C0-AVC-832

# Molecular Cloning of Hormone Genes

# Molecular Biology and Biophysics

**Molecular Cloning of Hormone Genes,** edited by ***Joel F. Habener,*** 1987

**Molecular Biophysics of the Extracellular Matrix,** edited by ***Struther Arnott, D. A. Rees,*** and ***E. R. Morris,*** 1984

# Molecular Cloning of Hormone Genes

edited by

Joel F. Habener

Humana Press · Clifton, New Jersey

**Library of Congress Cataloging in Publication Data**

Molecular cloning of hormone genes.

(Molecular biology and biophysics)
Includes bibliographies and index.
1. Molecular cloning. 2. Peptide hormones.
I. Habener, Joel F. II. Series.
QH442.2.M65 1986 574.19′256 86-19980
ISBN 0-89603-091-1

Crescent Manor
PO Box 2148
Clifton, NJ 07015



Printed in the United States of America

# Preface

The peptide hormones are small proteins that regulate cellular metabolism through their specific interactions with tissues of the endocrine, nervous, and immune systems, as well as in embryonic development. During the past ten years, refinements in the techniques of recombinant DNA technology have resulted in the cloning of genes encoding approximately 50 different hormonal and regulatory peptides, including those in which the peptides themselves and the mRNAs encoding the peptides are present in only trace amounts in the tissues of origin. In addition to providing the coding sequences of recognized hormonal and regulatory peptides, gene sequencing has uncovered new bioactive peptides encoded in the precursor prohormones that are then liberated along with the hormonal peptides during cellular cleavages of the precursors. The encoding of multiple peptides in a single monocistronic mRNA appears to be a genetic mechanism for the generation of biologic diversification without requiring amplification of gene sequences.

Two of the objectives in the assembly of this book are to present, in one volume, the known primary structures of the genes encoding several of the polypeptide hormones and related regulatory peptides, and to provide an account of the various approaches that have been used to identify and select the cloned genes encoding these polypeptides. The contents of the two introductory chapters are intended to provide the reader with a brief background of the approaches to gene cloning and the structure and expression of hormone-encoding genes. The remaining 16 chapters describe the conditions employed in the cloning of the specific genes, as well as certain of the unique structural aspects of the genes that are important in both their expression and their diversification of biologic informational content. The genes encoding growth hormone, glucagon, and the pituitary glycoprotein hormones are examples of biologic diversification at the level of gene duplication and multiplication. The calcitonin- and gastrin-releasing peptide genes provide specific examples of diversification at the level of alternate RNA splicing of primary gene transcripts. The proenkephalin, glucagon, and proopiomelanocortin gene products illustrate posttranslational diversification of information by way of tissue-specific processing of prohormones to yield distinct peptides. It is anticipated that this information will be helpful to workers who are entering the field of gene cloning and will thereby facilitate efforts to characterize additional hormone-encoding genes.

*Joel F. Habener*

# Contents

## CONTRIBUTORS

**SUSAN G. AMARA** • *Eukaryotic Regulatory Biology Program, University of California, San Diego, La Jolla, California*

**SHU JIN CHAN** • *Department of Biochemistry and Molecular Biology, The University of Chicago, Chicago, Illinois*

**WILLIAM C. CHIN** • *Section on Molecular Genetics, Joslin Diabetes Center and Howard Hughes Medical Institute, Boston, Massachusetts*

**PIERRE CORVOL** • *Inserm, Pathologie Vasculaire et Endocrinologie Renale, Paris, France*

**R. EGGER** • *Institute of Molecular Biology, Austrian Academy of Sciences, Salzburg, Austria*

**RONALD EVANS** • *Molecular Biology and Virology Laboratory, The Salk Institute, San Diego, California*

**HOWARD M. GOODMAN** • *Department of Molecular Biology, Massachusetts General Hospital and Department of Genetics, Harvard Medical School, Boston, Massachusetts*

**RICHARD H. GOODMAN** • *Department of Endocrinology, Tufts-New England Medical Center, Boston, Massachusetts*

**UELI GUBLER** • *Department of Molecular Genetics, Hoffman-LaRoche, Nutley, New Jersey*

**JOEL F. HABENER** • *Laboratory of Molecular Endocrinology and Howard Hughes Medical Institute, Massachusetts General Hospital, Harvard Medical School, Boston, Massachusetts*

**GERHARD HEINRICH** • *Laboratory of Molecular Endocrinology and Howard Hughes Medical Institute, Massachusetts General Hospital, Boston, Massachusetts*

**EDWARD HERBERT** • *Department of Chemistry, University of Oregon, Eugene, Oregon*

**W. HOFFMANN** • *Institute of Molecular Biology, Austrian Academy of Sciences, Salzburg, Austria*

**INGE HOLM** • *Unite de Genetique et Biochimie du Developpment, Institut Pasteur, Paris, France*

**G. KRIEL** • *Institute of Molecular Biology, Austrian Academy of Sciences, Salzburg, Austria*

**HENRY M. KRONENBERG** • *Endocrine Unit, Massachusetts General, Boston, Massachusetts*

**JANET KURJAN** • *Department of Biological Sciences, Columbia University, New York, New York*

**MALCOLM J. LOW** • *Department of Endocrinology, Tufts-New England Medical Center, Boston, Massachusetts*

**I. MALEC** • *Institute of Molecular Biology, Austrian Academy of Sciences, Salzburg, Austria*

**MARC R. MONTMINY** • *Department of Endocrinology, Tufts-New England Medical Center, Boston, Massachusetts*

**DAVID D. MOORE** • *Department of Molecular Biology, Massachusetts General Hospital and Department of Genetics, Harvard Medical School, Boston, Massachusetts*

**DANIEL S. ORY** • *Department of Molecular Biology, Massachusetts General Hospital and Department of Genetics, Harvard Medical School, Boston, Massachusetts*

**JEAN-JACQUES PANTHIER** • *Unite de Genetique et Biochimie du Developpment, Institut Pasteur, Paris, France*

**EDOUARD PROST** • *Department of Molecular Biology, Massachusetts General Hospital and Department of Genetics, Harvard Medical School, Boston, Massachusetts*

**DIETMAR RICHTER** • *Institut für Physiologische Chemie, Abteilung Zellbiochemie, Universität Hamburg, Hamburg, West Germany*

**K. RICHTER** • *Institute of Molecular Biology, Austrian Academy of Sciences, Salzburg, Austria*

**MICHAEL G. ROSENFELD** • *Eukaryotic Regulatory Biology Program, University of California, San Diego, La Jolla, California*

**FRANCOIS ROUGEON** • *Unite de Genetique et Biochimie du Developpment, Institut Pasteur, Paris, France*

**RICHARD F SELDEN** • *Department of Molecular Biology, Massachusetts General Hospital and Department of Genetics, Harvard Medical School, Boston, Massachusetts*

**FLORENT SOUBRIER** • *INSERM, Pathologie Vasculaire et Endocrinologie Renale, Paris, France*

**ELIOT R. SPINDEL** • *Section on Molecular Genetics, Joslin Diabetes Center and Howard Hughes Medical Institute, Boston, Massachusetts*

**MICHAEL UHLER** • *Department of Pharmacology, University of Washington, Seattle, Washington*

**R. VLASAK** • *Institute of Molecular Biology, Austrian Academy of Sciences, Salzburg, Austria*

# Chapter 1

# Approaches to Gene Cloning

JOEL F. HABENER

## 1. INTRODUCTION

Approaches to the molecular cloning of DNA complementary to mRNA and chromosonal genes have advanced greatly over the past 15 yr. These advances have consisted largely of (a) the development of enzymatic methods to synthesize full length, or near full length, cDNA copies of mRNA, (b) the preparation of recombinant genomic DNA in cosmid vectors that accommodate 30–40 kilobases (kb) of genomic sequence, (c) the development of bacteriophage vectors that express the genes at the level of their encoded proteins, allowing for the identification of recombinant DNA by immunostaining expressed fusion proteins, and (d) the application of synthetic oligonucleotides complementary to predicted RNA sequences encoding known amino acid sequences for the identification of various species of recombinant DNA.

The purpose of this chapter is to provide a brief review of the general approaches that are utilized in the construction of recombinant DNA libraries and the approaches used to identify specific DNA sequences within their bacterial host plasmids or viruses. More detailed descriptions of the recombinant DNA techniques are provided in the chapters that follow, as well as in several laboratory manuals and review articles (*1–5*).

## 2. CONSTRUCTION AND ANALYSIS OF RECOMBINANT DNA

The techniques for the construction of recombinant DNA make use of two classes of reagents. The first class is a battery of enzymes consisting of restriction endonucleases, polymerases, and ligases. These enzymes cleave DNA at specific sites, synthesize DNA from RNA and DNA templates, and join segments of DNA together. The second class includes bacterial plasmids and viruses; these are extrachromosomal organisms that replicate at high efficiency in their bacterial hosts. By the application of these two techniques, it is possible to synthesize DNA from cellular RNA and to engineer the enzymatically synthesized DNA in ways that permit its insertion into the genomes of the bacterial plasmids and viruses. The recombinant DNA can then be amplified by replication within bacterial hosts, thereby providing large quantities of the recombinant DNA for (a) structural analyses, (b) use as hybridization probes, and (c) introduction into the genomes of mammalian cell lines for purposes of analyzing the structural properties of the DNA involved in the regulation of gene expression.

The discovery of two enzymes facilitated the development of DNA technology: reverse transcriptase (*6,7*) and restriction endonuclease (*8*). Reverse transcriptase was found in the RNA of tumor viruses and is the means by which the virus makes DNA copies of the RNA templates. This enzyme permits a copy of DNA to be made from the RNA, an essential first step in the preparation of recombinant DNA for purposes of cloning. Restriction endonucleases cleave DNA at specific sequences, generally of 4–6 base pairs (bp). Each of the many endonuclease enzymes now isolated is specific for any given sequence of nucleotides; they reproducibly and predictably cleave DNA at specific sites, a property that is critical for the precise engineering of DNA segments.

Analysis of a particular gene begins by the preparation and cloning of cDNA complementary to the RNA of a particular cell (*1–5*) (Fig. 1). The cDNAs are prepared by "priming" the reverse transcription (reverse transcriptase) of the mRNA with short oligonucleotide fragments of oligodeoxyribothymidine, which preferentially binds to the 3′-polyadenylate [poly(A)] tract characteristic of cellular mRNA. Double-stranded DNA is then prepared

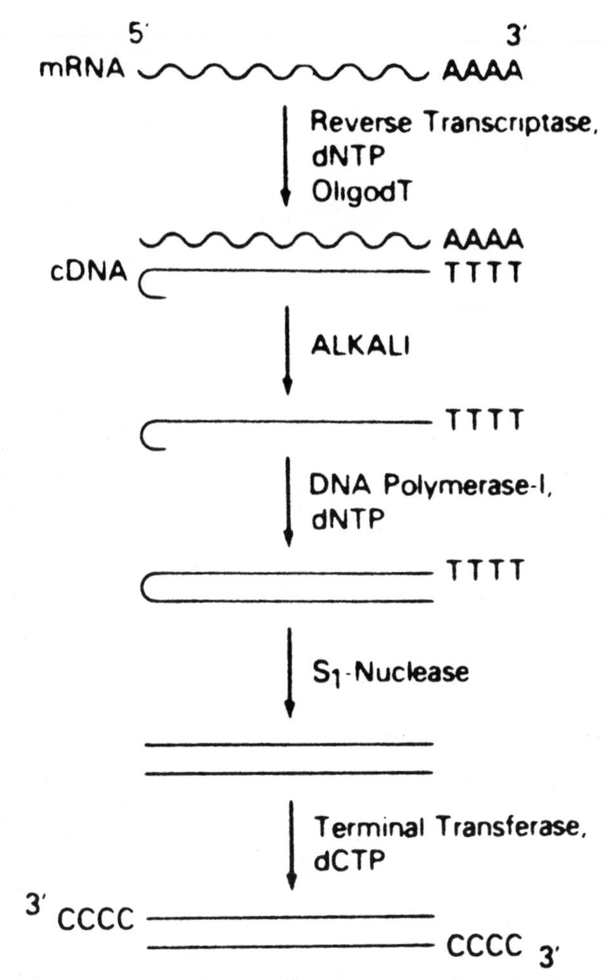


Fig. 1. Preparation of complementary DNA by enzymatic synthesis from a messenger RNA template. (1) Oligodeoxythymidylate (oligo dT) is hybridized to the 3′ polyadenylate tract characteristically found at the 3′ end of messenger RNAs; (2) The enzyme reverse transcriptase is used in the presence of the four deoxyribonucleotides (DNTPs) to synthesize the complementary cDNA copy of the messenger RNA; (3) The messenger RNA is selectively hydrolyzed by alkali; (4) DNA polymerase I is then used to synthesize the second strand of DNA, which is primed from intramolecular hybridization of a self-forming loop at the 3′ end of the cDNA; (5) The enzyme S-1 nuclease cleaves the loop structure, resulting in a double-stranded, flush-ended DNA sequence; (6) "Sticky" ends are created by the synthesis of polycytidylate tracts directed by the enzyme terminal transferase at the 3′ ends of the double-stranded DNA.

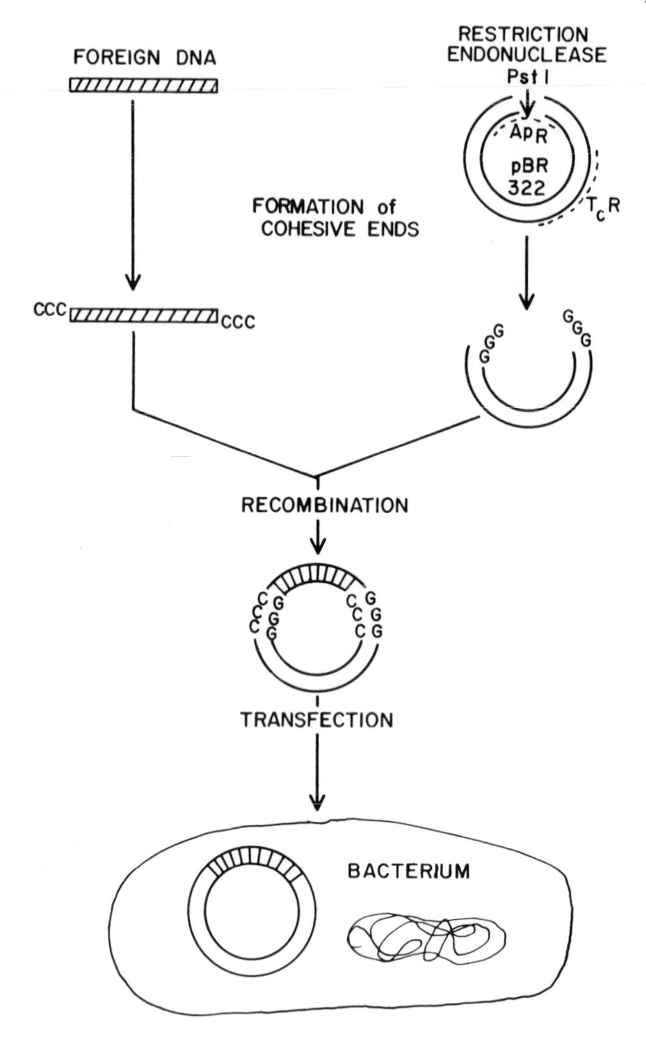


Fig. 2. Construction and molecular cloning of recombinant DNA. A circular bacterial virus (plasmid) is cleaved at a single site by a restriction endonuclease. In the example shown, the plasmid pBR322 is cleaved by the restriction endonuclease Pst-I, which, as a consequence of the cleavage, inactivates the gene encoding ampicillin resistance (ApR). The cleavage at the Pst-I site leaves intact the gene that confers tetracycline resistance (TcR). Poly(G) homopolymer extensions are synthesized on the 3′ ends of the linear plasmid, which are complementary to the corresponding Poly(C) homopolymer extensions that are created on the double-stranded cDNA of the foreign DNA (*see* Fig. 1). The foreign DNA is then ligated into the linearized plasmid and sealed with

from the single-stranded cDNA using DNA polymerase, and the cDNA is ligated into bacterial plasmids that have been linearized by cleavage at a single site in the plasmid with a restriction endonuclease (Fig. 2). To ensure a reasonably high efficiency of ligation of the foreign DNA into the plasmids, cohesive, or "sticky," ends are first prepared by adding short complementary DNA sequences to the ends of the foreign DNA and to the plasmids. A vector that is commonly used is the plasmid pBR322, engineered specifically for the purposes of cloning DNA fragments. Foreign DNA is commonly ligated into the unique *Pst*I of pBR322 site in which poly(dC) and poly(dG) homopolymers serve as the complementary, or cohesive, ends. The *Pst*I site is located within the gene that confers resistance to the antibiotic ampicillin. The plasmid also carries a gene for resistance to tetracycline. Thus, bacteria containing the plasmids can be selected by their sensitivity to ampicillin because the ampicillinase gene is inactivated by the inserted foreign DNA.

## 3. TECHNIQUES OF IDENTIFICATION

The recombinant plasmids containing DNA sequences complementary to the specific mRNA of interest are identified by any of a number of methods. One method involves hybridizing groups of recombinant plasmids to the initial mRNA preparations used in the cloning. The hybrid-selected mRNA is translated in a cell-free system in which the protein whose synthesis is programmed by the mRNA has been identified (*9*). Alternatively, specific inhibition of the translation of a mRNA–cDNA hybrid is detected in the

---

DNA ligase to form a colinear circle of recombinant DNA. The recombinant plasmid is then transfected into suitable bacterial hosts at low-density conditions, such that only a single plasmid will infect a given bacterium. The bacterial plasmids are allowed to amplify within the bacterium, which are then cloned by low-density plating and growth amplification. The bacterial clones containing plasmids with recombinant foreign DNA are selected from among the other clones by growth in media that contain tetracycline. All tetracycline-resistant bacteria are then replicated and tested for sensitivity to ampicillin. Those tetracyline-resistant, ampicillin-sensitive bacteria contain the plasmids with recombinant foreign DNA.

cell-free system (*10*). More recently, the techniques of hybridization selection and hybridization arrest and cell-free translation have been supplanted by hybridization of the bacterial colonies with synthetic oligonucleotide "probes" that have been labeled with P-32. Mixtures of oligonucleotides in the range of 14–17 bases are prepared complementary to the nucleotide sequences predicted from the known amino acid sequences of segments of the protein encoded by the mRNA. Because of the degeneracy in the genetic code (there are 61 amino acid codons and 20 amino acids), mixtures of from 24 to 96 oligonucleotides will ordinarily represent all possible sequences complementary to a particular 14-to-17-base region of the mRNA.

As a consequence of the development of automated instruments for the chemical synthesis of oligonucleotides of from 80 to 100 bases in length, single-sequence cDNA has been used successfully to identify by hybridization recombinant cDNA, as well as recombinant chromosonal genes (*11*). The sequence of the cDNA synthesized is based on the known amino acid sequence of the protein encoded by the desired recombinant DNA. In the selection of the specific base used in the position of the degenerate bases in the codon, consideration is given to three factors: (a) Whenever possible, a GT mismatch is selected because GT base pairs have little destabilizing effect on the DNA helix, (b) pyrimidine–pyrimidine mismatches are avoided because of their destabilizing effects, and (c) CG sequences are minimized because they appear at low frequency in the protein-coding sequences of genes, presumably because they serve as sites for methylation. An additional aid in the construction of such a cDNA probe is to utilize the codon preference for a given animal species. A recent technologic advance in the preparation of cDNA libraries is the utilization of lambda phage vectors instead of plasmid vectors. The advantage of the utilization of the lambda vector is the high efficiency and reproducibility of in vitro packaging of lambda DNA as a method for introducing DNA sequences into *E. coli*. Using this approach, it is readily possible to prepare cDNA libraries containing $10^5$–$10^7$ recombinants. Two of the lambda phage vectors that are suitable for cloning cDNA are the λ-GT10 and λ-GT11. Phage libraries cloned in λ-GT10 are useful for screening with nucleic acid probes. The phage vector λ-GT11 utilizes an insertion site for foreign DNA within the structural gene for beta galactosidase, thereby providing the potential for the foreign DNA sequences to be expressed as a fusion protein with beta-galactosidase. Such recombinant cDNA libraries constructed in λ-GT11 can be screened with antibody probes for antigen produced by specific recombinant clones. In a typical experiment, $10^5$–$10^6$ individual recombinants are screened in the form of phage plaques on a lawn of bacteria genetically deficient in certain proteases that might destroy the foreign fusion protein. The bacterial colonies are lysed, and proteins within the plaques are immobilized by a nitrocellulose filter placed over the bacterial lawn. Protein bound to the nitrocellulose filter is incubated with antibodies specific to the particular antigen that is to be detected, and the binding of antibody to specific antigen is detected by probing the filter with either a radioactive second antibody or radioiodinated protein A derived from *Staphylococcus aureus*.

The ability to prepare DNA by reverse transcription of mRNA was an important initial development in recombinant-DNA technology. This technique is important because in the cell there are many more copies of mRNA than there are genes that encode particular polypeptides. Usually cells contain only two copies of the genes, whereas in cells in which the gene is expressed, there may be from 10,000 to as many as 100,000 copies of the mRNA. Hence, it is much easier to isolate recombinant DNA prepared by reverse transcription of RNA templates extracted from these cells than it is to isolate specific gene sequences. In practice, the gene sequences themselves are isolated by hybridization with either [$^{32}P$]-labeled cloned recombinant cDNA or homologous complementary oligonucleotides.

## 4. CLONING TECHNIQUES

The technique used in the cloning of genomic DNA is similar to that used for cloning cDNA, except that the genomic sequences are much longer than cDNA sequences, and different cloning vectors are required. The most common vectors used are derivatives of the bacteriophage lambda that can accomodate DNA fragments of from 10 to 20 kilobase pairs (kbp). Recently, hybrids of bacteriophages and plasmids, called "cosmids," have been developed that can accommodate inserts of DNA of up to 40–50 kbp. In the cloning of genomic DNA, restriction fragments are prepared by partial digestion of unsheared DNA with a restriction endonuclease that cleaves the DNA frequently. DNA fragments of proper size are then prepared by a fractionation on agarose gels and are ligated to the bacteriophage DNA. The recombinant DNA so formed is then mixed with bacteriophage proteins, which results in the packaging of DNA and in the production of viable phage particles. The recombinant bacteriophages are grown on agar ,plates covered with lawns of growing bacteria. When the bacteria are infected by a phage particle, they are lysed and form visible plaques. Specific phage colonies are identified by their transfer to nitrocellulose filters and hybridization by complementary DNA probes labeled with P-32. In practice, different libraries of genomic DNA fragments of various animal species cloned in bacteriophage are available from a number of laboratories.

## 5. HYBRIDIZATION PROBES

In addition to providing the nucleotide sequences of the messenger RNA and the encoded amino acid sequence deduced from the nucleotide sequence, recombinant cDNA is highly valuable as a hybridization probe to measure cellular levels of mRNA and mRNA precursors, and as a probe to analyze the number of gene copies contained in the genomes of animals. The latter two procedures involve the separation of either the cellular RNA or restriction endonuclease digests of genomic DNA on agarose gels, followed by the transfer of the polynucleotide fragments to nitrocellulose filters and hybridization with [$^{32}$P]-labeled cDNA probes. These procedures are known as Northern transfer (RNA) and Southern transfer (DNA), respectively. In some instances, it is possible to use labeled probes for hybridization directly to tissue slices and to spreads of metaphase chromosomes. For example, labeled cDNA encoding proopiomelanocortin, the precursor to ACTH, localized specific neurons containing proopiomelanocortin mRNA by hybridization of the cDNA to histologic sections prepared from the medial basal hypothalamus of rats (*12*). In addition, the human insulin gene has been mapped to the distal end of the short arm of chromosome 11 by hybridization of mitotic chromosome preparations obtained from human peripheral blood lymphocyte cultures with a tritium-labeled recombinant plasma encoding preproinsulin (*13*). These new and powerful histohybridization techniques add a new dimension to molecular technology, allowing individual cells to be assessed for expression of specific genes.

## REFERENCES


1. Wu R., Grossman L., and Moldave K., eds. (1979–1983) in *Recombinant DNA: Methods in Enzymololgy, 1979–1983* (Colowick S. P. and Kaplan N. O., eds.-in-chief); vols. 68, 100-B, 100-C.
2. Maniatis T., Fritsch E. F., and Sambrook J. (1982) *Molecular Cloning—A Laboratory Manual.* Cold Springs Harbor Laboratory.
3. Williamson, S. R., ed. (1983) *Genetic Engineering.* Academic, London.
4. Huynh T. V., Young R. A., and Davis R. W. (1984) Constructing and Screening cDNA Libraries in λgt10 and λgt11, in *DNA Cloning Techniques: A Practical Approach* (David Glover, ed.), IRL, Oxford.
5. Young R. A. (1985) Immunoscreening λgt11 Recombinant DNA Expression Libraries, in *Genetic Engineering* (Setlow J. and Hollaender A., eds.), vol. 7, Plenum, New York.
6. Baltimore D. (1976) Viruses, polymers and cancer. *Science* **192,** 623–636.
7. Temin H. M. (1976) The DNA provirus hypothesis. *Science* **192,** 1075–1080.
8. Nathans D. and Smith H. O. (1975) Restriction endonucleases in the analysis and restructuring of DNA molecules. *Annu. Rev. Biochem.* **44,** 273–293.
9. Ricciardi R. P., Miller J. S., and Roberts B. E. (1979) *Proc. Natl. Acad. Sci. USA* **76,** 4927–4931.
10. Paterson B. M., Roberts B. E., and Kuff E. L. (1977) *Proc. Natl. Acad. Sci. USA* **74,** 4370–4374.
11. Seeburg, P. H. and Adelman J. P. (1984) Characterization of cDNA for precursor of human luteinizing hormone releasing hormone. *Nature* **311,** 666–668.
12. Gee C. E., Chen C. L. C., and Roberts J. L. (1983) Identification of proopiomelanocortin neurones in rat hypothalamus by *in situ* cDNA–mRNA hybridization. *Nature* **306,** 374–376.
13. Harper M. E., Ullrich A., and Saunders G. F. (1981) Localization of the human insulin gene to the distal end of the short arm of chromosome 11. *Proc. Natl. Acad. Sci. USA* **78,** 4458–4460.

and even bacteria (*6*). Examples of the primitive origins of certain regulatory peptides are the alpha-mating factor of the yeasts, somatostatin, insulin, and glucagon. The alpha-mating factor is a decapeptide secreted by the yeast *Saccharomyces cerevesiae* of the alpha-mating type, which interacts with the receptors on yeasts of the opposite, or A-mating, type to induce the two haploid organisms to fuse into a diploid form (*7*). The structure of the alpha-mating-factor-peptide is remarkably homologous with the mammalian hypothalamic gonadotropin-releasing hormone, a decapeptide produced in the hypothalamus that acts as a hormone by passage through the neurohypophyseal portal system and stimulates the release of gonadotropins from the pituitary (*8*).

It is interesting to contemplate that the reproductive functions of these peptides have been conserved in evolution over the great distances between unicellular organisms (yeasts) and mammals, including man. Calcitonin, a peptide hormone that regulates calcium metabolism in mammals, has been detected by immunochemical techniques in celenterates and tunicates (sea squirts). Both somatostatin-like and insulin-like immunoreactivity, as well as biological activity, have been detected in extracts of the ciliated protozoan, *Tetrahymena pyriformis.* Glucagon-like immunoreactivity is reported in insects (tobacco hornworm). Hence, one can conclude that the genes encoding these important peptides, as well as their regulatory functions, originated very early in the development of life forms and have been retained throughout a vast evolutionary time (*6*). These observations indicate that the genes were likely present and expressed in very early forms of life, and that the genes have been conserved to a remarkable degree throughout evolution. Moreover, it appears probable that one of the earliest, and most important, functions of these peptides is that of paracrine communication among adjacent cells by way of local cellular secretion and action on adjacent cells rather than by endocrine communication between organs, which requires the passage of the peptides through the circulatory system. In view of the widespread distribution of the many peptide hormones throughout lower organisms and the nervous system of higher organisms, it is tempting to speculate that in the course of evolution these peptides first served as local neurotransmitters before the appearance of their specific hormonal functions. With the growing complexity of the organism—namely, development of highly specialized organs and a circulatory system—the peptides were utilized as messenger molecules to convey information among the organs. Moreover, one can reason that the neuroregulatory functions of the peptides in mammals were maintained separately from the late-evolving endocrine functions through the development of the blood–brain barrier. The peptidergic neurons of the hypothalamus appear to represent a transition between the cell-to-cell communicator functions and the organ-to-organ regulatory functions of the peptides, since many neurocrine peptides of the hypothalamus, such as vasopressin, oxytocin, somatostatin, and the thyrotropin-, gonadotropin-, somatotropin-, and corticotropin-releasing hormones are secreted from peptidergic neurons directly into the circulatory system, through which they act on specific receptors on distant target organs to regulate metabolic processes. Thus, it is of interest to ascertain specific functions of these hormones and regulatory peptides: how they are synthesized, how their synthesis and release are regulated, and how their respective genes are related to one another.

### 1.3. Present Information Based on Molecular and Cellular Biology

The inauguration of the era of molecular endocrinology occurred in the 1950s with the determination of the amino acid sequences and the nonapeptides vasopressin and oxytocin by duVigneaud and his coworkers (*9,10*). In the ensuing 25 yr, the amino acid sequences of approximately 30 different polypeptide hormones and regulatory peptides were established through the tedious approach of the isolation and purification of scarce peptides from large amounts of endocrine tissues. The successful cloning of the structural genes for insulin (*11*) and growth hormone (*12*) in 1977 established the techniques of genetic engineering of recombinant-DNA molecules as a highly efficient approach for the determination of the structures of proteins by way of the decoding of nucleotide sequences. Refinements in the techniques of recombinant DNA over the past 6–8 yr have permitted the cloning of genes encoding 30–40 different hormonal and regulatory peptides, including those in which the peptides themselves and the mRNAs encoding them are present in only trace amounts in their tissues of origin. These refinements include techniques for the cloning of several hundred thousand DNA sequences complementary to cellular mRNAs (cDNAs) at a time, the preparation of full-length cDNAs, and the use of synthetic sequences (artificial genes) complementary to the coding regions of genes as hybridization probes to identify specific cloned cDNAs and chromosomal (genomic) DNAs.

The application of recombinant-DNA technology has accelerated markedly the gathering of information about the molecular workings of cells through the ability to analyze directly the structure and expression of genes. This powerful new technology has provided (a) information on the structure and organization of genes, (b) an indication of the levels of gene expression at which biologic diversification and amplification of information occur, and (c) insights into the molecular mechanisms by which specific genes are regulated.

## 2. THE STRUCTURE AND ORGANIZATION OF GENES

Analyses of the primary structures of both DNAs complementary in sequence to mRNAs (cDNAs) and the chromosomal genes encoding hormones and regulatory peptides have provided much new information about the structure, organization, and expression of these genes. This information includes: (a) the amino acid sequences of the proteins deduced from the encoded genes, resulting in the discovery that all of the known bioactive peptides are contained within larger precursors, frequently in accompaniment with other peptides, (b) the recognition that the coding regions of the genes (exons) are interrupted by intervening DNA sequences (introns) that are cleaved from the initial RNA transcripts during the processing and assembling of the specific mRNAs, (c) the discovery of new pathways for the diversification of genetic information in gene expression at the levels of tissue-specific alternative patterns of RNA splicing and posttranslational processing of the peptide precursors, and (d) the identification of regulatory sequences flanking the structural genes that determine the cellular specificity, accuracy, and rates of initiation of gene transcription.

Before considering the primary nucleotide structures of genes, it is important to recognize that the DNA of higher organisms is wound into a tightly and regularly packed chromosomal structure in association with a number of different proteins. The resultant nuclear protein is organized into elements called nucleosomes (*13*). Nucleosomes are composed of four or five different protein subunits, or histones, that form a core structure around which are wound approximately 140 base pairs (bp) of genomic DNA. Structures of histones are highly conserved throughout evolution, indicating their fundamental importance in the detailed architecture of the nucleosome. The nucleosomes are arranged as beads on a string along the DNA strand and become further organized into more highly ordered structures consisting of coils of many closely packed nucleosomes that in turn form the fundamental organizational units of the eukaryotic chromosome. Although the function of nucleosomal structure is incompletely understood, nucleosomes may serve one or more purposes. For example, nucleosomes may be a means to compact the large amount of DNA (approximately $2 \times 10^9$ bp) comprising the human genome. Nucleosomes may also be involved in the process of DNA replication and gene transcription—a loosening or obliterating of the nucleosome structure may render the DNA available for transcription by RNA polymerase. In addition to histones, other diverse proteins are associated with DNA, and it is likely that the complex nucleoprotein structure provides: (a) specific recognition sites resulting in access of specific regions of the DNA to the many regulatory proteins and enzymes involved in DNA replication, and (b) rearrangements of DNA segments and the transcriptional process of gene expression.

The expressed gene consists of two functional units: (a) a transcriptional region, and (b) a promoter or regulatory region. The transcriptional unit is the segment of the gene corresponding to the sequence that is transcribed into a mRNA precursor. The coding sequence of the gene consists of the exonic sequences that are spliced out from the primary transcript during the post-transcriptional processing of the precursor RNA; these exons consist of both the segment of the mRNA sequence that is translated into protein and the 5′- and 3′-flanking untranslated sequences. The 5′ sequence begins typically with a methylated guanine residue known as the cap site. The 3′ untranslated region contains within it a short sequence, AATAAA, that signals the site of cleavage of the 3′ end of the RNA and the addition of a polyadenylate tract of 100–200 nucleotides located approximately 20 bases from the AATAAA sequence. Although the functions of these modifications of the ends of messenger RNAs are poorly understood, there is some evidence that they enhance the stability of the RNAs, perhaps through providing resistance to degradation by exonucleases. Likewise, the nature of the enzymatic mechanisms that result in the precise excision of intronic sequences and the rejoining of exons is incompletely understood. Short consensus sequences of nucleotides reside at the splice junctions; for example, the bases GT and AG of the 5′ and 3′ ends of the introns, respectively (*14*) (Fig. 1). A population of small nuclear RNAs, known as the U1 RNAs, contains short nucleotide sequences that are complementary to the splice junctions. There is a growing body of information that indicates these small RNAs serve as templates that base pair with the splice junctions, providing secondary structure necessary for specific endonucleolytic cleavages (*15*). The protein-coding sequence of the messenger RNA begins with the codon AUG for methionine and ends with the codon immediately preceding one of the three nonsense, or stop, codons (TGA, TAA, and TAG). As will be discussed later, the protein-coding sequences of polypeptide hormones invariably encode precursors of prehormones (or preprohormones) that then undergo specific posttranslational cleavages during their passage through the secretory pathway.

The factors that are involved in the regulation of the expression of specific genes encoding polypeptides have been the focus of much attention over the past several years. Although our understanding of this area is incomplete, some conclusions can be made. As a result of experiments involving the selective deletion of 5′ sequences located upstream from structural genes encoding polypeptides, followed by the analyses of their expression after introduction into cell lines, some insights have been obtained about the identification of specific regulatory sequences in the DNA flanking the structural genes. The information gained thus far indicates that regulatory sequences, termed promoters in the broad sense, consist of unexpectedly short polynucleotide sequences. These sequences can be characterized into four groups with respect to their functions and distances from the transcriptional initiation site: First is the sequence TATAAA (TATA, or Goldberg-Hogness box), which is invariably present within 25–30

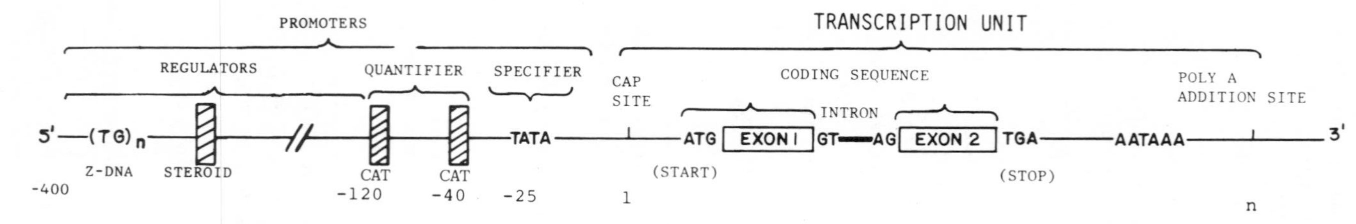


Fig. 1. Diagrammatic structure of a consensus sequence of a gene encoding a polypeptide hormone. Such a gene typically consists of a promoter and a transcription unit. The transcription unit is that region of the DNA composed of exons and introns that is transcribed into an RNA precursor. Transcription begins at a cap-site sequence in the DNA and extends several hundred bases beyond the poly(A) addition site in the 3′ region. During the posttranscriptional processing of the RNA precursor, the 5′ end of the messenger RNA is capped by addition of methyl guanine residues at the 3′ end. The transcript is then cleaved at the poly(A) addition site approximately 20 bases to the right of the AATAAA signal sequence and a polyadenylate tract is added to the 3′ end of the RNA. Introns are cleaved from the RNA precursor, and the exons are joined together. The dinucleotides GT and AG are invariably found at the 5′ and 3′ ends of introns—the so-called Chambon consensus sequence. Translation of the messenger RNA invariably starts with the codon ATG for methionine. Translation is terminated when the polyribosome reaches stop codons TGA, TAA, or TAG. The promoter region of the gene is less well-defined than is the transcriptional unit. It consists of several different sequences, some of which are involved in the regulation of the transcription of the gene. Approximately 30 nucleotides upstream (in the 5′ direction) from the cap site, the sequence TATAAA (Goldberg-Hogness box) specifies the site of initiation of transcription at the cap site. Further upstream, sequences consisting of some variation of the nucleotides CCATT are located and are involved in the amplitude of transcription. Recently, a number of different quantifier, or enhancer, sequences (designated by boxes enclosing E) have been identified, all of which are in some way involved in the rate of initiation of the transcription of genes. Further upstream in the region of 400–100 bases from the transcriptional initiation site, there are specific regulator sequences involved in the recognition of specific regulatory signals for the given gene, such as steroid-hormone-recognition sites.

nucleotides upstream from the point of transcriptional initiation. Site-specific mutation studies indicate that the integrity of the TATA sequence is required to ensure the accuracy of initiation of transcription at a particular site. Upstream from the TATA sequence resides a series of sequences, tissue-specific enhancers, that are important in the tissue-specific and quantitative expression of the structural gene. Even farther upstream reside sequences that are involved in the metabolic regulation of gene expression such as the glucocorticoid and the progesterone promoters. In addition, in the surrounding regions of genes sequences exist that consist of alternating pyrimidine and purine bases, such as thymidine and guanidine, (called TpG sequences), that form a Z-DNA (*16*), as opposed to B-DNA structure, and homopurine–homopyrimidine tracts that may determine tissue-specific expression in responses to a particular regulatory molecule (*17*).

## 3. PRECURSORS OF THE PEPTIDE HORMONES

By deciphering the nucleotide sequences of the coding regions of the genes, it has been possible to determine the primary amino acid sequences of the peptides. This information has revealed that the hormones and regulatory peptides are encoded within the structures of larger precursor molecules (prohormones or preprohormones), from which the peptides are cleaved during their co- and posttranslational processing. Studies of the co- and posttranslational processing of hormones and regulatory peptides have brought to light several functions that these precursors appear to fulfill in biologic systems. Two of these proposed functions for which there is considerable supporting evidence are (a) intracellular signaling, by which the cell distinguishes among specific classes of proteins and directs them to their specific sites of action, and (b) the generation of multiple biologic activities from a common gene product by regulated or cell-specific variations of posttranslational modifications (Fig. 2).

### *3.1. Precursor Sequences Signal Intracellular Segregation*

Without exception, all of the peptide hormones and regulatory peptides studied thus far contain amino-terminal sequences,

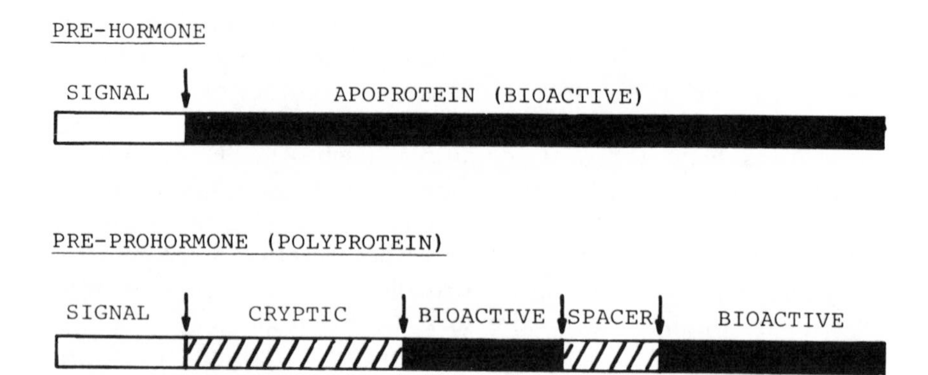


Fig. 2. Diagrammatic depiction of two configurations of precursors of polypeptide hormones. The diagrams represent the polypeptide backbones of the protein sequences encoded in the mRNA. One form of precursor consists of the amino-terminal signal, or presequence, followed by the apoprotein portion of the polypeptide that needs no further proteolytic processing for activity. The second form of the precursor is a preprohormone that consists of the amino-terminal signal sequence followed by a polyprotein, or prohormone, sequence consisting of several peptide domains linked together that are subsequently liberated by cleavages during posttranslational processing of the prohormone. The reason for the synthesis of polypeptide hormones in the form of precursors is only partly understood. Clearly, the amino-terminal signal sequences function in the early stages of transport of the polypeptide into the secretory pathway (*see* section 3.1 and Fig. 3). The prohormones, or polyproteins, often serve to provide a source of multiple bioactive peptides (*see* Fig. 4). However, many prohormones contain peptide sequences that are cleaved out and have no known biologic activity. These peptides are referred to as "cryptic" peptides. Other peptides may serve as spacer sequences betwen two bioactive peptides—for example, the C peptide of proinsulin. In instances in which the bioactive peptide is located at the carboxy-terminus of the prohormone, the amino-terminal prohormone sequence may simply facilitate the cotranslational translocation of the polypeptide in the endoplasmic reticulum (*see* Fig. 6).

termed signal, leader, or presequences, that are hydrophobic and recognize specific sites on the membranes of the rough endoplasmic reticulum, resulting in the transport of the nascent polypeptides into the secretory pathway of the cell (*18,19*) (Fig. 3). As a consequence of this specialized nature, the precursor proteins

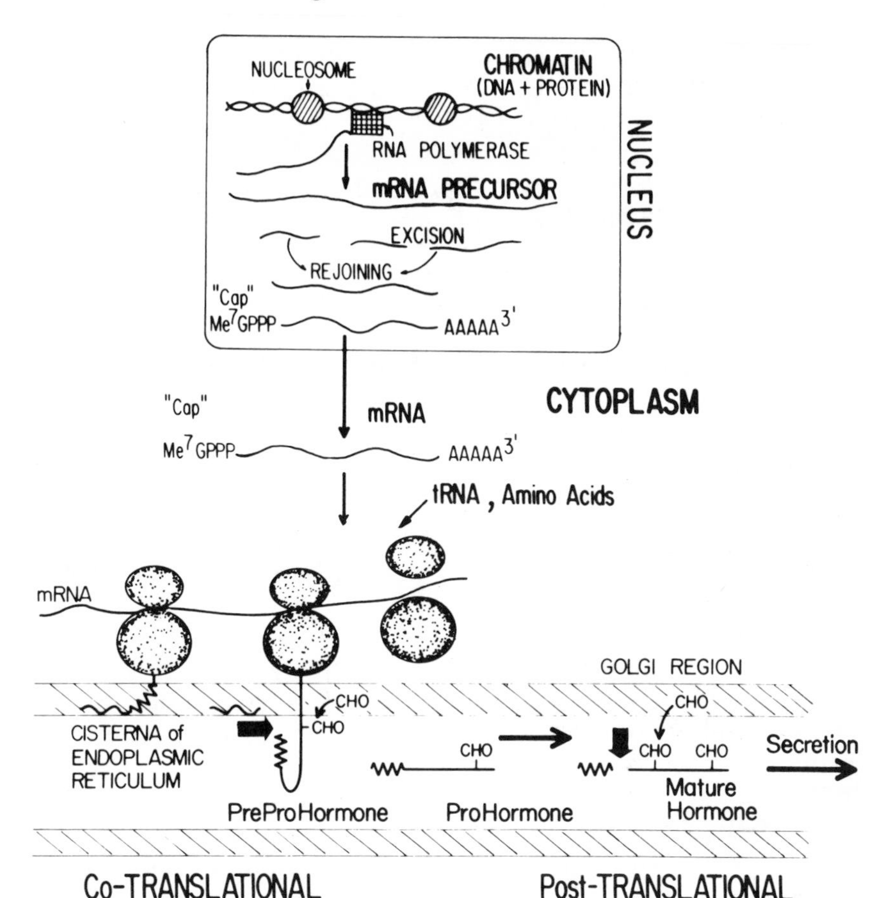


Fig. 3. Steps in the cellular synthesis of polypeptide hormones. Steps that take place within the nucleus include the transcription of genetic information into an RNA precursor (pre-mRNA) followed by posttranscriptional processing of pre-mRNA, a process that includes the steps of RNA cleavage, excision of introns, and rejoining of exons, resulting in the formation of mRNA. The ends of the mRNA are modified by the addition of methylguanine caps at the 5′ end and addition of poly(A) tracts at the 3′ ends. The cytoplasmic mRNA is assembled with ribosomes. Amino acids, carried by aminoacylated tRNAs, are then polymerized into the polypeptide chain. The final step in protein synthesis is that of posttranslational processing. These processes take place both during the growth of the nascent polypeptide chain (cotranslational) and after the release of the completed chain (posttranslational), and include proteolytic cleavages of the polypeptide chain (conversion of

*(continued)*

destined for secretion are selected from a great many other cellular proteins for sequestration and subsequent packaging into secretory granules and export from the cell.

### *3.2. Many Peptide Hormone Precursors Are Polyproteins That Encode Multiple Peptides*

In addition to the presence of amino-terminal signal sequences in the precursors encoding the hormones and regulatory peptides, most if not all of the smaller hormones and regulatory peptides are synthesized in the form of polyproteins, or prohormones, from which multiple peptides are produced as a consequence of the posttranslational cleavages of the precursors within the Golgi complex of the secretory cells (Fig. 4). The sites of cleavages in the precursors characteristically consist of short sequences of 2–3 basic (Arg, Lys) amino acids. Proteolysis occurs by way of combined trypsin-like (endopeptidase) and carboxypeptidase-B-like (exopeptidase) activities. Additional cleavages of certain precursors occur by way of sequential removal of dipeptides by exopeptidases when they are present in the sequences of alternating acidic (Asp or Glu) and hydrophobic (e.g., Gly, Ala) amino acids. Such dipeptidylpeptidase action is seen in the processing of the precursors of the yeast alpha-mating factor, melittin, and cerulein.

A function that appears to be fulfilled by these polyproteins is the provision of multiple biologic activities from a single gene product. Since all of the many mRNAs thus far identified in eukaryotes are monocistronic, i.e., code for only a single protein,

---

Fig. 3. (*cont.*) preprohormones or prohormones to the hormones), derivatizations of amino acids (glycosylation, phosphorylation), and cross-linking and assembly of the polypeptide chain into its conformed structure. The diagram depicts the posttranslational synthesis and processing of a typical secreted polypeptide, which requires the vectorial, or unidirectional, transport of the polypeptide chain across the membrane bilayer of the endoplasmic reticulum, resulting in the sequestration of the polypeptide in the cisterna of the endoplasmic reticulum, a first step in the export process of proteins destined for secretion from the cell (*see* Fig. 6). Most translational processing occurs within the cell as depicted (presecretory), and, in some instances, outside the cell, during which time further proteolytic cleavages or modifications of the protein may take place (postsecretory).

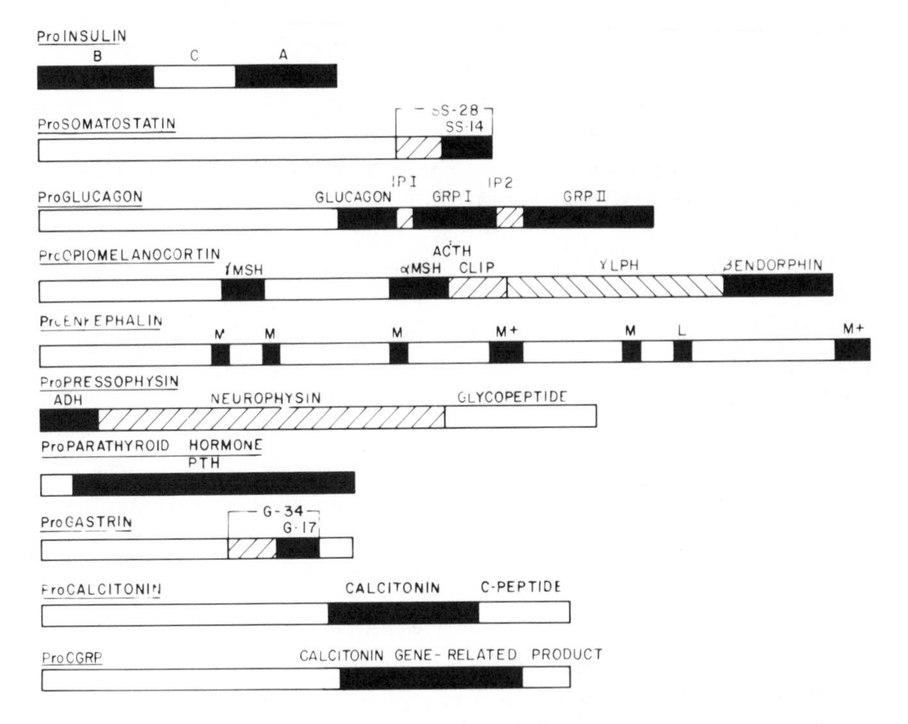


Fig. 4. Diagrammatic illustration of the primary structures of several prohormones. The dark shaded regions of the prohormones denote the regions of the sequence that constitute known biologically active peptides after their posttranslational cleavage from the prohormones. The sequences indicated by the cross-hatching denote regions of the precursor that alter the biologic specificity of that region of the precursor. For example, ACTH contains a sequence of α-MSH, but when it is covalently attached to the CLIP peptide, it constitutes adrenocorticotropin (ACTH). Somatostatin-28 is an amino-terminally extended form of somatostatin-14 that has a higher potency than has somatostatin-14 on certain receptors. The neurophysin sequence linked to the C-terminus of antidiuretic hormone functions as a carrier protein for the hormone during its transport down the axon of the neurons in which it is synthesized. The precursor proenkephalin represents a polyprotein that contains multiple similar peptides within its sequence, either metenkephalin (M) or leuenkephalin (L) or a carboxy-terminally extended form of metenkephalin (M+). Procalcitonin and procalcitonin-gene-related product (CGRP) share identical amino-terminal sequences, but differ in their carboxy-terminal regions as a result of alternative splicing during the posttranscriptional processing of the messenger-RNA precursor (*see* section 4).

multiple but separate biologic activities appear to be generated at the level of posttranslational cleavages of the precursor proteins. The encoding of multiple peptides within a single monocistronic mRNA appears to be a genetic mechanism for the generation of biologic diversification without the requirement for amplification of gene sequences (*vide infra*). Different polyprotein precursors may contain two or more peptides with different spectra of biologic activities; for example, proopiomelanocortin, which is the precursor of (a) ACTH, (b) MSH, and (c) beta-endorphin. Other polyproteins may contain several copies of the same peptide, as seen in the example of the common precursors of the yeast alpha-mating factor, thyrotropin-releasing hormone, and in proenkephalin, the precursor of the enkephalins. Alternatively, some polyproteins encode peptides of unknown functions, i.e., proglucagon, which contains, in addition to glucagon, two peptides of 30–40 amino acids related in their structures to glucagon, but for which no biological activity is yet known.

One possible explanation for the existence of cryptic peptides located at the amino terminus of prohormones is that the peptides may serve as spacers to provide a minimum length of polypeptide necessary for the signal sequence on the growing nascent chain to emerge from the large ribosomal subunit for interaction with the signal-receptor particle during the synthesis of the polypeptide. This appears to be a plausible explanation for the existence of amino-terminal precursor peptides in situations in which the bioactive peptide is located at or near the carboxy-terminus of the precursor, e.g., somatostatin and calcitonin.

### 3.3. Coding Sequences of Genes Are Interrupted

It came as a surprise to learn that the coding sequences of genes encoding the proteins and ribosomal RNAs in eukaryotes are separated by intervening DNA sequences (introns) into coding blocks (*20,21*). Analyses of the structures of bacterial genes had shown previously that the nucleotide sequences in the chromosomal genes matched precisely the corresponding sequences in the mRNAs. Thus, the interruption of the continuity of genetic information appears to be a unique property of nucleated cells. The colinear array of exons and introns is transcribed into an RNA precursor that is then processed within the nucleus into the mature mRNA by enzymatic removal of the introns, followed by ligation of the exons. The mRNAs so formed are modified by the addition of methylguanosine caps to the nucleotides located at the 5′ ends and by addition of polyadenine tracts to the 3′ ends.

The reasons for, and the functional significance of, the interruptions of gene sequences by introns is unknown. In many instances, however, introns appear to separate exons into functional domains with respect to the proteins that they encode. Examples of such a separation of functional domains include the genes for proglucagon, a precursor of glucagon in which three introns separate the four exons that encode the signal peptide, glucagon, and the two glucagon-related peptides contained within the precursor (22). Another example of the division by introns of the gene into functional domains is seen in the growth-hormone gene. The gene is divided into five exons by four introns that separate the promoter region of the gene from the protein-coding region and the protein-coding region into three partly homologous, repeated segments, two of which provide the growth-promoting activity of the hormone, and the third remaining segment providing the carbohydrate metabolic functions (23). As a general rule, the genes encoding most of the precursors of hormones and regulatory peptides contain introns at or about the region where the signal peptide joins the apoprotein or prohormone (propeptide), thus separating the signal sequence whose intracellular function lies in the segregation of the precursor into the secretory pathway (*vide supra*) from the components of the precursor that are exported from the cell in the form of hormones or peptides. It has been suggested that the topographic separation of exon-encoding regions of proteins of specific functions reflects the evolutionary processes of genetic recombination (*21*). The result of such recombination is the bringing together within a single gene of multiple-function components that are widely distributed within the genome and, as a consequence, the creation of chimeric proteins with new functions. Such recombination is evident in the "exon borrowing" seen in the gene encoding the LDL receptor, which contains exons with homology to those of the epidermal growth factor receptor and several of the blood-clotting factors (*23a*). The existence of specific functional coding blocks of DNA separated by noncoding DNA sequences would allow recombination to take place anywhere within the intronic noncoding DNA without interrupting the reading frames of the exonic coding DNA because, during the enzymic processing of the primary gene transcript, the introns within which recombination has occurred are excised and are not incorporated into the mature RNA sequence.

There are, however, exceptions to the theory of one exon–one function. The genes of many precursors of the polyprotein type are not interrupted by introns in a manner corresponding to the separation of the functional components of the precursor. Notable in this regard is the precursor, proopiomelanocortin, from which the peptides adreno-corticotropin, alpha-melanocyte-stimulating hormone, and beta-endorphin are cleaved out during the posttranscriptional processing of the precursor (*24*). Also, no introns interrupt the protein-coding region of the gene encoding proenkephalin, a precursor that contains seven copies of the enkephalins within its sequence (*25*). These exceptions, however, may be more apparent than real, since it is possible that introns that once separated each of the coding domains were simply lost during the course of evolution. The precedent for the selective loss of introns appears to be the rat insulin II gene (*26*). The rat contains two nonallelic insulin genes; insulin I contains two introns, and insulin II contains a single intron. The most probable explanation for this situation is that an ancestral gene containing two introns duplicated itself, and, in the process of duplication, or sometime thereafter, one of the introns was eliminated.

It seems reasonable to speculate that the processes of gene evolution involve a series of rearrangements of DNA sequences encoding specific but widely different functions, each separated by an intervening DNA sequence. These arrangements may or may not be random, but they could be random and still result in the formation of new, stable coding sequences. In the event that a new, functionally useful gene is created—that is, if the protein encoded in the gene provides a selective advantage to the organism—the gene would be maintained in the genome under strong selective pressures to conserve the coding assignment, tolerating only silent nucleotide changes that do not alter the encoded amino acid. This conservation would also allow only nucleotide changes that result in the substitution of an amino acid with chemical properties very similar to those of the original amino acid.

## 4. THE GENERATION OF BIOLOGIC DIVERSITY

In addition to providing specific control points for the regulation of gene expression, the various steps involved in the transfer of information encoded in the DNA of the gene to the final bioactive protein (via sequential processing of RNA precursors and protein precursors) provide the means for the diversification of the biologic information stored in the gene (Figs. 5 and 6).

At the level of DNA, diversification of genetic information comes about by way of gene duplication and amplification. This process of gene duplication is best appreciated in terms of evolution. Many of the polypeptide hormones are derived from the expression of families of multiple structurally related genes. Examples of structurally related hormones include the growth-hormone family, consisting of (1) growth hormone, (2) prolactin, and (3) chorionic somatomammotropin (placental lactogen); the glucagon family consisting of (1) glucagon, (2) vasoactive intestinal peptide, (3) secretin, (4) gastric inhibitory peptide, and (5) growth-hormone-releasing hormone; and the glycoprotein hormones, consisting of (1) thyrotropin, (2) luteotropin, (3) follicotropin, and (4) chorionic gonadotropin. It appears that over the course of evolution, an ancestral gene encoding a prototype polypeptide representative of each of these families duplicated itself one or more times and, through the processes of mutation and selection, the progeny of the ancestral gene assumed different biologic functions. The structural organization of the genome of higher animals lends itself to intergenic recombinational events, resulting in the rearrangement of transcriptional units and regulatory sequences within the genome (27). Mammalian genomes contain large numbers of DNA segments, called pseudogenes, that appear to consist of cDNAs that are partly mutated duplicates of structural genes, many of which lack introns and have 3′ poly(A)tracts. Their resemblance to mRNAs indicates that they have been reverse-transcribed from mRNA back into DNA and then reinserted into the genome. Such pseudogenes, or processed genes, have been observed for alpha-globin, immunoglobulins, alpha- and beta-tubulins, and others. It has been suggested that perhaps as much as 20% of the mammalian genome originated as RNA that was reverse transcribed back to DNA (27). If one assumes that pseudogenes, which generally differ from the

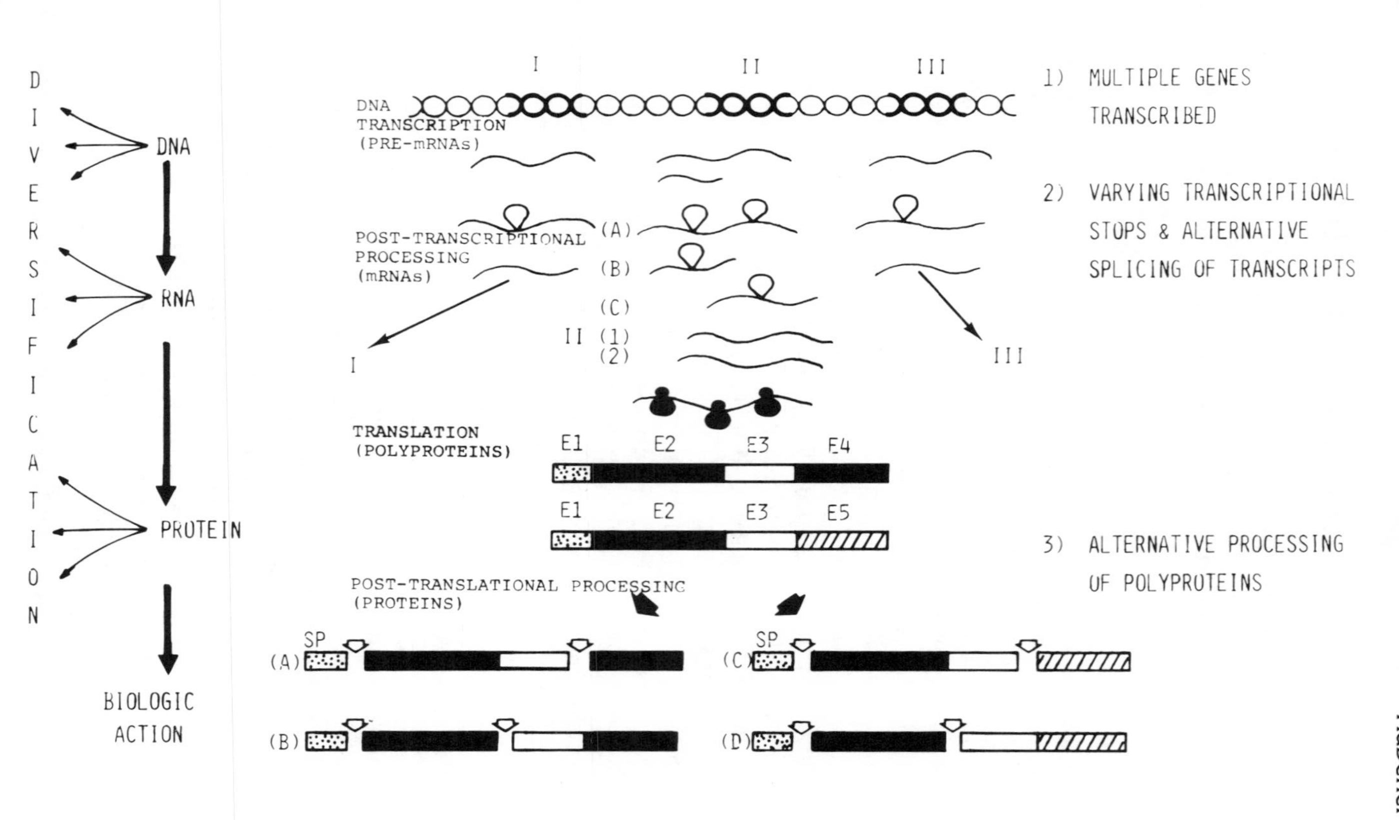
D I V E R S I F I C A T I O N
DNA
RNA
PROTEIN
BIOLOGIC
ACTION
I
II
III
DNA
TRANSCRIPTION
(PRE-mRNAs)
POST-TRANSCRIPTIONAL
PROCESSING
(mRNAs)
(A)
(B)
(C)
II (1)
(2)
I
III
TRANSLATION
(POLYPROTEINS)
E1
E2
E3
E4
E1
E2
E3
E5
POST-TRANSLATIONAL PROCESSING
(PROTEINS)
SP
(A)
(B)
SP
(C)
(D)
1) MULTIPLE GENES
TRANSCRIBED
2) VARYING TRANSCRIPTIONAL
STOPS & ALTERNATIVE
SPLICING OF TRANSCRIPTS
3) ALTERNATIVE PROCESSING
OF POLYPROTEINS

Fig. 5. Schema indicating the levels in the expression of genetic information at which diversification of information encoded in a gene may take place. The three major levels of genetic diversification are: (1) gene duplication, a process that occurs in terms of evolutionary time; (2) variation in the processing of RNA precursors, resulting in the formation of two or more messenger RNAs by way of alternative pathways of splicing of the transcript; and (3) use of alternative patterns in the processing of protein biosynthetic precursors (polyproteins, or prohormones). These three levels in gene expression provide a means for diversification of gene expression at the levels of DNA, RNA or protein. One or more of a combination of these processes leads to the formation of the final biologically active peptide, or hormone. In the diagram, the loops depicted in the transcripts denote introns, in the diagrammatic structures of proteins, the stippled, shaded, and unshaded areas denote exons. SP = signal peptide.

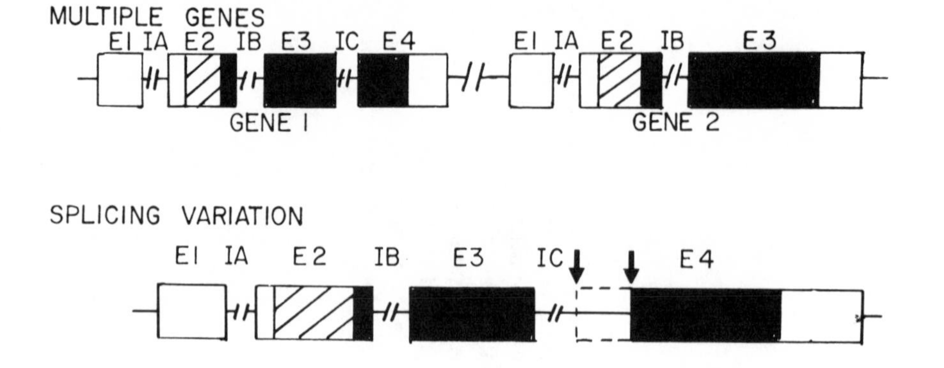


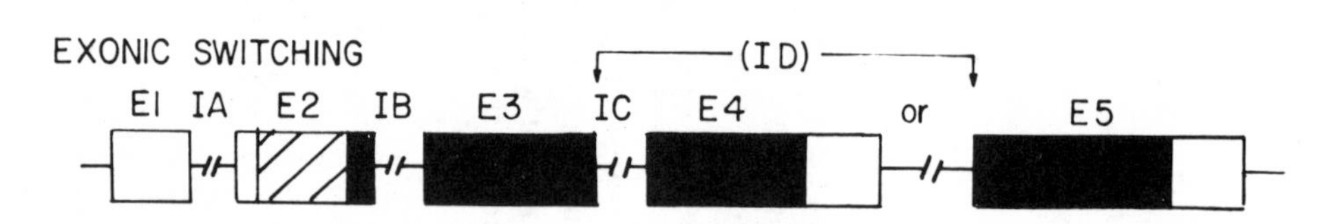


Fig. 6. Genetic and transcriptional origin of biologic diversity. This figure illustrates in greater detail the mechanisms of the alternative splicing of RNA transcripts. The initial level of diversification arises from the multiplication of genes, a process that occurs in a context of evolutionary time. Each of the genes consists of a series of exons (E1 through 4) and introns (IA through ID). Variations at the site at which introns are cleaved from the primary transcript can lead to the formation of two or more messenger RNAs with sequence deletions or insertions. When the variations of splicing occur in the protein-coding sequence of the RNA, they result in corresponding insertions of deletions of codons for amino acids. Such a splicing variation occurs in the human growth-hormone gene in which a minor transcribed mRNA lacks 15 codons in its central region, resulting in the translation of a growth hormone that is 2 kdaltons smaller than the normal growth hormone. A second splicing variation, exonic switching, occurs in the transcription of the calcitonin gene (*see* Fig. 18). By this mechanism of RNA splicing, an entire exon is substituted for another. In the illustration, exon 5 (E5) can be substituted for exon 4 (E4), and intron C (IC) and exon 4 become the functional equivalent of intron D (ID) of the second messenger. The shading of the exons denotes the untranslated regions of the transcript (unshaded), the signal-sequence region (crosshatched), and the prohormone sequence (shaded).

sequence of the known expressed genes in 15–20% of their bases, are present for 10% of genes, one can estimate that there has been a reintegration event in the germ line for about 10% of genes within the last 10–20 million yr. This corresponds to about 0.5–1% of genes per million yr., or once for each gene every 100–200 million yr. This is a relatively high frequency of introduction of new genetic information in evolutionary terms.

DNA that has been reverse-transcribed from mRNA may reinsert itself back into the genome by one or the other of at least two known mechanisms. First, the processed DNA may donate its RNA-based mutations to its parent gene by gene conversion, a process that appears to take place quite frequently in the genomes of higher cells (*27*). If a cDNA reverse transcribed from mRNA were matched against its original gene, then conversion in one direction could imprint the restructured intron-deficient, or intron-lacking, sequence into its genomic homolog, while leaving the promoter signal intact. This process would allow processed information to feed back into the genome in functional form.

The second possible mechanism is that RNA molecules transcribed from transposon-like elements (*28*) may be reverse-transcribed into a heterogeneous pool of DNA particles, as in the case of the Alu family of sequences. These DNA particles would be potentially genetically transmissible through the germ line in a nonmendelian fashion without necessarily integrating into chromosomal DNA. Thus, cells clonally derived from a common ancestor may become polymorphic in those molecules that have passed through an RNA phase, although retaining rigorous sequence identity in the original DNA genes. In this way, new adaptive possibilities could occur without erasing the existing phenotype. These mechanisms for providing genetic diversification are theoretically attractive because they allow nuclear information to occasionally pass through a noisy RNA copier resulting in changes in the genetic information that can then be introduced back into the parent gene. It is also possible, but not yet proved, that rearrangement and amplification of genetic sequences encoding polypeptide hormones could take place during the lifespan of an individual animal. It is well known that the enormous diversification of antibody molecules is a consequence largely of the rearrangement of genetic segments encoding regions of the immunoglobulins. Perhaps, as work progresses on the analysis of genes encoding receptor molecules, such a mechanism of genetic rearrangement coupled with somatic mutation may be involved in the generation of what must be a highly complex set of receptor proteins.

The identification of the mosaic structure of transcriptional units encoding polypeptide hormones and other proteins, consisting of exons and introns that are spliced during posttranscriptional processing, immediately raised the possibility that the use of alternative pathways in RNA splicing could provide informationally distinct molecules. One can imagine that a wide variety of different coding sequences could be generated through the use of alternative splicing mechanisms. Different coding sequences could arise by either inclusion or exclusion of specific exonic segments by utilization of parts of introns in one mRNA as exons in another mRNA. In addition, differences in the splice sites utilized would result in the expression of new codon translational reading frames. An example of an alternative splicing mechanism is the gene-encoding calcitonin, in which the production of a novel neuropeptide occurs via the translation of an mRNA arising from tissue-specific RNA processing (*29*). Alternative processing of the RNA transcribed from the calcitonin gene results in the production of an mRNA in neural tissues that is distinct from that formed in the C-cells of the thyroid. The mRNA found in the thyroid encodes a precursor to calcitonin, whereas the mRNA in the neural tissues generates a novel neuropeptide, known as the calcitonin-gene-related peptide. Although the biologic functions of this calcitonin-gene-related peptide are under investigation, immunocytochemical analyses of the distribution of the peptide in the brain and other tissues suggest functions for the peptide in nociperception, ingestive behavior, and modulation of the autonomic and endocrine systems.

There are other examples of genetic diversification that arise through the programmed flexibility in the splicing of coding regions, allowing an array of coding sequences (exons) to be put together in a number of possible useful combinations. For example, the coding sequences of immunoglobulin heavy chains can be brought together in two different ways—one to include, another to exclude an exonic coding sequence specifying part of the polypeptide chain that anchors an immunoglobulin molecule to the surface of a lymphocyte (*30*). If mRNA splicing occurs in a way excluding the anchor's peptide sequence, a circulating, rather than a surface, immunoglobulin is produced.

It appears that the splicing of the RNA precursor that encodes the tachykinins substance P and substance K can take place in at least two ways (*31*). One splicing pattern results in the mRNA that encodes substance P and another putative peptide called substance K in a common protein precursor. Other messenger RNAs are spliced to encode only substance K. An alternative RNA-splicing pattern also occurs in the processing of the transcripts arising from the gene encoding bradykinin (*32*). The high- and low-molecular weight kininogens are translated from mRNAs that differ by the alternate use of 3′ exons encoding the carboxy termini of the prohormones, a situation similar to that found in the transcription of the calcitonin gene. The primary RNA transcript from the human growth-hormone gene is processed in an alternative manner to provide 10–15% of messenger RNAs that lack a portion of the third exon of growth hormone, resulting in the translation of a growth hormone lacking 15 amino acids from its central region (*23*). Recently, analyses of the gene encoding the bombesin-like peptide, gastrin-releasing peptide, indicate that an alternative RNA-processing pathway actually changes the coding frame of the mRNA in a region corresponding to the carboxy-terminal amino acid sequence of the gastrin-releasing-hormone precursor. This event comes about by the deletion (or insertion) of a 19-base segment of the RNA during the processing of the RNA precursor (*32a*). It seems likely that, as investigations of other polypeptide-encoding genes continue, other examples of alternative splicing will be found in the processes that generate biologic diversification.

A third level in gene expression at which diversification of biologic information can take place is that of posttranslational processing. Many precursors of polypeptide hormones, particularly those encoding small peptides, contain multiple peptides that are cleaved during the posttranslational processing of the prohormones. Certain polyprotein precursors, however, appear to simply amplify the number of peptides produced, since they contain several copies of the identical peptide. Examples of prohormones that contain multiple identical peptides are the precursors encoding thyrotropin-releasing hormone (*33*) and the alpha-mating factor of yeast (*34*), each of which contains four copies of the respective peptide. Other polyproteins contain several distinct peptides, although they may be structurally related to each other. Representatives of such polyprotein are proenkephalins (*25*), proopiomelanocortin (*24*), and proglucagon (*22*).

In many instances, biologic diversification at the level of differential posttranslational processing occurs in a tissue-specific manner. The patterns of processing of proopiomelanocortin differ markedly in the anterior compared with the intermediate pituitary (*35*). In the anterior pituitary, the primary peptide products are ACTH and beta-endorphin, whereas, in the intermediate pituitary, one of the primary products is α-MSH. The smaller peptides produced are extensively modified by acetylation and phosphorylation of amino acid residues. The processing of proglucagon in the pancreatic A cells and in the intestinal L cells is quite different (*36*). In the pancreatic A cells, the predominant bioactive product of the processing of proglucagon is glucagon itself. Whether the two glucagon-like peptides are also individually processed from proglucagon in the A cells in unknown. If one assumes, however, that they are not processed, they would predictably be biologically inactive by virtue of having amino-terminal and carboxy-terminal extensions. On the other hand, in the intestinal L cell, the glucagon immunoreactive product has been shown to be a molecule, called glicentin, which consists of the amino-terminal extension of the proglucagon plus glucagon and the small C-terminal peptide known as intervening peptide I. The glicentin has no glucagon-like biologic activity, and therefore one may suppose that the bioactive peptide in the intestinal L cells is one or the other or both of the two glucagon-like peptides. The biologic activities of these glucagon-like peptides are not understood at present, but it is possible that they may function as insulinotropic and/or intestinal-growth factors.

This wide variety for the diversification of biologic information provided by the alternative pathways of gene expression is rather impressive when one considers that these pathways can occur in multiple combinations. One can reasonably expect that continued investigations of the patterns of the expression of genes encoding polypeptide hormones will reveal many more examples of the utilization of alternative pathways at the levels of the processing of RNA transcripts and protein precursors of the hormones.

# 5. GENE EXPRESSION

## *5.1. Regulation of Gene Expression*

The regulation of the expression of genes encoding polypeptide hormones can take place at one or more levels in the pathway of hormone biosynthesis (Fig. 7) (*37,38*). In addition to (1) DNA syn-

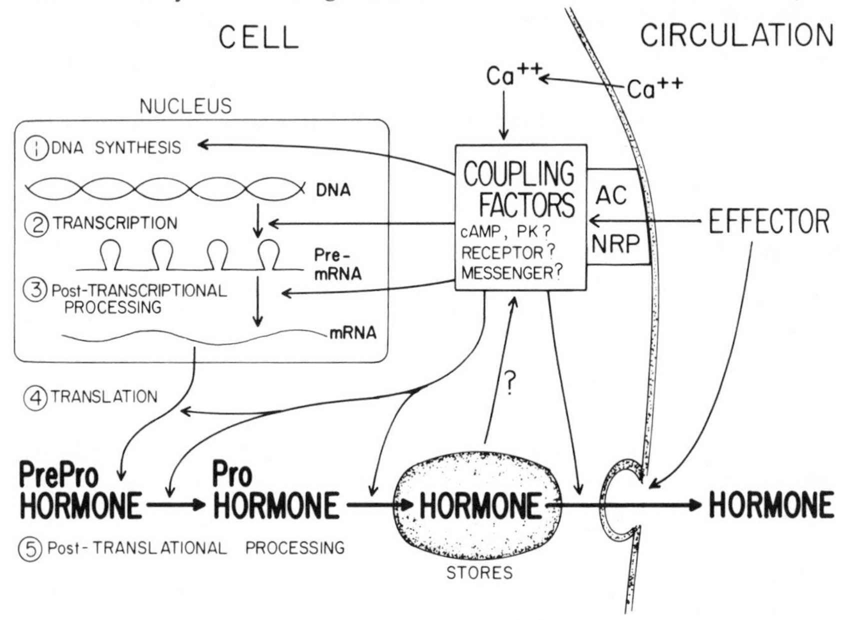


Fig. 7. Diagram of an endocrine cell showing the potential control points for the regulation of gene expression in hormone production. Specific effector substances bind to either plasma-membrane receptors (peptide effectors) or cytosolic receptors (steroids), leading to the initiation of a series of events that couple the effector signal with gene expression. Peptide–effector–receptor complex interactions appear to act initially via the activation of adenylate cyclase (AC) coupled with a nucleotide regulatory protein (NRP). The coupling factors through which other effectors act, such as glucose, thyroid hormones, and cations, are unknown but probably involve, at least in part, the activation of protein kinases and a series of phosphorylations of macromolecules. As discussed in the text, specific effectors for various endocrine cells appear to act at one or more of the indicated five levels of gene expression, with the possible exception of posttranslational processing of prohormones, for which no definite examples of metabolic regulation have yet been found.

thesis (cell growth and division), the levels include, (2) transcription, (3) posttranscriptional processing, (4) translation, and (5) posttranslational processing. In different endocrine cells, one or more levels may serve as specific control points for the regulation of the production of a particular hormone.

Although the emphasis of this review is on the control of peptide-hormone-gene expression at the level of the transcription of the genome, a brief consideration is given to control at the other levels of gene expression.

The regulation by calcium of the biosynthesis of parathyroid hormone appears to take place principally at the levels of DNA synthesis and cell division. Stimulation of the parathyroid glands by lowering levels of calcium appears to have little effect on the rates of RNA synthesis, but readily leads to hyperplasia of the glands (*39*). In addition, a decrease in the basal rates of intracellular turnover of parathyroid hormone takes place when the gland is stimulated by hypocalcemia.

Regulation of proinsulin biosynthesis appears to take place primarily at the level of translation (*40*). Within minutes after raising blood glucose levels, five- to tenfold increases are observed in the rate of proinsulin biosynthesis. It is believed that the glucose is acting either directly or indirectly through mechanisms that alter the efficiency of the initiation of translation of proinsulin mRNA.

Regulation at the level of posttranscriptional processing of mRNA precursors is not yet clearly established. However, the intriguing findings that the primary transcripts derived from the calcitonin gene and substance P are alternately spliced to provide two or more messenger RNAs encoding chimeric protein precursors, containing both common and different amino acid sequences, suggest that regulation might take place at the level of processing of the calcitonin-gene transcripts (*29*).

As yet, there are no examples of the specific physiologic regulations of polypeptide-hormone biosynthesis at the level of posttranslational processing. Although tissue-specific differences in the patterns of processing of prohormones appear to occur (as described in section 4 on generation of biologic diversification), rates of conversion of a prohormone to a hormone under physiologic circumstances in a given tissue have not been identified as a point of cellular control.

In many instances, the particular level of gene expression that is under regulatory control appears to correspond to a level that is best suited for meeting secretory and biosynthetic demands of the endocrine organ. For example, after a meal there is an immediate requirement for the release of large amounts of insulin. Because the large amounts of insulin released in the bloodstream over the course of several minutes deplete insulin stores of the pancreatic beta cells, the beta cell utilizes the immediate mechanism of increasing the translational efficiency of preformed proinsulin mRNA to provide additional hormone. In contrast, the release of parathyroid hormone remains almost constant at all times, since small fluctuations in the secretion rate are adequate to maintain the levels of serum ionized calcium within a narrow range. Because of the importance of ionized calcium in the over-all homeostatic maintenance of cells, a hormonal feedback system between ionized calcium and parathyroid-hormone secretion is tightly regulated (*41*). The consequence of this situation is that prolonged stimulation of the parathyroid gland by hypocalcemia, in chronic renal failure for example, results in a marked hyperplasia of the glands; below-normal levels of ionized calcium act as a "growth factor" for the parathyroid glands.

### 5.2. *cis* and *trans* Mechanisms of Gene Regulation

Studies of gene regulation carried out thus far, although still in the early stages of investigation, lead to a conceptual consideration of the possible mechanisms for the regulation of expressed genes. Such a consideration of gene control requires recognition of the likelihood that there are two qualitatively different modes of gene regulation. The first mode includes the factors and structural components of a gene that determine whether a gene can or cannot be expressed in a given tissue when the appropriate inducer is present. The second mode is the physiologic induction and regulation of a gene that can normally be expressed in a particular tissue.

Clearly, certain genes or certain sets of genes are expressed only in specific tissues. Two general but conceptually distinct mechanisms have been proposed for the activation of gene expression: These are *cis* mechanisms and *trans* mechanisms (Fig.

8). The separation of *cis*- and *trans*-acting regulatory mechanisms is probably more conceptual than real, since when the processes involved in gene regulation are understood, it will likely be found that both mechanisms are used. In the *cis* mechanism, the simpler of the two, a specific signaling factor interacts with a sensor, or receptor, region of the gene, resulting in the activation of transcription of the producer, or structural, gene and the formation of mRNA and specific protein. The *cis*-activation mechanism works by induction of structural or conformational changes in the chromatin through interaction of the signaling factor with the sensor region on the gene.

The *trans* mechanism of gene regulation requires the addition of a second step of gene activation in which a diffusible intermediary product of a regulatory gene is responsible for the activation of the transcription of a producer gene encoding a specific polypeptide. In this model of gene activation, the intracellular signal that arises as a consequence of the activation of the cell interacts with the sensor region of a regulatory gene, resulting in the transcription of an RNA from an associated integrator gene (acting in *cis*). The RNA either serves directly as an activator RNA or alternatively as the mRNA for the encodement of a protein that, in turn, interacts with the activator receptor responsible for initiating the transcription of the producer gene. The sum of these events represents the biologic response of the cell to the extracellular (intracellular) signal. The *cis* and *trans* models depicted in Fig. 8 reflect the ideas initially set forth by Britten and Davidson (*42,43*), who suggested that repetitive *cis* recognition sequences might provide the structural basis for the coordinate induction of unlinked structural genes. This hypothesis arose from the observations that sets of specific DNA sequences appear to exist in multiple copies in the genomes of eukaryotic cells, for example, the so-called Alu sequences, which consist of highly homologous sequences of approximately 300 bases and are present in 300,000–500,000 copies in the human genome (*44*). The proposed repetitive regulatory elements are designated as either sensor and/or receptor sequences, and may function in the coordinate expression of sets of structural genes. The sensor sequences are thought of as genomic sites that can induce activator synthesis in response to changes in external circumstances—for example, activation of a cell by a specific extracellular peptide ligand or steroid hormone. Recent studies indicate that the repeated DNA se-

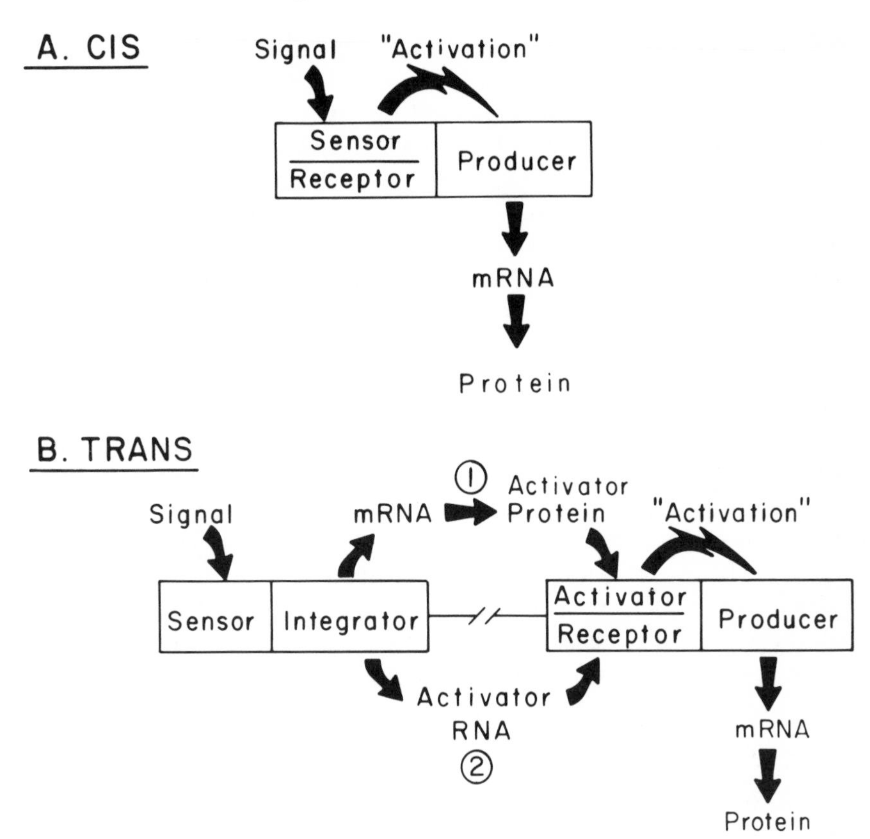


Fig. 8. Hypothetical *cis* (*A*) and *trans* (*B*) models for the activation of gene expression. In both of these models, the specific intracellular signal interacts with a sensor receptor on the gene. In the *cis* model, the sensor receptor (activator) is adjacent to the producer gene, which is the transcriptional unit, leading to the production of messenger RNA and protein. In the *trans* model, the sensor is separated from the activator receptor, which is adjacent to the producer gene, and the activator substance, originating from the sensor, acts in *trans* by transport to the activator receptor. This activator substance may either be an RNA molecule or a protein activator translated from the RNA. The activator binds or otherwise interacts with an activator-receptor sequence on the gene, resulting in initiation of transcription.

quences are scattered throughout the genome because many or all of them have, or have had, the capacity to transpose during evolution (*28,45*). This is an intriguing observation in the explanation of the evolutionary origin of new biologic forms. The dominant process in evolution is likely to have been the creation of novel, coordinately induced, structural-gene networks. One hypothetical mechanism for the evolution of new gene products is the diffusion within and around the genome of transposable *cis* regulatory sequences and their occasional insertion at productive locations in the vicinity of structural (producer) genes (*20,27,46,47*). One can envision that transposable DNA-sequence elements could occasionally carry with them and deposit specific control sequences in the right genomic environment, resulting in the expression of new gene functions.

In support of the possibility of mobile control sequences is the recent discovery of enhancer sequences located in the region of immunoglobulin genes (*48*). This discovery not only has profound implications with regard to the genetic elements involved in regulation of gene expression already described, but is the first unequivocal demonstration that enhancer sequences, originally identified in the genomes of viruses, are found in cellular genes. These enhancers consist of small sequences of 8–10 bp that markedly increase the transcription of genes into mRNA. Enhancers in immunoglobulin genes appear to be cell-specific because they induce expression of immunoglobulin genes when introduced into lymphocyte-derived cells that normally produce immunoglobulins, but not when introduced into other cell types such as fibroblasts. The enhancers can activate transcription when placed either upstream or downstream from the 5′ end of the immuoglobulin genes. Their action appears to be independent of their location or their orientation in the genome; that is, they enhance transcription when their orientation is inverted and is the reverse of its usual orientation. The discovery of these enhancer sequences may help in the elucidation of the signals needed for gene activation and may lead to a better understanding of the regulation of gene expression during development.

## 6. TISSUE-SPECIFIC GENE EXPRESSION

An understanding of the factors resulting in tissue-specific expression of genes will probably come about only when biologists gain an understanding of the factors involved in cellular differentiation. Differentiated cells possess a remarkable capacity for the selective expression of specific genes. In one cell type, a single gene may account for a large fraction of the total gene expression, yet in another cell type, the same gene may be expressed at undetectable levels.

The evidence currently available suggests that the chromatin is more loosely arranged in genes that are capable of expression than it is in those same genes residing in another tissue in which they are never expressed. For example, the DNA within the chromatin of genes and tissues in which they are expressed is more susceptible to cleavage by DNAse than it is in tissues in which the genes are quiescent (*49*). This looseness on the chromatin may facilitate access of RNA polymerase and other regulatory DNA-binding proteins to the gene for purposes of transcription. In addition, genes in tissues in which they are not expressed appear to have a higher content of methylated cytosine residues than do comparable genes in tissues in which they are expressed (*50*).

Intriguing evidence has been obtained recently that suggests the determinants for the tissues-specific transcriptional expression of genes may exist in control sequences residing within 300 bp of the 5′-flanking region of the transcriptional sequences of the genes (*51*). Recombinant-DNA techniques were used to link 5′-flanking regions of the insulin and chymotrypsin genes to the coding sequence of a chloramphenicol acetyltransferase gene—a gene that provides an enzymatic reporter function from which to detect the activity of putative gene-control regions. When these genes were introduced into cell lines prepared from pancreatic beta cells in which the insulin gene, but not the chymotrypsin gene, is expressed and into the cell line of pancreatic exocrine cells in which, conversely, the chymotrypsin gene, but not the insulin gene, is normally expressed, the insulin-gene recombinant elicited preferential expression of the chloramphenicol acetyltransferase only when introduced into the beta-cell line. Similarly, the chymotrypsin-gene recombinant elicited preferential expression only in the cell line derived from exocrine pancreatic cells. If these observations are borne out by studies of additional genes in other tissues, they will have profound implications for providing an understanding of the nature of the control elements involved in the tissue-specific expression of genes.

The simplest interpretation of these results is that a *trans*-acting positive regulatory factor, which can be termed a differentiator or activator, interacts with the upstream-control element flanking the insulin and chymotrypsin genes, resulting in the selective expression of the gene uniquely in its specific differentiated cell. It is possible that a relatively small number of differentiators may be sufficient to direct the many alterations in gene expression characteristic of the differentiated state of particular cell types. Several such *trans*-acting proteins that either activate or repress the expression of specific genes have been identified in viral genes. For example, the large T antigen of SV-40 and the EIA protein of adenovirus-2 activate not only their own respective viral encoded genes but specific genes involved in growth control within the cells that they infect as well (*52,53*). Certain products of oncogenes, such as the myc gene, appear to be DNA-binding proteins that most probably promote growth control by activation of the transcription of specific cellular genes (*54*). The EIA and myc proteins contain regions of similar amino acid sequences (*55*). The actions of EIA on gene expression can be either positive or negative; the transcription of certain genes is increased, whereas that of other genes is decreased (*56*). In certain instances, the sequences of the *cis* elements of DNA, at which gene activation DNA-binding proteins interact, are known. The SV-40 T-antigen binding is known to occur at two regions at least—a double 72-bp repeat and several 21-bp repeated segments, all located within 300 bp upstream from the site of the initiation of transcription of the early SV-40 genes (*57*). The regulatory sequences that have been identified, or are suspected of being regulatory elements, often have dyad symmetries. That is, they are palindromic sequences that have the potential capability for the formation of stem-loop structures. This property of these sequences raises the interesting possibilities that the loops may be sites of interactions of the specific binding proteins, perhaps by way of a destabilization of the chromatin or of the helical structure of the DNA caused by torsional forces on the helical DNA that comprises the loop or both.

## 7. COUPLING OF EFFECTOR ACTION TO CELLULAR RESPONSE

The second qualitative mode of gene control consists of the processes that are involved in the induction and suppression of the expression of genes that are normally expressed within a specific tissue. These processes are at work in the minute-to-minute and day-to-day regulation of the rates of production of the specific proteins produced by the cells; for example, the production of polypeptide hormones in response to their extracellular stimuli. As indicated below, at least two different classes of macromolecules, phosphoproteins and steroid-hormone receptors, appear to be involved in the physiologic regulation of hormone-gene expression.

Understanding of the precise mechanisms by which information encoded in a cellular effector is coupled to cellular response is incomplete, but there is a growing body of evidence that at least two distinct coupling processes exist in endocrine cells. First, peptide ligands bind to receptor complexes located on the plasma membrane, resulting in the activation of a series of steps that include hydrolysis of phosphatidylinositol, mobilization of calcium, formation of phosphorylated nucleotide intermediates, activation of protein kinases, and phosphorylation of specific regulatory proteins (*58*). Second, steroidal compounds, because of their hydrophobic composition, readily diffuse through the plasma membrane and bind to what are generally believed to be specific nuclear or cytosolic receptors that are then transported to the nucleus, in which they interact with other macromolecules, includinng specific domains on the chromatin located in and around the gene that is activated (*59,60*).

The cellular mechanism of protein phosphorylation often utilizes cyclic AMP as a second messenger (Fig. 9). In this model, the stimulatory factor (ligand) interacts with a receptor located within the plasma membrane and, in some manner, when bound to the plasma-membrane receptor, activates adenylate cyclase, resulting in the generation of 3′,5′-AMP, which, in turn, converts an inactive form of a phosphorylating enzyme, protein kinase, to an active form by way of dissociation of a regulatory (R) subunit from the active catalytic (C) subunit. The protein kinase (active subunit) catalyzes the phosphorylation of certain intracellular proteins, and the phosphoproteins thus formed are believed to function in the processes of gene inactivation. As indicated earlier, one demonstration of such a cascade of phosphorylated intermediates in the activation of gene expression by a peptide ligand appears to be the transcription of the prolactin gene stimulated by thyrotropin-releasing hormone (*61*). New gene transcription is detectable within minutes after interaction of the

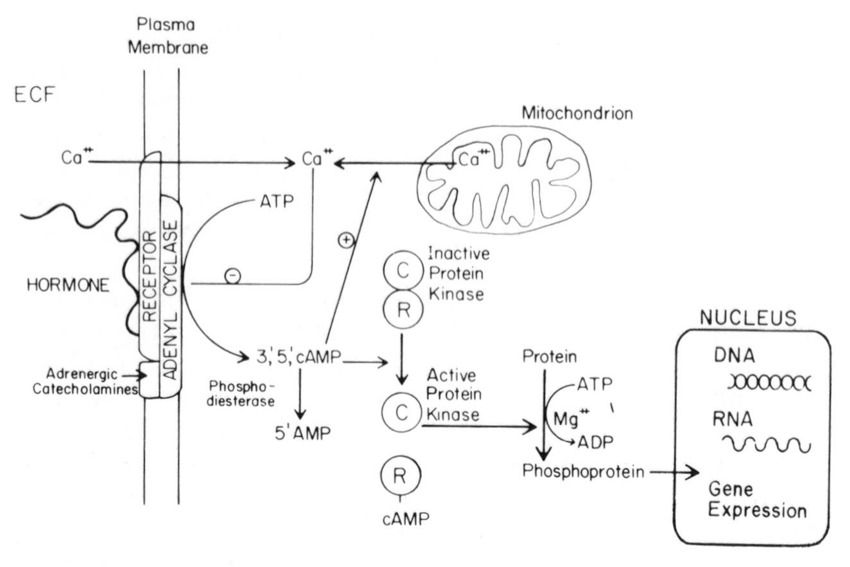


Fig. 9. Proposed cellular mechanism through which a peptide-hormone effector might activate gene expression of an endocrine cell. In the model shown, binding of the peptide to the plasma-membrane receptor activates adenylate cyclase, leading to the formation of 3′,5′-cyclic AMP and a cascade of reactions leading to the conversion of inactive to active protein kinases. The kinases then phosphorylate specific proteins. The presumed final active protein in this cascade of reactions is a phosphoprotein that interacts with regulatory sites on the gene, thereby activating gene transcription and expression. C and R refer to the catalytic and cyclic-AMP-receptor subunits of the protein kinase, respectively.

thyrotropin-releasing hormone with its plasma receptor and, concomitant with the activation of transcription, a specific subset of nuclear proteins is phosphorylated. Because of the early stage of these investigations, the exact mechanisms by which phosphoproteins activate gene transcription are unknown. One might speculate, however, that the phosphorylation of a protein substrate changes its conformation and thereby activates the protein, which in turn interacts with a chromatin "receptor," thereby allowing RNA polymerase to initiate gene transcription.

It should be noted that phosphorylated nucleotides, as well as calcium, appear to have important functions in secretory processes. In particular, fluxes of calcium from the extracellular fluid into the cell, as well as fluxes from intracellular organelles, e.g., mitochondria, into the cell sol, are closely coupled to the event of secretion (*62*).

In circumstances of the regulation of gene expression in endocrine cells that are responsive to steroids—for example, pituitary corticotropes, which produce ACTH and are regulated by cortisol, and pituitary gonadrotropes, in which the release of gonadotropins is related by gonadal steroids (estrogen and progesterone)—the steroids penetrate the plasma membrane by simple or facilitated diffusion. Upon entry into the cell, the steroid is bound to a specific hormone receptor (Fig. 10) (*60,63*). This hormone–receptor complex is then transferred to the nucleus in an "activated" form that is bound to the target-cell genome. Then, by as-yet-undefined processes, transcription of the gene takes place, followed by increased protein synthesis. Considerable progress has been made in the elucidation of the molecular structures involved in gene activation by the steroidal hormones progesterone and cortisol. The two subunits of the progesterone receptor have been isolated, receptor A and receptor B, but only receptor A contains strong DNA-binding activity. Utilizing recombinant-RNA techniques, investigators have been able to localize the specific DNA sequences in nuclei of cells from the chick oviduct to which the progesterone–receptor complex binds (*64*). These DNA sequences are undergoing intensive analysis to elucidate the properties that are responsible for the regulation of the expression of genes. It appears that these sequences are short, in the range of 10–30 bp, and that many of them form palindromes. Palindromic sequences have been noted in the putative control regions of a number of other eukaryotic genes as well (*47*).

Sequences to which the corticosteroid–receptor complex binds have also been identified. Of considerable interest in this regard has been the independent finding of three separate sequences that appear to be sites that are regulated by glucocorticoids (*47*). Each of these three sequences consists of approximately 25 bp and is homolgous in 70–90% of each bases. The sequences are located in their respective genes between 400 and 500 bases upstream, in the 5′ direction, from the site of transcriptional initiation. These three sequences identified thus far have been located in the genes encoding human proopiomelanocortin, the precursor of adrenocorticotropin and beta-endorphin, mouse mammary-tumor virus, and rat growth hormone. Although the mouse mammary-tumor virus is not a polypeptide hormone, the structure of the gene

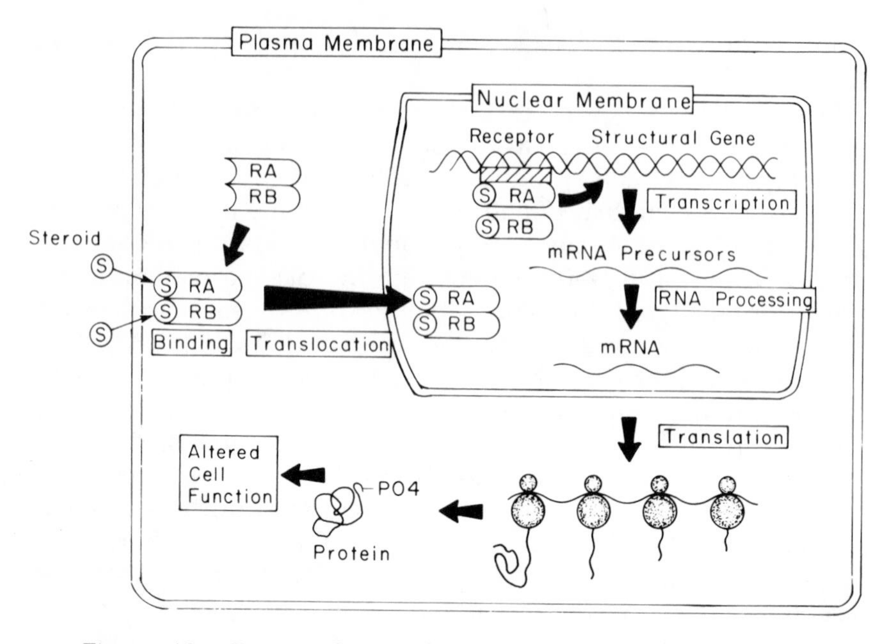


Fig. 10. Proposed mechanism of action of steroids (glucocorticoids, estrogens, progesterone) in the activation of specific gene transcription. In this model, steroid readily diffuses across the plasma membrane and binds to a cytosolic receptor consisting of two subunits, RA and RB. The steroid–receptor complex is translocated to the nucleus, in which one of the steroid–receptor subunit complexes, RA, binds to a chromatin receptor, activating the transcription of specific genes involved in steroid-hormone action. The RNA transcripts are translated into proteins that constitute the steroid-mediated response to the cell. Recent evidence suggests an alternative model in which the steroid receptor resides in the nucleus and not in the cytoplasm. Presumably in this model, the steroid diffuses through the cytoplasm into the nucleoplasm where it binds to the receptor before gene activation occurs (from ref. *59*).

encoding this virus has been of great interest to molecular biologists because the glucocorticoid-sensitive regulatory sequence is located in a long, terminally repeated segment of DNA. This gene has the characteristics of a transposable element similar to that found in yeasts and fruit flies. These observations from studies of viruses and lower eukaryotes suggest that regulatory sequences exist in mobile genetic elements that allow them to move around in the genome, perhaps providing the means to specifically activate new genes.

At present, an understanding of the participation of specific nucleotide sequences in the regulation of the expression of genes resides at the frontiers of investigative molecular biology. Major efforts are under way in an attempt to understand the precise molecular mechanisms involved in the regulation of the expression of specific genes. Once the processes of the regulation of gene expression are known, it should then be possible to plan and execute experiments that will result in the alteration of the expression of these genes. Information obtained from such experiments will, in turn, lead to a more complete understanding of gene-control mechanisms.

## REFERENCES

1. Habener J. F. (1985) Genetic Control of Hormone Formation in *Textbook of Endocrinology* (Wilson J. and Foster D., eds.), Saunders, Philadelphia.
2. Banting Best C. H., Collip J. B., Macleod J., Jr., and Noble E. C. (1933) The effect of pancreatic extract (insulin) on normal rabbits. *Am. J. Physiol.* **62,** 162–176.
3. Scharrer E. and Scharrer B. (1940) Secretory Cell Within the Hypothalamus, in *The Hypothalamus* (Assoc. Research in New Mental Disease) Hafner, New York.
4. Krieger D. T. (1983) Brain peptides: What, where and why *Science* **222,** 975–985.
5. Reichlin S. (1983) Somatostatin. *N. Eng. J. Med.* **309,** 1501–1563.
6. Habener J. F. (1985) Neuropeptides in Lower Forms of Life, in *Bioenergetics of Neurohormonal Peptide* (Hakanson R. and Thornell H., eds.), Academic, New York.
7. Kurjan J. and Herskowitz I. (1982) Structure of a yeast pheromone gene (MFx): A putative α factor precursor contains four tandem copies of mature α factor. *Cell* **30,** 933–943.
8. Loumaye E., Thorner J., and Catt K. J. (1982) Yeast-mating pheromone activates mammalian gonadotrophs: Evolutionary conservation of a reproductive hormone. *Science* **218,** 1323–1325.
9. Popenos N. A. and duVigneaud V. (1954) A partial sequence of amino acids in performic acid-oxidized vasopressin. *J. Biol. Chem.* **205,** 353–360.
10. Pierce J. G. and duVigneaud V. (1950) Preliminary studies on amino acid content of a high-potency preparation of the oxytocic hormone of the posterior lobe of the pituitary gland. *J. Biol. Chem.* **182,** 359–366.
11. Ullrich A., Shine J., Chirgwin J., Pictet R., Tischer E., Rutter W. J.,

and Goodman H. M. (1977) Rat insulin genes: Construction of plasmids containing the coding sequence. *Science* **196,** 1313–1319.

12. Seeburg P. H., Shine J., Martial J. A., Baxter J. D., and Goodman H. M. (1977) Nucleotide sequence and amplification in bacteria of the structural gene for rat growth hormone. *Nature* **270,** 486–494.
13. Kornberg R. D. and Klug A. (1981) The nucleosome. *Sci. Am.* **244,** 52–63.
14. Sharp P. A. (1981) Speculations on RNA processing. *Cell* **23,** 643–46.
15. Rogers J. and Wall R. (1980) A mechanism for RNA splicing. *Proc. Natl. Acad. Sci. USA* **77,** 1877–1879.
16. Nordheim A., Pardue M. L., Lafer E. M., Moller A., Stollar B. D., and Rich A. (1981) Antibodies to left-handed Z-DNA bind to interband regions of Drosophila polytene chromosomes. *Nature* **294,** 417–422.
17. Schon E., Evans T., Welsh J., and Efstratiatis A. (1983) Conformation of promoter DNA: Fine mapping of S1-hypersensitive sites. *Cell* **35,** 837–848.
18. Hand A. R., and Oliver C., eds. (1981) *Basic Mechanisms of Cellular Secretion.* Academic, New York.
19. Blobel G. (1980) Intracellular protein topogenesis. *Proc. Natl. Acad. Sci. USA* **77,** 1496–1500.
20. Crick F. (1979) Split genes and RNA splicing. *Science* **204,** 264–271.
21. Gilbert W. (1978) Why genes in pieces? *Nature* **271,** 501.
22. Heinrich G., Gros P., and Habener J. F. (1984) Glucagon gene sequences: Four of six genes encode separate functional domains of rat pre-proglucagon. *J. Biol. Chem.* **259,** 14082–14087.
23. Miller W. and Eberhardt N. L. (1983) Structure and evolution of the growth hormone gene family. *Endocr. Rev.* **4,** 97–130.

23a. Südhof T. C., Goldstein J. L., Brown M. S., and Russel D.W. (1985) The LDL receptor gene: A mosaic of exons shared with different proteins. *Science* **228,** 815–822.

24. Nakanishi S., Inoue A., Kita T., Nakamura M., Chang A. C. Y., Cohen S. N., and Numa S. (1979) Nucleotide sequence of cloned cDNA for bovine corticotropin-lipotropin precursor. *Nature* **278,** 423–427.
25. Noda M., Teranishi Y., Takahashi H., Toyasoto M., Notake M., Nakanishi S., and Numa S. (1982) Isolation and structural organization of the human prepro-enkephalin gene. *Nature* **297,** 431–434.
26. Perler F., Efstratiadis A., Lomedico P., Gilbert W., Kolodner R., and Dodgson J. (1980) The evolution of genes: The chicken preproinsulin gene. *Cell* **20,** 555–565.
27. Dover G. (1982) Molecular drive: A cohesive mode of species evolution. *Nature* **299,** 111–117.

28. Calos M. P. and Miller, J. H. (1980) Transposable elements. *Cells* **20,** 579–595.
29. Rosenfeld M. G., Mermod J. J., Amara S. G., Swanson L. W., Sawchenko P. E., Rivier J., Vale W. W., and Evans R. M. (1983) Production of a novel neuropeptide encoded by the calcitonin gene via tissue-specific RNA processing. *Nature* **304,** 129–136.
30. Leder P., Max E. E., and Seidman J. F. (1980) The Organization of Immunoglobulin Genes and the Origin of Their Diversity, in *Fourth International Congress of Immunology: Immunology 80.* (Fougerau M. and Dausset J., eds.), Academic, London.
31. Nawa H., Kotani H., and Nakanishi S. (1984) Tissue-specific generation of two preprotachykinin mRNAs from one gene by alternative RNA splicing. *Nature* **312,** 20–27.
32. Kitamura N., Takagaki Y., Furuto S., Tanaka T., Nawa H., and Nakanishi S. (1983) A single gene for bovine high molecular weight and low molecular weight kininogens. *Nature* **305,** 545–549.
32a. Spindel E. R., Zilberberg M. D., Habener J. F., and Chin W. W. (1986) Two prohormones for gastrin-releasing peptide are encoded by two mRNAs differing by 19 nucleotides. *Proc. Natl. Acad. Sci. USA* **83,** 19–23.
33. Richter K., Kawadshima E., Egger R., and Kreil G. (1981) Biosynthesis of thyrotropin releasing hormone in the skin of *Xenopus laevis;* partial sequence of the precursor deduced from cloned cDNA. *EMBO J.* **3,** 617–621.
34. Kurjan J. and Herskowitz I. (1982) Structure of a yeast pheromone gene (MF alpha): A putative alpha-factor. *Cell* **30,** 933–943.
35. Zakarian S. and Smyth D. G. (1982) Beta-endorphin is processed differently in specific regions of rat pituitary and brain. *Nature* **296,** 250–252.
36. Habener J. F., Lund P. K., and Goodman R. H. (1984) Complementary DNAs Encoding Precursors of Glucagon and Somatostatin, in *Biogenetics of Neurohormonal Peptides.* (Hakanson R. and Thorell J., eds.), Academic, London.
37. Darnell J. E. (1982) Variety in the level of gene control in eukaryotic cells. *Nature* **297,** 365–371.
38. Brown D. D. (1981) Gene expression in eukaryotes. *Science* **211,** 667–674.
39. Habener J. F. and Jacobs J. W. (1982) Biosynthesis and Control of Secretion of the Calcium-Regulating Peptides, in *Endocrinology of Calcium Metabolism* (Parsons, J. A., ed.), Raven, New York.
40. Itoh N. and Okamoto H. (1980) Translational control of proinsulin synthesis by glucose. *Nature* **283,** 100–102.

41. Habener J. F. (1981) Regulation of parathyroid hormone secretion and biosynthesis. *Ann. Rev. Physiol.* **43,** 211–223.
42. Britten R. J. and Davidson E. H. (1969) Gene regulation for higher cells: A theory. *Science* **165,** 349–357.
43. Davidson E. H. and Britten R. J. (1979) Regulation of gene expression: Possible role of repetitive sequences. *Science* **204,** 1052–1059.
44. Schmid C. W. and Jelinek W. R. (1982) The alu family of dispersed repetitive sequences. *Science* **216,** 1065–1070.
45. McClintock B. (1951) *Cold Spring Harbor Symp. Quant. Biol.* **16,** 13–47.
46. Davidson E. H., Jacobs H. T., and Britten R. J. (1983) Very short repeats and coordinate induction of genes. *Nature* **301,** 468–470.
47. Banerji J., Olson L., and Schaffner W. (1983) A lymphocyte-specific cellular enhancer is located downstream of the joining region in immunoglobulin heavy chain genes. *Cell* **33,** 729–774.
48. Marx J. F. (1983) Immunoglobulin genes have enhancers. *Science* **221,** 735–737.
49. Wu C. and Gilbert W. (1981) Tissue-specific exposure of chromatin structure at the 5 prime terminus of the rat preproinsulin II gene. *Proc. Natl. Acad. Sci. USA* **78,** 1577–1580.
50. Razin A. and Riggs A. D. (1980) DNA methylation and gene function. *Science* **210,** 604–610.
51. Walker M. D., Edlund T., Boulet A. M., and Rutter W. J. (1983) Cell-specific expression controlled by the 5 prime-flanking region of insulin and chymotrypsin genes. *Nature* **306,** 557–561.
52. Dynan W. S. and Tjian R. (1983) The promoter-specific transcription factor SP 1 binds to upstream sequences in the SV40 early promoter. *Cell* **79,** 87.
53. Berk A. J., Lee F., Harrison T., Williams S. J., and Sharp P. A. (1979) Pre-early adenovirus 5 gene product regulates synthesis of early viral messenger RNAs. *Cell* **17,** 935.
54. Persson H. and Leder P. (1984) Nuclear localization and DNA binding of proteins expressed by human C-myc oncogene. *Science* **225,** 718–721.
55. Ralston R. and Bishop J. M. (1983) The protein products of the *myc* and *myb* oncogenes and adenovirus E1a are structurally related. *Nature* **306,** 803.
56. Borrelli E., Hen R., and Chambon P. (1984) Adenovirus-2 EIA products repress enhancer-induced stimulation of transcription. *Nature* **312,** 608–612.
57. Gidoni D., Dynan W. S., and Tjian R. (1984) Multiple specific contacts between a mammalian transcription factor and its cognate promoters. *Nature* **312,** 409–413.

58. Cohen P. (1982) The role of protein phosphorylation in neural and hormonal control of cellular activity. *Nature* **296,** 613–620.
59. Chan L. and O'Malley B. W. (1976) Mechanism of action of the sex steroid hormones. *N. Eng. J. Med.* **294,** 1322–1437.
60. O'Malley B. (1984) Steroid hormone action in eukaryotic cells. *J. Clin. Invest.* **74,** 307–312.
61. Murdoch G. H., Franco R., Evans R. M., and Rosenfeld M. G. (1983) Polypeptide hormone regulation of gene expression. *J. Biol. Chem.* **258,** 15329–15336.
62. Rubin R. P. (1970) The role of calcium in the release of neurotransmitter substances and hormones. *Phar. Rev.* **22,** 389.
63. Baxter J. D. and Ivarie R. D. (1978) Regulation of gene expressed by glucocorticoid hormones: Studies of receptors and responses in cultured cells. *Receptors in Hormone Action* **2,** 251–284.
64. Compton J. G., Schrader W. T., and O'Malley B. W. (1983) DNA sequence preference of the progesterone receptor. *Proc. Natl. Acad. Sci. USA* **80,** 16–20.

# Chapter 3

# Insulin Genes

## Molecular Cloning and Analyses

Shu Jin Chan

## 1. INTRODUCTION

Insulin is an essential hormone in the higher-order animals, in which it functions to facilitate the assimilation of nutrients and to promote growth (for a review, *see* ref. *1*). The hormone is produced by endocrine cells in the pancreas. In humans, diabetes is often correlated with a relative or absolute deficiency in insulin production, although the precise defect(s) involved has not been determined (*2*). The primary structure of insulin, first determined by Ryle et al. in 1955 (*3*), consists of two polypeptide chains (A and B) linked by disulfide bonds. However, the discovery of proinsulin by Steiner and Oyer (*4*) and, more recently, the identification of preproinsulin (*5,6*) have clearly established that biosynthesis occurs via these larger precursor forms.

In the past few years, the development of recombinant DNA technology has extended studies on insulin production to the gene level. The objectives of this report are to review the strategies that have been used to clone insulin genes and mRNAs and to briefly describe the structural features of the insulin gene that have been revealed by DNA sequence analyses. In particular, the focus is on the occurrence of polymorphism and coding mutations in the human insulin gene and the possibility that these sequence alterations may be contributory factors in some forms of diabetes.

## 2. MOLECULAR CLONING STRATEGIES

In studies on the insulin gene, several strategies have been devised to clone both the chromosomal gene and complementary DNA (cDNA) copies of the preproinsulin mRNA. For cDNA cloning, an important initial consideration is to select a tissue that is relatively enriched in the mRNA of interest. In the case of insulin, isolated islets of Langerhans, prepared by collagenase digestion of whole pancreas, are an excellent choice because the insulin-producing beta cells comprise the majority of the cells in the islet (*5,7*). Alternatively, insulin-producing tumors, when available, have also been used successfully (*8–10*).

Total RNA is prepared by rapid homogenization and extraction of the tissue in hot SDS-phenol (*5*) or in the chaotropic salt, guanidinium isothiocyanate (*11*). Poly(A)-enriched mRNA is isolated by affinity chromatography on oligo(dT)-cellulose (*12*). The poly(A)-mRNA is converted into double-stranded (ds) cDNA in vitro by the action of reverse transcriptase and *E. coli* DNA polymerase I (*8*). After treatment with S1 nuclease to remove remaining single-stranded (ss) DNA regions, the ds-cDNA is ligated via homopolynucleotide tails or oligonucleotide linkers to a plasmid vector, such as pBR322, and transformed into an appropriate *E. coli* strain (*9*).

Several techniques have been used to identify preproinsuslin cDNA clones, including hybridization with an oligonucleotide primer (*8*) and hybrid-arrested, cell-free translation (*9*). The highly conserved sequence of insulin, however, renders it likely that genomic or cDNA clones of most mammalian species will crosshybridize under appropriately reduced stringency conditions. Thus, Perler et al. (*13*) obtained the chicken preproinsulin gene and Kwok et al. (*14*) isolated the dog preproinsulin gene, using cloned rat and human preproinsulin cDNAs as hybridization probes, respectively. Alternatively, it was found that in some species, such as the hagfish (*15*), in which the insulin sequence has diverged substantially, a useful procedure has been to hybridize all the transformed *E. coli* colonies with $^{32}$P-labeled cDNA reverse transcribed from the total islet mRNA and to analyze those clones that react most strongly after a short period of autoradiography (4–10 h). This technique is based on the finding that the proproinsulin mRNA is the predominant species in islet mRNA preparations and consequently yields a strong hybridization signal.

Although the techniques just mentioned yielded nearly full-length preproinsulin cDNA clones, an important caveat is that cDNA clones occasionally contain sequence errors (*9*). These occur predominantly near the 5′ end, and may be caused during second strand synthesis when self-priming may generate inverted sequences. Kemper observed similar errors during cloning of bovine preproparathyroid cDNA and has compiled other instances from the literature (*16*). The recent development of a procedure for cloning of full-length cDNA by Okayama and Berg (*17*), utilizing oligonucleotide primers for synthesis of both DNA strands, may help to circumvent this problem.

In contrast to the cloning of cDNA, fewer manipulations are involved in the isolation of the chromosomal gene for insulin. The standard isolation procedure was developed by Maniatis and his coworkers (*18*) and involves the creation of a genomic library containing DNA fragments cloned into bacteriophage-λ vectors. Briefly described, high molecular weight, genomic DNA is partly digested with restriction enzymes, such as *Hae*III and *Alu*I, to generate a random assortment of large overlapping DNA fragments. These fragments are ligated to isolated bacteriophage-λ-DNA arms via oligonucleotide linkers, and the recombinant λ-DNA is packaged in vitro into phage particles and amplified by growth on nutrient agar plates. The resultant genomic library is then screened for the gene of interest by *in situ* hybridization (*19*).

For a genome size of $10^9$ base pairs (bp), typically about one million plaques from the library are screened on 20 nutrient agar plates to isolate a single-copy gene, such as insulin. However, to facilitate studies on mutations in the human insulin gene, a more rapid and less tedious procedure has been developed for cloning the gene, based on the πVX recombinant system developed by Seed (*20*). In this procedure, which is outlined schematically in Fig. 1, we initially constructed plasmid πHnIns by ligating a unique 5′-upstream segment of the human insulin gene (HnIns) to the πVX plasmid (containing the supF gene), and cloned the recombinant plasmid into an *E. coli* strain, which is $sup^0$. Genomic DNA is then digested with *Eco*RI, ligated to λgtWES·EcoRI arms, packaged in vitro into recombinant phage, and grown on *E. coli* containing πHIn. Lytic growth is supported

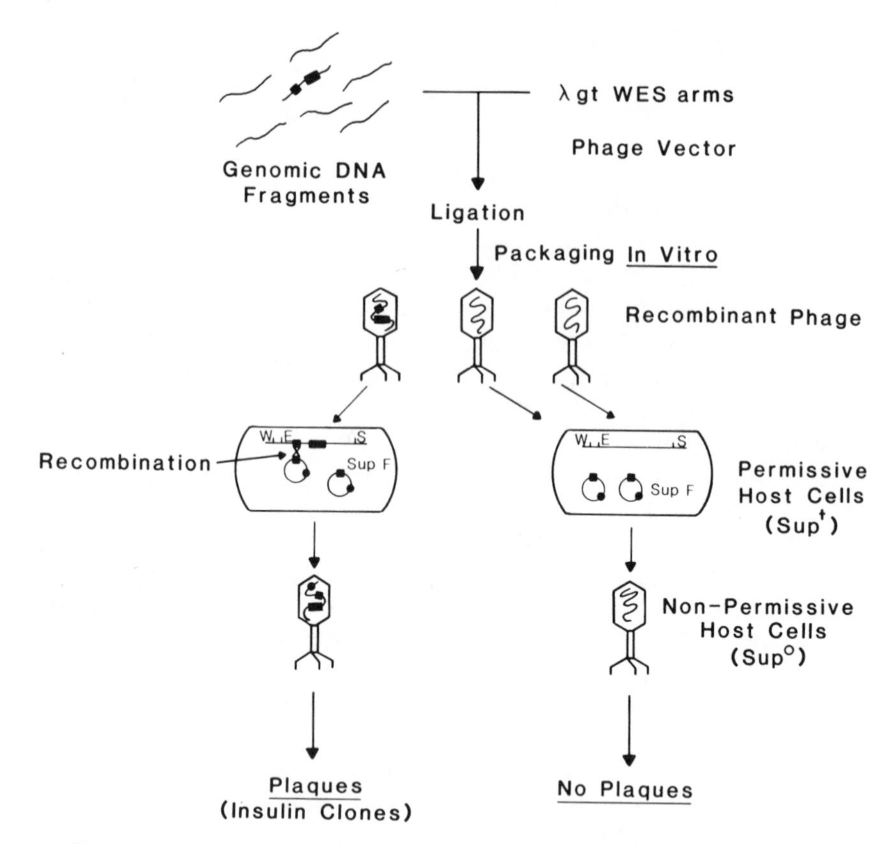


Fig. 1. Cloning of insulin genes using πVX in vivo recombination system. Genomic DNA fragment containing the insulin gene is represented by line with shaded boxes; small circle inside bacterium is πVX-HnIns. For details of cloning procedure, see text.

by the supF gene carried on πHIns, but only the phage containing the cloned insulin gene segment will recombine in vivo and thus incorporate the supF gene into the phage progeny. The recombinant insulin gene can then be selectively isolated by growth on nonpermissive ($sup^0$) *E. coli*.

## 3. STRUCTURE AND EXPRESSION OF THE INSULIN GENE

With the recombinant-DNA method described above, the insulin gene and preproinsulin mRNA have now been cloned and sequenced from a variety of species, including human (*11,21,22*), monkey (*23*), dog (*15*), rat (*8–10,24,24*), guinea pig (*26*), hamster (*27*), chicken (*14*), carp (*28*), anglerfish (*29*), and a primitive jawless vertebrate, the hagfish (*30*). In most species, insulin is represented in a single-copy gene; however, in rats (*31*) and a few species of fish (*32*), the insulin gene has been duplicated.

Comparison studies of the DNA sequence data demonstrate, as might be expected, that the insulin gene has slowly accumulated both silent and replacement nucleotide substitutions over evolutionary time (*26*). However, the structural organization of the gene has been well conserved. As illustrated in Fig. 2, all insulin genes, except rat I, contain three expressed segments (exons) separated by two intervening sequences (introns). The first intron is positioned within the 5′ untranslated region of the preproinsulin mRNA; the second intron interrupts the C-peptide coding segment between the sixth and seventh amino residues. The introns vary in size, ranging from 119 bp (*26*) to over 3500 bp (*14*). Interestingly enough, the second intron is deleted in the rat insulin I gene, whereas both introns are present in rat insulin II (*24*). However, both genes are expressed in approximately equal ratio within the beta cell (*32*).

In mammals, the mature preproinsulin mRNA is about 500 nucleotides, excluding the 3′ poly(A) tract with 330 nucleotides comprising the 110-residue coding sequence. The mRNA sequence and predicted primary structure of human preproinsulin are shown in Fig. 3. Biosynthesis studies, using radiolabeled amino acids, established that the 24-residue prepeptide segment

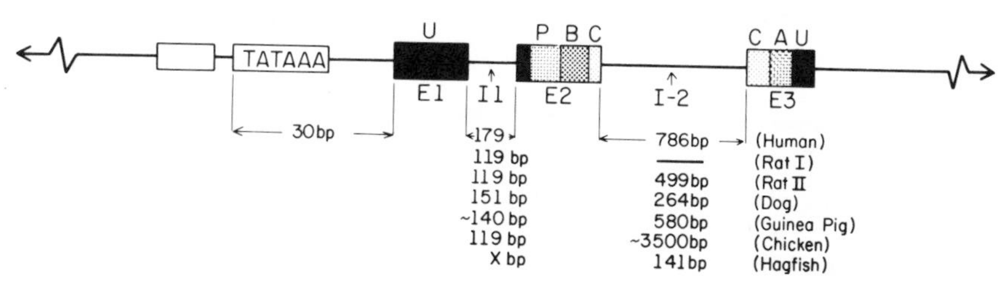


Fig. 2. Structural organization of the insulin gene. The exons (E1, E2, E3) transcribed into the mature preproinsulin mRNA encode the 5′ and 3′ untranslated regions (U), prepeptide (P), B chain (B), C-peptide (C), and A chain (A). The size of the two introns (I1, I2) varies in different species, as shown. The second intron is deleted in the rat insulin I gene; the length of the first intron in hagfish insulin gene has not been determined (S. J. Chan, unpublished results).

```
AGCCCUCCAGGACAGGCUGCAUCAGAAGAGGCCAUCAAGCAGAUCACUGUCCUUCUGCC

-24             -20                                                 -10
met ala leu trp met arg leu leu pro leu leu ala leu leu ala leu
AUG GCC CUG UGG AUG CGC CUC CUG CCC CUG CUG GCG CUG CUG GCC CUC

                                 1
trp gly pro asp pro ala ala ala phe val asn gln his leu cys gly
UGG GGA CCU GAC CCA GCC GCA GCC UUU GUG AAC CAA CAC CUG UGC GGC

     10                                      20
ser his leu val glu ala leu tyr leu val cys gly glu arg gly phe
UCA CAC CUG GUG GAA GCU CUC UAC CUA GUG UGC GGG GAA CGA GGC UUC

                         30                                 40
phe tyr thr pro lys thr arg arg glu ala glu asp leu gln val gly
UUC UAC ACA CCC AAG ACC CGC CGG GAG GCA GAG GAC CUG CAG GUG GGG

                                     50
gln val glu leu gly gly gly pro gly ala gly ser leu gln pro leu
CAG GUG GAG CUG GGC GGG GGC CCU GGU GCA GGC AGC CUG CAG CCC UUG

             60                                         70
ala leu glu gly ser leu gln lys arg gly ile val glu gln cys cys
GCC CUG GAG GGG UCC CUG CAG AAG CGU GGC AUU GUG GAA CAA UGC UGU

                             80                         86
thr ser ile cys ser leu tyr gln leu glu asn tyr cys asn AM
ACC AGC AUC UGC UCC CUC UAC CAG CUG GAG AAC UAC UGC AAC UAG ACG

CAGCCCGCAGGCAGCCCCCCACCCGCCGCCUCCUGCACCGAGAGAGAUGGAAUAAAGCCCUUG

AACCAGC
```

Fig. 3. Sequence of human preproinsulin mRNA. The sequence is modified from Bell et al. (*21*). The translated amino acid sequence for human preproinsulin is shown.

is rapidly removed after translation and sequestration of the nascent protein into the cisternal space of the endoplasmic reticulum (*7*). After transport through the Golgi apparatus and packaging into secretion granules, proinsulin is then proteolytically cleaved at the paired basic residues to release insulin and the C-peptide (*33*).

One of the interesting results that has emerged from analyses of the cloned insulin genes is the finding that the C-peptide is more highly conserved in length and overall sequence than was previously supposed. Thus, Kwok et al. (*15*) found that the dog insulin gene sequence predicted a normal 31-residue C-peptide instead of the 23-residue form originally isolated from dog pancreas and sequenced by Peterson et al. (*34*). Similarly, the guinea pig C-peptide sequence has been recently revised from 29 to 31 residues, based on the cloned gene sequence (*26*). These results suggest the possibility that conserved structural features in the C-peptide may play a functional role to facilitate the vectorial transport and correct folding of proinsulin and its ultimate proteolytic conversion to insulin.

Outside the coding sequence, the insulin genes contain the typical transcription, initiation, and termination signals found in most eukaryotic genes. A Goldberg-Hogness, or "TATA," promotor sequence is located about 30 nucleotides upstream from the transcription initiation site (*35*), whereas the hexanucleotide transcription termination signal, usually AATAAA, is found in the 3′ untranslated segment near the polyadenylation site (*36*).

In addition to these elements, there is evidence that the insulin gene may also contain tissue-specific regulatory sequences that may specifically activate or enhance transcription of the gene. Wu and Gilbert (*37*) found that chromatin isolated from rat insulinoma cells contained a DNAse I hypersensitive region that mapped near the 5′ end of the insulin gene, whereas chromatin isolated from tissue that does not express insulin, such as brain or liver, did not. This differential expression, however, did not correlate with the methylation pattern of the rat insulin genes in different tissues (*38*). Recently, Walker et al. (*39*) provided additional evidence for a cell-specific controlling element. These workers constructed plasmid vectors that contained 5′ flanking regions of the human or rat insulin genes fused to the coding sequence for chloramphenicol acetyltransferase (CAT). When these plasmids were transfected into hamster insulinoma cells, a rat exocrine pancreatic tumor, or hamster fibroblasts, only the insulinoma cells expressed significant amounts of CAT activity. From analysis of the amounts of CAT activity induced by different constructs, these workers concluded that an insulin cell-specific control element was localized between 150 and 300 bp upstream from the 5′ initiation site. Within this context, it is of interest that Lomedico et al (*24*) showed, from analysis of heteroduplex DNA formed between the rat insulin I and II genes, that a 500-bp 5′-flanking region of the two genes was highly conserved in sequence, whereas the 3′-flanking region diverged very quickly. Using computerized homology programs, significant regions of conserved homology in the 5′ flanking region of the different mammalian insulin genes were also found (MacKrell, A. and S. J. Chan, unpublished results). However, the definitive identification of regulatory DNA sequences will require additional experiments in an in vivo or in vitro trancription system combined with in vitro mutagenesis to generate altered forms of the putative activator or enhancer sequences.

## 4. POLYMORPHISM IN THE HUMAN INSULIN GENES

An unusual feature of the human insulin gene that has not been found in other cloned insulin genes was first identified by Bell and coworkers (*40,41*), who restriction mapped the gene in detail. These workers found that the human gene contained a unique polymorphic insert in the 5′-flanking region, about 1500 bp upstream from the transcription initiation site. The existence of this polymorphic region was confirmed by Ullrich et al (*42*), and DNA sequence analysis showed that it consisted of a family of tandemly repeated segments related to the sequence, ACAGGGGTGTGGGG (*43*). The size distribution of the polymorphic region is heterogeneous, but most allelic forms contain a small insert (< 600 bp); genes containing larger inserts (greater than 1600 bp) occur less frequently (*41*). Family studies confirmed that these polymorphic regions are inherited in a straightforward Mendelian fashion (*44*).

Several laboratories have investigated the distribution of the polymorphic insert in both normal and diabetic individuals. Rotwein et al. (*45*) initially reported in a study of 87 individuals that 66% of the patients with type II noninsulin-dependent diabetes contained at least one large allelic insert, whereas the corresponding frequency in normal controls and individuals with type I insulin-dependent diabetes was only 29%. In an independent study, Owerbach and Nerup obtained similar results (*46*). However, Bell et al. (*41*) reported finding no difference in allelic frequency between normals and diabetics in a survey of 56 individuals. In an attempt to resolve this discrepancy, Rotwein et al. (*47*) analyzed an additional number of subjects and also collated the data from the previous studies. The cumulative results showed that the frequency of small and large inserts was 73.4 and 26.6% in normal subjects and 65.6 and 34.4% in type II diabetics.

The above data suggest that there is no strong linkage to the insulin gene polymorphism in noninsulin-dependent diabetics. However, diabetes is known to be a heterogeneous set of disorders with multifactorial etiology. Thus, is is possible that the large insert allele may be an important contributory factor in a defined subpopulation of diabetics. Alternatively, the polymorphism may not be a causative factor *per se,* but may be loosely linked to the active locus. To investigate these possibilities, it will be important to perform further analyses on a carefully characterized and matched population of both normal and diabetic individuals.

## 5. ABNORMAL HUMAN INSULINS

In contrast to the inconclusive data on the 5′ polymorphic region, there is now strong evidence that some forms of diabetes may be a result of mutations in the insulin coding sequence that result in the production of structurally abnormal, biologically less active, insulin molecules. Patients with suspected structurally altered insulin are characterized clinically by: (1) hyperglycemia without evidence of insulin resistance, (2) normal response to exogenous insulin, (3) reduced biologic activity of endogenous insulin, (4) reduced plasma C-peptide/insulin molar ratio, and (5) hyperglycemia, which may be present or absent (*48*).

Recently, three patient-families with the just stated characteristics were studied in detail (*48–50*). In each case, insulin extracted from the patients' serum showed a markedly reduced biological activity when assayed in vitro (*48,49*). Analysis of the extracts by HPLC and radioimmunoassay revealed, in each case, the presence of an abnormal immunoreactive form, as well as a small amount of normal insulin (*50*). These results, coupled with family inheritance studies, demonstrated that the patients were heterozygous, with codominant expression of both insulin alleles. The reduced plasma concentrations of normal, relative to abnormal, insulin in these patients probably result from the reduced extraction and degradation of the mutant insulin via receptor-mediated endocytosis, a major pathway of insulin metabolism in vivo (*51*).

Both the normal and mutant insulin allelic genes have since been cloned from all three patients (*52–54*). From DNA sequence analyses, single-point mutations were identified within the coding sequence in each of the mutant genes, as shown in Fig. 4. In patient RC, a C (cytidylate) to G (guanylate) nucleotide transversion changed the codon for residue B25 from Phe to Leu (*52*). This is consistent with the earlier finding of Tager et al. (*55*), who had predicted a Leu for Phe substitution in residues B24 or B25, based on studies with semisynthetic insulin analogs.

Of the remaining two mutant genes, patient JF contained a T (thymidylate) to C transistion, which substituted Ser for Phe in residue B24, whereas patient KT had a Leu substituted for Val in the third residue of the A chain because of a G to T transversion (*53,54*). In all three mutant forms, the substitution occurred in invariant residues identified as being important for insulin biological activity (*56*). Interestingly enough, in two of the mutant genes, mutations in codons B24 or B25 destroyed an *Mbo*II restriction-

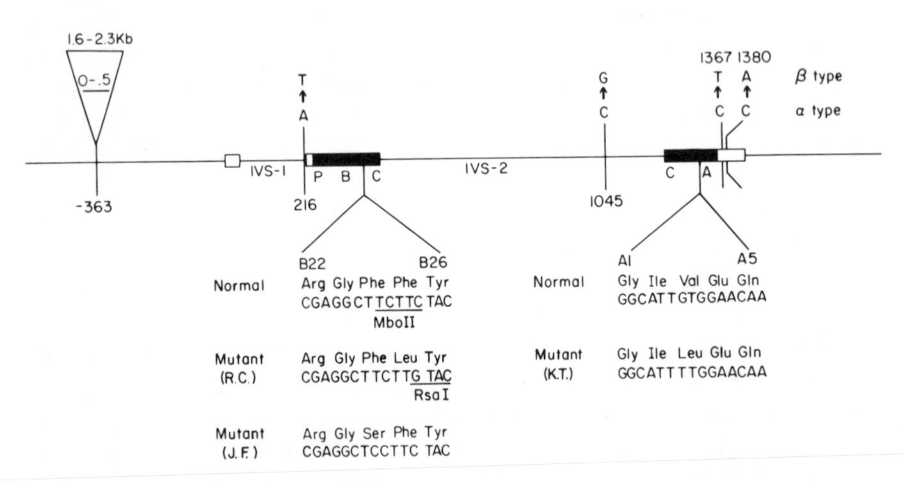


Fig. 4. Identification of point mutations in abnormal human insulin genes. The sites of single nucleotide substitutions and the predicted changes in amino acid sequence found in the abnormal insulin genes from patients RC, JF, and KT are shown. Also shown are the four nucleotide differences found in two allelic forms of the human gene as described by Ullrich et al. (22). All three mutant insulin genes are of the α type allele.

enzyme recognition site (TCTTC) that could be demonstrated on a Southern hybridization blot of genomic DNA (*57*). This diagnostic test might prove useful in analyzing other patients with suspected mutations in this important region of the insulin molecule.

In addition to the abnormal insulins just described, two patients with familial hyperproinsulinemia were also identified (*58,59*). In these individuals, alterations in the paired basic residues connecting the C-peptide to insulin inhibit the conversion of proinsulin to insulin. The sequence mutation in one patient was very recently identified by gene cloning and was shown to be a G to A (adenine) transition that substituted His and Arg at residue 65 of proinsulin (*60*).

From the examples just cited, it seems likely that additional mutations in the coding sequence of preproinsulin exist and that these mutations may affect the translation, folding, intracellular transport, or secretion of the (prepro)hormone. Moreover, analogous with the situation with the hemoglobinopathies (*61*), functionally significant mutations may also occur in the noncoding as well as coding sequences. In particular, noncoding mutations may have adverse effects on the transcription, processing, or translation of preproinsulin mRNA, but may be difficult to detect clinically because of the codominant expression of the abnormal allele. In seeking to identify these mutant insulin genes and to ascertain their relative importance in contributing to the incidence of diabetes, the further development of rapid cloning and analysis techniques, such as the πVX system, should prove useful.

## ACKNOWLEDGMENTS

I would like to express my gratitude to Dr. Donald F. Steiner for critical guidance and many helpful discussions, and in whose laboratory my research on the insulin genes was performed. Also, special thanks to Cathy Christopherson for assistance in preparing this manuscript. Support from the Lolly Coustan Memorial Fund is gratefully acknowledged.

## REFERENCES

1. Czeck M. P. (1981) Insulin action. *Am. J. Med.* **70,** 142–150.
2. Pyke D. A. (1979) Diabetes: The genetic connection. *Diabetologia* **17,** 333–343.

3. Ryle A. P., Sanger F., Smith L. F., and Kitai R. (1955) The disulfide bonds of insulin. *Biochem. J.* **60,** 541–556.
4. Steiner D. F. and Oyer P. (1967) The biosynthesis of insulin and a probable precursor of insulin by a human islet cell adenoma. *Proc. Natl. Acad. Sci. USA* **57,** 473–480.
5. Chan S. J., Keim P., and Steiner D. F. (1976) Cell-free synthesis of rat preproinsulins: Characterization and partial amino acid sequence determination. *Proc. Natl. Acad. Sci. USA* **73,** 1964–1968.
6. Patzelt C., Labrecque A., Duguid J., Carroll R., Keim P., Heinrikson R., and Steiner D. F. (1978) Detection and kinetic behavior of preproinsulin in pancreatic islets. *Proc. Natl. Acad. Sci. USA* **75,** 1260–1264.
7. Ullrich A., Shine J., Chirgwin J., Pictet R., Tischer E., Rutter W. J., and Goodman, H. M. (1977) Rat insulin genes: Construction of plasmids containing the coding sequences. *Science* **196,** 1313–1319.
8. Chan S. J., Noyes, B. E., Agarwal K. L., and Steiner D. F. (1979) Construction and selection of recombinant plasmids containing full-length complementary DNAs corresponding to rat insulins I and II. *Proc. Natl. Acad. Sci. USA* **76,** 5036–5040.
9. Villa-Komaroff L., Efstratiadis A., Broome S., Lomedico P., Tizard R., Naber S., Chick W., and Gilbert W. (1978) A bacterial clone synthesizing proinsulin. *Proc. Natl. Acad. Sci. USA* **75,** 3727–3731.
10. Bell G. I., Swain W. F., Pictet R., Cordell B., Goodman H. M., and Rutter W. J. (1979) Nucleotide sequence of a cDNA clone encoding human preproinsulin. *Nature* **282,** 525–527.
11. Chirgwin J. M., Przybyla A. E., MacDonald R. J., and Rutter W. J. (1979) Isolation of biologically active ribonucleic acid from sources enriched in ribonuclease. *Biochemistry* **18,** 5294–5299.
12. Aviv H. and Leder P. (1972) Purifiation of biologically active globin messenger RNA by chromatography on oligothymidylic acid-cellulose. *Proc. Natl. Acad. Sci. USA* **69,** 1408–1412.
13. Perler F., Efstratiadis A., Lomedico P., Gilbert W., Kolodner R., and Dodgson J. (1980). The evolution of genes: The chicken preproinsulin gene. *Cell* **20,** 555–566.
14. Kwok S. C. M., Chan S. J., and Steiner D. F. (1983) Cloning and Nucleotide sequence analysis of the dog insulin gene. *J. Biol. Chem.* **258,** 2357–2363.
15. Peterson J. D., Steiner D. F., Emdin S. O., and Falkmer S. (1975) The amino acid sequence of the insulin from a primitive vertebrate, the Atlantic hagfish (*Myxine glutinosa*). *J. Biol. Chem.* **250,** 5183–5191.
16. Weaver C. A., Gordon D. F., and Kemper B. (1981) Introduction by molecular cloning of artifactual inverted sequences at the 5′ terminus of the sense strand of bovine parathyroid hormone cDNA. *Proc. Natl. Acad. Sci. USA* **78,** 4073–4077.

17. Okayama H. and Berg P. (1982) High-efficiency cloning of full-length cDNA. *Mol Cell. Biol.* **2,** 161–170.
18. Maniatis T., Hardison R. C., Lacy E., Lauer J., O'Connell C., Quon D., Sim D. K., and Efstratiadis A. (1978) The isolation of structural genes from libraries of eucaryotic DNA. *Cell.* **15,** 687–701.
19. Benton W. D. and Davis R. W. (1977) Screening λgt recombinant clones by hybridizatin to single plaques *in situ*. *Science* **196,** 180–182.
20. Seed B. (1983) Purification of genomic sequences from bacteriophage libraries by recombination and selection in vivo. *Nucleic Acids Res.* **11,** 2427–2445.
21. Bell G. I., Pictet R. L., Rutter W. J., Cordell B., Tischer E., and Goodman H. M. (1980) Sequence of the human insulin gene. *Nature* **284,** 26–32.
22. Ullrich A., Dull T., Gray A., Brosius J., and Sures I (1980) Genetic variation in the human insulin gene. *Science* **209,** 612–615.
23. Wetekam W., Groneberg J., Leineweber M., Wengenmayer F., and Winnacker E. (1982) The nucleotide sequence of cDNA coding for preproinsulin from the primate *Macaca fascicularis*. *Gene* **19,** 179–183.
24. Lomedico P., Rosenthal N., Efstratiadis A., Gilbert W., Kolodner R., and Tizard R. (1979) The structure and evolution of the two nonallelic rat preproinsulin genes. *Cell* **18,** 545–558.
25. Cordell B., Bell G., Tischer E., DeNoto F., Ullrich A., Pictet R., Rutter W. J., and Goodman H. M. (1979) Isolation and characterization of a cloned rat insulin gene. *Cell* **18,** 533–543.
26. Chan S. J., Episkopou V., Zeitlin S., Karathanasis S., MacKrell A., Steiner D. F., and Efstratiadis A (1984) Guinea pig preproinsulin gene: An evolutionary compromise? *Proc. Natl. Acad. Sci. USA* **81,** 5046–5050.
27. Bell G. I. and Sanchez-Pescador R (1984) Sequence of a cDNA encoding syrian hamster preproinsulin. *Diabetes* **33,** 297–300.
28. Liebscher D., Coutelle C., Rapoport T., Hahn V., Rosenthal S., Prehn S., and Williamson R. (1980) Cloning of carp preproinsulin cDNA in the bacterial plasmid pBR322. *Gene* **9,** 233–245.
29. Hobart P. M., Shen L., Crawford R., Pictet R., and Rutter W. J. (1980) Comparison of the nucleic acid sequence of angler fish and mammalian insulin mRNAs from cloned cDNAs. *Science* **210,** 1360–1363.
30. Chan S. J., Emdin S. O., Kwok S. C. M., Kramer J., Falkmer S., and Steiner D. F. (1981) Messenger RNA sequence and primary structure of preproinsulin in a primitive vertebrate, the Atlantic Hagfish. *J. Biol. Chem.* **256,** 7595–7602.
31. Clark J. L. and Steiner D. F. (1969) Insulin biosynthesis in the rat: Demonstration of two proinsulins. *Proc. Natl. Acad. Sci. USA* **62,** 278–285.

32. Yamamoto M., Kotakii A., Okuyama T., and Satake K. (1960) Studies on insulin I. Two different insulins from Langerhans islet of Bonito fish. *J. Biol. Chem.* **48,** 84.
33. Steiner D. F., Kemmler W., Tager H. S., and Peterson J. D. (1974) Proteolytic processing in the biosynthesis of insulin and other proteins. *Fed. Proc.* **33,** 2105–2115.
34. Peterson J. D., Nehrlich S., Oyer P. E., and Steiner D. F. (1972) Determination of the amino acid sequence of the monkey, sheep, and dog proinsulin C-peptides by a semi-micro Edman degradation procedure. *J. Biol. Chem.* **247,** 4866–4871.
35. Gannon F., O'Hare K., Perrin F., LePennec J. P., Benoist C., Cochet M., Breathnach R., Royal A., Garapin A., Cami B., and Chambon P. (1979) Organization and sequences at the 5′ end of a cloned complete ovalbumin gene. *Nature* **278,** 428–434.
36. Proudfoot N. J. and Brownlee G. G. (1976) 3′ non-coding region sequences in eukaryotic messenger RNA. *Nature* (Lond.) **263,** 211–214.
37. Wu C. and Gilbert W. (1981) Tissue-specific exposure of chromatin structure at the 5′ terminus of the rat proproinsulin II gene. *Proc. Natl. Acad. Sci. USA* **78,** 1577–1580.
38. Cate R. L., Chick W., and Gilbert W. (1983) Comparison of the methylation patterns of the two rat insulin genes. *J. Biol. Chem.* **258,** 6645–6652.
39. Walker M. D., Edlund T., Boulet A. M., and Rutter W. J. (1983) Cell-specific expression controlled by the 5′ flanking region of insulin and chymotrypsin genes. *Nature* **306,** 557–561.
40. Bell G. I., Pictet R., and Rutter W. J. (1980) Analysis of the regions flanking the human insulin gene and sequence of an Alu family member. *Nucleic Acids Res.* **8,** 4091–4109.
41. Bell, G. I., Karam J., and Rutter W. J. (1981) Polymorphic DNA region adjacent to the 5′ end of the human insulin gene. *Proc. Natl. Acad. Sci. USA* **78,** 5759–5763.
42. Ullrich A., Dull T. J., Gray A., Philips III J. A., Peter S. (1982) Variation in the sequence and modification state of the human insulin gene flanking regions. *Nucleic Acids. Res.* **10,** 2225–2240.
43. Bell G. I., Selby M. J., and Rutter W. J. (1982) The highly polymorphic region near the human insulin gene is composed of simple tandemly repeating sequences. *Nature* **295,** 31–35.
44. Owerbach D., Poulsen S., Billesbølle P., and Nerup J. (1982) DNA insertion sequences near the insulin gene affect glucose regulation. *Lancet* **i,** 880.
45. Rotwein P., Chyn R., Chirgwin J., Cordell B., Goodman H. M., and Permutt M. A. (1981) Polymorphism in the 5′-flanking region of the human insulin gene and its possible relation to Type 2 Diabetes. *Science* **213,** 1117–1120.

46. Owerbach D. and Nerup J. (1982) Restriction fragment length polymorphism of the insulin gene in Diabetes Mellitus. *Diabetes* **31,** 275–277.
47. Rotwein P. S., Chirgwin J., Province M., Knowler W. C., Petitt D. J., Cordell B., Goodman H. M., and Permutt M. A. (1983) Polymorphism in the 5′ flanking region of the human insulin gene: A genetic marker for non-insulin dependent diabetes. *New Engl. J. Med.* **308,** 65–71.
48. Haneda M., Polonsky K., Bergenstal R., Jaspan J., Shoelson S., Blix P., Chan S., Kwok S., Wishner W., Zeidler A., Olefsky J., Freidenberg G., Tager H., Steiner D., and Rubenstein A. (1984) Familial hyperinsulinemia due to a structurally abnormal insulin. Definition of an emerging new clinical syndrome. *New Engl. J. Med.* **310,** 1288–1294.
49. Given B., Mako M., Tager H., Baldwin D., Markese J., Rubenstein A., Olefsky J., Kobayashi M., Kolterman O., and Poucher R. (1980) Diabetes due to secretion of an abnormal insulin. *New Engl. J. Med.* **302,** 129–135.
50. Shoelson S., Haneda M., Blix P., Nanjo A., Sanke T., Inouye K., Steiner D., Rubenstein A., and Tager H. (1983) Three mutant insulins in man. *Nature* **302,** 540–543.
51. Terris S. and Steiner D. F. (1975) Binding and degradation of $^{125}$I-insulin by rat hepatocytes. *J. Biol. Chem.* **250,** 8389–8398.
52. Kwok S. C. M., Steiner D. F., Rubenstein, A. H., and Tager H. S. (1983) Identification of a point mutation in the human insulin gene giving rise to a structurally abnormal insulin (Insulin Chicago). *Diabetes* **32,** 872–875.
53. Handeda M., Chan S. J., Kwok S. C. M., Rubenstein A. H., and Steiner D. F. (1983) Studies on mutant insulin genes: Identification and sequence analysis of a gene encoding [Ser-B24] insulin. *Proc. Natl. Acad. Sci. USA* **80,** 6366–6370.
54. Chan S. J., Kwok, S. C. M., Nanjo A., Sanke T., and Steiner D. F., unpublished results.
55. Tager H., Given B., Baldwin D., Mako M., Markese J., Rubenstein A., Olefsky J., Kobayashi M., Kolterman O., and Poucher R. (1979) A structurally abnormal insulin causing human diabetes. *Nature* **281,** 122–125.
56. Pullen R. A., Lindsay D. G., Wood S. P., Tickle I. J., Blundell T. L., Woolmer A., Krail G., Brandenberg D., Zahn H., Gliemann J., and Gammeltoft S. (1976) Receptor-binding region of insulin. *Nature* **259,** 369–373.
57. Kwok S. C. M., Chan S. J., Rubenstein A. H., Poucher R., and Steiner D. F. (1981) Loss of a restriction endonuclease cleavage site in the gene of a structurally abnormal human insulin. *Biochem. Biophys. Res. Comm.* **98,** 844–849.

58. Gabbay K. H., Bergenstal R. M., Wolff J., Mako M. E., and Rubenstein A. H. (1979) Familial hyperproinsulinemia: Partial characterization of circulating proinsulin-like material. *Proc. Natl. Acad. Sci. USA* **76,** 2881–2885.
59. Kanazawa Y., Hayashi M., Ikeuchi M., et al. (1979) Familial Proinsulinemia: A Rare Disorder of Insulin Biosynthesis in, *Proinsulin, Insulin, and C-peptide.* (Baba S., Kaneko T., Yanaihara N. (eds.)) Excerpta Medica, Amsterdam.
60. Shibasaki Y., Kanazawa Y., Akanuma Y., Kawakami T., and Takaku F. (1984) Abnormal proinsulin: Determination of amino acid substitution by recombinant DNA technology. *Diabetes* **33,** 335. (Abstract)
61. Orkin S. H. and Kazazian Jr., H. H. (1984) The mutation and polymorphism of the human β-globin and its surrounding DNA. *Ann. Rev. of Genet.* **13,** 131–180.

# Chapter 4

# The Glucagon Genes

GERHARD HEINRICH AND JOEL F. HABENER

## 1. INTRODUCTION

### *1.1. Background*

Glucagon is a peptide hormone of 29 amino acids produced and secreted by the A cells of the pancreatic islets (*1*). It is a member of a structurally related group of peptides that includes secretin (*2*), vasoactive intestinal peptide (*3*), gastric inhibitory peptide (*4*), and growth hormone releasing hormone (*5*) (Fig. 1). The secretion of glucagon is regulated by blood levels of glucose (*6*) and amino acids (*7*), as well as by a variety of hormonal stimuli (*8*). The action of glucagon on its target tissues, particularly the liver, is an important factor in protein and carbohydrate metabolism (*9,10*). Abnormal regulation of glucagon gene expression has been implicated in the pathogenesis of diabetes mellitus (*11*).

Peptides related immunologically to glucagon are produced in a number of extrapancreatic tissues that include brain (*12*), salivary glands (*13*), and intestine (*14*). The glucagon-related peptides helospectin (*15*) and helodermin (*16*) are produced and secreted by the salivary glands of a reptile, the gila monster.

Although the principal hormonal function of glucagon is to regulate carbohydrate, fat, and protein metabolism, it is likely that glucagon and glucagon-related peptides exercise regulatory functions both as endocrine agents carrying information from one organ to another, and/or as paracrine agents communicating with adjacent cells, for example within pancreatic islet cells and as neurotransmitters or growth factors within the nervous system.

| PEPTIDE | 1 5 10 15 20 25 30 35 40 45 |
|---|---|
| GLUCAGON | H-S-Q-G-T-F-T-S-D-Y-S-K-Y-L-D-S-R-R-A-Q-D-F-V-Q-W-L-M-N-T |
| VIP: | H-S-D-A-V-F-T-D-N-Y-T-R-L-R-K-Q-M-A-V-K-K-Y-L-N-S-I-L-N* |
| GIP: | Y-A-E-G-T-F-I-S-D-Y-S-I-A-M-D-K-I-R-Q-Q-D-F-V-N-W-L-L-A-Q-K-G-K-K-S-D-W-K-H-N-I-T-Q |
| SECRETIN: | H-S-D-G-T-F-T-S-E-L-S-R-L-R-D-S-A-R-L-Q-R-L-L-Q-G-L-V* |
| GRF: | H-A-D-A-I-F-T-S-S-Y-R-R-I-L-G-Q-L-Y-A-R-K-L-L-H-E-I-M-N-R-Q-Q-G-E-R-N-Q-E-Q-R-S-R-F-N |
| PHI: | H-A-D-G-V-F-T-S-D-F-S-R-L-L-G-Q-L-S-A-K-K-Y-L-E-S-L-I* |
| GLP I: | H-A-E-G-T-F-T-S-D-V-S-S-Y-L-E-G-Q-A-A-K-E-F-I-A-W-L-V-K-G-R-G-R-R-D |
| GLP II: | H-A-D-G-S-F-S-D-E-M-N-T-I-L-D-N-L-A-T-R-D-F-I-N-W-L-I-Q-T-K-I-T-D-K-K |
| FI GRCP: | H-A-D-G-T-F-T-S-D-V-S-S-Y-L-K-D-Q-A-I-K-D-F-V-D-R-L-K-A-G-0-V-R-R-E |
| FII GRCP: | H-A-D-G-T-Y-T-S-D-V-S-S-Y-L-Q-D-Q-A-A-K-D-F-V-S-W-L-K-A-G-R-G-R-R-E |
| HELOSPECTIN: | H-S-D-A-T-F-T-A-E-Y-S-K-L-L-A-K-L-A-L-Q-K-Y-L-E-S-I-L-G-S-S-T-S-P-R-P-P-S-S |
| HELODERMIN: | H-S-D-A-I-F-T-Q-Q-Y-S-K-L-L-A-K-L-A-L-Q-K-Y-L-A-S-I-L-G-S-R-T-S-P-P-P* |
| | 1 5 10 15 20 25 30 35 40 45 |

Fig. 1. The glucagon-related family of peptides. The sequences of known peptides related in structure to glucagon are shown. The underlined amino acids indicate identity with those found in glucagon. An asterisk at the COOH-terminal amino acid indicates that the last amino acid is amidated. Abbreviations: VIP, human vasoactive intestinal peptide; GIP, porcine gastric inhibitory peptide; secretin, porcine secretin; GRF, rat growth hormone releasing hormone; PHI, porcine peptide coencoded with VIP in the prohormone; GLP, rat glucagon-like peptides; FI and FII GRCP, anglerfish glucagon-related COOH-terminal peptides encoded by genes I and II.

Our interest in investigating the structure and expression of the glucagon and glucagon-related genes arose from several observations: (a) Glucagon is a member of a large multigene family in which each gene tends to encode multiple peptides in the form of polyprotein precursors, or prohormones. We are interested in the genetic mechanisms that have created the diversity of genes within the glucagon gene family, and the posttranscriptional mechanisms that generate the even greater diversity of biologically active peptides; (b) The various members of the glucagon gene family are expressed in diverse tissues and cell types, including brain, pancreas, intestine, and salivary glands. We are interested in the molecular mechanisms that determine the tissue and cell-specific expression of each of these closely related genes; (c) The multigenic character of the glucagon gene family raises the likelihood that there are additional glucagon-related family members to be discovered in the genome.

### 1.2. Recombinant cDNAs Encoding Preproglucagon

Analysis of mRNAs isolated from anglerfish islets has shown that at least two glucagon genes encode two separate and distinct mRNAs (*17*). Nucleotide sequence analyses of cloned cDNAs corresponding to these mRNAs reveal that the biosynthetic precursors of the two anglerfish glucagons both contain a glucagon and a second glucagon-like peptide arranged in tandem in polyproteins of $M_r$ = 12,500 and $M_r$ = 14,500 (*12*) (Fig. 2).

We have recently determined the nucleic acid sequence of cloned cDNAs corresponding to the mRNA encoding rat preproglucagon (*18*) (Fig. 3). The rat glucagon precursor of $M_r$ = 18,000 encoded by this mRNA is longer than the anglerfish preproglucagons, and contains the sequences of glucagon and two, rather than one, glucagon-like peptides arranged in tandem. Similar structures deduced from nucleic acid sequences have been reported for bovine (*19*), Syrian hamster (*20*), and human (*21*) glucagon precursors.

## 2. CLONING AND CHARACTERIZATION OF RAT PREPROGLUCAGON mRNAs AND cDNAs

### 2.1. Preproglucagon cDNAs

Total polyadenylated RNA was extracted from pancreas of newborn rats using the method described previously (*22*). Double-stranded cDNA with poly(dC) extensions at the 3′ termini was synthesized as described (*23*). Avian myeloblastosis virus reverse transcriptase was obtained from Life Sciences, St. Petersburg, FL, and the Klenow fragment of DNA polymerase I and S1 nuclease was obtained from New England Biolabs (Beverly, MA). Terminal transferase was obtained from PL Biochemicals (Milwaukee, WI).

The collection of cDNAs was inserted into the unique *Pst*I site of the plasmid pBR322 using the technique of poly(dG): (dC)

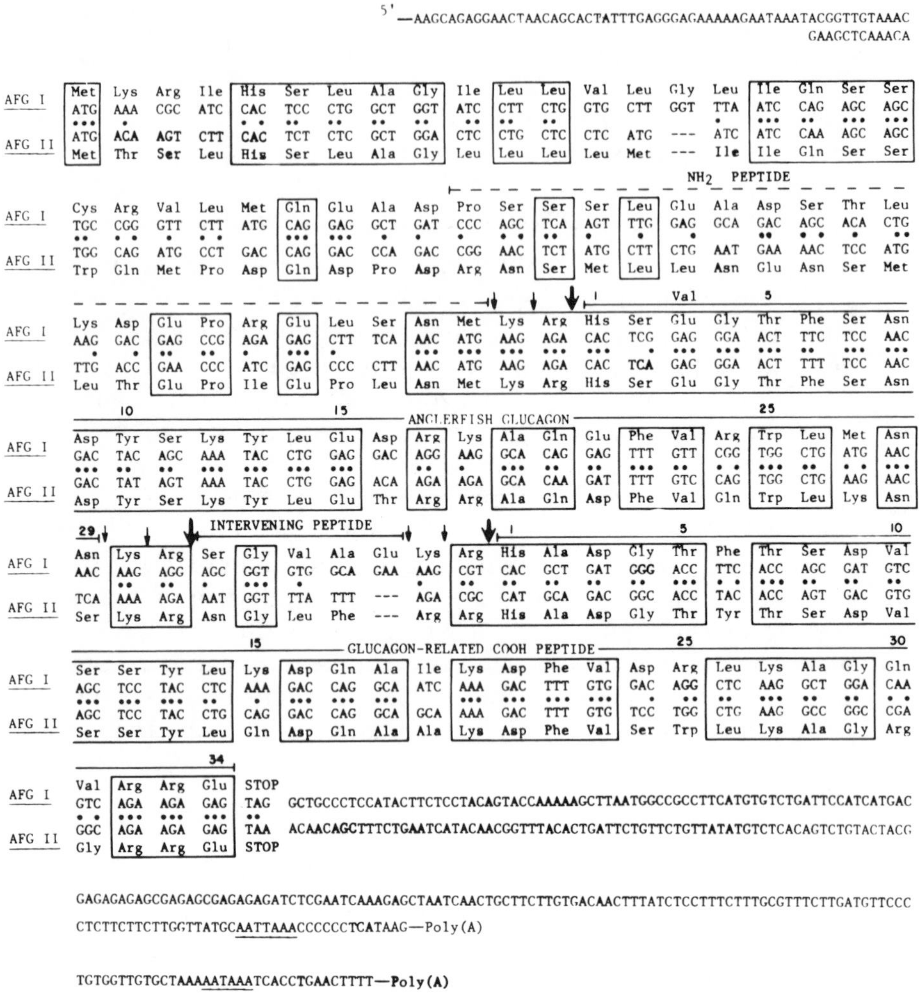

```
                                              5'—AAGCAGAGGAACTAACAGCACTATTTGAGGGAGAAAAAGAATAAATACGGTTGTAAAC
                                                                                              GAAGCTCAAACA

AFG I    Met Lys Arg Ile His Ser Leu Ala Gly Ile Leu Leu Val Leu Gly Leu Ile Gln Ser Ser
         ATG AAA CGC ATC CAC TCC CTG GCT GGT ATC CTT CTG GTG CTT GGT TTA ATC CAG AGC AGC
AFG II   ATG ACA AGT CTT CAC TCT CTC GCT GGA CTC CTG CTC CTC ATG --- ATC ATC CAA AGC AGC
         Met Thr Ser Leu His Ser Leu Ala Gly Leu Leu Leu Leu Met --- Ile Ile Gln Ser Ser

                                                                   NH2 PEPTIDE
AFG I    Cys Arg Val Leu Met Gln Glu Ala Asp Pro Ser Ser Ser Leu Glu Ala Asp Ser Thr Leu
         TGC CGG GTT CTT ATG CAG GAG GCT GAT CCC AGC TCA AGT TTG GAG GCA GAC AGC ACA CTG
AFG II   TGG CAG ATG CCT GAC CAG GAC CCA GAC CGG AAC TCT ATG CTT CTG AAT GAA AAC TCC ATG
         Trp Gln Met Pro Asp Gln Asp Pro Asp Arg Asn Ser Met Leu Leu Asn Glu Asn Ser Met

                                                     1       Val     5
AFG I    Lys Asp Glu Pro Arg Glu Leu Ser Asn Met Lys Arg His Ser Glu Gly Thr Phe Ser Asn
         AAG GAC GAG CCG AGA GAG CTT TCA AAC ATG AAG AGA CAC TCG GAG GGA ACT TTC TCC AAC
AFG II   TTG ACC GAA CCC ATC GAG CCC CTT AAC ATG AAG AGA CAC TCA GAG GGA ACT TTT TCC AAC
         Leu Thr Glu Pro Ile Glu Pro Leu Asn Met Lys Arg His Ser Glu Gly Thr Phe Ser Asn

          10                  15          ANGLERFISH GLUCAGON            25
AFG I    Asp Tyr Ser Lys Tyr Leu Glu Asp Arg Lys Ala Gln Glu Phe Val Arg Trp Leu Met Asn
         GAC TAC AGC AAA TAC CTG GAG GAC AGG AAG GCA CAG GAG TTT GTT CGG TGG CTG ATG AAC
AFG II   GAC TAT AGT AAA TAC CTG GAG ACA AGA AGA GCA CAA GAT TTT GTC CAG TGG CTG AAG AAC
         Asp Tyr Ser Lys Tyr Leu Glu Thr Arg Arg Ala Gln Asp Phe Val Gln Trp Leu Lys Asn

         29          INTERVENING PEPTIDE          1               5               10
AFG I    Asn Lys Arg Ser Gly Val Ala Glu Lys Arg His Ala Asp Gly Thr Phe Thr Ser Asp Val
         AAC AAG AGG AGC GGT GTG GCA GAA AAG CGT CAC GCT GAT GGG ACC TTC ACC AGC GAT GTC
AFG II   TCA AAA AGA AAT GGT TTA TTT --- AGA CGC CAT GCA GAC GGC ACC TAC ACC AGT GAC GTG
         Ser Lys Arg Asn Gly Leu Phe --- Arg Arg His Ala Asp Gly Thr Tyr Thr Ser Asp Val

                      15        GLUCAGON-RELATED COOH PEPTIDE         25              30
AFG I    Ser Ser Tyr Leu Lys Asp Gln Ala Ile Lys Asp Phe Val Asp Arg Leu Lys Ala Gly Gln
         AGC TCC TAC CTC AAA GAC CAG GCA ATC AAA GAC TTT GTG GAC AGG CTC AAG GCT GGA CAA
AFG II   AGC TCC TAC CTG CAG GAC CAG GCA GCA AAA GAC TTT GTG TCC TGG CTG AAG GCC GGC CGA
         Ser Ser Tyr Leu Gln Asp Gln Ala Ala Lys Asp Phe Val Ser Trp Leu Lys Ala Gly Arg

                  34
AFG I    Val Arg Arg Glu STOP
         GTC AGA AGA GAG TAG  GCTGCCCTCCATACTTCTCCTACAGTACCAAAAAGCTTAATGGCCGCCTTCATGTGTCTGATTCCATCATGAC
AFG II   GGC AGA AGA GAG TAA  ACAACAGCTTTCTGAATCATACAACGGTTTACACTGATTCTGTTCTGTTATATGTCTCACAGTCTGTACTACG
         Gly Arg Arg Glu STOP

         GAGAGAGAGCGAGAGCGAGAGAGATCTCGAATCAAAGAGCTAATCAACTGCTTCTTGTGACAACTTTATCTCCTTTCTTTGCGTTTCTTGATGTTCCC
         CTCTTCTTCTTGGTTATGCAATTAAACCCCCCTCATAAG—Poly(A)

         TGTGGTTGTGCTAAAAATAAATCACCTGAACTTTT—Poly(A)
```

Fig. 2. Nucleotide and corresponding deduced amino acid sequences of cDNAs encoding two anglerfish preproglucagons [AFG I (*17*) and AFG II]. The cDNA sequences shown were obtained by complete sequencing of both strands of each of five separate cDNAs. Dots between the nucleotide sequences indicate identical nucleotides (two single codon gaps shown as dashes were introduced into the AGF II sequence to maximize homology). The amino acid sequences of the two precursors, derived from the nucleotide sequences of the cDNAs, are shown above the nucleotide sequences. Boxed regions indicate regions of identity of the amino acid sequences in the two precursors. Arrows represent predicted prohormone cleavage sites (heavy arrows predict

homopolymer extension and ligation (*24*). The hybrid cDNAs were transfected into *Escherichia coli* strain MC 1061 (*25*). Bacterial clones containing recombinant plasmids were selected by their resistance to tetracycline and sensitivity to ampicillin. Plasmid pBR322 extended with poly(dG) at the 3′-OH termini of the *Pst*I site was obtained from New England Nuclear (Boston, MA).

Bacterial colonies containing recombinant plasmids were absorbed to nitrocellulose filters and screened by the colony hybridization method (*26*). The probe used for hybridization was a cDNA-encoding Syrian hamster preproglucagon that was kindly provided by G. I. Bell. The cDNA is a nearly full-length representation of the corresponding mRNA (*20*). The cDNA was labeled with [$\alpha^{32}$P]-dCTP by nick translation (*20*).

From approximately 30,000 transformants we obtained 27 bacterial clones (0.1%) containing recombinant plasmids that hybridized to the hamster preproglucagon cDNA. The surprisingly high frequency of hybridizing recombinants is consistent with the known preponderance of glucagon-producing cells within neonatal pancreas (0.7–1.5% v/v) compared to adult pancreas (0.18% v/v) (*27*). Twelve of the 27 hybridizing clones were selected for more detailed analysis of their resident plasmids. Preparation and restriction analysis of the plasmids (*28*) revealed that two plasmids, pRGlu5 and pRGlu16, had DNA inserts of 0.9 and 1.1 kb (kilobase), respectively.

## 2.2. Hybridization Analysis of Preproglucagon mRNA

The size of the mRNA-encoding rat preproglucagon was estimated by electrophoresis on a formaldehyde-agarose gel. Plasmid pRGlu16 was used as a hybridization probe. There appears a single hybridizing mRNA of 1400 ± 100 nucleotides in the neonatal pancreas (Fig. 4). Therefore, the longer of the two cDNAs rep-

---

Fig. 2. (*cont.*) sites of cleavage by trypsin-like activity; light arrows predict sites of cleavage by carboxypeptidase B-like activity). Horizontal lines delineate peptides formed by potential cleavages at the arrowed sites. The line indicating the $NH_2$-terminal peptide is dashed to indicate uncertainty as to the site of cleavage of the signal sequence. The underlined DNA sequences AATAAA and AATTAAA at the 3′ ends of AFG I and AFG II cDNAs, respectively, are characteristic of sites involved in addition of the poly(A) tract to eukaryotic mRNAs.

```
                                                                   -20        Asn Ile
5'                                                                  Met  Lys  Thr  Val  Tyr
aaaggagctccacctgtctacacctcctctcagctcagtcccacaaggcagaataaaaaaATG AAG ACC GTT TAC  75
                                                                              A    A
————— SIGNAL PEPTIDE ————————————|
             Phe       Cys  Gly  Ala  Gly                      -1   +1   Ser  Leu
Ile  Val  Ala  Gly  Leu  Phe  Val  Met  Leu  Val  Gln  Gly  Ser  Trp  Gln  His  Ala  Pro  Gln  Asp
ATC  GTG  GCT  GGA  TTG  TTT  GTA  ATG  CTG  GTA  CAA  GGC  AGC  TGG  CAG  CAT  GCC  CCT  CAG  GAC  135
 T                   T         TGT  GGT  GCT  GT                             T    T    A
———————————— NH2 - PEPTIDE ————————————
          Lys  Ser                            15                  Asp                          25
Thr  Glu  Glu  Asn  Ala  Arg  Ser  Phe  Pro  Ala  Ser  Gln  Thr  Glu  Pro  Leu  Glu  Asp  Pro  Asp
ACG  GAG  GAG  AAC  GCC  AGA  TCA  TTC  CCA  GCT  TCC  CAG  ACA  GAA  CCA  CTT  GAA  GAC  CCT  GAT  195
                A    T                                             C         C    G
——————————————|      |——— 35 ——— GLUCAGON ———————————————— 45
Gln  Ile  Asn  Glu  Asp  (Lys)(Arg) His  Ser  Gln  Gly  Thr  Phe  Thr  Ser  Asp  Tyr  Ser  Lys  Tyr
CAG  ATA  AAC  GAA  GAC  (AAA)(CGC) CAT  TCA  CAG  GGC  ACA  TTC  ACC  AGT  GAC  TAC  AGC  AAA  TAC  255
 A         T               G
——————————————————————————————————————————|          |
                                    55                                                    65
Leu  Asp  Ser  Arg  Arg  Ala  Gln  Asp  Phe  Val  Gln  Trp  Leu  Met  Asn  Thr  (Lys)(Arg) Asn  Arg
CTA  GAC  TCC  CGC  CGT  GCT  CAA  GAT  TTT  GTG  CAG  TGG  TTG  ATG  AAC  ACC  (AAG)(AGG) AAC  CGG  315
 G              A    C         G                        C                                  A
— IP - I —|      |——— 75 ——— GLP I ———————————————— 85
Asn  Asn  Ile  Ala  (Lys)(Arg) His  Asp  Glu  Phe  Glu  Arg  His  Ala  Glu  Gly  Thr  Phe  Thr  Ser
AAC  AAC  ATT  GCC  (AAA)(CGT) CAT  GAT  GAA  TTT  GAG  AGG  CAT  GCT  GAA  GGG  ACC  TTT  ACC  AGT  375
                          C    C         G                   C                                  C
————————————————————————————————————————————————
                                      95                                                  105
Asp  Val  Ser  Ser  Tyr  Leu  Glu  Gly  Gln  Ala  Ala  Lys  Glu  Phe  Ile  Ala  Trp  Leu  Val  Lys
GAT  GTG  AGT  TCT  TAC  TTG  GAG  GGC  CAG  GCA  GCA  AAG  GAA  TTC  ATT  GCT  TGG  CTG  GTG  AAA  435
           C                                  T
——————|          |——————————— IP - II ——————————————|
                                       115       Thr         Val                                125
Gly  Arg  Gly  (Arg)(Arg) Asp  Phe  Pro  Glu  Glu  Val  Ala  Ile  Ala  Glu  Glu  Leu  Gly  (Arg)(Arg)
GGC  CGA  GGA  (AGG)(CGA) GAC  TTC  CCG  GAA  GAA  GTC  GCC  ATA  GCT  GAG  GAA  CTT  GGG  (CGC)(AGA)  495
      A                G              A                   A    T    T    A         C    C
———————————————————————— GLP II ————————————————————————
                                       135                                Ser              145
His  Ala  Asp  Gly  Ser  Phe  Ser  Asp  Glu  Met  Asn  Thr  Ile  Leu  Asp  Asn  Leu  Ala  Thr  Arg
CAT  GCT  GAT  GGA  TCC  TTC  TCT  GAT  GAG  ATG  AAC  ACG  ATT  CTC  GAT  AAC  CTT  GCC  ACC  AGA  555
      G    C    C              C                                            GT                   G
———————————————————————————————————————————|
                                       155
Asp  Phe  Ile  Asn  Trp  Leu  Ile  Gln  Thr  Lys  Ile  Thr  Asp  (Lys)(Lys) End
GAC  TTC  ATC  AAC  TGG  CTG  ATT  CAA  ACC  AAG  ATC  ACT  GAC  (AAG)(AAA) TAG  gaatatttcaccatt  618
                                               A                                A

cacaaccatcttcacaacatctcctgccagtcacttgggatgtacatttgagagcatatccgaagctatactgctttgc  697

atgcggacgaatacatttccctttagcgttgtgtaacccaaaggttgtaaatggaataaagtttttccagggtgttgat  776

aaagtaacaactttacagtatgaaaatgctggattctcaaattgtctcctcgttttgaagttaccgccctgagattact  855

tttctgtggtataaattgtaaattatcgcagtcacgacacctggattacaacaacagaagacatggtaacctggtaacc  933

gtagtggtgaacctggaaagagaacttcttccttgaaccctttgtcataaatgcgctcagctttcaatgtatcaagaat  1012

agatttaaataaatatctcat                                                            1024
                     3'
```

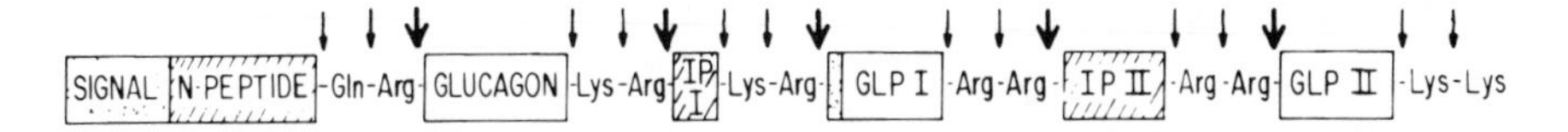


Fig. 3B. Schematic diagram of the preproglucagon encoded in the rat pancreatic mRNA. The two amino acids linking the boxed peptide domains are indicated. Stippled box, signal peptide; hatched box, $NH_2$-terminal peptide and two intervening peptides (IP, I, and II); open boxes, glucagon and two glucagon-like peptides (GLP, I, and II).

resents 90% of the mRNA encoding rat pancreatic preproglucagon and the remaining 100–200 nucleotides of the mRNA is consistent with the presence of a 3′ poly(A) tract of approximately 200 bases.

## 3. GLUCAGON GENE CLONING AND STRUCTURAL ANALYSIS

### *3.1. Hybridization Analysis of the Glucagon Chromosomal Gene*

To determine the representation of the cDNA in the rat genome we performed Southern blot hybridization (*29*) with DNA extracted from a rat liver and recombinant plasmid pRGlu16 labeled

Fig. 3A. (*on opposite page*) Composite nucleotide sequence of the sense strands of two rat preproglucagon cDNAs, with the derived amino acid sequence. Two recombinant cDNAs hybridizing to the Syrian hamster probe were selected for sequence analysis. In combination, the two cDNAs were sequenced on both strands. Restriction endonuclease fragments were prepared, labeled at the 3′ and 5′ ends with $^{32}P$, and sequenced according to Maxam and Gilbert (*39*). The signal sequence includes amino acids −20 to −1, the amino terminal peptide (glicentin) of the proglucagon amino acids +1 to +33, glucagon amino acids +33 to +61, and the COOH terminal peptide, which contains glucagon-like peptides I and II, as well as corresponding intervening peptides I and II, amino acids +62 to +160. Within the coding region, the appended nucleotides and amino acids shown below and above the continuous rat sequences represent the comparative changes in the corresponding sequences of the preproglucagon cDNA obtained from the Syrian hamster (*20*).

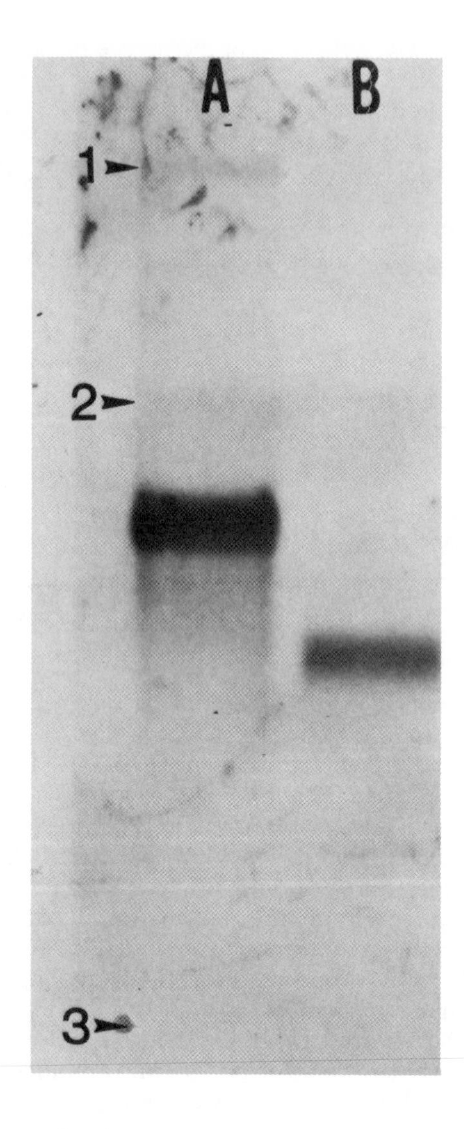
A
B
1
2
3

by nick translation with [$^{32}$P]-dCTP (Fig. 5). The recombinant plasmid hybridized strongly to only 2–3 restriction fragments in each enzymic digest, indicating that there is probably only one, or at most two, chromosomal genes with sequences homologous to the rat preproglucagon cDNA. At lower stringencies of hybridization, however, several fainter, weakly hybridizing bands were observed in the digest, suggestive of the presence of other genes with partial sequence homology to the preproglucagon cDNA (data not shown).

### 3.2. Structural Analyses of the Rat Glucagon Gene

To further analyze the expression and regulation of the gene, we determined the structure of the rat gene encoding preproglucagon.

### 3.3. Experimental Procedures Used for the Isolation of a Recombinant Bacteriophage Containing the Rat Glucagon Gene

A rat genomic DNA library prepared from rat liver DNA partially digested with restriction endonuclease *Hae*III and constructed in bacteriophage Charon 4A (*29*) was screened by filter hybridization (*30*). The $^{32}$P-labeled hybridization probe was a nearly full-length cloned cDNA encoding rat preproglucagon mRNA, plasmid pRGlu16, previously cloned and sequenced in our laboratory

Fig. 4. (*on opposite page*) Determination of the size of the mRNA encoding the rat pancreatic preproglucagon. Lane A, rat pancreatic RNA. Lane B, polyadenylated RNA prepared from the pancreatic islets of the anglerfish. Numbers indicate (1) 28S, (2) 18S ribosomal RNA, and (3) tRNA size markers. Ten micrograms of polyadenylated mRNA prepared from rat pancreas were size fractionated on a 1.5% agarose gel containing 6% formaldehyde, 25 m*M* MOPS (morpholinopropane sulfonic acid), pH 7.0, and 1 m*M* EDTA (ethylenediamine tetracetic acid), and blotted onto a nylon membrane (Gene Screen, New England Nuclear, Boston, MA). The blot was then hybridized with rat preproglucagon cDNA labeled with $^{32}$P by nick translation (*31*). Hybridization was performed in 50% formamide, 0.75*M* NaCl at 42°C for 12 h. The blot was washed at room temperature in 0.3*M* NaCl and 0.5% SDS. Hybridizing mRNA was then visualized by autoradiography. Size markers were 18S and 28S ribosomal RNA and tRNA (PL Biochemicals, Milwaukee, WI).

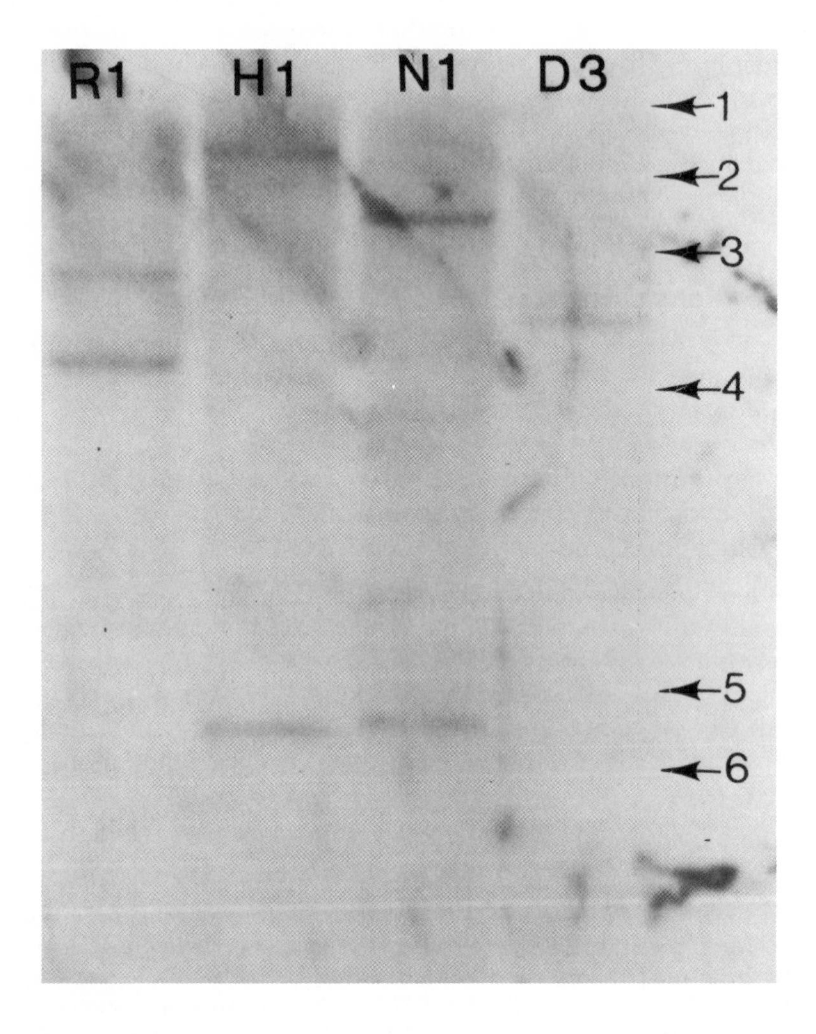


Fig. 5. Southern hybridization blot of rat liver genomic DNA with $^{32}$P-labeled recombinant plasmid containing a preproglucagon cDNA. The DNA was subjected to complete digestion with the restriction endonucleases *Eco*R1 (R1), BamH1 (H1), *Hin*dIII (D3), and NcoI (N1). Ten micrograms of the digested DNA were separated by electrophoresis on an 0.8% agarose gel containing 25 m*M* Tris-borate, pH 8.3, and 1 m*M*

(*18*). The cDNA was radioactively labeled with [$\alpha^{32}P$]-dCT by nick translation (*31*).

One of 300,000 recombinant bacteriophages hybridized specifically to the rat glucagon cDNA. Purified phage DNA was digested either with a single or sequentially with two or three restriction endonucleases. The digests were then analyzed by agarose gel electrophoresis and blot hybridization using the cDNA as a probe. Digestion of the recombinant phage with restriction endonuclease *Eco*RI yielded four distinct fragments of genomic DNA of sizes 1.4, 2.6, 4.3, and 6 kb. Each of the *Eco*RI restriction fragments was subcloned into the unique *Eco*RI site of plasmid pBR322. The four genomic subclones, each containing a unique *Eco*RI fragment, were then analyzed by restriction endonuclease digestion and blot hybridization. From the results, a restriction map of the recombinant gene and of the four subclones was constructed. The genomic insert of the recombinant bacteriophage DNA is approximately 15 kb long, and sequences that hybridize to the rat cDNA are dispersed over at least 10 kb pairs.

### 3.4. Partial Nucleotide Sequence Analysis of Gene Subclones

Knowledge of the nucleotide sequence of rat glucagon cDNA was used to select subfragments of the genomic subclones for nucleotide sequence analysis. This approach ensured that most sequencing would be started within exons of the glucagon gene. Exon and intron junctions, and the deduced amino acid sequence of the encoded glucagon precursor, were assigned according to the known rat glucagon cDNA nucleotide sequence (*see* Fig. 6).

---

Fig. 5. (*cont.*) EDTA, and transferred to a nylon membrane (Gene Screen Plus, New England Nuclear, Boston, MA) (*40*). The membrane prepared from the gel was hybridized with the recombinant plasmid containing the cDNA representing the entire mRNA encoding rat pancreatic preproglucagon that had been nick translated with [$^{32}P$]-dCTP (*31*). The conditions of hybridization were 65°C for 18 h in the presence of 1*M* NaCl, 10% dextran sulfate, and 1% SDS. Autoradiograms (24-h exposure) were prepared from the nylon membrane following hybridization. Arrows 1–5 represent size markers from a *Hin*dIII digest of λ phage DNA and point to DNA fragments of 23.6, 9.4, 5.5., 4.2, 2.3, and 1.9 kb, respectively. Exposure of the autoradiogram was for 48 h.

```
E-1
5'  gatgacgagagtgggcgagtgaaatcatttgaacaaaaccccattatttacagatgagaaatttatattgtcagcgtaatatctgcaaggctaaacagcttggagactatataaa   -19
gccacagcaccttggtgc  AGAAGGGCAGAGCTTGGGCGCAGAACACACTCAAAGTTCCCAAAGGAGCTCCACCTGTCTACACCTCCTCTCAGCTCAGTCCCACAAG  gtaagaggca   98
cactgtggggtggggatgtctggaggacattggaca...

                              Intron A  (3.0kb)

E-2
                                          -20                                          -10
                                          Met Lys Thr Val Tyr Ile Val Ala Gly Glu Phe Val Met Leu Val Gln Gly Ser
..ccccctaccccccactctgtgtccaacag  GCAGAATAAAAAA  ATG AAG ACC GTT TAC ATC GTG GCT GGA TTG TTT GTA ATG CTG GTA CAA GGC AGC   3200

    -1  +1                                   10
Trp Gln His Ala Pro Gln Asp Thr Glu Glu Asn Ala Ar
TGG CAG CAT GCC CCT CAA GAC ACG GAG GAG AAC GCC AG  gtactaaacagtagagcagcctgtcctgtggttagagtgag...   3278
     ↑

                              Intron B  (1.6kb)

E-3
              ...ttataaagcgtatactcctggagtatttaccctttgtgtcctgctcctacagttgtacttatgctacttaattactgcctctcaca   4800
                                 20                                30                     40
  g Ser Phe Pro Ala Ser Gln Thr Glu Pro Leu Glu Asp Pro Asp Gln Ile Asn Glu Asp Lys Arg | His Ser Gln Gly Thr Phe Pro Ser
agA TCA TTC CCA GCT TCC CAG ACA GAA CCA CTT GAA GAC CCT GAT CAG ATA AAC GAA GAC AAA CGC CAT TCA CAG GGC ACA TTC ACC AGT   4890
                                                                                   ↑
------------------------------ 50 ---- GLUCAGON ---------------- 60 --|
Asp Tyr Ser Lys Tyr Leu Asp Ser Arg Arg Ala Gln Asp Phe Val Gln Trp Leu Met Asn Thr Lys Arg Asn Ar
GAC TAC AGC AAA TAC CTA GAC TCC CGC CGT GCT CAA GAT TTT GTG CAG TGG TTG ATG AAC ACC AAG AGG AAC CG  gtaggagtctgaagttgtt   4983
                                                                                  ↑
gtacacaaatgcacattaagaaattctcctgactctcctaggttatcatgctcaccaaacaggcttattcatttgggtgttcgtgaggttgt...   5076

                              Intron C  (1.6kb)

E-4
                                                                                          ...gagatgaact   6700
                                      70                               80
                          g Asn Asn Ile Ala Lys Arg | His Asp Glu Phe Glu Arg His Ala Glu Gly Thr Phe Thr Ser Asp
tcattcaacacaactcccacatttctttcag  G AAC AAC ATT GCC AAA CGT CAT GAT GAA TTT GAG AGG CAT GCT GAA GGG ACC TTT ACC AGT GAT   6794
                                                        ↑
------------ 90 -------- GLP-I ---------------- 100 ----------------------------| 110
Val Ser Ser Tyr Leu Glu Gly Gln Ala Ala Lys Glu Phe Ile Ala Trp Leu Val Lys Gly Arg Gly Arg Arg As
GTG AGT TCT TAC TTG GAG GGC CAG GCA GCA AAG GAA TTC ATT GCT TGG CTG GTG AAA GGC CGA GGA AGG CGA GA   gtaagtctctggtttaa   6885
                                                                                       ↑
tatctgatgattttcctagatgagaaattacaaagaaaaaaagactatcctatggaaagctatactaaaacttaatattt...   6967
```

```
                                                     Intron D  (1.6kb)

                                                                                                              120
                                                                      p Phe Pro Glu Glu Val Ala Ile Ala Glu Glu Leu Gly Arg Arg
     ...cagcagttcacacgaatttatttaagataaacactgactccaagccttctttgcttag C TTC CCG GAA GAA GTC GCC ATA GCT GAG GAA CTT GGG CGC AGA ↑  8700

      ------------ 130 ----------------------------- 140 ------ GLP II ---------------- 150 --------------
E-5  His Als Asp Gly Ser Phe Ser Asp Glu Met Asn Thr Ile Leu Asp Asn Leu Ala Thr Arg Asp Phe Ile Asn Trp Leu Ile Gln Thr Lys
     CAT GCT GAT GGA TCC TTC TCT GAT GAG ATG AAC ACG ATT CTC GAT AAC CTT GCC ACC AGA GAC TTC ATC AAC TGG CTG ATT CAA ACC AAG  8790

     Ile Thr Asp Ly
     ATC ACT GAC AA   gtaagggtttttagttgattttgaaatctagtaaaacttt...                                                     8842

                                                     Intron E  (0.6kb)

                                                                                           ...aattattctgaacaattcctgcttttgacc  9500

               —160
             s Lys END
     tcatag  G AAA TAG GAATATTTCACCATTCACAACCATCTTCACAACATCTCCTGCCAGTCACTTGGGATGTACATTTGAGAGCATATACCGAAGCTATACTGCTTGGCATGCG  9613

     GACGAATACATTTCCCTTTAGCGTTGTGTAACCCAAAGGTTGTAAATGGAATAAAGTTTTTCCAGGGTGTTGATAAAGTAACAACTTTACAGTATGAAAATGCTGGATTCTCAAATTGT  9732

E-6  CTCCTCGTTTTGAAGTTACCGCCCTGAGATTACTTTTCTGTGGTATAAATTGTAAATTATCGCAGTCACGACACCTGGATTACAACAACAGAAGACATGGTAACCTGGTAACCGTAGTG  9851

     GTGAACCTGGAAAGAGAACTTCTTCCTTGAACCCTTTGTCATAAATGCGCTCAGCTTTCAATGTATCAAGAATAGATTTAAATAAATATCTCATCctttgttattatgctcctttctc  9970

     ttttattag... 3'                                                                                                       9979
```

Fig. 6A. Rat glucagon gene. Nucleotide sequence of specific regions of λrGlu1. The nucleotide sequences of λrGLu1 5′ flanking region, exons, and exon/intron junctions were determined using the chemical sequencing method (25). Exon sequences are in capital letters, and flanking and intronic sequences are in lower case letters and are given from 5′ to 3′ ends of the genomic insert, as indicated. Nucleotides are numbered starting with +1 at the most 5′ nucleotide of exon I, and running through each intron whose length, i.e., number of nucleotides, was estimated by restriction enzyme and blot hybridization analysis. The nucleotides of the 5′ flanking regions are numbered from −1 beginning with the nucleotide immediately 5′ to exon I, and continuing in the 5′ direction. The promoter of the glucagon gene is boxed, and the two polyadenylation sites are underlined. Amino acids encoded by exons are indicated above each codon, and are numbered −20 to −1 for the signal peptide, and +1 to +160 for the remainder of preproglucagon. Sites of known and presumptive proteolytic processing of preproglucagon are indicated by arrows.

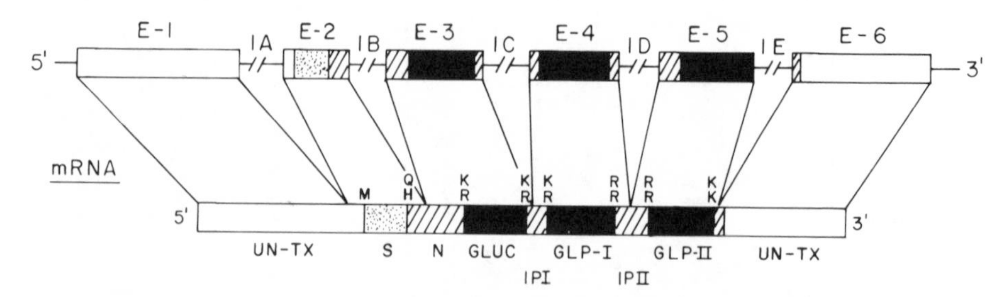


Fig. 6B. Structure of the encoded rat preproglucagon mRNA. The mRNA encoded by the glucagon gene, and the regions of preproglucagon corresponding to a given exon of the gene, are indicated by lines between gene and mRNA diagrams to illustrate the correspondence between exon and polyprotein functional domain structures. Exons (boxed) are designated E1–E6, and intervening sequences (interrupted lines) IA–IE. Open boxes represent untranslated regions of preproglucagon mRNA (UN-TX). The stippled boxes represent signal peptides (S), crosshatched boxes represent N-terminal extension (N) and C-terminal extension (IPI) of glucagon and the intervening peptides (IPI and IPII). Closed boxes represent glucagon and the two glucagon-like peptides I and II (GLPI and GLPII).

Accordingly, there are six exons and five introns. Exon 1 contains most of the 5′ untranslated region of glucagon mRNA. Exon 6 contains the entire 3′ untranslated region and the last four nucleotides of the coding region. Exon 2 encodes the signal peptide and a portion of the $NH_2$-terminal extension of glucagon. Exons 3, 4, and 5 contain glucagon, glucagon-like peptide I (GLPI), and glucagon-like peptide II (GLPII), respectively. Thus, each of the functional domains of preproglucagon is encoded by a separate exon. Introns B, C, and E in the coding sequence are located in intervening or connecting peptide regions (Fig. 2). The sizes of the introns were estimated by restriction endonuclease mapping and blot hybridization. Intron A is the longest and consists of approximately 3.0 kb. Introns B, C, and D are all of similar length of approximately 1.6 kb. Intron E is the shortest and consists of about 600–700 bp (base pairs). Introns B, C, and D are located in positions corresponding to those in the human glucagon gene (*21*).

### 3.5. Promoter and Transcriptional Initiation Sites of the Rat Glucagon Gene

Three techniques were used to identify the site of transcriptional initiation. First, hybridization analysis using a synthetic oligonucleotide complementary to 24 nucleotides of the 5′ region of glucagon mRNA (oligonucleotide I) revealed that a DdeI restriction fragment of one of the genomic subclones hybridized to the oligonucleotide. Sequence analysis of this fragment showed that it contained the 5′ untranslated region of the known rat glucagon cDNA (*18*), and extended upstream by an additional 150 nucleotides. The extension contained a classic TATAAA sequence. Moreover, transcription from this TATAAA sequence would yield a 5′ untranslated region of the encoded mRNA of approximately 90–100 bases. Therefore, this restriction fragment was chosen for S1 nuclease mapping of the 5′ end of glucagon mRNA. Two clusters of S1 nuclease resistant bands are seen (Fig. 7A). The upper band is clearly mRNA-dependent, whereas the lower and more prominent one is also present when no mRNA was included in the experiment. The main band of the upper cluster was localized with the aid of the sequencing ladder to a position of 26 bp downstream from the TATAAA sequence. The degeneracy of the transcriptional site as revealed by the S1 mapping could represent true degeneracy of the cap site, mRNA degradation, or instability of the RNA–DNA heteroduplex caused by the proximity of the mRNA cap. To further substantiate these results and the conclusion that the TATAAA sequence identified by the S1 mapping is the promoter of the glucagon gene, the oliogonucleotide used for hybridization analysis (oligonucleotide I) and two additional oligonucleotides complementary to 22 and 21 nucleotides of exons II and III (Fig. 7A), respectively, were labeled with $^{32}$P at the 5′ ends and each used separately to prime the reverse transcription of glucagon mRNA. The resulting $^{32}$P-labeled cDNAs were displayed by polyacrylamide gel electrophoresis. When oligonucleotide I was used, a rather prominent cluster of bands was seen (Fig. 7B) that was reproducible in two independent experiments (Fig. 7B, lanes E1). If derived from glucagon mRNA, the 3′ ends of these transcripts would correspond exactly with the site of transcriptional initiation suggested by the S1 nuclease mapping. With oligonucleotide II and III priming, major

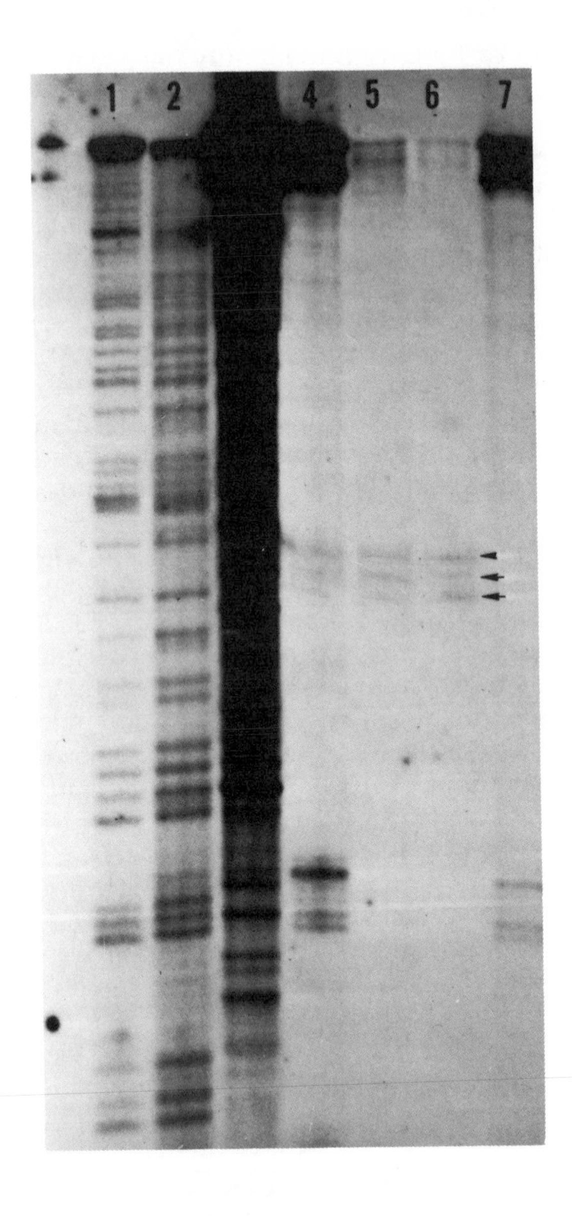
1
2
4
5
6
7

reverse transcripts were seen that correspond in length precisely to cDNAs originating at the complementary sequences on glucagon mRNA and extending to its presumptive 5′ end. Thus, these cDNAs are likely to represent glucagon mRNA reverse transcripts. These results suggest that transcription in the pancreas of the glucagon gene is initiated from the unique TATAAA sequence at position −26.

## 4. IMPLICATIONS DERIVED FROM THE STRUCTURE OF THE GLUCAGON GENE

### *4.1. Evolution of the Glucagon Gene*

The glucagon gene is a member of a larger gene family that includes secretin, vasoactive intestinal peptide, gastric inhibitory peptide, growth hormone releasing factor, PHI, and prealbumin genes. The glucagon gene appears to be a distinct and separate member of this family, inasmuch as we could detect only weakly cross-hybidizing sequences on the blot of rat genomic DNA that had been hybridizing with the glucagon cDNA (*see* Fig. 5), even under conditions of low hybridization stringency (data not shown). How did this gene family evolve? Clearly, gene duplication has been an important factor: There are at least two distinct glucagon genes in the anglerfish and each of these genes encodes

Fig. 7A. (*on opposite page*) Localization of the 5′ end of preproglucagon mRNA by S1 nuclease mapping and oligonucleotide primer extension. A. Pancreatic mRNA was hybridized to a DdeI restriction fragment of the glucagon gene labeled with $^{32}P$ at the 5′ ends and containing the presumptive 5′ end of glucagon mRNA. The length of the hybrid was then determined by S1 nuclease digestion and electrophoresis through an 8% polyacrylamide gel in the presence of 8*M* urea. For size reference, that strand of the DdeI fragment that is complementary to glucagon mRNA was degraded in base-specific chemical reactions (*39*), and the resulting fragments subjected to electrophoresis in parallel with the S1 nuclease-resistant hybrids. Lanes 1 and 2, chemical degradations of DdeI fragment specific for guanine (lane 1) and both adenine and guanine (lane 2); Lane 3, DdeI fragment without mRNA or S1 nuclease; Lane 4–6, mRNA/DdeI fragment hybrid digested with 5, 15, and 25 units of S1 nuclease; Lane 7, DdeI fragment alone digested with 5 units of S1 nuclease.

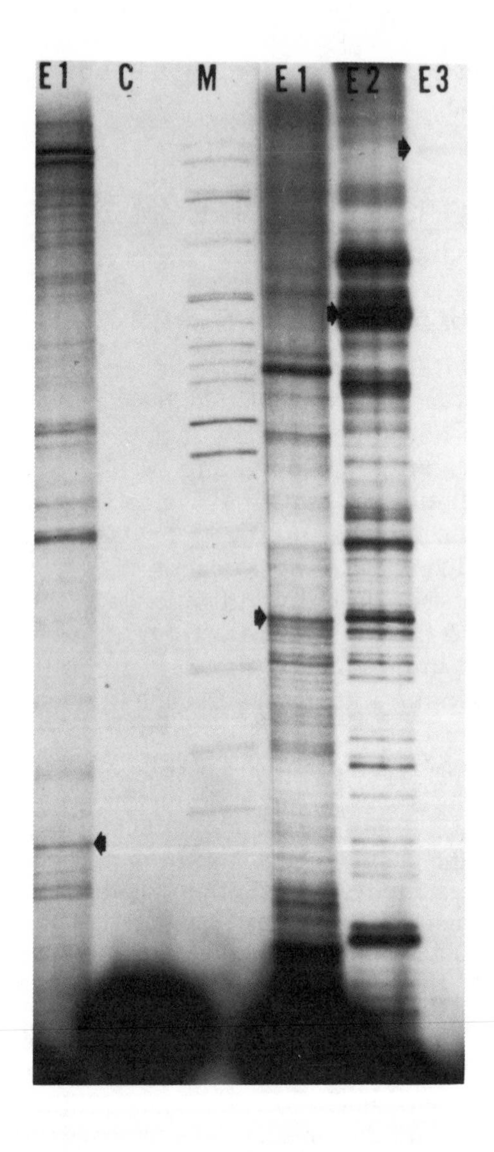
E1
C
M
E1
E2
E3

two closely related peptides arranged in tandem, suggesting duplication of an ancestral glucagon exon. Further duplication(s) must have produced the third glucagon-like sequences of the mammalian glucagon gene. Moreover, at least one additional member of the gene family, the gene-encoding vasoactive intestinal peptide and PHM-27, contains two closely related peptide sequences arranged in tandem (*32*). Thus, duplication appears to be a strong factor in the evolution of the glucagon gene family. On the other hand, it is interesting that the duplicated and strongly homologous exon 3, 4, and 5 of the human and rat glucagon genes are distinct and have not undergone mutual gene conversions.

### 4.2. Tissue-Specific Processing of Preproglucagon

A site of the expression of the rat glucagon gene is the pancreas. However, the mechanisms of the co- and posttranslational proteolytic processing of preproglucagon involved in the generation of pancreatic glucagon are still incompletely understood. A scheme based on analogy with the known processing of other secretory protein precursors at specific processing sites, for example, at pairs of basic amino acid residues, is compatible with data obtained from biosynthetic peptide-labeling and pulse-chase experiments, and has been proposed by us and others (*18,19,20,21,33*). Another site of glucagon gene expression may be the gut. The strongest evidence for this supposition is the colinearity and strong homology (>95% bovine vs porcine) of

Fig. 7B. (*on opposite page*) For additional mapping of the 5′ end of glucagon mRNA, three oligonucleotides complementary to nucleotides of the glucagon gene and synthesized by the phophotriester method (*41*) were labeled with $^{32}P$ at the 5′ ends and each hybridized separately to pancreatic mRNA. The hybrids were extended by reverse transcriptase and the resulting cDNAs separated according to size by gel electrophoresis through an 8% polyacrylamide gel in the presence of 8*M* urea. Lanes E1, E2, and E3, reverse transcription of pancreatic mRNA primed with oligonucleotide I, II, and III, respectively. Each E1 lane represents an independent experiment; Lane C, control using liver mRNA and oligonucleotide I; Lane M, marker restriction endonuclease HpaII digest of plasmid pBR322. Restriction fragments were labeled with $^{32}P$ at the 5′ ends. Arrows point to cDNAs whose length corresponds to transcription from the TATAAA sequence determined by S1 nuclease mapping.

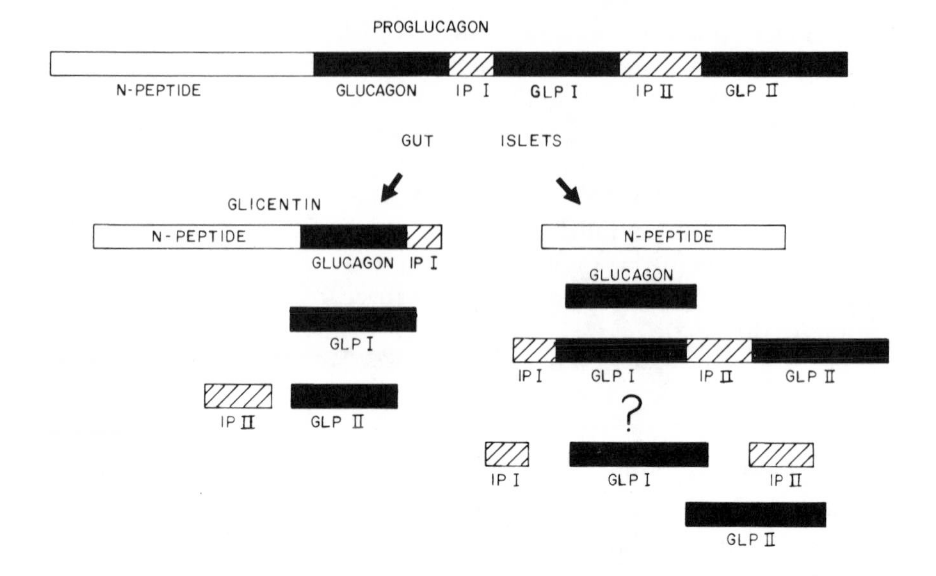


Fig. 8. Alternative pathways of the processing of proglucagon in the gut and pancreatic islets. The pathways shown on the left and right represent the predicted patterns of processing of proglucagon in the gut and islets, respectively. In the intestine, the major glucagon-containing peptide is glicentin, which consists of glucagon in covalent linkages with the N-terminal and the short C-terminal extensions. Although not proven, it is likely that the two glucagon-related peptides are the major biologically active peptides liberated in the gut by the processing of proglucagon. In the pancreatic islets, the major glucagon peptide is glucagon itself. Whether the C-terminal peptide resulting from the cleavages that liberate glucagon are further processed to the glucagon-related peptides is unknown.

glicentin, a glucagon-related gut peptide, with a fragment of pancreatic proglucagon (amino acids 1–61) that consists of glucagon, an $NH_2$-terminal extension, and a short COOH terminal extension that corresponds to intervening peptide I (*34*). In view of the presence of what appears to be a single glucagon gene in the rat genome, it is likely that glicentin and the gut glucagon-related peptides are derived from the same gene that gives rise to pancreatic glucagon. However, presently it is unknown whether identical mRNAs encode pancreatic and gut preproglucagon. It is conceivable that separate 5′ untranslated regions are utilized for gut and pancreatic glucagon mRNAs, as has been described for salivary and hepatic α-amylase mRNAs (*35*). Nevertheless, although they may differ at their 5′ and 3′ ends, or both, gut and pancreatic glucagon mRNAs are likely to encode the same glucagon precursor, which is then processed differentially in the two tissues. A known difference in preproglucagon processing is the preponderance of glucagon in the pancreas (*36*), whereas glicentin is the preponderant peptide detected in the intestine (*37*). The biologic activity of glicentin is poorly defined (*38*). It is therefore reasonable to speculate that the biologically important processing products of gut preproglucagon are the glucagon-like peptides (Fig. 8). The strong evolutionary conservation of the glucagon-like peptides in man, Syrian hamster, cow, and rat (90%) strengthens this speculation (*19,20*). Knowledge of the primary structure of the GLPs will not permit experimental testing of this hypothesis.

## REFERENCES

1. Unger R. H. and Orci L. (1981) Glucagon and the A cell: physiology and pathophysiology. *N. Eng. J. Med.* **304,** 1518–1524.
2. Mutt V., Jorpes, and Magnusson S. (1970) Structure of porcine secretin. The amino acid sequence. *Eur. J. Biochem.* **15,** 513–519.
3. Mutt V. and Said S. I. (1974) Structure of the porcine vasoactive intestinal octacosapeptide. The amino-acid sequence. Use of kallikrein in its determination. *Eur. J. Biochem.* **42,** 581–589.
4. Brown J. C. and Dryburgh J. R. (1971) A gastric inhibitory polypeptide. I. The amino acid composition and the tryptic peptides. *Can. J. Biochem.* **49,** 255–261.
5. Spiess J., Rivier J., Thorner M., and Vale W. (1982) Sequence analysis of a growth hormone releasing factor from a human pancreatic islet tumor. *Biochemistry* **21,** 6037–6040.
6. Gerich J., Schneider V., Dippe S., Langlois M., Noacco C., Karam T., and Forsham P. (1974) Characterization of the glucagon response to hypoglycemia in man. *J. Clin. Endocrinol. Metab.* **38,** 77–82.
7. Assan R., Attali J. R., Ballerio G., Boillot J., and Girard J. R. (1977) Glucagon secretion induced by natural and artificial amino acids in the perfused rat pancreas. *Diabetes* **26,** 300–307.
8. Samols E., Weir G. C., and Bonner-Weir S. (1983) Intra-Islet Insulin-Glucagon-Somatostatin Relationships, in *Handbook of Experimental Pharmacology* vol. 66/II. (P. J. Lefebvre, ed.), Springer Verlag, New York.

9. Aoki T. T., Muller W. A., Brennan M. F., and Cahill G. F. (1974) Effect of glucagon on amino acid and nitrogen metabolism in fasting man. *Metabolism* **23,** 805–814.
10. Cherrington A. D., Chiasson J. L., Liljenquist J. E., Hemmings A. S., Keller V., and Lacey W. W. (1976) The role of insulin and glucagon in the regulation of basal glucose production in the postabsorptive dog. *J. Clinic. Invest.* **58,** 1407–1418.
11. Dobbs R. E., Saki H., Faloona G. R., Valverde I., Baeteus D., Orci L., and Unger R. H. (1975) Glucagon: Role in the hyperglycemia of diabetes mellitus. *Science* **187,** 544–547.
12. Tager H., Hohenboken M., Markese J., and Dinerstein R. J. (1980) Identification and localization of glucagon-related peptides in rat brain. *Proc. Natl. Acad. Sci. USA* **77,** 6229–6233.
13. Lawrence A. M., Tan S., Hojvat S., and Kirsteins L. (1976) Salivary gland hyperglycemic factor: An extra pancreatic source of glucagon-like material. *Science* **195,** 70–72.
14. Conlon J. (1980) The glucagon-like peptides—order out of chaos. *Diabetologia* **18,** 85–88.
15. Parker D. S., Raufman J-P., O'Donohue T. L., Bledsoe M., Yoshida H., and Pisano J. J. (1984) Amino acid sequence of helospectins, new members of the glucagon superfamily, found in gila monster venom. *J. Biol. Chem.* **259,** 11751–11755.
16. Hoshino M., Yanaihara C., Hong Y-M., Kishida S., Katsumaru Y., Vandermeers A., Vandermeers-Piret M-C., Robberecht P., Christophe J., and Yanaihara N. (1984) Primary structure of helodermin, a VIP-secretin-like peptide isolated from Gila monster venom. *FEBS Let.* **178,** 233–239.
17. Lund P. K., Goodman R. H., Montminy M. R., Dee P. C., and Habener J. F. (1983) Anglerfish islet pre-proglucagon II; nucleotide and corresponding amino acid sequence of the cDNA. *J. Biol. Chem.* **258,** 3280–3284.
18. Heinrich G., Gros P., Lund P. K., Bentley R. C., and Habener J. F. (1984) Pre-proglucagon messenger ribonucleic acid: nucleotide and encoded amino acid sequences of the rat pancreatic complementary deoxyribonucleic acid. *Endocrinology* **115,** 2176–2181.
19. Lopez L. C., Fragier M. L., Chung-Jey S., Kumar A., and Saunders G. F. (1983) Mammalian pancreatic preproglucagon contains three glucagon-related peptides. *Proc. Natl. Acad. Sci. USA* **80,** 5485–5489.
20. Bell G. I., Santerre R. F., and Mullenbach G. T. (1983) Hamster pre-proglucagon contains the sequence of glucagon and two related peptides. *Nature* **302,** 716–718.
21. Bell I. G., Sanchez-Pescador R., Laybarron P. J., and Najarian R. C. (1983) Exon duplication and divergence in the human glucagon gene. *Nature* **304,** 368–371.

22. Heinrich G., Kronenberg H. M., Potts J. T., Jr., and Habener J. F. (1983) Parathyroid hormone messenger ribonucleic acid: Effect of calcium on cellular regulation *in vitro*. *Endocrinology* **112,** 449–458.
23. Goodman R. H., Jacobs J. W., Chin W. W., Lund P. K., Dee P. C., and Habener J. F. (1980) Nucleotide sequence of a cloned structural gene coding for a precursor of pancreatic somatostatin. *Proc. Natl. Acad. Sci. USA* **77,** 5869–5873.
24. Villa-Komaroff L., Efstratiadis A., Broomer S., Lomedico R., Tizard, Naber S. P., Chick W. L., and Gilbert W. (1978) A bacterial clone synthesizing proinsulin. *Proc. Natl. Acad. Sci. USA* **75,** 3727–3731.
25. Dagert M. and Erlich S. D. (1979) Prolonged incubation in calcium chloride improves the competence of *Escherichia coli* cells. *Gene* **6,** 23–28.
26. Grunstein M., Dickman M., and Hogness D. S. (1975) Colony hybridization: A method for the isolation of cloned cDNAs that contain a specific gene. *Proc. Natl. Acad. Sci. USA* **72,** 3961–3965.
27. McEvoy R. C. and Madson K. L. (1980) Pancreatic insulin, glucagon and somatostatin-positive islet cell populations during perinatal development of the rat I morphometric quantitation. *Biol. Neonate* **38,** 248–254.
28. Birnboim H. C. and Doty J. (1979) *Nucleac Acids Res.* **7,** 1513–1520.
29. Southern E. M. (1975) Detection of specific sequences among DNA fragments separated by gel electrophoresis. *J. Mol. Biol.* **98,** 503–517.
30. Benton W. D. and Davis R. W. (1977) Screening λgt recombinant clones by hybridization to single plaques in situ. *Science* **196,** 180–182.
31. Rigby P. W. J., Dieckman M., Rhodes C., and Berg P. (1977) Labeling deoxyribonucleic acid to high specific activity *in vitro* by nick-translation with DNA polymerase. *J. Mol. Biol.* **133,** 237–251.
32. Itoh N., Obata K., Yanaihara N., and Okamoto H. (1983) Human preprovasoactive intestinal polypeptide contains a novel PHI-27-like peptide, PHM-27. *Nature* **304,** 547–549.
33. Tager H. S. and Markese J. (1979) Intestinal and pancreatic glucagon-like peptides. Evidence for identity of higher molecular weight forms. *J. Biol. Chem.* **254,** 2229–2233.
34. Thim L. and Moody A. J. (1981) The primary structure of porcine glicentin (proglucagon). *Regul. Peptds.* **2,** 139–150.
35. Schibler U., Hagenbuchle O., Wellauer P. K., and Pittet A. C. (1983) Two promoters of different strengths control the transcription of the mouse alpha-amylase gene Amy-1a in the parotid gland and the liver. *Cell* **33,** 501–508.
36. Ravazzola M. and Orci L. (1980) Glucagon and glicentin immunoreactivity are topologically segregated in the alpha granule of the human pancreatic A cell. *Nature* **284,** 66–67.

37. Patzelt C., Tager H. S., Carroll R. J., and Steiner D. F. (1979) Identification and processing of proglucagon in pancreatic islets. *Nature* **282,** 260–6.
38. Moody A. J. and Thim L. (1983) Glucagon, Glicentin, and Related Peptides. in *Handbook of Experimental Pharmacology* vol. 66/I. (P. J. Lefebvre, ed.), Springer Verlag, New York.
39. Maxam A. and Gilbert W. (1980) Sequencing end-labeled DNA with base specific chemical cleavages. *Meth. Enzymol.* **65,** 499–560.
40. Grunstein M. and Hognes D.S. (1975) Colony hybridization: A method for the isolation of cloned cDNAs that contain a specific gene. *Proc. Natl. Acad. Sci. USA* **72,** 3961–3965.
41. Hirose T., Crea R., and Itakura K. (1978) Rapid synthesis of trideoxynucleotide blocks. *Tetrahedron Lett.* **28, 2449–2452.**

# Chapter 5

# The Somatostatin Genes

Richard H. Goodman, Marc R. Montminy, Malcolm J. Low, and Joel F. Habener

## 1. INTRODUCTION

Somatostatin (somatotropin-release inhibiting factor; SRIF) was originally named for its ability to inhibit the release of growth hormone from cultured rat pituitary cells (*1*). Like the names of many other regulatory peptides, the designation SRIF proved to be far too limited. In addition to inhibiting the release of pituitary growth hormone (*2,3*), somatostatin inhibits the release of thyrotropin (*3–6*) and, in certain circumstances, prolactin (*7*) and adrenocorticotropin (*8*). Somatostatin appears to regulate the secretion of the pancreatic-islet hormones glucagon (*9*) and insulin (*10*) and to have a number of effects on gastrointestinal function. These gastrointestinal effects include inhibition of the secretion of gut hormones (gastrin, pancreozymin, pancreatic polypeptide, vasoactive intestinal peptide, glucagon, motilin, gastric inhibitory peptide, secretin), exocrine secretion (gastric acid, pancreatic bicarbonate, gastric and pancreatic enzymes), gastrointestinal motility (gastric emptying, gallbladder contraction), absorption, and blood flow (reviewed in *11*). Finally, somatostatin influences the release of the neurotransmitters acetylcholine (*12*), norepinephrine (*13*), and serotonin (*14*) and acts as a neurotransmitter itself (*15*). Although somatostatin does not affect all aspects of pituitary, pancreatic, and gastrointestinal function, it is clear that it plays a central role in the physiology of these systems.

Soon after Brazeau and coworkers determined the amino acid sequence of ovine hypothalamic somatostatin (*1*), Schally et al. demonstrated that an identical substance could be isolated from porcine hypothalamic extracts (*16*). It was subsequently found that the sequence of the somatostatin tetradecapeptide was conserved in porcine intestine (*17*), pigeon pancreas (*18*), and anglerfish islets (*19*). A substantially different form of somatostatin was isolated from catfish islets (*20*) and an even more distantly related form, urotensin II, from the caudal neurosecretory system of teleost fish (*21*). The structures of somatostatin-like peptides from the lower phylogenetic species *Ciona intestinalis, Locust migratorius* (*22*), and *Tetrahymena pyriformis* (*23*) have not yet been determined.

Elucidation of the amino acid sequence of somatostatin allowed the generation of antibodies that recognized larger forms of the hormone (*24,25*). It was postulated that these larger forms, detected initially in porcine hypothalamus and stomach, represented precursor forms of the tetradecapeptide. In some instances, the larger forms were biologically active (*26*), suggesting that the precursors might have some physiological role in addition to generating the smaller form of somatostatin.

The best characterized of these higher molecular weight forms of somatostatin is somatostatin-28, an amino-terminally extended form of the tetradecapeptide initially found in porcine intestine (*27*). Somatostatin-28 was later identified in extracts of porcine (*28*) and ovine (*29*) hypothalamus. Several studies suggested that somatostatin-14 and somatostatin-28 may have distinct biological functions (*30–36*), and that specific tissues vary in their ability to produce one or the other peptide. In addition to having its own biological activity, somatostatin-28 may serve as an intermediate in the formation of somatostatin-14 by virtue of the dibasic sequence of amino acids adjacent to the tetradecapeptide sequence.

Still larger forms of somatostatin, with a molecular weight of 11,000–15,000, have been isolated from hypothalamus, cerebral cortex, and jejunum (*16,25,37–41*). These forms were also detected in hepatic (*42*) and pituitary (*43*) portal blood. Pulse-chase studies using radioactive amino acids demonstrated a transfer of radioactivity from the high molecular weight forms to somatostatin-14 and -28, suggesting that the larger forms are, in fact, precursors of the smaller somatostatin peptides. These studies were done in anglerfish and rat pancreatic islets (*41,44*) and in rat hypothalami (*38*).

The concept that somatostatin is synthesized as a precursor is consistent with the current understanding of the processes involved in the biosynthesis of protein hormones. Precursors are processed to their final protein products by a series of proteolytic cleavages and other modifications during their transport within specialized subcellular organelles (for review, *see 45*). Heterologous cell-free translation systems have been particularly useful for the study of the early events in the biosynthesis of polypeptide hormones. The initial translation products invariably have hydrophobic amino-terminal extensions (signal sequences) and have been termed prehormones or preprohormones in instances where intermediate precursor forms exist. Cell-free translation reactions, supplemented with microsomal membranes, result in the removal of the amino-terminal signal sequences and, in the case of neuropeptide precursors, in the generation of prohormones (*46*). In some instances, microsomal membranes can produce other modifications as well, such as glycosylation (*47*). Although the subsequent proteolytic cleavages of the hormone precursors cannot be faithfully reproduced in cell-free reactions, these reactions do serve to identify the initial products of translation.

## 2. ANGLERFISH SOMATOSTATINS

In our studies, to identify preprosomatostatin, we used mRNA isolated from anglerfish pancreatic islets. This source of mRNA had several advantages: (a) the anglerfish islets are anatomically distinct from pancreatic exocrine tissue and are relatively free of ribonuclease, (b) mRNA from the anglerfish islets generates a limited number of cell-free translation products, (c) somatostatin-14 from anglerfish islets has the same sequence as that found in mammals and crossreacts with antisera to the mammalian peptide, and (d) nearly 30% of the cells in the anglerfish islets contain immunoreactive somatostatin (*48*).

Polyadenylated RNA prepared from anglerfish islets directed the synthesis of at least four major discernible proteins, as analyzed on SDS-polyacrylamide gels (Fig. 1). The apparent molecu-

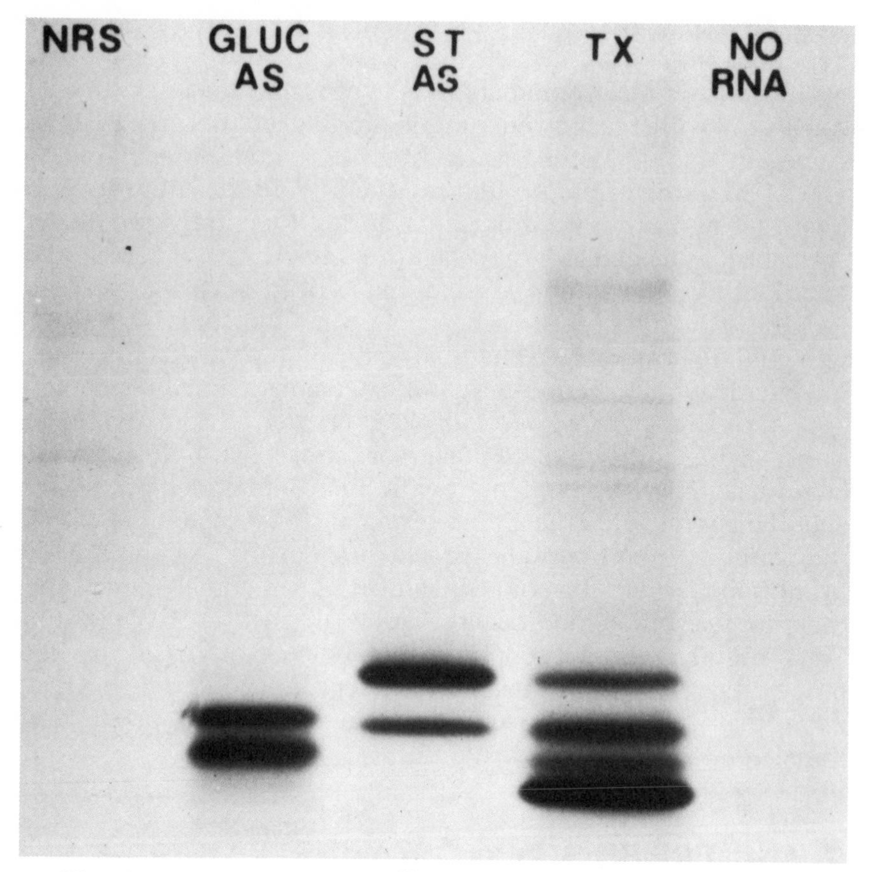


Fig. 1. Autoradiogram of [$^{35}$S]-methionine-labeled cell-free translation products of mRNA isolated from anglerfish islets analyzed on an SDS-polyacrylamide gel. TX represents total translation products, ST-AS represents immunoprecipitation using antiserum to synthetic somatostatin-14, GLUC-AS represents immunoprecipitation using antisera to purified porcine glucagon, NRS represents nonimmune control. Proteins range in size from 16,000 to 11,000 $M_r$.

lar weights of these proteins ranged from 11,000 to 16,000. Immunoprecipitations, using antisera to synthetic somatostatin and purified porcine glucagon, revealed that the larger two products, 16 and 14 kdaltons, were specifically related to somatostatin (*49*). Two different products, 14 and 13 kdaltons, were found to be related to glucagon (*50*). The smallest product had previously been identified as a precursor of insulin (*51*). When the cell-free translations were performed in the presence of microsomal membranes, smaller somatostatin-related products of 14 and 13 kdaltons were identified. These products were resistant to limited proteolysis with trypsin and chymotrypsin, suggesting that the membrane-processed forms were sequestered within the microsomes (*49*). It is likely, therefore, that the 14.5- and 13.5- kdaltons peptides represent prosomatostatin forms. These molecular weight estimates are slightly higher than those determined from the actual amino acid sequences of the precursors, based on the nucleotide sequences of the cDNA (*52*). These differences are probably a result of anomalous migration of the prosomatostatins through SDS-polyacrylamide gels.

Studies by Shields et al., again using immunoprecipitation techniques, confirmed the existence of two forms of anglerfish preprosomatostatin (*53*). These workers later analyzed cell-free translation products derived from islet mRNA on two-dimensional polyacrylamide gels and observed as many as nine separate proteins (*54*). Similar observations were made by P. K. Lund, R. H. Goodman, and J. F. Habener (unpublished results). Whether this multiplicity of prosomatostatin forms is a result of (a) the transcription of nine separate genes, (b) an unusually high error rate in fish gene transcription, (c) the existence of various base modifications in the fish mRNAs, or (d) artifacts of the cell-free translation assays, is uncertain. Similarly, the biological consequences of such variability in prohormone sequences are unknown. Subsequent studies confirmed the existence of two major forms of anglerfish somatostatin, termed preprosomatostatin I and II (*52,55*). The 16-kdaltons form that was identified on SDS-polyacrylamide gels appears to represent preprosomatostatin I. The identify of the 14-kdaltons form has not been determined, but is probably preprosomatostatin II.

Cell-free translations, using mRNA isolated from anglerfish gastrointestinal tract, gave somewhat different results (*56*). Immunoprecipitations of cell-free translation products from GI tract mRNA revealed only one preprosomatostatin form, identical on gel electrophoresis to the 16-kdaltons precursor. The smaller 14-kdaltons form could not be identified, even on long exposures of the autoradiograms. These studies raise the possibility that the different fish preprosomatostatins may be expressed in a tissue-specific fashion.

To characterize the structures of the somatostatin precursors, a complementary DNA library from the anglerfish islet mRNAs was prepared, using standard dC·dG tailing techniques to insert the cDNA into the *Pst*I site of the plasmid pBR322 (*57*). Recombinant clones encoding the 16-kdaltons somatostatin precursor were identified by hybridization-arrest (*58*) and hybridization-selection (*59*) techniques. The nucleotide sequence of anglerfish preprosomatostatin I is shown in Fig. 2. The coding sequence of the tetradecapeptide somatostatin was found at the carboxy-terminus of a 121-amino acid precursor. This structure had been previously predicted on the basis of pulse-labeling stud-

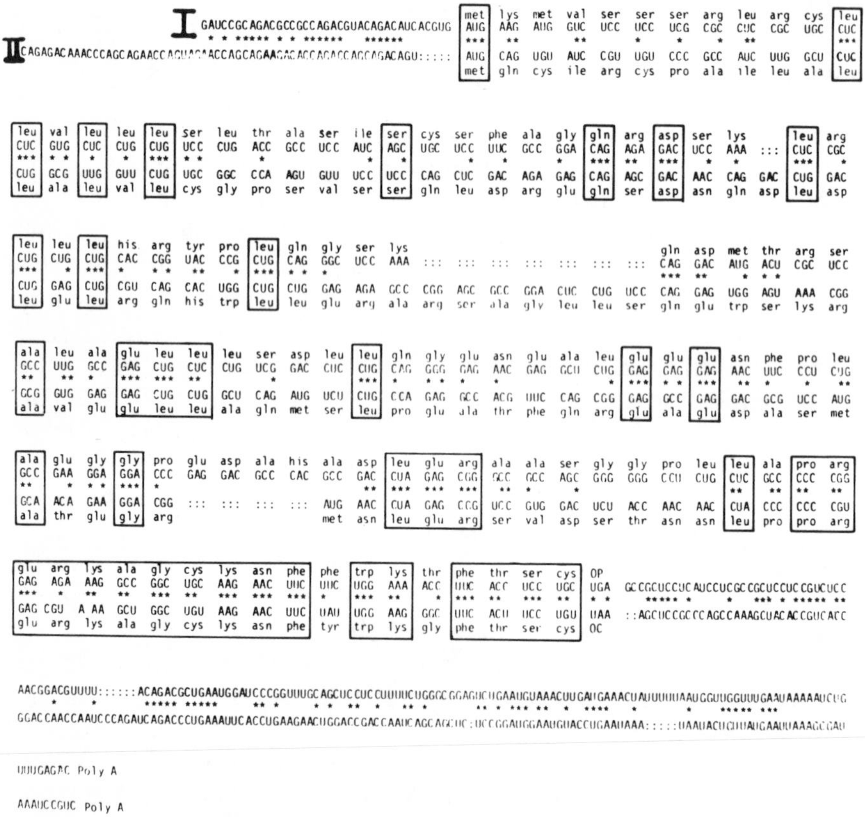

```
                                         met lys met val ser ser ser arg leu arg cys leu
 I  GAUCCGCAGACGCCGCCAGACGUACAGACAUCACGUG AUG AAG AUG GUC UCC UCC UCG CGC CUC CGC UGC CUC
    * * ***** * * ******  ******          *** **      **       *   *   *  **          ***
 II CAGAGACAAACCCAGCAGAACCAGUAGAACCAGCAGAAGACACCAGACCAGCAGACAGU::::: AUG CAG UGU AUC CGU UGU CCC GCC AUC UUG GCU CUC
                                         met gln cys ile arg cys pro ala ile leu ala leu

leu val leu leu leu ser leu thr ala ser ile ser cys ser phe ala gly gln arg asp ser lys     leu arg
CUC GUG CUC CUG CUG UCC CUG ACC GCC UCC AUC AGC UGC UCC UUC GCC GGA CAG AGA GAC UCC AAA ::: CUC CGC
*** * *  *   *  *** * *      *          *   *       *   *       *  *** **  ***  *   *      ***  *
CUG GCG UUG GUU CUG UGC GGC CCA AGU GUU UCC UCC CAG CUC GAC AGA GAG CAG AGC GAC AAC CAG GAC CUG GAC
leu ala leu val leu cys gly pro ser val ser ser gln leu asp arg glu gln ser asp asn gln asp leu asp

leu leu leu his arg tyr pro leu gln gly ser lys                             gln asp met thr arg ser
CUG CUG CUG CAC CGG UAC CCG CUG CAG GGC UCC AAA ::: ::: ::: ::: ::: ::: ::: CAG GAC AUG ACU CGC UCC
***   * *** *   * *  **   * *** * * *                                       *** **    *  * *
CUG GAG CUG CGU CAG CAC UGG CUG CUG GAG AGA GCC CGG AGC GCC GGA CUC CUG UCC CAG GAG UGG AGU AAA CGG
leu glu leu arg gln his trp leu leu glu arg ala arg ser ala gly leu leu ser gln glu trp ser lys arg

ala leu ala glu leu leu leu ser asp leu leu gln gly glu asn glu ala leu glu glu glu asn phe pro leu
GCC UUG GCC GAG CUG CUC CUG UCG GAC CUC CUG CAG GGG GAG AAC GAG GCU CUG GAG GAG GAG AAC UUC CCU CUG
**   ** *   *** *** **       *          *** *   * * *   *            * *** * * *** **  *    *   **
GCG GUG GAG GAG CUG CUG GCU CAG AUG UCU CUG CCA GAG GCC ACG UUC CAG CGG GAG GCC GAG GAC GCG UCC AUG
ala val glu glu leu leu ala gln met ser leu pro glu ala thr phe gln arg glu ala glu asp ala ser met

ala glu gly gly pro glu asp ala his ala asp leu glu arg ala ala ser gly gly pro leu leu ala pro arg
GCC GAA GGA GGA CCC GAG GAC GCC CAC GCC GAC CUA GAG CGG GCC GCC AGC GGG GGG CCU CUG CUC GCC CCC CGG
**    * * * *** *                        ** *** *** *** **  *     *                 **  ** **  **
GCA ACA GAA GGA CGG ::: ::: ::: ::: AUG AAC CUA GAG CGG UCC GUG GAC UCU ACC AAC AAC CUA CCC CCC CGU
ala thr glu gly arg                 met asn leu glu arg ser val asp ser thr asn asn leu pro pro arg

glu arg lys ala gly cys lys asn phe phe trp lys thr phe thr ser cys OP
GAG AGA AAG GCC GGC UGC AAG AAC UUC UUC UGG AAA ACC UUC ACC UCC UGC UGA GCCGCUCCUCAUCCUCGCCGCUCCUCCGUCUCC
***  *  **  **  *** **  *** *** *** *   *** **    * *** **  *** **  * *   ***** *   *   *** * ***** **
GAG CGU A AA GCU GGC UGU AAG AAC UUC UAU UGG AAG GGC UUC ACU UCC UGU UAA ::AGCUCCGCCCAGCCAAAGCUACACCGUCACC
glu arg lys ala gly cys lys asn phe tyr trp lys gly phe thr ser cys OC

AACGGACGUUUU::::::ACAGACGCUGAAUGGAUCCCGGUUUGCAGCUCCUCCUUUUCUGGCCGGAGUCUGAAUGUAAACUUGAUGAAACUAUUUUUAAUGGUUGGUUUGAAUAAAAAUCUG
    *     *      ***** *****     ** *    * * ** *  ** *        ** * *** ** *  ****   *      *    ***** ***
GGACCAACCAAUCCCAGAUCAGACCCUGAAAUUCACCUGAAGAACUGGACCGACCAAUCAGCAGCUC:UCCGGAUGGAAUGUACCUGAAUAAA::::UAAUACUCUUAUGAAUUAAAGCGAU

UUUGAGAC Poly A

AAAUCCGUC Poly A
```

Fig. 2. Comparison of anglerfish preprosomatostatin I and II [adapted from Hobart et al. (*52*)]. Conserved nucleotides are indicated by asterisks. Conserved amino acids are enclosed in boxes. Gaps are introduced in the sequence to maximize homology.

ies by Patzelt et al. (*41*). At the amino-terminus of the precursor was a sequence encoding a region of hydrophobic amino acids characteristic of a signal sequence. The exact cleavage site of the signal sequence from the remainder of the molecule is not known for certain, but probably occurs after the Gly residue at amino acid position 29.

Adjacent to the sequence encoding somatostatin-14 is an Arg-Lys sequence that serves as a prohormone cleavage site. This cleavage site falls within the sequence Ala-Pro-Arg-Glu-Arg-Lys, which is completely conserved between the anglerfish somatostatin precursors and precursors from mammalian species, including rodents (*60–62*) and humans (*63*). The complete conservation of this hexapeptide between fish and mammals, species that diverged over 400 million yr ago, suggests that this region is important in some essential biological function. Whether this function is to allow cleavage at the Arg-Lys sequence preceding somatostatin-14 or to participate in the biological activity of another secreted portion of the somatostatin precursor is unknown. Recent findings by Noe and Spiess (*64*) indicate that conservation of the hexapeptide is not sufficient for cleavage at the Arg-Lys sequence preceding somatostatin-14.

Between the signal sequence and the sequence of somatostatin-14 lies a peptide of 78 amino acids, the "proregion" of prosomatostatin I. Little is known about the metabolic fate and biological activity of this portion of the hormone precursor. The proregion of anglerfish prosomatostatin I is very similar to the corresponding region in the rat somatostatin precursor (*60*) and will be discussed. Overall, 32 of 78 amino acids (41%) are conserved between the two proregions. The degree of homology is even greater toward the carboxy-termini of these sequences.

The sequence depicted in Fig. 2 as anglerfish preprosomatostatin I closely resembles that reported by Hobart et al. (*52*). There are a few codons in the two sequences that differ, however. These differences could result from reverse-transcription errors or from the sequencing of distinct polymorphic cDNA. We have recently sequenced a number of cDNAs encoding preprosomatostatin I from an anglerfish cDNA library and have noted single nucleotide differences among several of the cloned cDNAs. Because this cDNA library was derived from pooled anglerfish islets, it is possible that these alterations represent polymorphic differences between individual fish. Alternatively, it is possible that these single nucleotide differences could represent the expression of separate somatostatin genes. The latter explanation would support the findings of Warren et al. (*54*).

Also depicted in Fig. 2 is the sequence of anglerfish preprosomatostatin II determined by Hobart et al. (*52*). The somatostatin-14 sequence encoded by this precursor has two amino acid changes from that found in anglerfish somatostatin I or the mammalian somatostatin—a Tyr residue at position seven and a Gly at position ten. The hexapeptide Ala-Pro-Arg-Glu-Arg-Lys just described is also found in the anglerfish somatostatin II precursor, but the remainder of the proregion is quite different from prosomatostatin I. This difference in amino acid sequence is reflected by marked differences in the tertiary structures of the two precursors estimated by determination of the levels of hydrophobicity, α-helix, and β-turn of various regions of the two molecules. Comparison of the anglerfish preprosomatostatins with mammalian somatostatin precursors suggests that the mammalian somatostatins were derived from a form closely related to preprosomatostatin I. This in turn suggests that the biological actions of the somatostatin II tetradecapeptide and proregion may be important for physiological functions that are different from those of the corresponding regions of prosomatostatin I. The functions served by these somatostatin II sequences may either be nonessential in mammals or have been taken over by completely separate mechanisms. Other "variant" forms of somatostatin have also been isolated from catfish islets (*20*) and goldfish neurosecretory glands (*21*).

Recent studies by Noe and Spiess (*64*) suggested that the two anglerfish preprosomatostatins undergo a differential process of proteolytic cleavage. Characterization of pulse-labeled peptides synthesized by the anglerfish islet indicated that the tetradecapeptide form of somatostatin is generated from preprosomatostatin I by cleavage at the Arg-Lys sequence immediately adjacent to somatostatin-14. Prosomatostatin II is proteolytically cleaved at a site nearer the amino-terminus of the precursor to generate a somatostatin-28-like peptide. As described above, the two potential cleavage sites are identical in the two precursors, and it is not known how the differential processing is controlled.

It is possible that the two somatostatin precursors are synthesized in different cell types and are exposed to a different set of processing enzymes.

## 3. MAMMALIAN SOMATOSTATIN

The structural similarity of anglerfish preprosomatostatin I to mammalian somatostatin precursors made it possible to obtain cDNA encoding rat preprosomatostatin (*60*). Immunoreactive somatostatin had been previously identified in extracts of rat medullary carcinomas of the thyroid (*65*), but identifying somatostatin precursors in cell-free translations of mRNA from such tumors proved to be unsuccessful. By using a cloned cDNA encoding the carboxy-terminal portion of anglerfish somatostatin I as a hybridization probe (*55*), identification of a somatostatin-related cDNA in a cDNA library prepared from a rat medullary carcinoma of the thyroid was possible (*66*). Initially, only one cDNA, which was 400 base pairs (bp), was identified among the 10,000 colonies that were screened. The paucity of clones encoding somatostatin reflected the low level of the neuropeptide in this particular tumor. Another cDNA library was subsequently prepared from a somatostatin-rich medullary carcinoma of the thyroid. By screening this second library with the original rat prosomatostatin cDNA, identification of 55 additional somatostatin-related cDNA clones was made. The nucleotide and corresponding amino acid sequence of rat preprosomatostatin I are depicted in Fig. 3. An identical sequence for rat preprosomatostatin was reported by Funckes et al. (*62*).

The rat cDNA sequences encode a 116-amino acid protein of $M_r$ 12,737, somewhat smaller than the rat hypothalamic preprosomatostatin identified in cell-free translations by Joseph-Bravo et al. (*67*). As in the anglerfish precursor, the tetradecapeptide somatostatin is located at the carboxy-terminus of the molecule, preceded by an Arg-Lys sequence, and followed by a stop codon. The 28 amino acids at the carboxy-terminus of rat preprosomatostatin are identical to those found in ovine and porcine somatostatin-28 (*60*). Adjacent to the sequence of somatostatin-28 is a sequence Leu-Gln-Arg that presumably serves as an additional

TGCGGACCTGCGTCTAGACTGACCCACCGCGCTCAAGCTCGGCTGTCTGAGGCAGGGGAG

ATG CTG TCC TGC CGT CTC CAG TGC GCG CTG GCC GCG CTC TGC ATC
*Met Leu Ser Cys Arg Leu Gln Cys Ala Leu Ala Ala Leu Cys Ile* 15

GTC CTG GCT TTG GGC GGT GTC ACC GGG GCG CCC TCG GAC CCC AGA
*Val Leu Ala Leu Gly Gly Val Thr Gly Ala Pro Ser Asp Pro Arg* 30

CTC CGT CAG TTT CTG CAG AAG TCT CTG GCG GCT GCC ACC GGG AAA
*Leu Arg Gln Phe Leu Gln Lys Ser Leu Ala Ala Ala Thr Gly Lys* 45

CAG GAA CTG GCC AAG TAC TTC TTG GCA GAA CTG CTG TCT GAG CCC
*Gln Glu Leu Ala Lys Tyr Phe Leu Ala Glu Leu Leu Ser Glu Pro* 60

AAC CAG ACA GAG AAC GAT GCC CTG GAG CCT GAG GAT TTG CCC CAG
*Asn Gln Thr Glu Asn Asp Ala Leu Glu Pro Glu Asp Leu Pro Gln* 75

GCA GCT GAG CAG GAC GAG ATG AGG CTG GAG CTG CAG AGG TCT GCC
*Ala Ala Glu Gln Asp Glu Met Arg Leu Glu Leu Gln Arg Ser Ala* 90

AAC TCG AAC CCA GCC ATG GCA CCC CGG GAA CGC AAA GCT GGC TGC
*Asn Ser Asn Pro Ala Met Ala Pro Arg Glu Arg Lys Ala Gly Cys* 105

AAG AAC TTC TTC TGG AAG ACA TTC ACA TCC TGT TAG CTTTAATATT
*Lys Asn Phe Phe Trp Lys Thr Phe Thr Ser Cys Stop* 116

GTTGTCTCAGCCAGACCTCTGATCCCTCTCCTCCAAATCCCATATCTCTTCCTTAACTCC

CAGCCCCCCCCCCAATGCTCAACTAGACCCTGCGTTAGAAATTGAAGACTGTAAATACAA

AATAAAATTATGGTGAAATTATGpoly(A)

Fig. 3. Nucleotide and corresponding amino acid sequence of the rat preprosomatostatin cDNA. Signal sequence, potential glycosylation site, and somatostatin-28 sequences are underlined. Somatostatin-14 sequence is enclosed in box at the carboxy-terminus of the precursor.

prohormone cleavage site. This region is reminiscent of the sequence Leu-Glu-Arg at the analogous position in the anglerfish preprosomatostatins (*52*). It has been suggested that cleavage of prohormones at single Arg residues, rather than at dibasic sequences, occurs in other hormonal precursors, including relaxin (*68*), propressophysin (*69*), somatomedin (*70*), and pancreatic polypeptide (*71*). Amino-terminal to somatostatin-28 in the precursor is the sequence Asn-Gln-Thr (amino acid positions *61–63*), which represents a potential N-glycosylation site (*72*). Several investigators have suggested that the rat somatostatin precursor may be glycosylated (*41,73*), but studies in our laboratory (*61*) using cell-free translation systems supplemented with microsomal membranes have not supported this observation. The absence of glycosylation in cell-free systems does not exclude the possibility that prosomatostatin may be glycosylated in specific tissues, however.

A region of hydrophobic amino acids characteristic of a signal sequence is present at the amino-terminus of the rat somatostatin precursor. Nucleotides encoding this hydrophobic region are preceded by an AUG codon that codes for the initiator methionine. A purine-rich sequence AGGGGAG, reminiscent of the prokaryotic Shine-Delgarno sequence (*74*), is found adjacent to the coding region of rat preprosomatostatin.

To determine the site of cleavage of the signal sequence from the rat precursor, the somatostatin prohormone generated in cell-free translations supplemented with dog pancreas microsomal membranes was microsequenced (*61*). These studies indicated that rat preprosomatostatin is cleaved between Gly and Ala, at positions 24–25, to produce a prosomatostatin molecule of 92 amino acids. Cleavage at this position is similar to that seen in other presecretory proteins. Such cleavage commonly occurs after amino acids with small neutral side chains, such as Gly, Cys, or Ser (*75*). After removal of the signal sequence, a prosomatostatin molecule of 10,389 daltons is formed. The size of this prohormonal form suggests that the largest form of immunoreactive somatostatin identified in numerous earlier studies represents the intact prosomatostatin molecule.

The biological function and metabolic fate of the proregion of the somatostatin precursor (i.e., the portion that lies amino-terminal to somatostatin-28) have not been determined. As already mentioned, this region contains areas that are highly conserved between the rat and anglerfish somatostatin I precursors. Recently, Shen et al. reported the nucleotide and corresponding amino acid sequence of a cDNA encoding human preprosomatostatin (*63*). The human cDNA sequence is strikingly similar to that of the rat precursor.

To compare the cDNA sequences of the rat and human preprosomatostatins, the nucleotide sequences were analyzed by a dot matrix computer program (*76*). Each dot in Fig. 4 denotes an identical nucleotide triplet in the two DNA sequences. Perfect homology between the two sequences would be represented by a

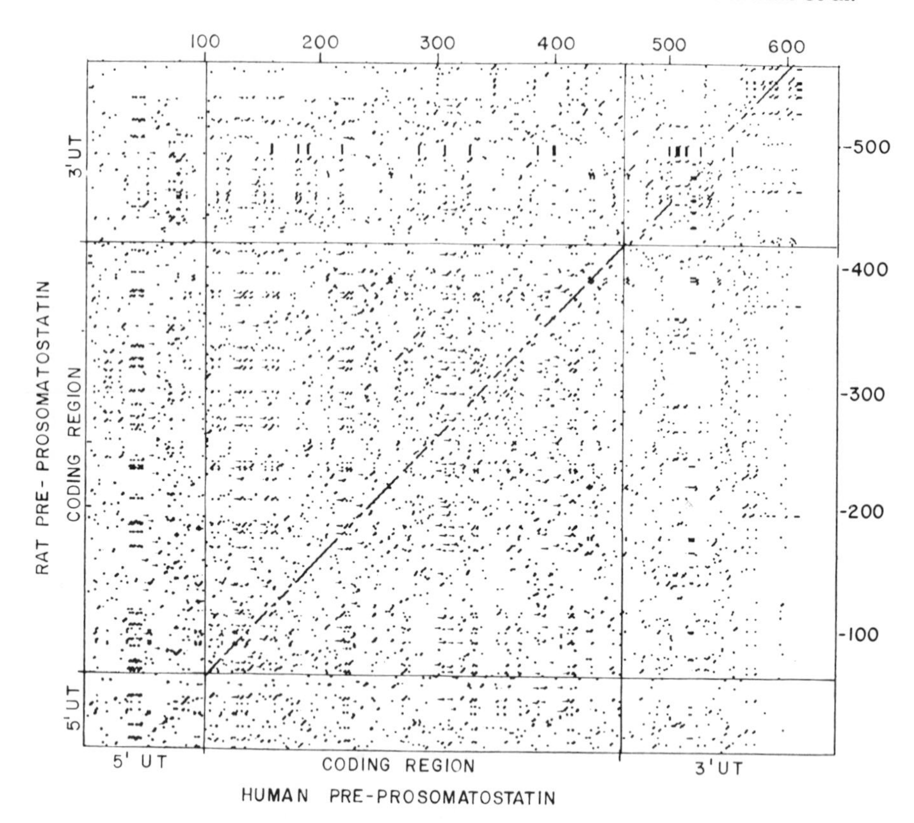


Fig. 4. Comparison of nucleotide sequences of rat and human cDNAs, using a dot matrix computer program. Each dot in the matrix denotes identical triplet of nucleotides between the two sequences. Coding and untranslated regions of the cDNA are indicated.

straight diagonal line. Gaps in the line denote differences in the sequences, parallel displacements represent insertions or deletions of DNA sequences. As shown in Fig. 4, the DNA sequences corresponding to the coding region of the preprosomatostatins are highly conserved. There is also a high degree of conservation within the 3′ untranslated regions, especially in the region adjacent to the polyadenylate tail. Insertion or deletion of two short DNA sequences accounts for much of the difference between the two 3′ untranslated regions. By contrast, less homology is seen in the 5′ untranslated regions.

At the protein level, only four amino acids were found to differ between the rat and human precursors. These differences include a Cys for Ser at position 14, Gly for Cys at position 21, Thr for Ala at position 43, and a Pro for Ser at position 74. The first two of these substitutions lie within the signal sequence of the precursor; the last two are within the proregion. The cysteine substitutions, although not usually considered conservative changes, maintain the hydrophobic character of the signal sequence, and are therefore probably not functionally significant. The remarkably high level of conservation between rodents and humans of the amino acid sequence of the entire prosomatostatin molecule (90/92 amino acids) is probably the result of strong biological pressures to conserve this sequence during evolution. The importance of the proregion could be that it allows proper processing of the precursor or serves as a substrate for the generation of additional biologically active peptides.

To determine the neuroanatomical distribution and fate of the somatostatin prohormone, we prepared an antiserum to a 15-amino acid peptide corresponding to amino acids 63–77 of rat preprosomatostatin (Fig. 5) (77). This fragment is referred to as rat somatostatin cryptic peptide, RSCP. Using this antiserum in immunohistochemical studies, it has been demonstrated that immunoreactive RSCP is present in diverse regions of the central nervous system, in both cell bodies and axon terminals, in a distribution indistinguishable from that of somatostatin-14 (Fig. 6). The appearance of immunoreactive RSCP in cell bodies of the hypothalamic periventricular system, their descending projections to the median eminence, and the presence of immuno-

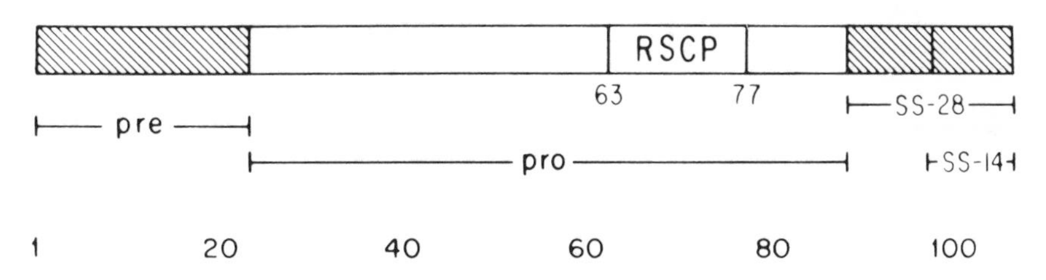


Fig. 5. Schematic representation of rat preprosomatostatin. The shaded region at the amino-terminus represents the pre- or signal sequence. The shaded region of the carboxyl-terminus represents somatostatin-28 and -14. A proregion of 64 amino acids separates the signal sequence from the somatostatin-28 sequence. Rat somatostatin cryptic peptide (RSCP) is indicated within the proregion. Numbering refers to amino acid positions within the precursor.

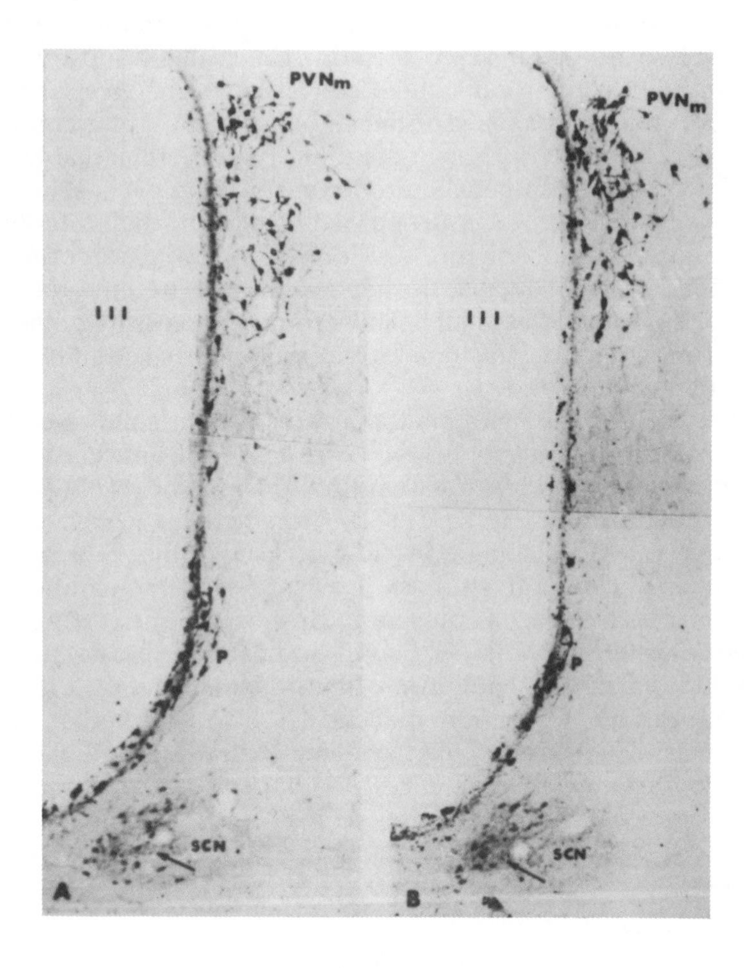


Fig. 6. Serial sections through the anterior hypothalamus reacted with antiserum to RSCP (A) and somatostatin-14 (B). Neuronal cell bodies in the parvocellular subdivision of the paraventricular ($PVN_m$), periventricular (P), and suprachiasmatic nucleus (SCN) that stain using both antisera are indicated.

reactive material within the terminals in the median eminence are compatible with the transport of immunoreactive RSCP throughout the course of this neuronal system. The close proximity of immunoreactive material to the capillary plexus in the external zone of the median eminence makes it possible that this material is secreted into portal vessels as well. Similarly, immunoreactive terminal fields in extrahypothalamic regions suggest that transport of the immunoreactive RSCP from the cell bodies of origin to the axon terminals is a general phenomenon in the central nervous system and not limited exclusively to the hypothalamic tuberoinfundibular system.

Although there are no prototypical cleavage sites within the proregion of the somatostatin precursor, it is possible that this region undergoes further proteolytic cleavage. When we chromatographed acetic acid extracts of rat median eminence on Sephadex G50 and assayed the fractions using an antibody to synthetic RSCP, the major immunoreactive fraction migrated with an apparent size of 7 kilodaltons (*78*). This fraction also contained somatostatin-14 immunoreactivity. These results imply that the prosomatostatin molecule is cleaved at a site amino-terminal to residue 63. Similar conclusions have been reached by other investigators, using an antiserum directed toward the amino-terminus of prosomatostatin (*78a*). Benoit et al. sequenced one such fragment generated by amino-terminal cleavage of prosomatostatin (*79*). The sequence of this fragment indicates that proteolytic cleavage occurs between two Leu residues at position 56–57 (Fig. 3). Production of the resultant 7260-dalton fragment would require that the somatostatin prohormone be cleaved at a somewhat unusual site. A similar site is used, however, in the generation of angiotensin from angiotensinogen. The putative cleavage region is highly conserved between the human, rat, and anglerfish somatostatin I precursors (Figs. 2 and 3). The biological activity of this 7260-dalton peptide has not been tested.

Control of the choice of cleavage sites used in the processing of the somatostatin precursor could occur at several levels. First, there could be multiple somatostatin genes that differ in regions encoding specific cleavage sites. These individual genes could differ by as little as a single nucleotide. Alternatively, the precise cleavage sites of the individual precursors could be maintained, but may lie within different flanking regions that alter the tertiary structure at the site. In either of these two instances, control of processing would occur at the transcriptional level. Transcription of one or another gene would determine the processing pattern of the precursor. Posttranslational modification of a single somatostatin gene product could also influence processing. Addition of carbohydrate residues to the Asn at position 61, possibly masking the adjacent Leu-Leu cleavage site, would be an example of such a mechanism. Finally, it is possible that the ability of various tissues to produce somatostatin-14, -28, or other forms may depend on the presence of particular enzymatic processing systems within specific cell types.

As a first step in determining which of these mechanisms is important in controlling prosomatostatin processing, we examined genomic DNA isolated from a single rat spleen, using the method of Southern (*80*). Rat genomic DNA was digested with the restriction enzymes *Eco*RI, *Hin*dIII, or *Pvu*II, electrophoresed on agarose gels, transferred to nitrocellulose, and probed with a nick-translated cDNA encoding rat preprosomatostatin. Two fragments were identified in each digestion. Probing individually with 5′ and 3′ portions of the cDNA identified a single fragment in each digestion. These results are most consistent with the presence of a single somatostatin gene in the rat. Therefore, it is likely that the differential processing of prosomatostatin within various tissues is a result of posttranslational modifications or cell-specific processing enzymes.

The rat preprosomatostatin cDNA was subsequently used as a hybridization probe to screen two genomic libraries cloned in Charon-4A bacteriophage. The first library, prepared from a partial *Eco*RI digest of rat liver genomic DNA (*81*), contained a recombinant bacteriophage that represented the 5′ exon of the rat somatostatin gene and extended 15 kb in the 5′ direction. No sequences representing the 3′ portion of the gene were detected in this library, despite the ten genome equivalents that were screened. A second recombinant bacteriophage was isolated from a partial *Hae*III genomic library (*81*). This bacteriophage contained a representation of the entire preprosomatostatin gene and extended 10 kb in the 3′ direction. Southern blotting analysis of rat liver genomic DNA revealed hybridizing fragments identical to those obtained from digests of the two recombinant bacteriophages, indicating that the bacteriophage libraries contained an accurate representation of the somatostatin gene.

A restriction map of the 25-kb region encompassed by the two bacteriophages indicated that the somatostatin gene is encoded by two *Eco*RI fragments of 4.6 and 9.7 kb, which represent the 5′ and 3′ regions of the gene (Fig. 7). The two hybridizing *Eco*RI fragments isolated from the bacteriophage clones were subcloned

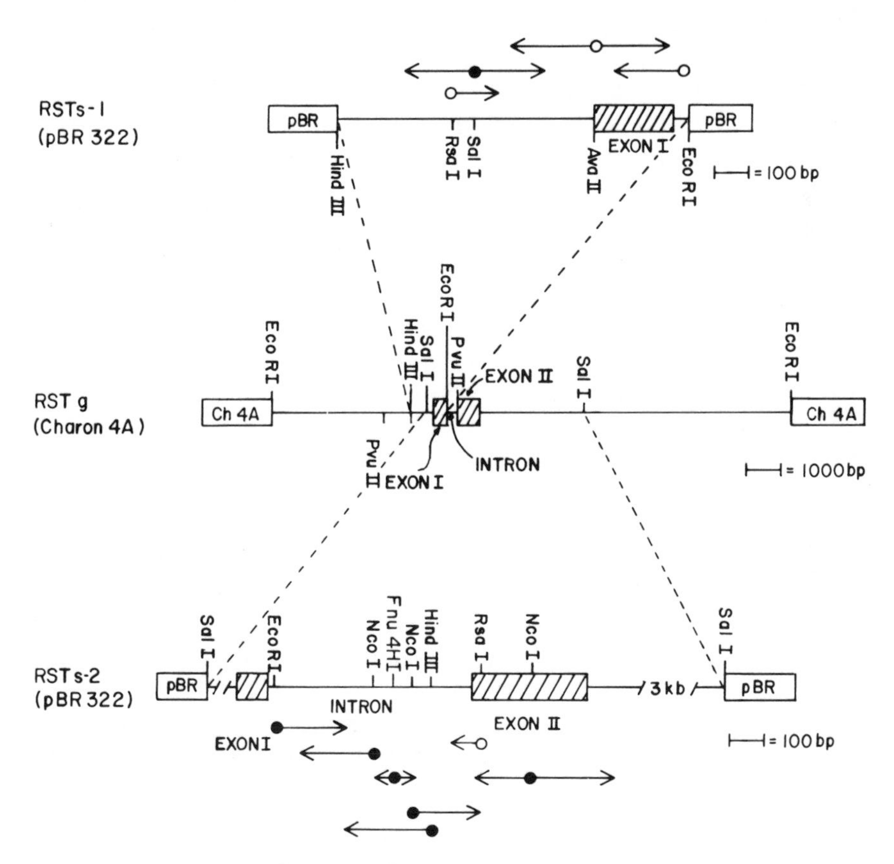


Fig. 7. Recombinant phages containing the rat preprosomatostatin gene (RST g) and two subclones of the genomic insert (RSTs-1 and -2) in plasmid pBR322. Orientation of the rat preprosomatostatin gene is from left to right. Open boxes represent DNA of the cloning vehicles, phage Charon-4A and plasmid pBR322. Hatched boxes indicate the two exons of the gene. Lines represent the remaining rat DNA, including the single intron.

into the plasmid pBR322 and sequenced by the method of Maxam and Gilbert (*82*).

Sequencing of the subclones revealed a somatostatin gene of approximately 1.2 kb (*83*). Similar findings were reported by Tavianini et al. (*84*). Comparison of the nucleotide sequences of the genomic subclones with that of the preprosomatostatin cDNA revealed a single intron of 630 bases that separates the glutamine and glutamic acid codons at amino acid positions 46–47 (Fig. 8). The intervening sequence within the human somatostatin gene occurs at the identical position (*85*). Characteristic GT–AG donor–acceptor sites are present adjacent to the coding sequences. The biological function of the amino-terminal region of prosomatostatin is not known. It is not possible, therefore, to assess whether the intervening sequence separates functional domains within the precursor. The Glu-Gln sequence lies amino-terminal to any known or postulated cleavage sites within prosomatostatin.

The transcriptional initiation point of the rat somatostatin gene was determined by S-1 nuclease mapping. Polyadenylated RNA from the somatostatin-rich medullary carcinoma of the thyroid was hybridized to a $^{32}$P-labeled fragment of the 5′ region of the somatostatin gene. The hybrids were digested with S-1 nuclease (*86*), and the length of the nuclease-resistant fragment was determined on denaturing polyacrylamide gels. A single band was observed on autoradiography of the gels, migrating at a size that placed the transcriptional initiation site 102 bases upstream from the translational initiator methionine codon. The 5′ untranslated region is slightly longer, therefore, than the corresponding region

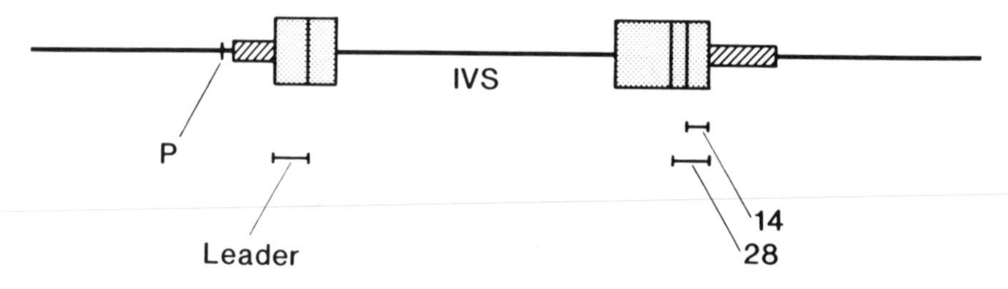


Fig. 8. Structure of the rat somatostatin gene. P refers to promoter, IVS to intervening sequence, 14 and 28 to somatostatin-14 and -28 peptides. The repetitive element is located 650 bases upstream from the promoter.

in most other eukaryotic genes (*87*). The size of the somatostatin mRNA detected on Northern blotting analysis is consistent with this assignment of the transcriptional initiation site, however. The unique band seen in the S-1 nuclease experiment suggests that the transcription of the somatostatin gene occurs at a single site in the medullary carcinoma of the thyroid. Whether this is true of somatostatin gene transcription in other tissues as well is unknown. Thirty-one bases upstream from the transcriptional initiation site lies an AT-rich sequence, TTTAAAA, surrounded by a GC-rich sequence. This AT-rich sequence is likely to represent a variant form of the Goldberg-Hogness (TATA) box.

Comparison of the nucleotide sequences of the rat and human somatostatin genes between positions -27 and -153 reveals that 90% of the nucleotides are identical (Fig. 9). This level of conservation is higher than that seen in any other region of the gene, including the protein coding region, and suggests that these sequences may be particularly important for gene expression.

When DNA isolated from the bacteriophage clones was digested with *Eco*RI, transferred to nitrocellulose, and probed with nick-translated rat genomic DNA, the 4.7-kb band representing the 5′ portion of the gene was detected. These experiments suggested that a repetitive sequence is located near the 5′ end of the

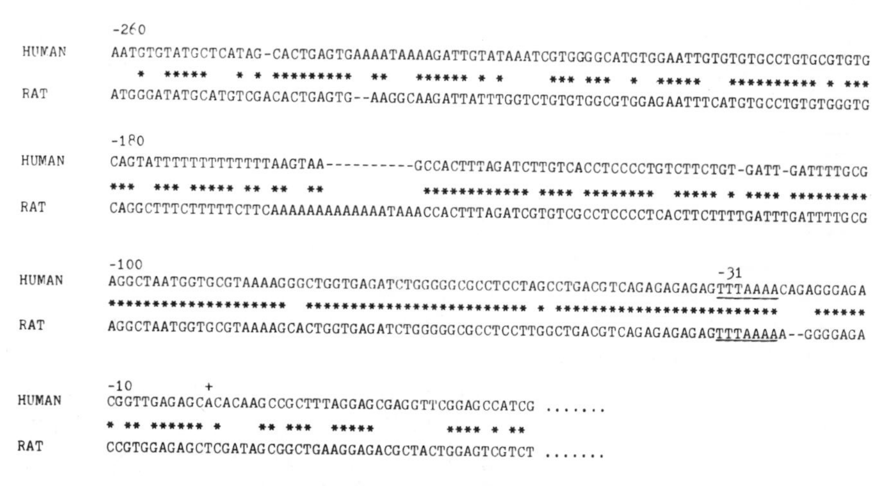


Fig. 9. Comparison of the rat and human somatostatin gene promoter regions. Identical nucleotides are indicated by an asterisk. Atypical Goldberg-Hogness box is underlined, and + indicates transcriptional initiation point.

somatostatin gene. Further restriction-endonuclease mapping of this portion of the gene localized the repetitive sequence to a region approximately 650 bp upstream from the promotor. To estimate the frequency of this repetitive element, the 5′ flanking region of the gene was used as a hybridization probe to screen the rat *Hae*III genomic library. Approximately 10% of the plaques hybridized strongly to the probe. Assuming that bacteriophages containing the repetitive element are not selectively amplified within the library, these findings suggest that the element occurs at a frequency of approximately 10,000 copies/genome.

The nucleotide sequence of the repetitive element included a 42-bp alternating purine and pyrimidine region. Such sequences are capable of entering a Z-configuration and perhaps influence transcriptional activity of distant sites (*88*). H. Hamada and coworkers (personal communication) determined that a synthetic GT sequence, similar in length to that seen in the somatostatin gene, acts as a transcriptional enhancer in CHO cells. It is possible, therefore, that the Z-DNA region within the rat somatostatin gene may also influence transcriptional activity.

Similar examples of other long purine-pyrimidine regions have been identified in other eukaryotic genes, including the human somatostatin gene (*85*). The positioning of the repetitive element relative to the promoter region in most other examples does not appear to be favorable for transcriptional regulation, however. The rat somatostatin gene may therefore provide a model system for determining whether Z-DNA switching has an influence on gene transcriptional activity.

## 4. FUTURE DIRECTIONS

Because of the wealth of structural information now available on somatostatin, it is feasible to ask specific questions about the regulation of somatostatin expression and processing in different cell types. For example, despite knowledge of the complete primary sequence of prosomatostatin, the role of the proregion in determining specific peptide cleavage has remained elusive. By using molecular techniques of site-directed mutagenesis, it is now possible to assay the effects of changes in tertiary structure on precursor cleavage. In addition, putative regulatory regions, such as the Z-DNA region discussed above, can be assayed for their influence on promoter activity in different cell types. Information obtained from these types of experiments should greatly increase our understanding of somatostatin biosynthesis and physiology.

## REFERENCES

1. Brazeau P., Vale W., Burgus R., Ling N., Butcher M., Rivier J., and Guillemin R. (1973) Hypothalamic peptide that inhibits the secretion of immunoreactive pituitary growth hormone. *Science* **129,** 77.
2. Arimura A., Smith W., and Schally A. V. (1976) Blockade of the stress-induced decrease in blood GH by anti-somatostatin serum in rats. *Encodrinology* **98,** 540.
3. Ferland L., Labrie F., Jobin M., Arimura A., and Schally A. V. (1976) Physiological role of somatostatin in the control of growth hormone and thyrotropin secretion. *Biochem. Biophys. Res. Commun.* **68,** 149.
4. Vale W., Rivier C., Brazeau P., and Guillemin R. (1975) Effects of somatostatin on the secretion of thyrotropin and prolactin. *Endocrinology* **95,** 968.
5. Chihara K., Arimura A., Chihara M., and Schally A. V. (1978) Studies on the mechanism of growth hormone and thyrotropin responses to somatostatin antiserum in anesthesized rats. *Endocrinology* **103,** 1916.
6. Weeke J., Hansen A. P., and Lundaek K. (1975) Inhibition by somatostatin of basal levels of serum thyrotropin in normal men. *J. Clin. Endocrinol. Metab.* **41,** 168.
7. Drovin J., DeLean A., and Rainville D. (1976) Characteristics of the interaction between thyrotropin releasing hormone and somatostatin for thyrotropin and prolactin release. *Endocrinology* **98,** 514.
8. Tyrrell J., Lorenzi M., and Gerich J. (1975) Inhibition by somatostatin of ACTH secretion in Nelson's syndrome. *J. Clin. Endocrinol. Metab.* **40,** 1125.
9. Gerich J. E. (1981) Somatostatin and diabetes. *Am. J. Med.* **70,** 619.
10. Gerich J. E., Lovinger R., and Grodsky G. (1975) Inhibition by somatostatin of glucagon and insulin release from the perfused rat pancreas in response to arginine, isoproterenol, and theophylline: Evidence for a preferential effect on glucagon secretion. *Endocrinology* **96,** 749.
11. Gerich J. E. and Patton G. S. (1978) Somatostatin-physiology and clinical applications. *Med. Clin. North Am.* **62,** 375.
12. Guillemin R. (1976) Somatostatin inhibits the release of acetylcho-

line induced electrically in the myenteric plexus. *Endocrinology* **99,** 1653.

13. Gothert M. (1980) Somatostatin selectively inhibits noradrenaline release from hypothalamic neurons. *Nature* **288,** 86.
14. Tanaka S. and Tsujimoto A. (1981) Somatostatin facilitates the serotonin release from rat cerebral cortex, hippocampus, and hypothalamus slices. *Brain Res.* **208,** 219.
15. Barker J. L. (1976) Peptides: Role in neuronal excitability. *Physiol. Rev.* **56,** 435.
16. Schally A. V., Dupont A., Arimara A., Redding T. W., Nishi N., Linthicym G. L., and Schlesinger D. H. (1976) Isolation and structure of growth hormone-release inhibiting hormone from porcine hypothalami. *Biochemistry* **15,** 509.
17. Pradayrol L., Jornvall H., Mutt V., and Ribet A. (1980) N-terminally extended somatostatin: The primary structure of somatostatin-28. *FEBS Lett.* **109,** 55.
18. Spiess J., Rivier J., Rodkey J. A., Bennett C. D., and Vale W. (1979) Isolation and characterization of somatostatin from pigeon pancreas. *Proc. Natl. Acad. Sci. USA* **76,** 2974.
19. Noe B. D., Spriss J., Rivier J., and Vale W. (1979) Isolation and characterization of somatostatin from anglerfish pancreatic islet. *Endocrinology* **105,** 1410.
20. Oyama H., Bradshaw R. A., Bates O. J., and Permutt A. (1980) Amino acid sequence of catfish pancreatic somatostatin I. *J. Biol. Chem.* **225,** 2251.
21. Pearson D., Shively J. E., Clark B. R., Geschwind I. I., Barkley M., Nishioka R. S., and Bern H. A. (1980) Urotensin II: A somatostatin-like peptide in the caudal neurosecretory system of fishes. *Proc. Natl. Acad. Sci. USA* **77,** 5021.
22. Reichlin S. (1983) Somatostatin, in *Brain Peptides* (Krieger D., Brownstein M., and Martin J., eds.) Wiley, New York.
23. Berelowitz M., LeRoith D., Von Schenk H., Newgard C., Szabo M., Frohman L. A., Shiloach J., and Roth J. (1982) Somatostatin-like immunoreactivity and biological activity is present in tetrahymena pyriformis, a ciliated protozoan, *Endocrinology* ***110,*** 1939.
24. Arimura A., Sato H., Coy D., and Schally A. V. (1975) Radioimmunoassay for GH-release inhibiting hormone. *Proc. Soc. Exper. Biol. Med.* ***148,*** 784.
25. Spiess J. and Vale W. (1978) Investigations of larger forms of somatostatin in pigeon pancreas and rat brain. *Metabolism* ***27,*** 1175.
26. Rorstad O. P., Epelbaum J., Brazeau P., and Martin J. B. (1979) Chromatographic and biological properties of immunoreactive

somatostatin in hypothalmic and extrahypothalamic brain regions of the rat. *Endocrinology* **105,** 1083.

27. Pradayrol L., Jornvall H., Mutt V., and Ribet A. (1980) N-terminally extended somatostatin: The primary structures of somatostatin-28 *FEBS Lett.* **109,** 55.
28. Schally A. V., Huang W.-Y., Chang R. C. C., Arimura A., Redding T. W., Millar R. P., Hunkapiller M. W., and Hood L. E. (1980) Isolation and structure of pro-somatostatin: A putative somatostatin precursor from pig hypothalamus. *Proc. Natl. Acad. Sci. USA* **77,** 4489.
29. Esch F., Bohlen P., Ling N., Benoit R., Brazeau P., and Guilleman R. (1980) Primary structure of ovine hypothalamic somatostatin-28 and somatostatin-25. *Proc. Natl. Acad. Sci. USA* **77,** 6827.
30. Mandarino L., Stenner D., Blanchard W., Nissen S., Gerich J., Ling N., Brazeau P., Bohlen P., Esch F., and Guillemin R. (1981) Selective effects of somatostatin-14, -25, and -28 on in vitro insulin and glucagon secretion. *Nature* **291,** 76.
31. Brown M., Rivier J., and Vale W. (1980) Somatostatin 28: Selective action on the pancreatic B cell and brain. *Endocrinology* **108,** 2391.
32. Konturek S. J., Tasler J., Jaworek J., Pawlik W., Walus K. M., Schusdziarra V., Meyers C. A., Coy D. H., and Schally A. V. (1981) Gastrointestinal, secretory, motor, circulatory, and metabolic effects of prosomatostatin. *Proc. Natl. Acad. Sci. USA* **78,** 1967.
33. Brazeau P., Ling N., Bohlen P., Benoit R., and Guillemin R. (1981) High biological activities of the synthetic replicates of somatostatin-28 and somatostatin-25. *Regul. Pept.* **1,** 255.
34. Susini C., Esteve J. P., Vayesse W., Pradayrol L., and Ribet A. (1980) Somatostatin-28; Effect on exocrine pancreatic secretion in conscious dogs. *Gastroenterology* **79,** 720.
35. Vayesse N., Pignal F., Esteve J. P., Pradayrol L., Susini C., and Ribet A. (1980) Effects of somatostatin-14 and somatostatin-28 on bombesin-stimulated release of gastrin, insulin, and glucagon in dog. *Gastroenterology* **78,** 1284.
36. Meyers C. A., Murphy W. A., Redding T. W., Coy D. H., and Schally A. V. (1980) Synthesis and biological actions of prosomatostatin. *Proc. Natl. Acad. Sci. USA* **77,** 6171.
37. Arimura A., Sato H., Dupont A., Nishi N., and Schally A. V. (1975) Somatostatin: Abundance of immunoreactive hormone in rat stomach and pancreas. *Science* **189,** 1007.
38. Lauber M., Camier M., and Cohen P. (1979) Higher molecular weight forms of immunoreactive somatostatin in mouse hypothalamic extracts: Evidence of processing in vitro. *Proc. Natl. Acad. Sci. USA* **76,** 6004.

39. Trent D. F., and Weir G. G. (1981) Heterogeneity of somatostatin-like peptides in rat brain, pancreas, and gastrointestinal tract. *Endocrinology* **108,** 2033.
40. Spiess J. and Vale W. (1980) Multiple forms of somatostatin-like activity in hypothalamus. *Biochemistry* **19,** 2861.
41. Patzelt C., Tager H. S., Carroll R. J., and Steiner D. F. (1980) Identification of prosomatostatin in pancreatic islets. *Proc. Natl. Acad. Sci. USA* **77,** 2410.
42. Schusdziarra V., Zyzner E., Rouiller D., Harris V., and Unger R. H. (1980) Free somatostatin in the circulation: Amounts and molecular sizes of somatostatin-like immunoreactivity in portal, aortic, and vena caval plasma of fasting and meal-stimulated dogs. *Endocrinology* **107,** 1572.
43. Chihara K., Minamitani N., Kaj H., Arimura A., and Fujita T. (1981) Intraventricularly injected growth hormone stimulates somatostatin release into rat hypophysial portal blood. *Endocrinology* **109,** 2279.
44. Fletcher D. J., Noe B. D., Bauer G. E., and Quigley J. P. (1980) Characterization of the conversion of a somatostatin precursor to somatostatin by islet secretory granules. *Diabetes* **29,** 593.
45. Habener J. F., and Potts J. T., Jr. (1978) Biosynthesis of parathyroid hormone. *N. Engl. J. Med.* **229,** 580.
46. Loh Y. P., Brownstein M. J., and Gainer H. (1984) Proteolysis in neuropeptide processing and other neural functions. *Annu. Rev. Neurosci.* **7,** 189.
47. Jacobs J. W., Lund P. K., Potts J. T., Jr., Bell N. H., and Habener J. F. (1981) Procalcitonin is a glycoprotein. *J. Biol. Chem.* **256,** 2803.
48. Johnson D. E., Torrence J. L., and Elde R. P. (1976) Immunohistochemical localization of somatostatin, insulin and glucagon in the principal islets of the anglerfish (*Lophius americanus*) and the channel catfish (*Ictalarus punctatus*). *Am. J. Anat.* **147,** 119.
49. Goodman R. H., Lund P. K., Jacobs J. W., and Habener J. F. (1980) Preprosomatostatins: Products of cell-free translations of messenger RNAs from anglerfish islets. *J. Biol. Chem.* **255,** 6549.
50. Lund P. K., Goodman R. H., Jacobs J. W., and Habener J. F. (1980) Glucagon precursors identified by immunoprecipitation of products of cell-free translation of messenger RNA. *Diabetes* **29,** 583.
51. Shields D. and Blobel G. (1977) Cell-free synthesis of fish preproinsulin and processing by heterologous mammalian microsomal membranes. *Proc. Natl. Acad. Sci. USA* **74,** 5300.
52. Hobart P., Crawford R., Shen L., Pictet R., and Rutter W. (1980) Cloning and sequence analysis of cDNAs encoding two distinct

somatostatin precursors found in endocrine pancreas of anglerfish. *Nature* **288,** 137.

53. Shields D. (1980) In vitro biosynthesis of somatostatin. *J. Biol. Chem.* **255,** 11625.
54. Warren T. G., and Shields D. (1982) Cell-free biosynthesis of somatostatin precursors: Evidence for multiple forms of pre-prosomatostatin. *Proc. Natl. Acad. Sci. USA* **77,** 4074.
55. Goodman R. H., Jacobs J. W., Chin W. W., Lund P. K., Dee P. C., and Habener J. F. (1980) Nucleotide sequence of a cloned structural gene coding for a precursor of pancreatic somatostatin. *Proc. Natl. Acad. Sci. USA* **77,** 5869.
56. Goodman R. H., Lund P. K., Barnett F. H., and Habener J. F. (1982) Intestinal pre-prosomatostatin: Identification of mRNA coding for a precursor by cell-free translations and hybridization with a cloned islet cDNA. *J. Biol. Chem.* **256,** 1499.
57. Maniatis T., Fritsch E. F., and Sambrook J. (1982) Molecular Cloning - A Laboratory Manual. New York: Cold Spring Harbor Laboratory.
58. Paterson B. M., Roberts B. E., and Kuff E. L. (1977) Structural gene identification and mapping by DNA-mRNA hybrid-arrested cell-free translation. *Proc. Natl. Acad. Sci. USA* **74,** 4370.
59. Ricciardi R. P., Miller J. S., and Roberts B. E. (1979) Purification and mapping of specific mRNAs by hybridization selection and cell-free translation. *Proc. Natl. Acad. Sci. USA* **76,** 4927.
60. Goodman, R. H., Jacobs J. W., Dee P. C. and Habener J. F. (1982) Somatostatin-28 encoded in a cloned cDNA obtained from a rat medullary thyroid carcinoma. *J. Biol. Chem.* **257,** 1156.
61. Goodman R. H., Aron D. C., and Roos B. A. (1983) Rat pre-prosomatostatin: Structure and processing by microsomal membrane. *J. Biol. Chem.* **257,** 1156.
62. Funckes M., Minth D., Deschenes R., Maganin M., Tavianini M. A., Sheets M., Collier M., Weith H. L., Aron D. C., Roos B. A., and Dixon J. (1984) Cloning and characterization of a mRNA encoding rat pre-prosomatostatin. *J. Biol. Chem.* **258,** 8781.
63. Shen L., Pictet R. L., and Rutter W. J. (1982) Human somatostatin I: Sequence of cDNA. *Proc. Natl. Acad. Sci. USA* **79,** 4575.
64. Noe B. D., and Spiess J. (1983) Evidence for biosynthesis and differential post-translational proteolytic processing of different (pre) prosomatostatins. *J. Biol. Chem.* **258,** 1121.
65. Berelowitz M., Cibelius M., Szabo M., Frohman L. A., Epstein S., and Bell N. H. (1980) Somatostatin-like immunoreactivity in a transplantable medullary carcinoma of rat thyroid: Partial chromatographic and biologic characterization. *Endocrinology* **107,** 1418.

66. Jacobs J. W., Goodman R. H., Chin W. W., Dee P. C., Bell N. H., Potts J. T., and Habener J. F. (1981) Calcitonin structural gene codes for multiple polypeptides. *Science* **213,** 457.
67. Joseph-Bravo P., Charli J. L., Sherman T., Boyer H., Bolivar F., and McKelvy J. F. (1980) Identification of a putative hypothalamic mRNA coding for somatostatin and of its product in cell-free translation. *Biochem. Biophys. Res. Commun.* **94,** 1004.
68. Hudson P., Haley J., John M., Cronk M., Crawford R., Haralambides J., Tregear G., Shine J., and Niall H. (1983) Structure of a genomic clone encoding biologically active human relaxin. *Nature* **301,** 628.
69. Land H., Schultz G., Schmale H., and Richter D. (1982) Nucleotide sequence of a cloned cDNA encoding arginine vasopressin-neurophysin II precursor. *Nature* **295,** 299.
70. Jansen M., Van Schaik F. M. A., Ricker A. T., Bullock B., Woods D. E., Gabbay K. E., Nussbaum A. L., Sussenbach J. S., and Van den Brande J. L. (1983) Sequence of cDNA encoding human insulin-like growth factor I precursor. *Nature* **306,** 609.
71. Leiter A. B., Keutmann H. T., and Goodman R. H. (1984) Structure of a precursor to human pancreatic polypeptide. *J. Biol. Chem.* **259,** 14702.
72. Pless D., and Lennarz W. (1977) Enzymatic conversion of proteins to glycoproteins. *Proc. Natl. Acad. Sci. USA* **74,** 134.
73. Aron D. C., O'Neil J., Birnbaum R. S., Muszynski M., Sabo S. W., Fleming A. A., and Roos B. A. (1982) Somatostatin-producing rat medullary thyroid carcinoma—identification and preliminary characterizations of high molecular weight forms of somatostatin. *Program of the 69th Annual Meeting of the Endocrine Society,* San Francisco, Abs. 246.
74. Shine J. and Delgarno L. (1974) The 3′ terminal sequence of *Escherichia coli* 16S ribosomal RNA: Complementarity to nonsense triplets and ribosome binding sites. *Proc. Natl. Acad. Sci. USA* **71,** 1342.
75. Docherty K., and Steiner D. F. (1982) Post-translational proteolysis in polypeptide hormone biosynthesis. *Annu. Rev. Physiol.* **44,** 625.
76. Novotny J. (1982) Matrix program to analyze primary structure homology. *Nucleic Acids Res.* **10,** 127.
77. Lechan R. M., Goodman R. H., Rosenblatt M., Reichlin S., and Habener J. F. (1983) Prosomatostatin specific antigen in rat brain: Localization by immunocytochemical staining with an antiserum to a synthetic sequence of pre-prosomatostatin. *Proc. Natl. Acad. Sci. USA* **80,** 2790.
78. Low M. J., Lechan R. M., Rosenblatt M., and Goodman R. H.

(1983) Distribution and content of prosomatostatin-specific antigen in rat brain. *Program of the 65th Annual Meeting of the Endocrine Society,* San Antonio, Abs. 281.

78a. Aron D. C., Andrews P. C., Dixon J. E., and Roos B. A. (1984) Identification of cellular prosomatostatin and nonsomatostatin peptides derived from its amino terminals. *Biochem. Biophys. Res. Commun.* **124,** 450–456.

79. Benoit R., Bohlen R., Esch R., and Ling N. (1984) Neuropeptides derived from prosomatostatin that do not contain the somatostatin-14 sequence. *Brain Res.* **311,** 23.

80. Southern E. (1975) Detection of specific sequences among DNA fragments separated by gel electrophoresis. *J. Mol. Biol.* **98,** 503.

81. Sargent T. D., Wu J. R., Sala-Trepat J. M., Wallace R. B., Reyes A. A., and Bonner J. (1979) The rat serum albumen gene: Analysis of cloned sequences. *Proc. Natl. Acad. Sci. USA* **76,** 3256.

82. Maxam A. M. and Gilbert W. (1977) A new method for sequencing DNA. *Proc. Natl. Acad. Sci. USA* **74,** 560.

83. Montminy M. R., Goodman R. H., Horovitch S., and Habener J. F. (1984) Primary structure of the gene encoding rat pre-prosomatostatin. *Proc. Natl. Acad. Sci. USA* **83,** 3337.

84. Tavianini M. A., Hayes T. E., Magazin M. D., Minth C. D., and Dixon J. E. (1985) Isolation, characterization and DNA sequence of the rat somatostatin gene. *J. Biol. Chem.* **259,** 11798.

85. Shen L.-P. and Rutter W. J. (1984) Sequence of the human somatatostatin I gene. *Science* **224,** 168.

86. Weaver R. F. and Weissman C. (1979) Mapping of RNA by a modification of the Berk-Sharp procedure: The 5′ termini of 15S globin mRNA precursor and mature 10S globin in RNA have identical map coordinates. *Nucleic Acids. Res.* **7,** 1175.

87. Kozak M. (1984) Compilation and analysis of sequences upstream from the translational startsite in eukaryotic mRNAs. *Nucleic Acids Res.* **12,** 857.

88. Wang A., Quigley G. J., Kuplack F. J., Crawford J. L., Van Boon J. H., VanderMarel G., and Rich A. (1979) Molecular structure of a left-handed double helical DNA fragment at atomic resolution. *Nature* **282,** 680.

# Chapter 6

# The Human Growth-Hormone Gene Family

DAVID D. MOORE, RICHARD F SELDEN, EDOUARD PROST, DANIEL S. ORY, AND HOWARD M. GOODMAN

## 1. INTRODUCTION

Human growth hormone (GH) is encoded by one member of a small, highly homologous gene family. The expression of these genes is under stringent tissue-specific control. The amino acid sequence of growth hormone, which is synthesized in the pituitary, is 85% homologous to chorionic somatomammotropin (CS, also known as placental lactogen), which is synthesized in the placenta (*1*), and the mRNA sequences are even more similar (93%) (*2,3*). Both proteins are major products of the differentiated cells that secrete them. In addition to this tissue-specific level of control, growth-hormone expression is further modulated by a variety of other factors. This is best characterized in the rat, for example, in which glucocorticoids and thyroid hormone synergistically induce growth-hormone expression (*4*).

As an initial approach to unraveling these different levels of control of expression, the genes for the various members of the human GH family have been isolated in several laboratories (*5–12*). Initially there was confusion over the total number of genes as a result of the unusually frequent rearrangements of cloned segments of the locus. However, a minimal map with five genes accounts for all the data from Southern blots and is now generally accepted (*10*). The sequences of all five of the family members have been determined, and show extensive homology of introns and flanking sequences, as well as coding regions (*6,7,11,13*). Here we discuss the overall structure of the locus and some details of the sequences of the family members from the points of view of evolution of the locus as a whole and of the elements that control the expression of its members.

## 2. MAP OF THE GH LOCUS

Figure 1 shows a restriction map of the GH gene cluster. The map was constructed by analysis of the four independent cosmid clones and 11 independent lambda-phage clones indicated. These clones are the result of the screening of three independent genomic libraries with an hGH cDNA probe, and one with a 21-nucleotide synthetic oligonucleotide specific for CS-type genes (*13*). This map confirms and extends that presented by Barsh et al. (*10*), and is fully consistent with extensive genomic Southern blotting results. One cosmid clone (not shown) and one phage clone (as indicated) contained different small rearrangements. Not

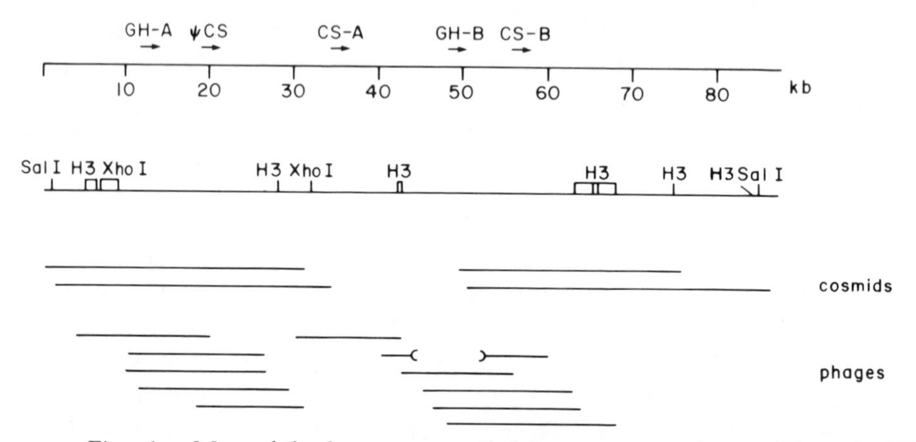


Fig. 1. Map of the human growth-hormone gene cluster. The relative positions of the members of the human growth-hormone gene family are shown over a scale that indicates the size of the cluster in kilobase pairs (kbp). The positions of cleavage sites for several restriction enzymes that cut the locus infrequently are given (H3 = *Hin*dIII). The relative positions and extents of four cosmid clones and 11 phages used to compile the map of the locus are given.

shown are one cosmid and one phage that contained sequences hybridizing specifically to GH probes, but did not yield maps that corresponded to results of geonomic Southern blots. This class of apparent rearrangements includes the gene that we previously called 2.9-1 (*9,14*), as well as several reported by others (*8,12*). The reason for the apparently high instability of this gene segment is not known; however, the high number of Alu family repeats that it contains, and the fact that the segment is composed of several large direct repeats, or both circumstances, may be relevant (see below).

Examination of patterns of restriction-enzyme cleavage sites and direct analysis of homology between flanking regions suggest that the GH locus is composed of two repeats of a basic GH/CS two-gene unit, separated by a fractional unit containing a single CS gene. This arrangement is diagrammed in Fig. 2, along with two specific and simple evolutionary pathways (out of many possibilities) that could generate such a pattern. The first of these is nearly identical to one proposed by Barsh et al. (*10*). The two pathways are basically similar, but are potentially resolvable. They differ in that the first generates an extra, central CS gene by unequal recombination between loci consisting of two homologous dimeric units, whereas, in the second, the central CS gene is a result of a similar deletion, starting from loci with three dimeric units. As diagrammed in Fig. 2, the first pathway requires that the two leftmost CS genes be derived from a common progenitor and, therefore, clearly predicts that they should be more similar to each other than to the rightmost gene. The second suggests that the central CS gene could be derived from either the leftmost or rightmost gene, depending on the location of the crossover generating a triplication of the locus. (A crossover of the latter type is shown.) This pattern is consistent with sequence analysis of the three CS-type genes (described below), which shows that the two rightmost genes are much more homologous to each other than to the leftmost gene. The probable existence of a recent gene-conversion event involving the left halves of the structural portions of this pair of CS genes raises some doubts about the significance of their similarity, but the increased homology does extend beyond this recently converted area. A detailed examination of the structure of the GH locus in other primates could provide further evidence to confirm or reject the proposed pathways.

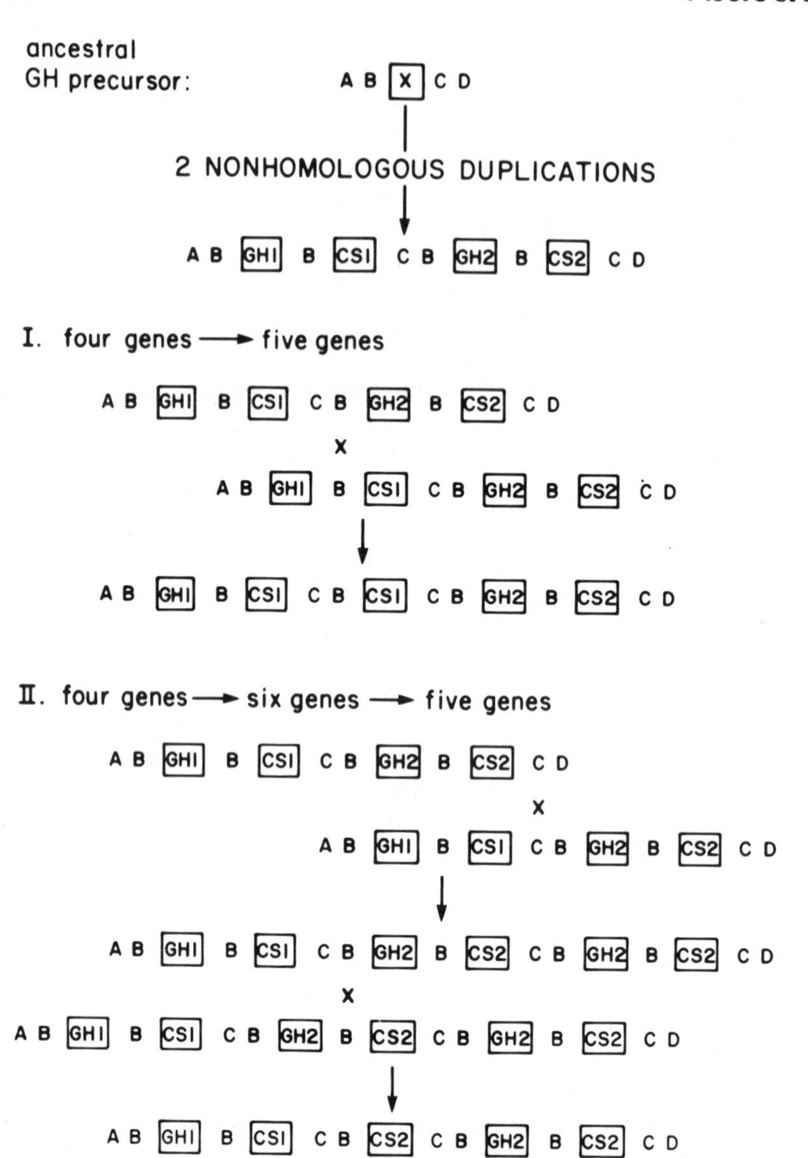


Fig. 2. Possible modes of evolution of the human growth-hormone gene cluster. Two different means for generating the currently observed 5-member family from an ancestral 4-gene set are indicated. A, B, C, and D indicate different blocks of sequence homology in the dispersed elements of the cluster. Pathways I and II differ in that I generates a central CS gene that is similar to CS1, whereas II generates a central CS gene similar to CS2.

Only three of the five genes in the cluster are known to be expressed. GH-A encodes the expressed growth-hormone protein. GH-B encodes a protein that can be produced in vitro (*15*), but there is no evidence that it is ever expressed in vivo. CS-A and CS-B encode CS proteins differing at only a single amino acid in the signal peptide (*7,11*). Evidence shows that both these genes are expressed as mRNA. For example, Barrera-Saldena et al. isolated CS cDNA clones corresponding to both CS-A and CS-B, and demonstrated that the ratio of the two mRNAs was approximately 3/2 (*16*). As shown by sequence analysis, PsCS, the third CS-type gene, is apparently a pseudogene that could not express a CS protein because of a splice-site mutation (*13*).

## 3. SEQUENCES OF THE GH FAMILY MEMBERS

DNA sequences for all the members of the family have been determined (*6,7,11,13*). Complete sequences for three different alleles of the CS-B gene have been obtained, which are basically in agreement with previously published sequences (*7*), except for the presence of a four- rather than five-base insert into the 3′ untranslated region, relative to CS-A. Such a four-base insert has been observed in CS cDNA sequences (*16*). Three different alleles of the GH-A gene and one allele of the CS pseudogene have been sequenced.

Figure 3 shows the sequences of all the members of the family in comparison to the sequence of GH-A. With the exception of the CS-A/CS-B pair, all of the family members are approximately equally divergent from each other. A more quantitative analysis confirms this. For example, GH-A and GH-B are 93.1% homologous overall, GH-A and CS-A are 92.5%, but CS-A and CS-B are 98.9% homologous. As shown in Fig. 3, the homology of CS-A and CS-B is not randomly distributed; they are nearly identical from the 5′ flanking region through the fourth-exon (99.5% homology), whereas the fourth intron, fifth exon, and 3′ flanking region are somewhat less similar (96.4%). Such stepwise variation in homology strongly implies that these two genes have been homogenized by a gene conversion event in the very recent past. One end of this conversion event must be near the junction of the fourth exon and fourth intron, whereas the other must be beyond

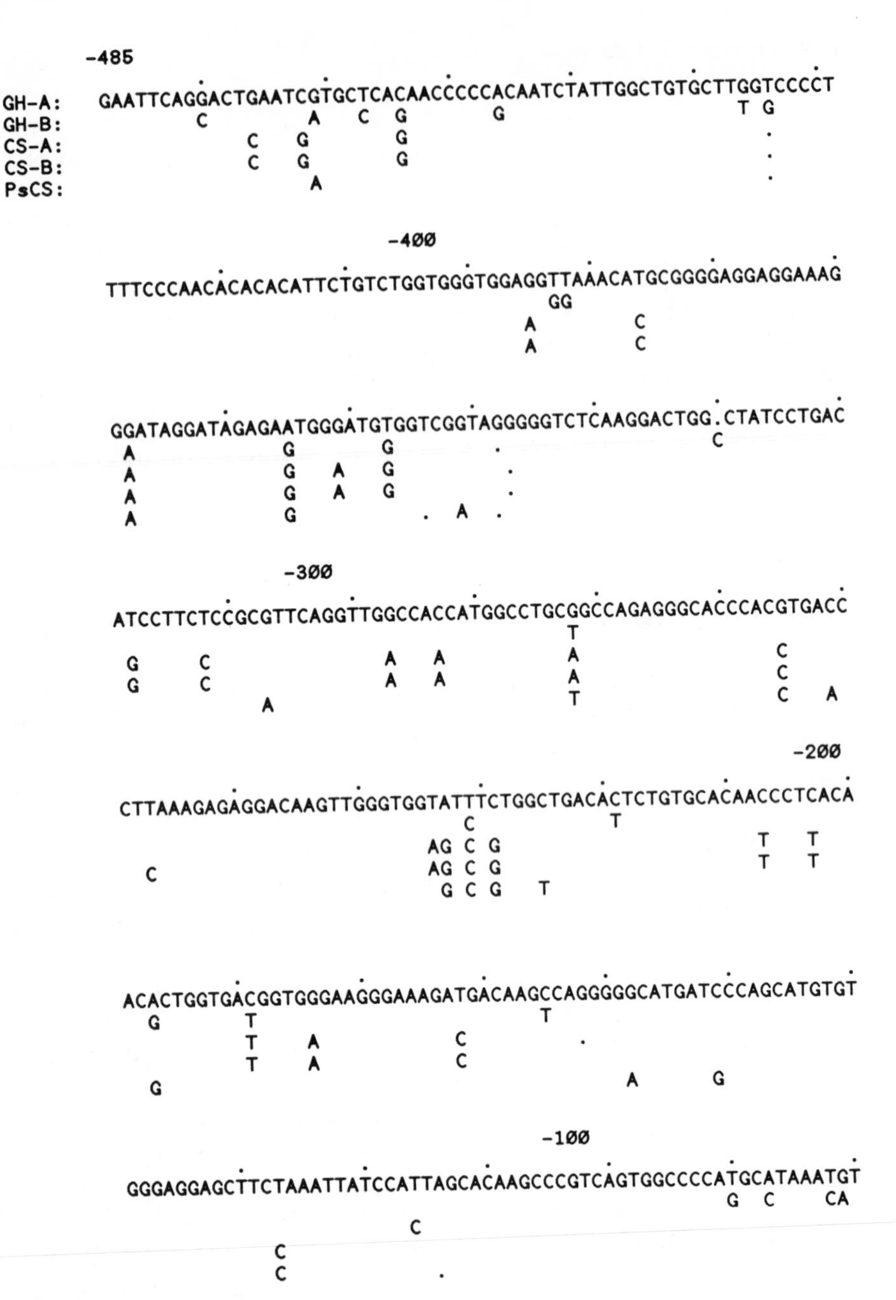


Fig. 3. Sequences of the members of the human growth-hormone gene family. The sequences of all the members of the family are compared with the sequence of the GH-A gene. Only differences from GH-A are indicated (. = deletion). The amino acid sequence encoded by GH-A is shown.

```
         .         .         .         .         .         .
ACACAGAAACAGGTG.GGGGCAACAGTGGGAGAGAAGGGGCCAGGGTATAAAAAGGGCCC
G  G           A  A A G   C A              .
               G  TCA G   G A       CT
               G  TCA G   G A       CT
G              G          C A

                1 ----->
         .         .         .         .         .         .
ACAAGAGACCGGCTCAAGGATCCCAAGGCCCAACTCCCCGAACCACTCAGGGTCCTGTGG
          A
               T
               T
               CG C            G        C   G   G

                                                         100
         .      (-26)       .         .         .         .
                  MetAlaThrG
ACAGCTCACCTAGCTGCAATGGCTACAGGT.AAGCGCCCCTAAAATCCCTTTGGGCACAA
        .     G         G                           .
             TG         C
              G         G
       G      G         G     G G

         .         .         .         .         .         .
TGTGTCCTGAGGGGAGAGGCAGCGACCTGTAGATGGGACGGGGGCACTAACCCTCAGGTT
                    G   T
C                       .
C                       C
C                   G   C    C                    .

                                     200
         .         .         .         .         .         .
TGGGGCTTCTGAATGTGAG.TATCGCCATGTAAGCCCAG.TATTTGGCCAATCTCAGAAA
        A       T  C         C                         T   T
        A                    C    G    A               T   T
        A                    C    G    A               T   T
                  .A         C    G    A               T   T

         .         .         .         .         .         .
GCTCCTGGTTCCTGGAGG.........GATG..GAGAGAGAAAAACAAA...CAGCTCCT
 T       C          AGGCAGA  GAGA            A   ACC
 T       CT                    GA            A  CAAA
 T       CT                    GA            .  CAAA
 T       C                     GA            G     A

                                 300
         .         .         .         .         .         .
GGAGCAGGGAGAGTGCTGGCCTCTTGCTCTCCGGCTCCCTCTGTTGCCCTCTGGTTTCTC
   A         C                  A             .    C
             C           C               CA        C
             C           C               CA        C
   A                          G         . C        C
```

Fig. 3. *(cont.)*

```
         .         .         .         .         .         .
TyrSerPheLeuGlnAsnProGlnThrSerLeuCysPheSerGluSerIleProThrPro
TATTCATTCCTGCAGAACCCCCAGACCTCCCTCTGTTTCTCAGAGTCTATTCCGACACCC
                                   C                 A     T

              TG  T           T    C        C
              TG  T           T    C        C
              TG  T           T    C        C            T

        800

         .         .         .         .         .         .
SerAsnArgGluGluThrGlnGlnLysSer
TCCAACAGGGAGGAAACACAACAGAAATCCGTGAGTGGATGCCTTCTCCCC.AGGCGGGG
          T A    G  G        T                      .A GT
       T         G                         G       T
       T         G                         G       T
       T         G  G                     TG       T

                                                900

         .         .         .         .         .         .
ATGGGGGAGACCTGTAGTCAGAGCCCCCGGGCAGCACAGCCAATGCCCGTCCTTCCCCTG
      T        G                          C    G
               G     G  T                 C    G
               G     G  T                 C    G
               G                          C    G

 .         .         .         .         .         .
   AsnLeuGluLeuLeuArgIleSerLeuLeuLeuIleGlnSerTrpLeuGluProVal
CAGAACCTAGAGCTGCTCCGCATCTCCCTGCTGCTCATCCAGTCGTGGCTGGAGCCCGTG
                                            A
     T                                 G
     T                                 G
      TC           A                   G     C

                             1000

         .         .         .         .         .         .
GlnPheLeuArgSerValPheAlaAsnSerLeuValTyrGlyAlaSerAspSerAsnVal
CAGTTCCTCAGGAGTGTCTTCGCCAACAGCCTGGTGTACGGCGCCTCTGACAGCAACGTC
   C          C                       T        G
 G             A G          A         T A A    G      G T A
 G             A G          A         T A A    G      G T A
 G             AC    A      A         T A A    G      G T A

         .         .         .         .         .         .
TyrAspLeuLeuLysAspLeuGluGluGlyIleGlnThrLeuMetGly
TATGACCTCCTAAAGGACCTAGAGGAAGGCATCCAAACGCTGATGGGGGTGAGGGTGGCG
   CG  A   G                                 T             A
   C
   C
   C                                 T                     A

         1100

         .         .         .         .         .         .
CCAGGGGTCCCCAATCTTGGAGCCCCACTGACTTTGAGAGCTGTGTTAGAGAAACACTGC
     ...AT      C   G         G   CC  G ACTG GG
         A           A        G   C   G    G GG       T
         G      C    A        G   A   G    G GG
                C    AG       G   C   G    G GG
```

Fig. 3. (*cont.*)

```
         .         .         .         .         .         .
     lySerArgThrSerLeuLeuLeuAlaPheGlyLeuLeuCysLeuProTrpLeuGl
CCCAGGCTCCCGGACGTCCCTGCTCCTGGCTTTTGGCCTGCTCTGCCTGCCCTGGCTTCA
                                                 T
                                   C
                                   C
                                   C

              400
         .  (1)    .         .         .         .         .
nGluGlySerAlaPheProThrIleProLeuSerArgLeuPheAspAsnAlaSerLeuAr
AGAGGGCAGTGCCTTCCCAACCATTCCCTTATCCAGGCTTTTGACAACGCTATGCTCCG

     CTG     G   A    G                       C            A
     CTG     G   A    G    G                  C            A
     CTG     G   A    G                    A AG G          A

                                                    500
         .         .         .         .         .         .
gAlaHisArgLeuHisGlnLeuAlaPheAspThrTyrGlnGluPhe
CGCCCATCGTCTGCACCAGCTGGCCTTTGACACCTACCAGGAGTTTGTAAGCTCTTGGGG
     G   C   T          A A         T                      T
A        CGC             A                         T
A        CGC             A                         T
A        CGCA            A                    A

         .         .         .         .         .         .
AATGGGTGCGCATCAGGGGTGGCAGGAAGGGGTGACTTTCCCCCGCT.GGGAAATAAGAG
           T    A                  A                 G   TG
          GG            A                   A  G     G   TG
          GG            A                   A  G     G   TG
          GG            A                   A  G     G   TG

                                  600
         .         .         .         .         .         .
GAGGAGACTAAGGAGCTCAGGGTT.TTTCCCGAAGCGAAAATGCAGGCAGATGAGCACAC
                        G   T T    T                     T
                        G   T T                          T G
                        G   T T                          T G
                        G   T T                          T G

         .         .         .         .         .         .
GCTGAGTGAGGTTCCCAGAAAAGTAACAATGGGAGCTGGTCTCCAGCGTAGA........
                                    A          A
      .C                C                      A    AACCAGCA
      CC                C                     ,A    AACCAGCA
      CT             G       G                 A    AAGCAGTG

                            700
         .         .         .         .         .         .
                              GluGluAlaTyrIleProLysGluGlnLys
.....CCTTGGTGGGCGGTCCTTCTCCTAGGAAGAAGCCTATATCCCAAAGGAACAGAAG
                                              TG     G
GTCCTT         G                    A                C
GTCCTT         G                    A                C
GTCCTT         G        C                    A
```

Fig. 3. (*cont.*)

```
                                                 1200
         .         .         .         .         .         .
TGCCCTCTTTTTAGCAGTCAGGCCCTGACCCAAGAGAACTCACCTTATTCTTCATTTCCC
                       G                    G
                  A    G                                  G
          G            G                                  G
          CC           G


         .         .         .         .         .         .
CTCGTGAATCCT.CAG.GCCTTTCTCTACACCCTGAAGGGGAGGGAGGAAAATGAATGAA
            C                 A    G                  G
  G         CT                                        G  A
  G         C       C                                 G
            C T A                       .    A        G  A


                               1300
         .         .         .         .         .         .
TGAGAAAGGGAGGGAACAGTACCCAAGCGCTTGGCCTCTCCTTCTCTTCCTTCACTTTGC
     G              G
     G              G
     G              G
     G              G             T             G


         .         .         .         .         .         .
  ArgLeuGluAspGlySerProArgThrGlyGlnIlePheLysGlnThrTyrSerLysP
AGAGGCTGGAAGATGGCAGCCCCCGGACTGGGCAGATCTTCAAGCAGACCTACAGCAAGT
                                           T   T
             C       G                C
             C       G                C
             C       A  T           C C


           1400
         .         .         .         .         .         .

heAspThrAsnSerHisAsnAspAspAlaLeuLeuLysAsnTyrGlyLeuLeuTyrCysP
TCGACACAAACTCACACAACGATGACGCACTACTCAAGAACTACGGGCTGCTCTACTGCT
 T        A  G                 G
 T           G      C          G
 T                  C          G
 T           G      C          G                     C


                                                   1500
         .         .         .         .         .         .
heArgLysAspMetAspLysValGluThrPheLeuArgIleValGlnCysArgSerValG
TCAGGAAGGACATGGACAAGGTCGAGACATTCCTGCGCATCGTGCAGTGCCGCTCTGTGG

                                        G
                                        G                 A
                                        G


          .  (191) .         .         .         .         .
luGlySerCysGlyPheEND
AGGGCAGCTGTGGCTTCTAGCTGCCCGGGTGGCATCCCTGT....GACCCCTCCCCAGTG

                    G      A  A  . .
    T               G      C     ˙ .     GACC
                    GG     C C     .
```

Fig. 3. (*cont.*)

```
                            1600
         .         .         .         .         .         .
CCTCTCCTGGCCTTGGAAGTTGCCACTCCAGTGCCCACCAGCCTTGTCCTAATAAAATTA
          T G      G   T
      T     C .    G
            C      G                 T
            C .    G
```

```
         .         .         .         .         .
AGTTGCATCATTTTGTCTGACTAGGTGTCCTTC.TATAATATTATGGGGTGGAGGGGG
                T               .G                     C
     T       CA      C       A                       A  T
     T       CA              A                       A  T
     T  TG   CAG C               A          G A CG A   C
```

Fig. 3. (*cont.*)

the *Eco*RI site, which is the beginning of the sequenced region. This analysis of the complete sequence basically supports our previous proposal (*9,14*) that the CS-B allele is a product of gene conversion. However, the earlier conclusion that the first exon of CS-B was more similar to GH than to CS was based on a CS cDNA sequence that does not reflect the mRNA sequence (*17*), and is incorrect.

Comparison of the sequences of the fourth intron plus fifth exon among the three CS-type genes shows that CS-A and CS-B are somewhat more similar to each other (96.8%) than they are to the pseudogene (95.2 and 94.9%, respectively). Detailed examination of those positions where CS-A and CS-B differ shows that neither has significantly more homology with the pseudogene. This implies that the CS-A and CS-B genes were derived from a common progenitor after it had diverged from the PsCS gene. As already described, this is consistent with the notion that the current five genes are the result of a deletion of a GH-type gene from a six-gene cluster.

It is remarkable that the introns and flanking regions of the genes are, in general, as highly homologous as the coding regions. It is extremely unlikely that the introns and flanking regions are subject to as much conservation pressure as are the protein-coding regions, and therefore the genes must have been diverging for only a very short time. Two extreme scenarios could account for this situation: Either the genes observed were all generated very recently, or much older genes have been exchanging information by gene conversion or other homogenizing processes. It is not clear which of these two possibilities is correct. However, as pointed out before (*9,14*), the apparently recent emergence of the difference between GH and CS implied by the sequence comparisons (approximately 55 million yr ago) conflicts with the estimate (greater than approximately 75 million yr ago) suggested by the general presence of CS function in various placental mammals. In this regard, it is particularly interesting that some evidence suggests that some nonprimate CS genes may have arisen by duplication of a prolactin, rather than a GH, gene (*18*). Since the prolactin and GH genes were separated at least 300 million yr before the mammalian radiation (*12*), the idea that the CS genes of the different mammalian species arose by distinct duplications of separate (albeit distantly related) genes seems plausible. Thus, in primates at least, some functions of a preexisting CS gene may have been subsumed by the newly generated gene derived from GH. One prediction of this convergent model is that a relic of the ancient, prolactin-derived CS gene could still be present in primate genomes. Perhaps the activity of such putative prolactin-derived CS functions accounts for the fact that deletion of both CS-A and CS-B does not cause any obvious phenotype (*19*).

Whether the current CS genes are the product of conversion or convergence, it is remarkable that their pattern of tissue-specific expression is quite different from that of GH-A, although the sequences of their promoter and 5′ flanking regions are virtually identical to those of GH-A. There are other examples of highly homologous genes that are not expressed in similar patterns. In given cell types of the yeast *Saccharomyces cerevisiae,* only one of two sets of mating-type genes that differ in position, but not sequence, is expressed (*20*). Similarly, independent insertions of identical retroviruses into different positions in the mouse genome can result in widely divergent, heritable patterns of tissue-specific expression (*21*). In a more conventional mammalian gene family, the promoter and flanking regions of the three goat β-globin genes are very similar, but the genes show strong temporal control or expression (*22*). All of these cases require regulatory elements that can govern expression over distances of as much as up to several thousand base pairs (bp) away from the regulated promoter.

The broadly defined class of enhancer elements, originally discovered in viruses, show several of the properties required for such regulators, including action at considerable distances (*23*). For example, a number of immunoglobulin genes have enhancers located in introns that stimulate transcription from promoters located up to several thousands of bp away (*24,25*). Furthermore, the action of these immunoglobulin enhancers is cell-type specific: They only function in the cell types appropriate for immunoglobulin expression. Although it is not clear whether such enhancer elements are present in the growth-hormone gene cluster, some constraints on their potential locations are implied by the large repeated segments characteristic of the locus. A growth-hormone specific element might be expected to be located at least several kb upstream from GH-A in an area not repeated near other genes, whereas CS regulatory elements could be guessed to be several kb downstream from the expressed genes, in repeated segments specific to CS genes. This and similar schemes point out a particularly puzzling aspect of the expression of the locus: The apparent lack of placental expression of the GH-B gene, even though it is flanked by two very strong placental-expression units. The explication of the mechanism of tissue-specific expression of the GH locus should be particularly interesting and rewarding.

## REFERENCES


1. Niall H. D., Hogan M. L., Sayer R., Rosenblum I. Y., and Greenwood P. C. (1971) Sequences of pituitary and placental lactogenic and growth hormones: Evolution from a primordial peptide by gene duplication. *Proc. Natl. Acad. Sci. USA* **68,** 866–870.
2. Shine J., Seeburg P. H., Martial J. A., Baxter J. D., and Goodman H. M. (1979) Cloning and analysis of recombinant DNA for human chorionic somatomammotropin. *Nature* **270,** 494–498.
3. Martial J. A., Hallewell R. A., Baxter J. D., and Goodman H. M. (1979) Human growth hormone: Complementary DNA cloning and expression in bacteria. *Science* **205,** 602–606.
4. Diamond D. J. and Goodman H. M. (1985) Regulation of growth hormone messenger RNA synthesis by dexamethasone and triiodothyronine: Transcriptional rate and mRNA stability changes in pituitary tumor cells. *J. Mol. Biol.* **181,** 41–58.

5. Fiddes J. C., Seeburg P. H., DeNoto F. M., Hallewell R. A., Baxter J. D., and Goodman H. M. (1979) Structure of the genes for human growth hormone and chorionic somatomammotropin. *Proc. Natl. Acad. Sci. USA* **76,** 4294–4298.
6. DeNoto F. M., Moore D. D., and Goodman H. M. (1981) Human growth hormone DNA sequence and mRNA structure: Possible alternative splicing. *Nucleic Acids Res.* **9,** 3719–3728.
7. Seeburg P. H. (1982) The human growth hormone gene family: Nucleotide sequences show recent divergence and predict a new polypeptide hormone. *DNA* **1,** 239–246.
8. Kidd V. J. and Saunders G. F. (1982) Linkage arrangement of human placental lactogen and growth hormone genes. *J. Biol. Chem.* **257,** 10673–10677.
9. Moore D. D., Walker M. D., Diamond D. J., Conkling M. A., and Goodman H. M. (1982) Structure, expression and evolution of growth hormone genes. Recent *Prog. Horm. Res.* **38,** 197–223.
10. Barsh G. S., Seeburg P. H., and Gelinas R. E. (1983) The human growth hormone gene family: Structure and evolution of the chromosomal locus. *Nucleic Acids Res.* **11,** 3939–3948.
11. Selby M. J., Barta A., Baxter J. D., Bell G. L., and Eberhardt N. L. (1984) Analysis of a major human chorionic somatomammotropin gene: Evidence for two functional promoter elements. *J. Biol. Chem.* **259,** 13131–13137.
12. Miller W. L. and Eberhardt N. L. (1983) Structure and evolution of the growth hormone gene family. *Endocrine Rev.* **4,** 97–141.
13. Moore D. D., Selden R. F., Conkling M. A., Fiddes J. C., Hallewell R. A., Prost E., DeNoto F. M., and Goodman H. M. Structure of the human growth hormone gene cluster: sequence of a chrorionic somatomammotropin pseudogene, in preparation.
14. Moore D. D., Conkling M. A., and Goodman H. M. (1982) Human growth hormone: A multigene family. *Cell* **29,** 285–287.
15. Pavlakis G. N., Hizuka N., Gorden P., Seeburg P. H., and Hamer D. H. (1981) Expression of two human growth hormone genes in monkey cells infected by simian virus 40 recombinants. *Proc. Natl. Acad. Sci. USA* **78,** 7398–7402.
16. Barrera-Saldana H. A., Seeburg P. H., and Saunders G. F. (1983) Two structurally different genes produce the same secreted human placental lactogen hormone. *J. Biol. Chem.* **258,** 3587–3592.
17. Goodman H. M., DeNoto F. M., Fiddes J. C., Hallewell R. A., Page G. S., Smith S., Tischer E. (1980) Structure and Evolution of Growth Hormone-Related Genes, in *Mobilizational Reassembly of Genetic Information,* (Scott W. A., Werner R., Joseph D. R., Schultz J., eds.), Academic, New York.

18. Robertson M. C., Gillespie B., and Friesen H. G. (1982) Characterization of the two forms of rat placental lactogen: rPL-I and rPL-II. *Endocrinology* **111,** 1860–1870.
19. Wurzel J. M., Parks J. S., Herd J. E., and Nielsen P. V. (1982) A gene deletion is responsible for absence of human chorionic somatomammotropin. *DNA* **1,** 251–259.
20. Astell C. R., Ahlstrom-Jonasson L., Smith M., Tatchell K., Naysmyth K. A., and Hall B. D. (1981) The sequence of the DNAs coding for the mating-type loci of *Saccharomyces cerevisiae. Cell* **27,** 15–24.
21. Stewart C., Harbers K., Jahner D., and Jaenisch R. (1983) X chromosome linked transmission and expression of retroviral genomes micro-injected into mouse zygotes. *Science* **221,** 760–765.
22. Townes T. M., Shapiro S. G., Wernke S. M., and Lingrel J. B. (1984) Duplication of a four gene set during the evolution of the goat beta-globin locus produced genes now expressed differentially in development *J. Biol. Chem.* **259,** 1896–1902.
23. Gluzman Y. and Shenk T. (eds) (1983) Enhancers and Eukaryotic Gene Expression. Cold Spring Harbor Laboratory, Cold Spring Harbor, New York.
24. Banerji J., Olsen L., and Schaffner W. (1983) A lumphocyte-specific cellular enhancer is located downstream of the joining region in immunoglobulin heavy chain genes. *Cell* **33,** 729–739.
25. Gillies S., Morrison S., Oi V., and Tonegawa S. (1983) A tissue specific transcription enhancer element is located in the major intron of a rearranged immunoglobin heavy chain gene. *Cell* **33,** 717–728.

# Chapter 7

# Glycoprotein Hormone Genes

William W. Chin

## 1. INTRODUCTION

### 1.1. General Considerations

The glycoprotein hormone family is composed of four structurally related polypeptide hormones primarily produced by two specific endocrine tissues. Three such hormones, thyrotropin (TSH), lutropin (LH), and follitropin (FSH), are produced in the anterior pituitary gland in most mammals. A fourth hormone, chorionic gonadotropin (CG), is produced by the placenta in primates and possibly other mammals (*1*). In addition, it is well known that CG is produced by trophoblastic and nontrophoblastic tumors (2). In particular, malignant conditions of the lung, breast, gastrointestinal system, and central nervous system may produce CG or its components, which therefore serve as possible cancer markers (*3–5*). Also, recent reports suggest that pituitary glycoprotein hormones may be produced in the central nervous system (*6,7*).

The glycoprotein hormones serve important functions in the regulation of several critical physiologic processes. TSH is the major regulator of the activity of the thyroid gland and is responsible for the stimulation of production of thyroid hormones—thyroxine (T4) and triiodothyronine (T3). These hormones are ultimately responsible for general cellular development and function, and for regulation of the metabolic rate (*8*). The pituitary gonadotropins, LH and FSH, serve to stimulate the gonads to produce gonadal sex-steroid hormones and gametes. The former are responsible for normal sexual development and expression of secondary sexual characteristics, and the latter for reproduction (*9*). Placental

CG is produced in early pregnancy and stimulates the ovary, in particular the corpus luteum, to produce progesterone and estrogen, which are critical for the maintenance of the feto-placental unit during early and midpregnancy (*2,10*).

The synthesis and secretion of glycoprotein hormones are controlled by complex regulatory systems. TSH stimulates the thyroid gland to produce thyroid hormones that, in turn, enter the bloodstream to act on the thyrotrope to decrease the production and secretion of TSH. These thyroid hormones may also influence the production of thyrotropin-releasing hormone (TRH) and other factors by the hypothalamus that may further regulate the production of TSH by the thyrotrope. In this classic endocrine negative-feedback regulatory system, the blood levels of thyroid hormone are thus kept nearly constant (*8,11*). Similarly, LH stimulates the gonadal tissues to produce testosterone and estrogen in males and females, respectively. These gonadal sex-steroid hormones enter the bloodstream to act on the production of LH. These sex-steroid hormones, like thyroid hormones, may also affect the production of gonadotropin-releasing hormone (GnRH) and other factors by the hypothalamus that further regulate the production of LH by the gonadotrope (*9*). FSH may also be involved in a similar negative-feedback regulatory system. However, the target-organ product that may mediate this negative-feedback relation is not clearly understood (*12*). Finally, the regulation of CG is not well known. There have been reports of the isolation of a placental GnRH that may have an important role in the regulation of CG (*13*). However, for general purposes it may be considered that the production of CG by the placenta in early pregnancy is constitutive in nature and, hence, is under minor regulation. In addition, CG produced by trophoblastic and nontrophoblastic tumors is probably not under hormonal regulation (*2*).

Hence, we see that the glycoprotein hormones serve important functions in endocrine physiology. Without these hormones, individuals may lack normal metabolic function or be unable to reproduce. Considering the importance of the glycoprotein hormones in normal body function, it is imperative to understand how the biosynthesis of these hormones may be regulated. In this chapter, I will describe recent work involved in the isolation and characterization of the genes that encode the proteins that form the glycoprotein hormone family. It is hoped that these studies will establish the basis for our future understanding of the mechanisms involved in the hormonal regulation of the production of these hormones.

### *1.2. Structure of Glycoprotein Hormones*

The glycoprotein hormones share important structural features. Each glycoprotein hormone is composed of two different, noncovalently associated subunits, α- and β- (Fig. 1). The α-, or common, subunit is identical in protein sequence among the hormones within a species. The β-subunits are different and impart biologic specificity to each hormone. Each subunit contains carbohydrate moieties that may be important for the biosynthesis and ultimate expression of biologic activity of the active dimer. In ad-

β SUBUNITS

COMMON α SUBUNIT

$+\beta_{TSH}$ → $\alpha\beta_{TSH}$ TSH

$+\beta_{LH}$ → $\alpha\beta_{LH}$ LH

$+\beta_{FSH}$ → $\alpha\beta_{FSH}$ FSH

$+\beta_{CG}$ → $\alpha\beta_{CG}$ CG

Fig. 1. The glycoprotein hormones are composed of two different subunits, α and β, associated in a noncovalent fashion. The α-, or common, subunit is nearly if not completely identical among hormones within a species and may combine with any one of the β-subunits to yield the biologically active dimers. Abbreviations: TSH, thyrotropin or thyroid-stimulating hormone; LH, lutropin or luteinizing hormone; FSH, follitropin or follicle-stimulating hormone; CG, chorionic gonadotropin.

dition, each subunit contains multiple cysteine residues (α, 10; β, 12) that combine to form multiple disulfide linkages that are critical for achieving the proper conformations that enable dimerization. Apparently, only the α–β dimer is biologically active, whereas separate, or free (i.e., unassociated), subunits are biologically inactive. Reconstitution experiments show that the α-subunit from any glycoprotein hormone may combine with another β-subunit to produce a biologically active dimer that corresponds to the specificity of the β-subunit (*1,14*).

Several investigators have studied the possible evolution of the α- and β-subunits from a common ancestral gene (*15,16*). The primary structure of the α-subunit is generally well conserved from species to species, with homology levels from 74 to 95% (*1,17*). The β-subunits are slightly more variable in that the homology level of human TSHβ and FSHβ with the LHβ–CGβ subunits is 30–40% (*1,14*). However, this limited protein sequence homology, along with the strict conservation of the cysteine residues (number and position), suggests that the β-subunits are derived from a common ancestral gene. The evolutionary relationship of the α- to the β-subunit genes, however, is less clear, although conservation of several sequence features has suggested the past existence of yet another shared ancestral gene.

The subunits of the glycoprotein hormones are encoded by separate mRNAs (18–23) (Fig. 2). The product of the α gene is a messenger RNA (mRNA) of approximately 800 bases. This mRNA encodes a protein that is the precursor of the α-subunit and consists, in general, of 116–120 amino acids. This precursor consists of a 24-amino-acid leader, or signal, peptide followed by an apoprotein of 92–96 amino acids. The signal peptide is located at the $NH_2$-terminal end of precursors of secretory proteins and is hydrophobic. The translation product, as analyzed by polyacrylamide–gel electrophoresis, has an apparent molecular mass of 14 kdaltons (*18,19,21*). This α-subunit precursor is converted, in a cotranslational fashion, to a glycosylated product containing N-linked carbohydrate moieties and an apoprotein devoid of its leader peptide. This intermediate form of the α-subunit is then further processed by several posttranslational steps, including modification of the sugar moieties to yield complex carbohydrates, protein folding, and subunit dimerization to form the biologically active hormone (*24–26*).

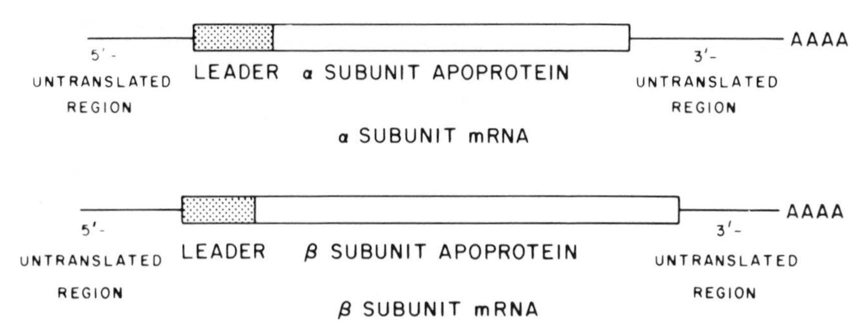


Fig. 2. Schematic diagram of mRNAs encoding the α- and β-subunits of the glycoprotein hormones. The boxes represent the coding regions for the subunit precursors; the stippled box denotes the leader, or signal, peptide, which is cleaved and removed cotranslationally, and the open box denotes the apoprotein. The lines at the left and right of the coding box represent the 5′- and 3′-untranslated regions, respectively, of the mRNAs. The AAAA denotes the polyadenylic acid tail at the 3′ end of the mRNAs characteristic of most mammalian mRNAs.

The genes for the β-subunit are transcribed to produce mRNAs of approximately 700 bases. These mRNAs encode precursor proteins containing 138–165 amino acids, in general. Each precursor consists of a 20-amino-acid leader, or signal, peptide followed by an apoprotein of 118–145 amino acids. Again, like the α-subunit, the β-subunits consist of an apoprotein attached to N-linked complex carbohydrate moieties. The initial translation products are approximately 17–20 kdaltons in molecular mass and are also processed cotranslationally with signal-peptide cleavage and N-glycosylation (*18,21,27*). Further posttranslational processes similar to those involved in the α-subunit occur, beginning in the rough endoplasmic reticulum (*28,29*). The formation of the appropriate secondary and tertiary structures of the β-subunits then allows dimerization with the α-subunit to yield the biologically active products. In addition, the COOH-terminal extension of the CGβ-subunit is modified by the acquisition of four O-linked carbohydrate chains that may affect its metabolic stability (*2,30*). The final products are then sorted into secretory granules in which these hormones await the putative secretory signals to allow their release into the general circulation (*31*).

Furthermore, an important feature of most cells that produce glycoprotein hormones is the production of the α-subunit in excess (*32,33*). The unassociated, or free, α-subunit thus formed is apparently biochemically distinct from the α-subunit that is found in the dimer (34–36). Parsons and coworkers have shown that bovine free α-subunit contains an extra carbohydrate moiety linked via an O-glycosidic bond that is located in the known α–β contact region (*37*). This carbohydrate moiety is sufficient to inhibit the dimerization process. Chemical removal of the O-linked carbohydrate moiety restores the ability of the free α-subunit to combine with the β-subunit, thus proving that this additional modification may modulate the ability of the subunits to recombine (*38*). Recent data also indicate that the β-subunit may be found in excess in some choriocarcinoma cell lines (*39*). This finding is significant, inasmuch as the synthesis of the β-subunit, the minority intracellular subunit, has been implicated as the rate-limiting step in the production of glycoprotein hormones. However, the finding of free, or unassociated, β-subunits now suggests that other regulatory situations may pertain. In particular, it is possible that glycosylation of the α-subunit may be the rate-limiting step. Thus, the O-glycosylation of the free α-subunit might dictate ultimately whether the α-subunit may combine with available β-subunits.

### 1.3. Major Questions

Thus, a major question that arises in the biosynthesis of the glycoprotein hormones is: What is the mechanism (or mechanisms) of expression of subunits in a coordinate fashion? It is now known that the subunits are encoded by separate mRNAs and, hence, separate genes that are located on separate chromosomes. How are these genes activated in a tissue-specific fashion so that the α-subunit, that is present in mammals as a single gene is expressed along with the particular β-subunit, which is located on a separate chromosome? In addition, how are the levels of expression of each gene modulated so that there is nearly balanced production of the subunits? Finally, and perhaps most important, how do the hormones produced by the target end-organs, such as thyroid hormone and gonadal sex-steroid hormones, act on these genes to decrease their expression and therefore decrease production of these hormones? These tantalizing questions can now be approached because of our recent progress in the understanding of the genes encoding the subunits of the glycoprotein hormones. In particular, molecular cloning has allowed detailed characterization of these subunit genes. The major objective of this chapter is to document the results of these studies, placing emphasis on the recombinant DNA techniques used to isolate the complementary DNAs (cDNAs) and genes encoding these subunits. This is a particularly challenging problem because of the relatively low abundance of the mRNAs that encode the subunits of the glycoprotein hormones in some of these tissues. The mRNAs encoding prolactin and growth hormone represent 1–5% of the total mRNAs (*40*) in the anterior pituitary gland, whereas the mRNAs encoding the separate subunits of the glycoprotein hormones in the pituitary gland represent only 0.1–0.5% in normal, and 0.5–1% in the stimulated, pituitary gland (*21,40*).

It is our hope that the molecular cloning of the genes encoding the subunits of the glycoprotein hormones will eventually allow us to understand the molecular basis for the effect of various hormones on the regulation of the activity of these genes. In addition, in vitro changes of these subunit genes may be made using recombinant DNA technology and allow introduction of these altered genes into foreign cells. Such DNA transfection studies will then permit the evaluation of the structure–function relation of the subunits and the effect of gene location on the ultimate expression of these hormones in various cells. Finally, such information may ultimately allow us to understand the basis for the tissue-specific expression of these genes.

## 2. MOLECULAR CLONING OF GENES ENCODING THE SUBUNITS OF GLYCOPROTEIN HORMONES

### *2.1. Molecular Cloning of α-Subunit cDNAs*

Inasmuch as the α-subunit mRNA is, in general, produced in excess of the β-subunit mRNAs in thyrotropic and gonadotropic tissues (*41*), the molecular cloning of the α-subunit cDNA is relatively straightforward.

I will first describe the molecular cloning of cDNAs encoding the α-subunit of the mouse glycoprotein hormones. The tissue source was a transplantable mouse thyrotropic tumor (MGH 101) developed in the Thyroid Unit at the Massachusetts General Hospital, Boston, Massachusetts. The amount of TSH produced by this tumor is relatively large compared with that produced by the pituitary gland. In fact, approximately 1–5% of the total protein produced by this tissue is TSH (*19*). Also, the tumor growth and associated production of TSH are regulated by thyroid hormones and hence are exhibited only in hypothyroid animals. The growth of the tumor is completely inhibited in animals with normal, functioning thyroid glands. The prototype of this thyrotropic tumor was initially developed by Jacob Furth in the early 1950s (*42*).

In our initial studies, poly(A)+ RNA from the mouse thyrotropic tumor was isolated and translated in a cell-free wheatgerm system (*19*). The major translation product was the precursor to the α-subunit with $M_r$ = 14 kdaltons. The identity of this 14-kdalton protein as the α-subunit precursor was confirmed by immunologic and biochemical techniques. The immunologic techniques included precipitation of the [$^{33}$S]-methionine-labeled translation product using antibody against reduced and carboxymethylated bovine α-subunit, followed by analysis by polyacrylamide-gel electrophoresis. The biochemical techniques involved comparison of tryptic peptides of labeled mature and precursor α-subunit proteins. Hence, we possessed an indirect but reliable method to identify mRNA that encoded the α-subunit (*19*). The next step involved the production of a cDNA library derived from mRNA from the mouse thyrotropic tumor. The cDNA library was produced by ligating double-stranded cDNA, synthesized using reverse transcriptase, *E. coli* DNA polymerase I, and S1-nuclease digestion, to the bacterial plasmid pBR322. The cDNA inserts were introduced at the unique *Pst*I site (*43*), using dG · dC homopolymeric tails. Bacterial colonies containing plasmids bearing cDNAs (sensitivity and resistance to ampicillin and tetracycline, respectively) were then screened for their ability to hybridize to α-subunit mRNA using the method of hybridization–selection–translation described by Ricciardi *et al.* (*44*). In this technique, cDNAs from different bacterial colonies were isolated and attached to small pieces of nitrocellulose filter paper. Total mRNA from the mouse thyrotropic tumor were allowed to hybridize with the cDNAs, and nonhybridizing mRNAs were washed away. The hybridized RNAs were then eluted and translated in a cell-free wheatgerm translation system. Presumably, if a cDNA contained information encoding the α-subunit, then the mRNA selected would be translated into the 14-kdalton α-subunit precursor. The results of such an experiment (*43*) are shown in Fig. 3. As you can see in this figure, two cDNAs were isolated and purified that contained sequences complementary to mRNA encoding the α-subunit precursor. Nucleotide-sequence analyses of these cDNAs indicated that indeed these cDNAs contained the major portion of the coding region for the mouse α-subunit mRNA. However, because these various cDNAs did

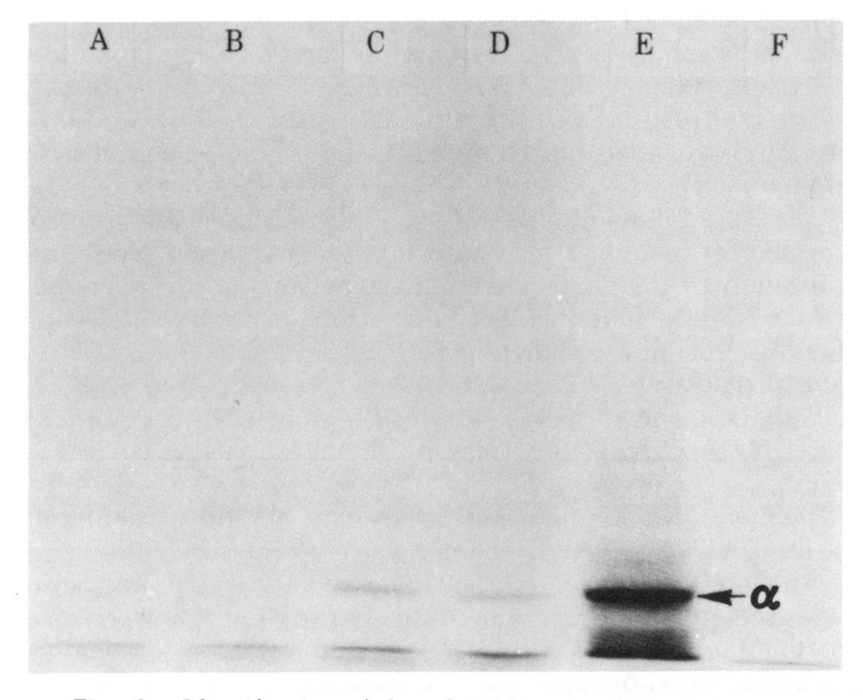


Fig. 3. Identification of cloned cDNAs encoding the precursor of mouse α-subunit. Autoradiogram of labeled proteins in the cell-free translations of thyrotropic mRNA using [$^{35}$S] methionine in the wheat-germ system is shown. Proteins are analyzed by electrophoresis on a gradient 10–20% polyacrylamide slab gel containing sodium dodecyl sulfate. Tumor mRNAs, hybridized to pBR322 (lane A), no DNA (lane B), pTSHα-1 (lane C), and pTSHα-2 (lane D) that were immobilized on nitrocellulose filter paper, were translated. pTSHα-1 and pTSHα-2 are two cDNAs that encode portions of the mouse α-subunit mRNA. Lane E, tumor mRNA; lane F, no added mRNA. The arrow points to the precursor of mouse α-subunit ($M_r$ 14,000) [details given in ref. (*43*), with permission].

not correspond to the extreme 5′ end of the α-subunit mRNA, a small DNA fragment was obtained from the 5′ end of the cDNA that represented the furthest 5′ extent of the cDNAs. This fragment was used as a primer to initiate the synthesis of α-subunit-specific cDNAs from tumor mRNA template to yield a transcript that was further analyzed by chemical sequencing techniques to yield the nucleotide sequence of the 5′ end of the mRNA. The composite nucleotide sequence of these cDNAs resulted in the almost-complete nucleotide sequence of the mRNA encoding the mouse α-subunit (*43*). These results are shown in Fig. 4.

The cDNA encoding the α-subunit of the rat glycoprotein hormone was obtained by Godine et al. (*17*). A cDNA library was prepared as just described using mRNA from the anterior pituitary gland of ovariectomized rats (*17*). These animals were chosen for their increased production of LH and presumed increased levels of LH subunit mRNAs (*21,45*). The rat cDNA library was screened for α-subunit cDNAs using a high-density-colony hybridization technique developed by Hanahan and Meselson (*46*) and the $^{32}$P-labeled mouse α-subunit cDNA as probe. The results of an experiment are shown in Fig. 5. The rat α-subunit cDNAs were also subjected to nucleotide-sequence analysis to yield results that are shown in Fig. 4, which compares the sequence of the cDNA encoding the α-subunit in the rat with that of the mouse.

The molecular cloning of the cDNA encoding the α-subunit of the human glycoprotein hormones was performed by Fiddes and Goodman (*47*). They noted that the α-subunit was produced in large amounts in the placenta in the first trimester of pregnancy and that the α-subunit mRNA was a major mRNA species in this tissue. They first produced double-stranded cDNA using reverse transcriptase and showed that digestion of the cDNAs with several restriction enzymes produced characteristic DNA fragment patterns. These fragments were also detected after digestion with the same enzymes of a cDNA obtained from a library produced from placental mRNA. This result suggested that the cDNA encoded an abundant mRNA in the placenta. The subsequent DNA sequence analysis of this cDNA revealed that it indeed encoded the α-subunit (Fig. 4).

Finally, the molecular cloning of cDNAs encoding the α-subunit of bovine glycoprotein hormones was performed by

```
                                                  -24         -20                                     -10
                                                  Met Asp Cys Tyr Arg Arg Tyr Ala Ala Val Ile Leu Val Met Leu Ser Met Val
RAT      CACATCCTTCCAAGATCCAGAGTTTGCAGGAGAGCT     ATG GAT TGC TAC AGA AGA TAT GCG GCT GTC ATT CTG GTC ATG CTG TCC ATG GTC
MOUSE          C  A  AG      C      A                     A       A           A                                         T
COW      G  TA  A   TGCA AAATCCAGAG AC A      C           A           A       A                       C   T T     T C   T T
MAN      A    CATCCTGC A AAGCCCAGAGAA      C  C           A           A       A     A   T             CA  T       G G   T T

                         -1 ▽ +1                                 +10                                     +20
         Leu His Ile leu His Ser Leu Pro Asp Gly Asp Leu Ile Ile Gln Gly Cys Pro Glu Cys Lys Leu Lys Glu Asn Lys Tyr Phe
RAT      CTG CAT ATT CTT CAT TCT CTT CCT GAT GGA GAC CTT ATT ATT CAG GGT TGT CCA GAA TGT AAA CTA AAG GAA AAC AAA TAC TTC
MOUSE                                                T                                                        T
COW            A       C       C T               G T   CA   G       C       T       C   G       A
MAN              G     C       C GC          TG C G @@@@@@@@@@@@@@@ A     C         C  CG     C           CC  T

                                    +30                                         +40                                     +50
         Ser Lys Leu Gly Ala Pro Ile Tyr Gln Cys Met Gly Cys Cys Phe Ser Arg Ala Tyr Pro Thr Pro Ala Arg Ser Lys Lys Thr
RAT      TCC AAG CTG GGT GCC CCC ATC TAT CAG TGT ATG GGC TGT TGC TTC TCC AGG GCA TAC CCG ACT CCC GCA AGG TCC AAG AAG ACA
MOUSE            A   A           C                                       T   C       C
COW              CA  A     T   A           C       G   C                     C     A   G       T
MAN          C   C             A   A CT    C           C         T   A     T   C     A CT                           G

                                           +60                                           +70
         Met Leu Val Pro Lys Asn Ile Thr Ser Glu Ala Thr Cys Cys Val Ala Lys Ser Phe Thr Lys Ala Thr Val Met Gly Asn Ala
RAT      ATG TTG GTT CCA AAG AAT ATT ACC TCG GAG GCC ACG TGC TGT GTG GCC AAA TCA TTT ACT AAG GCC ACA GTG ATG GGA AAC GCC
MOUSE      C                                   A                   G                       A           T
COW            C   C       C   C         A   T   A                 G       C                           T T
MAN            C   A       C G C     A     T     T         A   T       A   AC  G   T       A       G GGT TT

          +80                                    +90            +96
         Arg Val Glu Asn His Thr Asp Cys His Cys Ser Thr Cys Tyr Tyr His Lys Ser  ***
RAT      AGA GTG GAG AAC CAC ACG GAC TGC CAC TGT AGC ACT TGT TAC TAC CAC AAG TCG  TAG  CTTCCATGTGTGCCAAGGGCTGCGCTGACGACT
MOUSE                  T   T   G                   C                                 TGA C C  AG       T      G
COW                          C   G       C             T   T       A   C   A  TAGTTTGCA   GGCCTT    ATGATGGCTGA
MAN        A                   CG        C   T         T   T       A   T   A  A GTTT ACCAAGTCCT T  TGATGACT CTG
```

Fig. 4. Nucleotide and deduced amino acid sequences of cDNA encoding the α-subunits of rat, mouse, and bovine and human glycoprotein hormones. The third line in each section represents the DNA sequence of rat α cDNA *(17)*. The fourth, fifth, and sixth lines indicate the sequence of the α-subunit cDNAs for mouse *(43)*, cow *(48,50)*, and humans *(47)*, respectively. Only bases that differ from the rat cDNA sequence are shown. Bases in these sequences that result in an amino acid substitution are underlined. The second line in each section represents the deduced amino acid sequence of rat α-subunit precursor. The first line refers to the number of the codons in the precursor of the rat α-subunit; −24 to −1 represents the leader peptide, whereas +1 to +96 represents the apoprotein of the α-subunit. The inverted open triangle between −1 and +1 indicates the site of leader-peptide cleavage; @@@@@@ represents the putative deletion of a sequence from the α-subunit mRNA in humans; *** represents termination of translation.

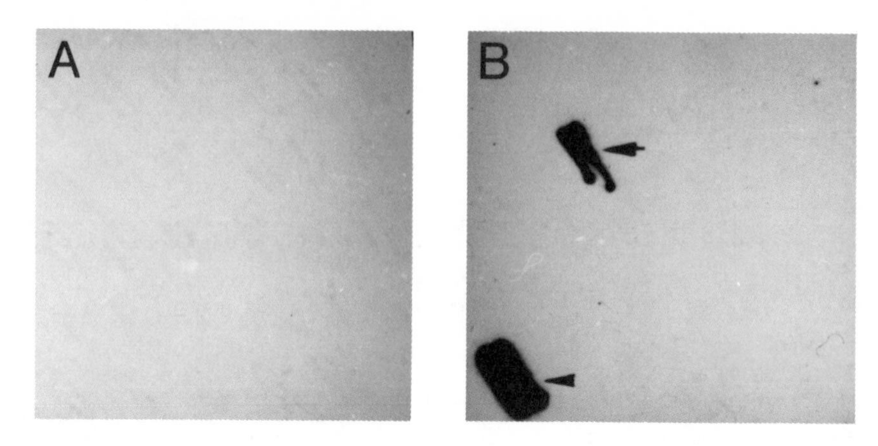


Fig. 5. Isolation of cDNA encoding the α-subunit of rat glycoprotein hormones. Several bacterial colonies containing cDNAs derived from the pituitary glands of ovariectomized rats were screened by colony hybridization using a labeled cDNA encoding the precursor of the mouse α-subunit. Panel A shows the bacterial colonies in streaks. Panel B shows the autoradiogram of colonies hybridizing to $^{32}P$-labeled pTSHα-1 cDNA (mouse). The arrowhead points to mouse α cDNA colony control. The arrow points to a colony containing rat α-subunit cDNAs.

two groups using cDNA libraries obtained from the anterior pituitary glands of cows. One group (*48*) synthesized oligodeoxyribonucleotides that corresponded to the known amino acid sequence of a region of the bovine α-subunit (Tyr-Gln-Cys-Met-Gly). Several such nucleotides were synthesized to encompass the natural ambiguity in these codons. The synthetic DNAs were then labeled, and those probes that detected α-mRNA by blot hybridization were used to detect the putative α-subunit cDNAs by colony hybridization (*49*). The other group (*50*) produced a bovine pituitary cDNA library that was enriched in α-mRNA. The mRNA was enriched for α-mRNA by sucrose gradient centrifugation of total RNA from bovine pituitary glands and assay of fractions for the ability to direct the synthesis of the precursor of α-subunit in a cell-free translation system. Then, cDNAs encoding prolactin and growth hormone were identified and discarded. The remaining cDNAs were screened with a labeled "α-mRNA-enriched" cDNA, and putative α cDNAs were randomly subjected to DNA sequence analysis. Only 0.3% of all pituitary cDNAs encoded the α-subunit (*50*). The consensus nucleotide sequence of the bovine α-subunit cDNA is also shown in Fig. 4.

The comparisons of the nucleotide and deduced amino acid sequences of the α-subunits of rat, mouse, cow, and humans are shown in Fig. 4. They indicate high levels of protein and DNA homologies. The levels of homology at the DNA and protein levels range from 74 to 95%. The mRNAs encode a precursor for the α-subunit in each species. The precursor is, in general, 116–120 amino acids in size and includes a 24-amino-acid leader peptide and a 96- or 92-amino-acid α-subunit apoprotein. Most mammals possess a 96-amino-acid α-subunit; however, a 92-amino-acid residue α-subunit apoprotein is found in humans. Computer analysis and comparison of the α-cDNA sequences suggest that the four-amino-acid difference is the result of a deletion of four codons (+6 to +9) of the α-subunits of the rat, mouse, and cow (Fig. 4) (*50,51*). It is interesting that the site of the putative deletion of this four-codon stretch of DNA from the human gene corresponds well to the position of the second intron that has been described in the genes encoding the α-subunit of the glycoprotein hormone in cow and humans *(vide infra)*. A mechanistic etiology for the loss of this particular DNA stretch may involve splicing mechanisms (*50,51*). Another interesting observation is that the 3′-untranslated regions of the bovine and human α-mRNAs share 70% homology, which is the level of similarity seen between the two coding regions. This finding is contrasted with the marked sequence divergence between these regions and the 5′ untranslated regions between the rodent and both the bovine and human α-mRNAs, respectively. Whether there is functional significance to this finding is still unclear (*48*).

### *2.2. Molecular Cloning of the Genes Encoding the α-Subunit of Glycoprotein Hormones*

The genes encoding the α-subunit in humans (*52,53*) and cow (*54*) were isolated and characterized from bacteriophage-λ genomic DNA libraries. From Southern genomic DNA-blot hybridization and from gene-copy analyses, it is clear that a single gene exists in humans and cow. The strong homology is shown in Fig. 6. Each gene consists of four exons, or coding regions, containing sequences present in mature α-mRNA, and three introns located in

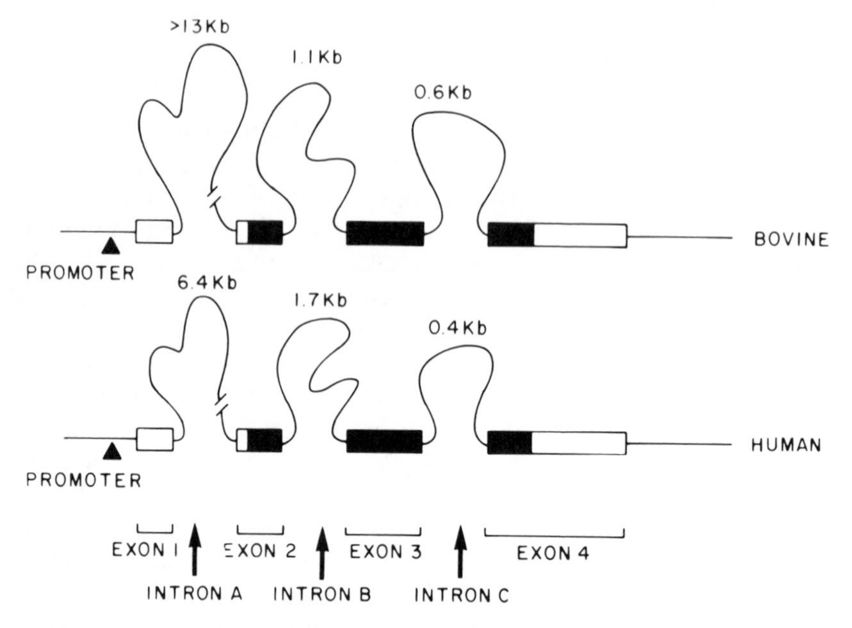


Fig. 6. Structures of the genes encoding the α-subunits of the glycoprotein hormones in cow and humans. The bovine (*54*) and human α (*52,53*) genes are 15 and 9.45 kb, respectively. Each gene contains four exons (boxes): the open boxes denote the coding regions of the α RNA transcripts. In between the exons are three introns with the sizes indicated. Intron A is located seven bases upstream of the initiation codon, ATG. Intron B is located between the first and second bases of codon +10 (cow) and codon +6 (human); note that there is a four-codon deletion in the human α cDNA when compared with bovine α cDNA at this point. Intron C is located between codons +71 and +72 (cow) and codons +67 and +68 (human). Solid triangles denote promoter elements.

identical positions (*55*). The first intron is located in the 5′ untranslated region and is very large (>6 kb), whereas the other two introns are located in the coding region of the α-subunit apoprotein and are, in general, smaller. Hence, the human and bovine α-genes are 9.4 and >16 kb, respectively.

Analysis of the bovine α-subunit gene indicates the presence of two 18-bp AT-rich sequences in the 5′-flanking and 5′-untranslated regions that are similar to a portion of the chick ovalbumin gene and was suggested as the possible target for the progesterone–receptor complex (*56*). Restriction-enzyme polymorphisms involving *Hin*dIII and *Eco*RI sites in the 3′-flanking region of the human α-subunit gene were reported (*52,53*). One such polymorphic pattern appears to be associated with choriocarcinoma cells, suggesting that particular genetic alterations may be associated with an increased risk for the development of this malignancy in certain trophoblastic tissues (*57*). The organization (number of genes and structure) of the α-subunit gene in various human tissues, including first and third trimester placental tissues, several trophoblastic and nontrophoblastic cell lines, and normal tissues is, in general, identical (*53*). Finally, the expression of the α-subunit gene in two tissues, pituitary gland and placenta, is qualitatively identical. That is, the α-mRNA appears to be identical by blot-hybridization analysis (*52*). Recently, the human α-subunit was expressed in a mouse fibroblast cell line. The α-subunit cDNA was controlled by the mouse metallothionein promoter in a bovine papilloma virus vector (*58*). The production of a nearly authentic glycosylated α-subunit was observed. The success of such experiments indicates the feasibility of these DNA transfection experiments to learn about structure–function relations, as well as the regulation of the expression of the subunit genes by thyroid and sex-steroid hormones.

### 2.3. Molecular Cloning of cDNA Encoding the β-Subunit of Human CG

Fiddes and Goodman (*59*) used an interesting approach to clone the cDNA encoding the β-subunit of human CG (hCGβ). They noted that a restriction enzyme, *Sau*96I, cleaves the DNA sequence CC*X*GG (where *X* = A, C, G, or T), which corresponds to the codons for amino acids Gly and Pro, which are CC*X* and GG*X*. Hence, any Gly-Pro sequence in a protein would be cleaved in its corresponding cDNA by *Sau*96I at those sites. Knowing the sequence of human CGβ and knowing that two Gly-Pro dipeptides are present in the carboxy-terminal end of the hCGβ, they predicted that *Sau*96I would cut putative CGβ cDNA into fragments of certain sizes. In this fashion, they were able to select cDNAs that contained the appropriate *Sau*96I sites and that were therefore candidates for hCGβ cDNAs.

One such cDNA encodes an open reading frame of 165 codons that represents the precursor for the hCGβ that includes a

20-amino-acid leader peptide and a 145-amino-acid hCGβ apoprotein (Fig. 7). The 5′-untranslated region is of normal and reasonable size; however, the 3′-untranslated region is unusually short, consisting of only 16 bp. The canonical polyadenylation recognition signal, AA*TAA*A, is present at the extreme end of the coding region and contains the translation-termination codon,

```
                          -20                                     -10
                          Met Glu Met Phe Gln Gly Leu Leu Leu Leu Leu Leu Leu Ser Met Gly Gly Thr Trp
AGACAAGGCAGGGGACGCACCAAGG ATG GAG ATG TTC CAG GGG CTG CTG CTG TTG CTG CTG CTG AGC ATG GGC GGG ACA TGG

-1 ▽ +1                         +10                                     +20
Ala Ser Lys Glu Pro Leu Arg Pro Arg Cys Arg Pro Ile Asn Ala Thr Leu Ala Val Glu Lys Glu Gly Cys Pro Val
GCA TCC AAG GAG CCG CTT CGG CCA CGG TGC CGC CCC ATC AAT GCC ACC CTG GCT GTG GAG AAG GAG GGC TGC CCC GTG

                +30                                     +40                                     +50
Cys Ile Thr Val Asn Thr Thr Ile Cys Ala Gly Tyr Cys Pro Thr Met Thr Arg Val Leu Gln Gln Val Leu Pro Ala
TGC ATC ACC GTC AAC ACC ACC ATC TGT GCC GGC TAC TGC CCC ACC ATG ACC CGC GTG CTG CAG GGG GTC CTG CCG GCC
                                                                                      ▲           ▲

                            +60                                     +70
Leu Pro Gln Val Val Cys Asn Tyr Arg Asp Val Arg Phe Glu Ser Ile Arg Leu Pro Gly Cys Pro Arg Gly Val Asn
CTG CCT CAG GTG GTG TGC AAC TAC CGC GAT GTG CGC TTC GAG TCC ATC CGG CTC CCT GGC TGC CCG CGC GGC GTG AAC

        +80                                     +90                                     +100
Pro Val Val Ser Tyr Ala Val Ala Leu Ser Cys Gln Cys Ala Leu Cys Arg Arg Ser Thr Thr Asp Cys Gly Gly Pro
CCC GTG GTC TCC TAC GCC GTG GCT CTC AGC TGT CAA TGT GCA CTC TGC CGC CGC AGC ACC ACT GAC TGC GGG GGT CCC
                                                                                                ▲

                    +110                                    +120
Lys Asp His Pro Leu Thr Cys Asp Asp Pro Arg Phe Gln Asp Ser Ser Ser Ser Lys Ala Pro Pro Pro Ser Leu Pro
AAG GAC CAC CCC TTG ACC TGT GAT GAC CCC CGC TTC CAG GAC TCC TCT TCC TCA AAG GCC CCT CCC CCC AGC CTT CCA
  ▲                                                                        ▲

+130                                    +140                +145
Ser Pro Ser Arg Leu Pro Gly Pro Ser Asp Thr Pro Ile Leu Pro Gln ***
AGC CCA TCC CGA CTC CCG GGG CCC TCG GAC ACC CCG ATC CTC CCA CAA TAA AGGCTTCTCAATCCGC AAAAA (POLY A)
                          ▲                                   ⇧
```

Fig. 7. Nucleotide and deduced amino acid sequences of cDNA encoding the β-subunit of hCG. The third line in each section represents the nucleotide sequence of hCGβ cDNA (*59*). The second line indicates the deduced amino acid sequence of the human CGβ precursor. The first line refers to the number of the codons in the precursor of hCGβ; −20 to −1 represents the leader peptide, and +1 to +145 represents the apoprotein of the CGβ subunit. The inverted triangle between codons −1 and +1 indicates the site of leader-peptide cleavage. The arrowheads point to *Sau*96I restriction-enzyme sites (_____) present in the cDNA. The open arrow indicates the polyadenylation signal sequence in which TAA, the stop codon for hCGβ mRNA, is located.

TAA. Inasmuch as the human LHβ subunit (hLHβ) was thought to contain 115 amino acids, it was suggested that residue +116, Gln of the hCGβ precursor, resulted from a point mutation in its codon from TAG, termination (stop) → CAG, Gln. This mutation would allow read-through of this codon into the 3′-untranslated region of the putative LHβ mRNA until the stop codon, TAA, was reached in the polyadenylation signal (AA*TAA*A) site. As we shall see, this picture is only partly correct.

### 2.4. Molecular Cloning of the Genes Encoding the Human CGβ–LHβ Subunits

Using the hCGβ cDNA, Fiddes and coworkers (*60–63*) screened a bacteriophage-λ human genomic DNA library and isolated several different DNA fragments that contained hCGβ-like sequences. The analysis of these DNAs revealed a surprise. In humans, there are at least seven hCGβ genes, along with a unique hLHβ gene in several clusters of tandem and inverted repeats (*63,64*) (Fig. 8). It is not clear, however, whether there is close

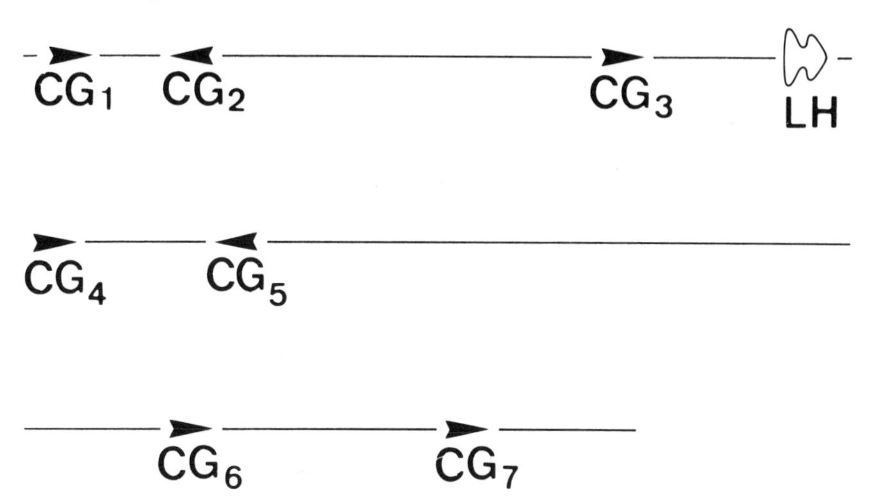


Fig. 8. Organization of the human CGβ and LHβ genes. There are at least seven CGβ genes and one LHβ gene in the haploid genome of humans. They have been isolated as part of three segments of DNA that are located on human chromosome 19 (*63,77*). The genes are organized as sets of tandem and inverted repeats. $CG\beta_1$–$CG\beta_7$ are the hCGβ genes represented by solid arrowheads that point in the putative direction of transcription. The open arrow denotes the hLHβ gene and also points in the supposed direction of transcription.

linkage of these genes. Later, data will be presented that suggest that the β-subunits of CG and LH in humans are located on the same chromosome.

All of the hCGβ–LHβ genes have similar structures. Each gene is 1.45 kb and contains three exons interrupted by two introns located between codons −16 and −15 and codons +41 and +42. Hence, it is probable that the multiple hCGβ genes evolved from a single hLHβ gene and arose from subsequent gene duplication and rearrangement. An important question is whether all hCGβ genes are expressed or whether some of them are pseudogenes (genes that may contain mutations that lead to aberrant RNA splice patterns, or premature translation termination). Detailed DNA sequence analyses of several hCGβ and hLHβ genes, as well as transient expression experiments, have helped to elucidate these and other points.

First, the human LHβ-subunit contains 121 amino acids, which is six more than previously reported. These data suggest that either a terminal COOH peptide was inadvertently missed during the protein isolation and sequence studies, or that the LHβ-subunit may be processed at the COOH terminus in vivo. Second, comparison of the DNA sequences of the hLHβ genes reveals the precise nature of the evolution of the hCGβ gene with a COOH-terminal extension (Fig. 9). Apparently, a single bp *deletion* occurred at the third base of codon, +114, which resulted in a translational frame shift that bypasses the *stop* codon, +122 of the hLHβ gene. The open reading frame continues for another 16 codons, at which point two bases are *inserted*. The open reading frame continues for another eight codons and stops with the TAA of the polyadenylation signal site, as noted previously (*59*). Thus, the hCGβ is 24 amino acids longer than the hLHβ with its extension at the COOH-terminus. Third, none of the hCGβ/LHβ genes are identical (*63*). For instance, the DNA sequence of the exons of $hCG\beta_5$ corresponds exactly to the hCGβ cDNA, whereas $hCG\beta_6$ contains a single amino acid substitution, Asp +117 to Ala +117. Asp +117, however, was the residue reported in the amino acid sequence of the hCGβ-subunit (*65*). Fourth, characteristic restriction enzyme sites, present in hCGβ cDNA, and the hCGβ genes, are found only in $hCG\beta_1$, $hCG\beta_3$, and $hCG\beta_5$. These data suggest that only these genes, if expressed, would yield the authentic CGβ mRNAs. Finally, transient expression of the hCGβ genes in

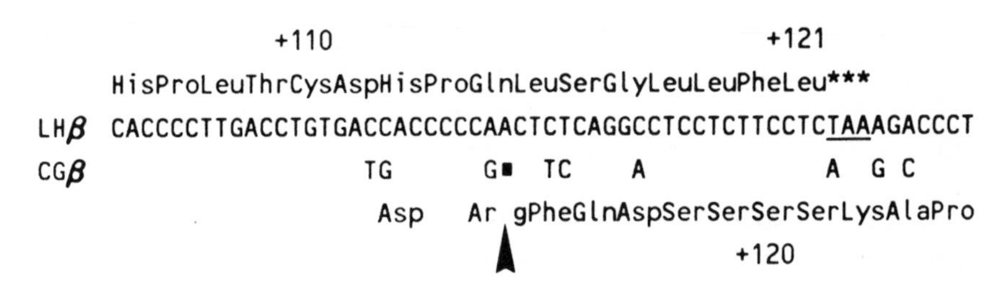


```
LHβ  CCCCGCAGCCTTCCAAGTCCATCCCGACTCCTGGAGCCCT■■GACACCCCGATCCTC
CGβ      C                            C  G    CG
     ProProSerLeuProSerProSerArgLeuProGlyProSerAspThrProIleLeu
                +130                              +140
```

```
LHβ  CCACAATAAAGGCTTCTCAATCCGCACTCTGGCAGTATC
CGβ       ___                        AG  G
     ProGln***
       +145
```

Fig. 9. The evolution of the hCGβ gene from the hLHβ gene: comparison of human LHβ and CGβ subunits. The third line indicates the partial cDNA sequence of the 3′ end of the hLHβ mRNA. The fourth line indicates the corresponding hCGβ sequence in which only bases that differ are shown. The second line indicates the deduced amino acid sequence of hLHβ. The first line indicates the number of the codons in hLHβ. The fifth line indicates the number of the amino acid sequence of hCGβ. The sixth line indicates the number of the amino acids with reference to the CGβ subunit. *** denotes the termination of translation, and _____ indicates a termination codon, TAA, in two locations. The small boxes indicate base deletions in one cDNA relative to the other cDNA. The arrowhead points to the single base deletion that allows a shift in the reading frame to bypass the first stop codon. Translation continues and would ordinarily stop at codon +138 if it were not for a two-base insertion in the hLHβ mRNA sequence that allows further read-through to the last stop codon in the polyadenylation recognition site. Thus the CGβ carboxy terminus probably evolved from a series of base deletions and insertions in the 3′ end of the mRNA encoding the LHβ subunit.

COS cells was evaluated to determine the potential function of each gene. Each gene was placed under the control of the SV40 early-region promoter and introduced into COS cells, which originated from African green monkey kidney cells that permit SV40 replication. Only a part of hCGβ$_1$ was cloned, and therefore was not tested. The studies indicated that only hCGβ$_3$ and hCGβ$_5$ were expressed. Thus, hCGβ$_3$ and hCGβ$_5$ (and possibly hCGβ$_1$) are likely to be expressed in placental tissue (normal or malignant) (*63*). Whether the other genes are expressed in other tissues or at different times in development is not known.

Another interesting feature of the DNA sequence comparisons of the hCGβ–LHβ genes is the nature of the nucleotide changes in hCGβ in its evolution from the LHβ gene. It appears that rate of nucleotide changes between the comparable introns of hCGβ–hLHβ is similar to that of the coding regions (both replacement and silent changes). These data suggest that the hCGβ genes have very recently evolved from the hLHβ gene and that the relatively high rate of nucleotide changes that result in amino-acid replacements (compared to intron nucleotide changes) indicates selection of the hCGβ gene on the basis of function, with alterations in the hCGβ–hLHβ apoprotein, rather than in the CGβ-COOH-terminal extension, being important.

### *2.5. Molecular Cloning of cDNA Encoding the Rat LHβ-Subunit*

Using the hCGβ cDNA, we isolated and characterized the cDNA encoding the β-subunit of rat LH (rat LHβ) from the same cDNA library derived from the pituitary glands of ovariectomized rats used for the isolation of the rat α-subunit cDNA (*17*). Inasmuch as the homology of CGβ to LHβ in humans is at the level of 82% in comparable regions, we reasoned that perhaps the hCGβ cDNA might cross-hybridize at low stringencies with the putative rat LHβ cDNA. We used the hCGβ cDNA insert, labeled to high specific activity by nick translation (*66,67*), to select bacterial colonies that seemed likely to contain the cDNA encoding the rat LHβ. The DNA and deduced protein sequences of rat LHβ cDNA are shown in Fig. 10. The sequences are compared with those of the presumed hLHβ cDNA reconstructed from the DNA sequence of the hLHβ gene. Note that each LHβ precursor contains 141 amino acids, with 20 amino acids comprising the leader peptide and 121 amino acids constituting the apoprotein. The rat LHβ

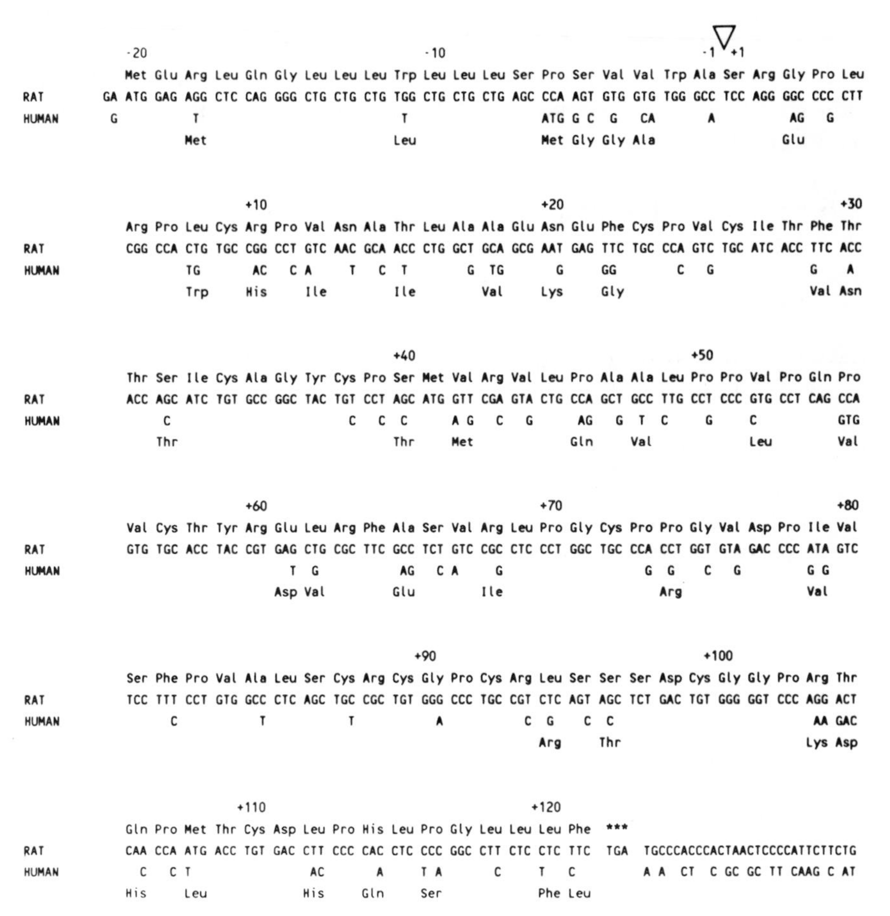

Fig. 10. Nucleotide and deduced amino acid sequences of the β-subunits of rat and human LH. The third line contains the DNA sequence of the cDNA encoding the precursor of rat LHβ (*67*). The fourth line represents the nucleotide sequence of the hLHβ cDNA (*61,63*), in which only bases that differ are shown. The fifth line indicates the amino acids present in the hLHβ as substitutions from that of the rat. The first line represents the codon number: −20 to −1, leader peptide and +1 to +121, the apoprotein of LHβ subunit. The inverted open triangle points to the site of leader-peptide cleavage. The *** indicates the site of translation termination.

mRNA is approximately 700 bases in extent, and the level of nucleotide sequence homology between coding regions of rat LHβ and hLHβ mRNAs is 80%. In addition, dot-matrix comparisons of the rat LH and hCGβ cDNA sequences show that the only region of extended homology (>18 bp) is at the Cys-Ala-Gly-Tyr-Cys sequence located at codons +34 to +38. It is probable that this small region allowed this hybridization–selection technique to succeed.

### 2.6. Molecular Cloning of the Gene Encoding the Rat LHβ-Subunit

The chromosomal gene encoding the β-subunit of rat LH was recently isolated and characterized (*68*). The rat LHβ and the human CGβ and human LHβ genes are similar in structure. Like the CGβ and LHβ genes in humans, the rat LHβ gene contains only three exons separated by two introns located in analogous positions. The entire gene is small in rat, encompassing 0.98 kb, whereas the hCGβ and hLhβ genes are 1.5 kb. The rat LHβ and human CGβ genes are compared in schematic form in Fig. 11. According to Southern analyses of rat genomic DNA, there is a

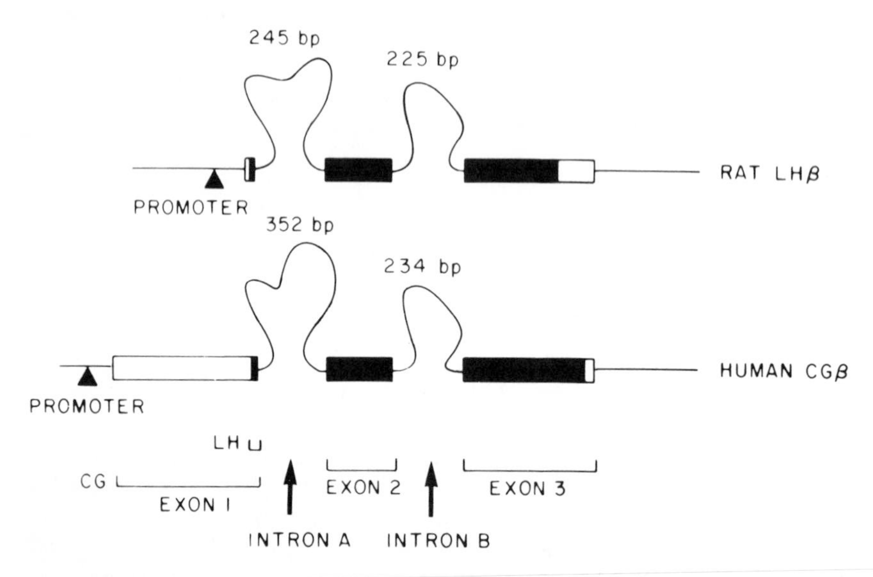


Fig. 11. Structures of the genes encoding the rat LHβ and hCGβ genes. The rat LHβ (*68*) and human CGβ (*61,73*) genes are 0.95 and 1.45 kb, respectively. Each gene contains three exons (boxes): the open boxes denote untranslated regions of the β transcripts. Intron A is located between codons −16 and −15, and intron B is located between codons +41 and +42. The solid triangles denote the putative promoter elements. Note that the 5′-untranslated region of hCGβ mRNA is larger than that of the rat LHβ mRNA.

single gene for rat LHβ and no evidence for a rat CGβ gene in hybridization studies at low stringency (*68–70*). This result is consistent with the recent findings of the absence of subunit mRNAs in rat placentas at all stages of development (*69–71*). Of interest is the 5′-untranslated region of hCGβ subunit mRNA that is over 300 bases (*61*), whereas, according to our studies using S1-nuclease protection and read-out transcription, the 5′-untranslated region of the rat LHβ-subunit gene is only seven bases (*68*). If one compares the corresponding 5′-flanking regions (bounded by the initiator ATG) of the rat LHβ and human CGβ genes, one finds striking homologies up to the hCGβ promoter, which is farther upstream than is the rat LHβ gene promoter because of the different sizes of the 5′-untranslated regions. These data suggest that the hCGβ gene was altered in a subtle manner, with the result that the promoter region that is used for the LHβ gene has shifted farther upstream to another promoter. This is interesting in light of the difference in hormonal regulation of these two genes. It should be noted that estrogen and progesterone may decrease LHβ-subunit gene expression, whereas they have relatively little effect on the expression of the hCGβ gene. These changes in the 5′-flanking regions and use of different promoters in the two genes may play major roles in the differential hormonal regulation and tissue-specific expression of the CGβ and LHβ genes in humans.

### 2.7. Molecular Cloning of cDNA Encoding the TSHβ-Subunit

The molecular cloning of cDNAs encoding the β-subunit of TSH has been more difficult than the cloning of the cDNAs discussed previously. This may be accounted for by the low abundance of TSHβ mRNA relative to the α- and LHβ-subunit mRNAs. Gurr *et al.* (*72*) were, however, successful in obtaining a cDNA encoding the β-subunit of mouse TSH (TSHβ). They utilized a cDNA library derived from poly(A)+ RNA obtained from a mouse thyrotropic tumor. The cDNA encoding the TSHβ was selected by using hybridization–selection–translation techniques utilizing information that the apparent molecular mass of the translation product for mouse TSHβ is 17 kdaltons. Using a bovine pituitary gland cDNA library, Maurer et al. (*73*) were successful in cloning a bovine TSHβ cDNA by eliminating colonies containing cDNAs encoding the α-subunit, prolactin, and growth hormone, and randomly sequencing cDNAs that remained. One of these coded for the bovine TSHβ. In addition, we obtained the rat TSHβ cDNA from a rat pituitary gland cDNA library by colony hybridization (*74*). The probe was a mouse TSHβ cDNA obtained from a mouse thyrotropic tumor cDNA library screened with a synthetic DNA corresponding to the sequence of a region of the mouse TSHβ-subunit cDNA isolated by Gurr et al. (*72*). Similar rat TSHβ cDNAs were obtained by Croyle and Maurer (*75*) using colony-hybridization techniques and the bovine TSHβ cDNA.

The nucleotide and deduced amino acid sequences of the cDNAS encoding the TSHβ-subunit are shown in Fig. 12. The cDNAs encode the precursor of the β-subunit of TSH in various species. Each contains 138 amino acids with a 20-amino-acid leader peptide and a 118-amino-acid apoprotein. The mRNAs are approximately 700 bases. Comparisons of these cDNA sequences show high levels of homology, varying from 85 to 95% between the rat, mouse, and bovine TSHβ cDNAs and protein.

### *2.8. Molecular Cloning of the Genes Encoding the TSHβ- and FSHβ-Subunits*

The gene encoding the TSHβ-subunit has been only partly characterized. There is evidence for a single rat TSHβ gene (*74*), and perhaps two mouse TSHβ genes (*76*). The molecular cloning of the cDNA and gene encoding the FSHβ-subunit has not yet been reported.

### *2.9. Chromosomal Localization of the Subunit Genes of Glycoprotein Hormones*

It is interesting to determine the genomic organization of the genes that encode the α- and β-subunits of the glycoprotein hormones and whether they are located on separate or the same chromosomes. Naylor et al. (*77*) used mouse–human somatic-cell hybrids containing varying numbers of human chromosomes characterized by karyotype and enzyme markers and DNA blot-hybridization techniques to localize the α gene to human chromosome 6q14→21 and the LHβ–CGβ gene cluster to human chromosome 19. Similar data for the hLHβ–hCGβ gene cluster and its localization on chromosome 19 were recently obtained by Julier et al. (*78*) using different cell lines. Hardin et al., using human choriocarcinoma–mouse cell hybrids, performed similar studies.

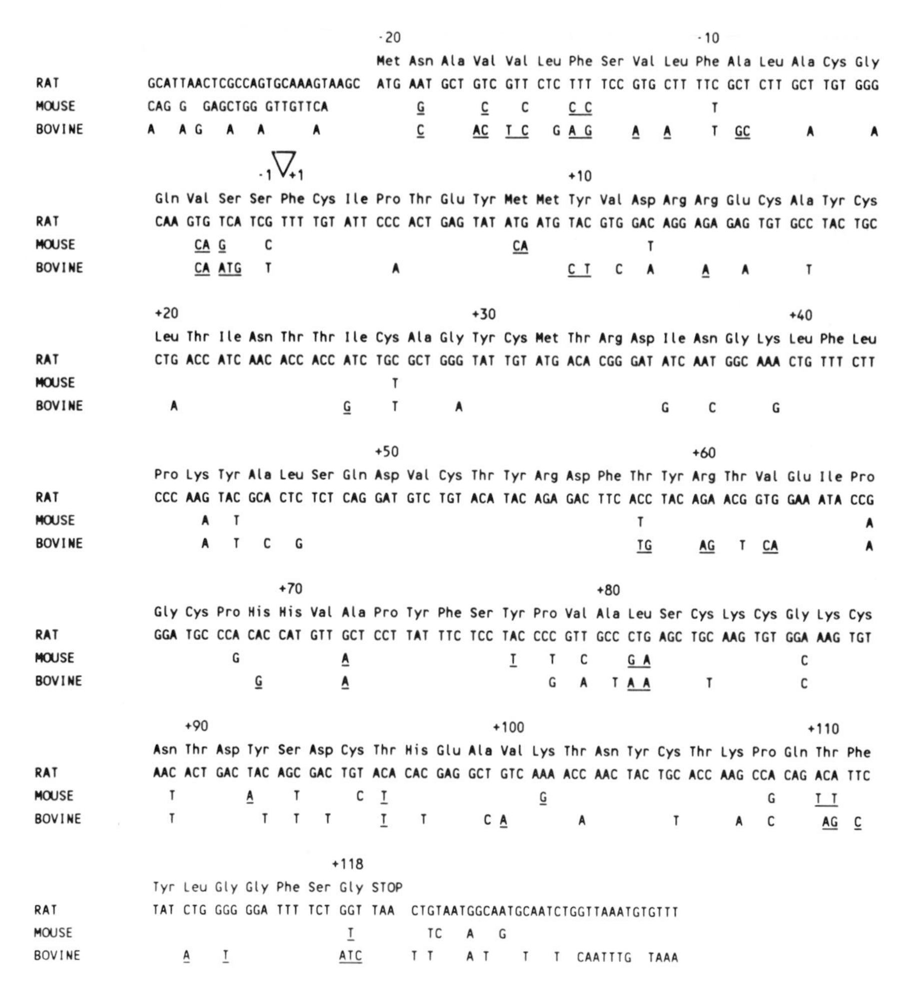

Fig. 12. Nucleotide and deduced amino acid sequences of cDNAs of TSHβ in rat, mouse, and cow. The third line in each section represents the DNA sequence of the cDNA encoding the β-subunit of rat TSH (*74,75*). The sequences of the corresponding cDNAs for mouse (*72*) and cow (*73*) are shown in the fourth and fifth lines. Only bases that differ from those of the rat sequence are shown. Nucleotides that result in amino acid substitutions are underlined. The second line indicates the deduced amino acid sequence of the precursor of the rat TSHβ subunit, and the first line indicates the codon number; −20 to −1 represents the leader peptide; and +1 to +118 denotes the TSHβ apoprotein. The inverted open triangle depicts the site of leader peptide-sequence cleavage, and STOP indicates the site of translation termination.

They concluded, however, that the α-subunit gene is located on human chromosome 18 (*79*) and that the CGβ-subunit cluster is located on human chromosome 10 (*80*). These data appear to confirm their original assignment of the α- and CGβ-subunit genes to human chromosomes 10 and 18 by the correlation of chromosomes with expression of CG in these somatic-cell hybrids (*81*). However, one should interpret these data with some caution. It is well known that the chromosomes in cancer lines may be rearranged, and it is possible that segments of DNA encoding the α–β-subunits have been placed in other chromosomes. If the observations of Naylor et al. and Julier et al. for the chromosomal localization of the α- and CGβ-subunit genes in normal cells are correct, then the observations of Hardin et al. are indeed interesting. Also, the question of why and how these CG subunits are expressed in the tumor remains a mystery. The possible translocation of such genes to other chromosomes may be a clue in the understanding of the process of tumorigenesis in these cells. Other studies using hamster–mouse somatic-cell hybrids have located the mouse α, LHβ, and TSHβ genes to chromosomes 4 (*77*), 7 (*77,82*), and 3 (*82*), respectively. Hence, the subunit genes are located on separate chromosomes in man and mouse (Fig. 13).

## 3. EXPRESSION OF THE SUBUNIT GENES

Over the past several years, much has been learned about the hormonal regulation of synthesis of glycoprotein hormones at the

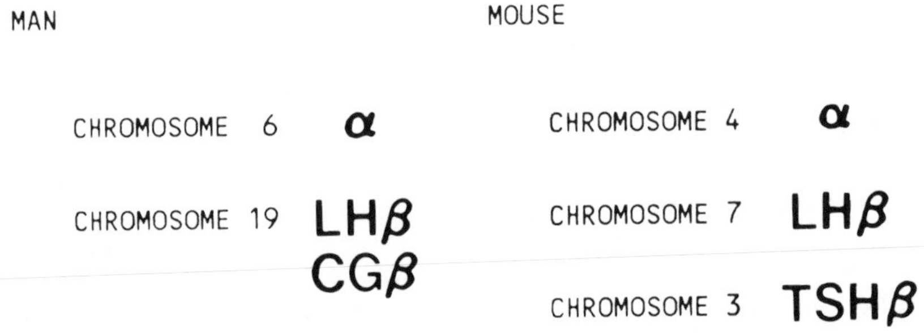

Fig. 13. Chromosomal localization of the α- and β-subunit genes in mouse and humans. The subunits are encoded by separate genes that are, in general, located on different chromosomes. Note, however, that the hLHβ–hCGβ gene cluster is located on a single chromosome.

pretranslational level. The availability of cDNAs to specific subunit mRNAs has allowed assay of the levels of these RNAs under the influence of various hormones. The thyroid hormones affect the α- and TSHβ-subunit genes by decreasing their expression. In several experiments, mouse thyrotropic tumors in hypothyroid mice were examined. These animals were treated with physiologic doses of T3 or T4, and a time-course study of the effects of thyroid hormones on α- and TSHβ-protein as measured by RIA and subunit mRNAs by blot-hybridization techniques, was performed. These data indicate that the thyroid hormones decrease the levels of TSH subunit mRNAs in a rapid and parallel fashion. However, the β-subunit mRNA seems to be affected much faster and to a greater extent than is the α-subunit. In addition, the α-subunit mRNA is never completely suppressed, even after long periods of thyroid hormone treatment *(75,83–85)*. Indeed, this is of particular interest, considering the fact that this tumor is essentially a pure population of thyrotropes and contains no gonadotropes (*86*). Theoretically, the study of the anterior pituitary gland with regard to the expression of the α-subunit could be confounding because the α-subunit may be expressed in both thyrotrope and gonadotrope cell populations. Hence, any effect of thyroid hormones on the α-subunit may reflect changes in the gonadotrope as well as the thyrotrope.

Shupnik et al. (*87*) proceeded further to analyze the rate of mRNA synthesis under the influence of thyroid hormones in the same system. They showed that thyroid hormones rapidly decrease the rate of transcription of both α- and TSHβ-subunit genes. The effects occur as rapidly as 15 min. Again, there is a greater effect on the β-subunit gene than on the α-subunit gene. By 4 h, the levels of transcription of the α and β genes are less than 14 and 7%, respectively, of the starting transcriptional rates. These data indicate that thyroid hormones rapidly and coordinately regulate TSH production at the pretranslational level, with the major effects at the transcriptional level (Fig. 14). However, there are some data to suggest that, in addition, thyroid hormones may decrease the half-lives of the TSH subunit mRNAs, allowing for even more rapid declines in the steady-state levels of these mRNAs (*84*). This finding is consistent with recent findings in other systems concerning the effects of hormones on RNA stability (*88–90*).

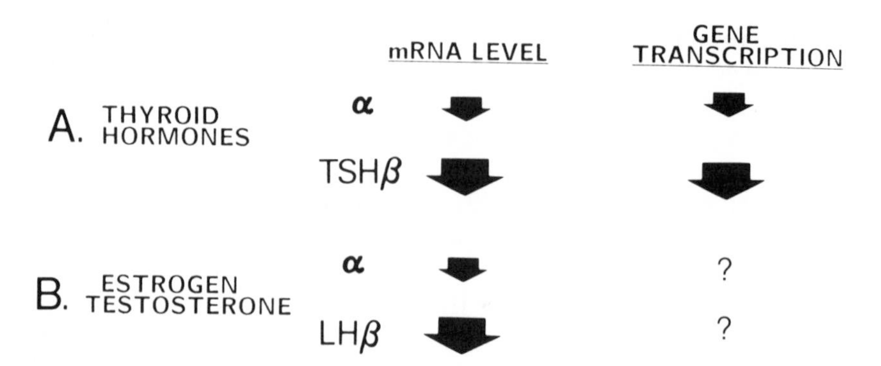


Fig. 14. Regulation of the α, TSHβ, and LHβ genes by thyroid and gonadal sex-steroid hormones at the pretranslational level. Thyroid hormones rapidly decrease the levels of α and TSHβ mRNAs, with an effect greater for TSHβ mRNA (*83–85*). Recent data indicate that this regulation is, at least in part, mediated at the transcriptional level (*87*). Estrogen and testosterone in vivo rapidly decrease the levels of α and LHβ mRNAs with a more pronounced effect on the LHβ mRNA (*40, 91–97*). It is not known whether these effects are exerted at the mRNA synthesis level.

In a similar fashion, we and others performed studies analyzing the effects of testosterone and estrogen on LH biosynthesis at the pretranslational level in male and female rats, respectively. These data are also reviewed in Fig. 14. These data indicate that, as in the case of TSH, estrogens and testosterone rapidly decrease the α and LHβ mRNAs in a rapid and parallel fashion (40, 91–97). Again, the effect on the LHβ- is greater than it is on the α-subunit mRNA. Transcriptional studies are in progress. Thus, the hormones from target organs appear to regulate the expression of the glycoprotein-hormone subunit genes at the level of mRNA synthesis or transcription, with possible additional effects at the level of the stability of the subunit mRNA. These effects appear to be rapid and raise the possibility that these hormones, in conjunction with their intracellular receptors, act directly on the subunit genes. Nevertheless, these data do *not* exclude the possibility that important posttranslational events may also modulate the amount and bioactivity of TSH and LH produced in these cells.

## 4. SUMMARY

The advent of recombinant DNA techniques has allowed the molecular cloning of the genes encoding the subunits of the glycoprotein hormones. The analysis of the structure and genomic organization of these genes has provided insight into their function and evolution. The availability of such DNAs will allow in vitro mutagenesis, reintroduction into foreign mammalian cells, and expression of the subunit genes in these cells, which, in turn, will allow study of structure–function relations. In addition, the isolation and characterization of 5′-flanking, or promoter, regions of these genes will, we hope, allow us to understand the molecular mechanisms of hormonal regulation by thyroid and gonadal sex-steroid hormones and their receptors. We look forward to an exciting period of interest and activity in which to understand further these important processes.

## ACKNOWLEDGMENTS

I thank Drs. John E. Godine, Larry Jameson, Soheyla Gharib, and Frances Carr, and Albert Chang, Steven Bowers, Lee Tan, and Deborah Klein for their valuable assistance in many aspects of this work. I also thank Dr. Joel F. Habener for his sound advice and support throughout the period of this work. I also wish to thank Nancy Patterson for her excellent assistance in the preparation of this manuscript.

## REFERENCES

1. Pierce J. G. and Parsons T. F. (1981) Glycoprotein hormones: structure and function. *Ann. Rev. Biochem.* **50,** 465–495.
2. Hussa R. O. (1980) Biosynthesis of human chorionic gonadotropin. *Endocr. Rev.* **1,** 268–294.
3. Borkowski A. and Muquardt C. (1979) Human chorionic gonadotropin in plasma of normal, non-pregnant subjects. *N. Eng. J. Med.* **301,** 298–302.
4. Braunstein G. D., Rasor J., and Wade M. E. (1975) Presence in normal human testes of a chorionic gonadotropin-like substance dis-

tinct from human luteinizing hormone. *N. Eng. J. Med.* **292,** 1339–1343.
5. Braunstein G. D., Vaitukaitis J. L., Carbone P. P., and G. T. Ross (1973) Ectopic production of human chorionic gonadotropin by neoplasms. *Ann. Intern. Med.* **78,** 39–45.
6. Emanuele N. V., Anderson J., Anderson E., Connick E., Baker G., Kirsteins L., and Lawrence A. M. (1983) Extra-hypothalamic brain luteinizing hormone: Characterization by radioimmunoassay, chromatography, radioligand assay and bioassay. *Neuroendocrinology* **36,** 254–260.
7. Hojvat S., Baker G., Kirsteins L., and Lawrence A. M. (1982) TSH in the rat and monkey brain: Distribution, characterization and effect of hypophysectomy. *Neuroendocrinology* **34,** 327–332.
8. Hershman J. M. and Pekary A. E. (1985) Regulation of Thyrotropin Secretion, in *The Pituitary Gland* (Imura H., ed.), Raven, New York.
9. Vaitukaitis J. L., Ross G. T., Braunstein G. D., and Rayford P. L. (1976) Gonadotropins and their subunits: Basic and clinical studies. *Recent Prog. Horm. Res.* **32,** 289–331.
10. Vaitukaitis J. L. (1974) Changing placental concentrations of human chorionic gonadotropin and its subunits during gestation. *J. Clin. Endocrinol. Metab.* **38,** 755–760.
11. Surks M. I. and Lifschitz B. M. (1977) Biphasic thyrotropin suppression in euthyroid and hypothyroid rats. *Endocrinology* **101,** 769–781.
12. Chappel S. C., Ulloa-Aguirre A., and Coutifaris C. (1983) Biosynthesis and secretion of follicle-stimulating hormone. *Endocr. Rev.* **4,** 179–211.
13. Clayton R. N. and Catt K. J. (1981) Gonadotropin-releasing hormone receptors: Characterization, physiological regulation, and relationship to reproductive function. *Endocr. Rev.* **2,** 186–209.
14. Pierce J. G., Faith M. R., Giudice L. C., and Reeve J. R. (1976) Structure and structure–function relationships in glycoprotein hormones. *Ciba Found. Symp.* **41,** 225–250.
15. Acher R. (1980) Molecular evolution of biologically active polypeptides. *Proc. R. Soc. Lond.* **200,** 21–43.
16. Stewart M. and Stewart F. (1977) Constant and variable regions in glycoprotein beta subunit sequences: Implications for receptor binding specificity. *J. Mol. Biol.* **116,** 175–179.
17. Godine J. E., Chin W. W., and Habener J. F. (1982) α-Subunit of rat pituitary glycoprotein hormones: Primary structure of the precursor determined from the nucleotide sequence of cloned cDNAs. *J. Biol. Chem.* **257,** 8368–8371.
18. Daniels-McQueen S., McWilliams D., Birken S., Canfield R., Landefeld T., and Boime I. (1978) Identification of the mRNAs encoding the α and β subunits of human choriogonadotropin. *J. Biol. Chem.* **253,** 7109–7114.

19. Chin W. W., Habener J. F., Kieffer J. D., and Maloof F. (1978) Cell-free translation of the messenger RNA coding for the α-subunit. *J. Biol. Chem.* **253,** 7985–7988.
20. Vamvakopoulos N. C. and Kourides I. A. (1979) Identification of separate mRNAs coding for the α and β subunits of thyrotropin. *Proc. Natl. Acad. Sci. USA* **76,** 3809–3813.
21. Godine J. E., Chin W. W., and Habener J. F. (1980) Luteinizing and follicle-stimulating hormones: Cell-free translations of mRNAs coding for subunit precursors. *J. Biol. Chem.* **255,** 8780–8783.
22. Godine J. E., Chin W. W., and Habener J. F. (1981) Cell-free synthesis and processing of the precursors to the subunits of luteinizing hormone. *J. Biol. Chem.* **256,** 2475–2479.
23. Alexander D. C. and Miller W. L. (1981) mRNA for ovine follicle-stimulating hormone β-chain: An in vitro ovine translation assay. *J. Biol. Chem.* **256,** 12628–12631.
24. Ruddon R. W., Bryan A. H., Hanson C. A., Perini F., Ceccorulli L. M., and Peters B. P. (1981) Characterization of the intracellular and secreted forms of the glycoprotein hormone chorionic gonadotropin produced by malignant cells. *J. Biol. Chem.* **256,** 5189–5196.
25. Chin W. W., Maloof F., and Habener J. F. (1981) Thyroid-stimulating hormone biosynthesis: Cellular processing, assembly, and release of subunits. *J. Biol. Chem.* **256,** 3059–3066.
26. Weintraub B. D., Stannard B. S., Linnekin D., and Marshall M. (1980) Relationship of glycosylation to de novo thyroid-stimulating hormone biosynthesis and secretion by mouse pituitary tumor cells. *J. Biol. Chem.* **255,** 5715–5723.
27. Giudice L. C. and Weintraub B. D. (1979) Evidence for conformational differences between precursor and processed forms of thyroid-stimulating hormone β-subunit. *J. Biol. Chem.* **254,** 12679–12683.
28. Magner J. A. and Weintraub B. D. (1982) Thyroid-stimulating hormone (TSH) subunit processing and combination in microsomal subfractions of mouse pituitary tumor. *J. Biol. Chem.* **257,** 6709–6715.
29. Hoshina H. and Boime I. (1982) Combination of rat lutropin subunits occurs early in the secretory pathway. *Proc. Natl. Acad. Sci. USA* **79,** 7649–7653.
30. Van Hall E. V., Vaitukaitis J. L., Ross G. T., Hickman J. W., and Ashwell G. (1971) Immunological and biological activity of hCG following progressive desialylation. *Endocrinology* **88,** 456–464.
31. Farquhar M. G. (1983) Multiple pathways of exocytosis, endocytosis, and membrane recycling: Validation of a Golgi route. *Fed. Proc.* **42,** 2407–2413.
32. Blackman M. R., Gershengorn M. C., and Weintraub B. D. (1978) Excess production of free alpha subunits by mouse pituitary thyrotropic tumor cells in vitro. *Endocrinology* **102,** 499–508.

33. Chin W. W., Maloof F., Martorana M. A., Pierce J. G., and Ridgway, E. C. (1981) Production and release of thyrotropin and its subunits by monolayer cultures containing bovine anterior pituitary cells. *Endocrinology* **108,** 387–394.
34. Chin W. W., Habener J. F., Martorana M., Keutmann H. T., Kieffer J. D., and Maloof F. (1980) Thyroid-stimulating hormone: Isolation and partial characterization of hormone and subunits from a mouse thyrotrope tumor. *Endocrinology* **107,** 1384–1392.
35. Weintraub B. D., Krauth G., Rosen S. W., and Rabson A. S. (1975) Difference between purified ectopic and normal alpha subunits of human glycoprotein hormones. *J. Clin. Invest.* **56,** 1043–1052.
36. Benveniste R., Linder J., and Rabin D. (1979) Human chorionic gonadotropin alpha-subunit from cultured choriocarcinoma (JEG) cells; Comparisons of the subunit secreted free with that prepared from secreted human chorionic gonadotropin. *Endocrinology* **105,** 581–587.
37. Parsons T. H., Bloomfield G. A., and Pierce J. G. (1983) Purification of an alternate form of the α subunit of the glycoprotein hormones from bovine pituitaries and identification of its O-linked oligosaccharide. *J. Biol. Chem.* **258,** 240–244.
38. Parsons T. H. and Pierce J. G. (1983) Free α-like material from bovine pituitaries: Restoration of the ability to reassociate with native LH-β. *Fed. Proc.* **42,** 1799.
39. Cole L. A., Hartle R. J., Laferla J. J., and Ruddon R. W. (1983) Detection of the free beta subunit of human chorionic gonadotropin (hCG) in cultures of normal and malignant trophoblast cells, pregnant sera, and sera of patients with choriocarcinoma. *Endocrinology* **113,** 1176–1178.
40. Nilson J. H., Nejedlik M. T., Virgin J. B., Crowder M. E., and Nett T. M. (1983) Expression of α subunit and luteinizing hormone β genes in the ovine anterior pituitary. *J. Biol. Chem.* **258,** 12087–12090.
41. Gurr J. A. and Kourides I. A. (1984) Ratios of α to TSHβ mRNA in normal and hypothyroid pituitaries and TSH-secreting tumors. *Endocrinology* **115,** 830–832.
42. Furth J. P., Moy J., Hershman J., and Ueda G. (1973) Thyrotropic tumor syndrome. *Arch. Pathol.* **96,** 217–226.
43. Chin W. W., Kronenberg H. M., Dee P. C., Maloof F., and Habener J. F. (1981) Nucleotide sequence of the mRNA encoding the pre-α-subunit of mouse thyrotropin. *Proc. Natl. Acad. Sci. USA* **78,** 5329–5333.
44. Ricciardi R. P., Miller J. S., and Roberts B. E. (1979) Purification and mapping of specific mRNAs by hybridization-selection and cell-free translation. *Proc. Natl. Acad. Sci. USA* **76,** 4927–4931.

45. Counis R., Ribot G., Corbani M., Poissonnier M., and Jutisz M. (1981) Cell-free translation of the rat pituitary messenger RNA coding for the precursors of α and β subunits of lutropin. *FEBS Lett.* **123,** 151–155.
46. Hanahan D. and Meselson M. (1980) Plasmid screening at high colony density. *Gene* **10,** 63–67.
47. Fiddes J. C. and Goodman H. M. (1979) Isolation, cloning and sequence analysis of the cDNA for the α-subunit of human chorionic gonadotropin. *Nature* **281,** 351–355.
48. Nilson J. H., Thomason A. R., Cserbak M. T., Moncman C. L., and Woychik R. P. (1983) Nucleotide sequence of a cDNA for the common α subunit of the bovine pituitary glycoprotein hormones. *J. Biol. Chem.* **258,** 4679–4682.
49. Erwin C. R., Croyle M. L., Donelson J. E., and Maurer R. A. (1983) Nucleotide sequence of cloned complementary deoxyribonucleic acid for the α subunit of bovine pituitary glycoprotein hormones. *Biochemistry* **22,** 4856–4860.
50. Grunstein M. and Hogness D. S. (1975) Colony hybridization: A method for the isolation of cloned DNAs that contain a specific gene. *Proc. Natl. Acad. Sci. USA* **72,** 3961–3965.
51. Chin W. W., Maizel J. V. Jr, and Habener J. F. (1983) Differences in sizes of human compared to murine alpha subunits of the glycoprotein hormones arises by a four-codon gene deletion or insertion. *Endocrinology* **112,** 482–485.
52. Fiddes J. C. and Goodman H. M. (1981) The gene encoding the common alpha subunit of the four human glycoprotein hormones. *J. Mol. Appl. Genet.* **1,** 3–18.
53. Boothby M., Ruddon R. W., Anderson C., McWilliams D., and Boime I. (1981) A single gonadotropin α-subunit gene in normal tissue and tumor-derived cell-lines. *J. Biol. Chem.* **256,** 5121–5127.
54. Goodwin R. G., Moncman C. L., Rottman F. M., and Nilson J. H. (1983) Characterization and nucleotide sequence of the gene for the common α subunit of the bovine pituitary glycoprotein hormones. *Nucleic Acids Res.* **11,** 6873–6882.
55. Gilbert W. (1978) Why genes in pieces? *Nature* **271,** 501.
56. Compton J. G., Schrader W. T., and O'Malley B. W. (1983) DNA sequence preference of the progesterone receptor. *Proc. Natl. Acad. Sci. USA* **80,** 16–20.
57. Hoshina M., Boothby M. R., Hussa R. D., Pattillo R. A., Camel H. M., and Boime I. (1984) Segregation patterns of polymorphic restriction sites of the gene encoding the α subunit of human chorionic gonadotropin in trophoblastic disease. *Proc. Natl. Acad. Sci. USA* **81,** 2504–2507.
58. Ramabhadran T. V., Reitz B. A., and Tiemeier D. C. (1984) Synthe-

sis and glycosylation of the common α subunit of human glycoprotein hormones in mouse cells. *Proc. Natl. Acad. Sci. USA* **81,** 6701–6705.

59. Fiddes J. C. and Goodman H. M. (1980) The cDNA for the β-subunit of human chorionic gonadotropin suggests evolution by read-through into the 3′-untranslated region. *Nature* **286,** 684–687.
60. Boorstein W. R., Vamvakopoulos N. C., and Fiddes J. C. (1982) Human chorionic gonadotropin β-subunit is encoded by at least eight genes arranged in tandem and inverted pairs. *Nature* **300,** 419–422.
61. Talmadge K., Vamvakopoulos N. C., and Fiddes J. C. (1984) Evolution of the genes for the β subunits of human chorionic gonadotropin and luteinizing hormone. *Nature* **307,** 37–40.
62. Talmadge K., Boorstein W. R., and Fiddes J. C. (1983) The human genome contains seven genes for the beta-subunit of chorionic gonadotropin but only one gene for the beta-subunit of luteinizing hormone. *DNA* **2,** 281–289.
63. Fiddes J. C. and Talmadge K. (1984) Structure, expression, and evolution of the genes for the human glycoprotein hormones. *Recent Prog. Horm. Res.* **40,** 43–78.
64. Policastro P., Ovitt C. D., Hoshina M., Fukuoka H., Boothby M. R., and Boime I. (1983) The β subunit of human chorionic gonadotropin is encoded by multiple genes. *J. Biol. Chem.* **258,** 11492–11499.
65. Keutmann H. T. and Williams R. M. (1977) Human chorionic gonadotropin: Amino acid sequence of the hormone-specific COOH-terminal region. *J. Biol. Chem.* **252,** 5393–5397.
66. Rigby W. J., Dieckmann M., Rhodes C., and Berg P. (1977) Labeling of deoxyribonucleic acid to high specific activity in vitro by nick translation with DNA polymerase I. *J. Mol. Biol.* **113,** 237–251.
67. Chin W. W., Godine J. E., Klein D. R., Chang A. S., Tan L. K., and Habener J. F. (1983) Nucleotide sequence of the cDNA encoding the precursor of the β subunit of rat lutropin. *Proc. Natl. Acad. Sci. USA* **80,** 4649–4653.
68. Jameson J. L., Chin W. W., Hollenberg A. N., Chang A. S., and Habener J. F. (1984) The gene encoding the β subunit of rat luteinizing hormone: Analysis of gene structure and evolution of nucleotide sequence. *J. Biol. Chem.* **259,** 15474–15480.
69. Tepper M. A. and Roberts J. L. (1984) Evidence for only one β-luteinizing hormone and no β-chorionic gonadotropin gene in the rat. *Endocrinology* **115,** 385–391.
70. Carr F. E. and Chin W. W. (1985) Absence of detectable chorionic gonadotropin mRNAs in rat placenta during gestation. *Endocrinology,* **116,** 1151–1157.
71. Wurzel J. M., Curatola L. M., Gurr J. A., Goldschmidt A. M., and

Kourides I. A. (1983) The luteotropic activity of rat placenta is not due to a chorionic gonadotropin. *Endocrinology* **83,** 1854–1857.

72. Gurr J. A., Catterall J. F., and Kourides I. A. (1983) Cloning of cDNA encoding the pre-β subunit of mouse thyrotropin. *Proc. Natl. Acad. Sci. USA* **80,** 2122–2126.
73. Maurer R. A., Croyle M. L., and Donelson J. E. (1984) The sequence of a cloned cDNA for the β subunit of bovine thyrotropin predicts a protein containing both $NH_2$-and COOH-terminal extensions. *J. Biol. Chem.* **255,** 5024–5025.
74. Chin W. W., Muccini J. A., and Shin L. (1985) Evidence for a single rat thyrotropin-β-subunit gene: Thyroidectomy increases its mRNA. *Biochem. Biophy. Res. Commun.* **128,** 1152–1158.
75. Croyle M. L. and Maurer R. A. (1984) Thyroid hormone decreases thyrotropin β subunit mRNA levels in rat anterior pituitary. *DNA* **3,** 231–236.
76. Kourides I. A., Gurr J. A., and Wolf D. (1984) The regulation and organization of thyroid stimulating hormone genes. *Recent Prog. Horm. Res.* **40,** 79–120.
77. Naylor S. L., Chin W. W., Goodman H. M., Lalley P. A., Grzeschik K. H., and Sakaguchi A. Y. (1983) Chromosomal assignment of genes encoding the α and β subunits of glycoprotein hormones in man and mouse. *Somatic Cell Genet.* **9,** 757–770.
78. Julier C., Weil D., Couillin P., Cote J. C., Nguyen V. C., Foubert C., Boue A., Thirion J. P., Kaplan J. C., and Junien C. (1984) The beta chorionic gonadotropin-beta luteinizing hormone gene cluster maps to human chromosome 19. *Hum. Genet.* **67,** 174–177.
79. Hardin J. W., Riser M. E., Trent J. M., and Kohler P. O. (1983) The chorionic gonadotropin α subunit gene is on human chromosome 18 in JEG cells. *Proc. Natl. Acad. Sci. USA* **80,** 6282–6285.
80. Hardin J. W., Riser M., and Kohler P. O. (1984) Chorionic gonadotropin β-subunit gene is on chromosome 10 in JEG cells. *Clin. Res.* **32,** 549A.
81. Kohler P. O., Riser M., Hardin J., Boothby M., Boime I., Norris J., and Siciliano M. J. (1981) Chorionic gonadotropin synthesis and gene assignment in human: mouse hybrid cells. *Adv. Exp. Med. Biol.* **138,** 405–418.
82. Kourides I.A., Baker P. E., Gurr J. A., Pravtcheva D. D., and Ruddle F. H. (1984) Assignment of the genes for the α and β subunits of thyrotropin to different mouse chromosomes. *Proc. Natl. Acad. Sci. USA* **81,** 517–519.
83. Shupnik M. A., Chin W. W., Ross D. S., Downing M. S., Habener J. F., and Ridgway E. C. (1983) Regulation of alpha mRNA levels in the thyrotrope by thyroxine (T4). *J. Biol. Chem.* **258,** 15120–15124.

84. Chin W. W., Shupnik M. A., Ross D. S., Habener J. F., and Ridgway E. C. (1985) Regulation of the α and β TSH subunit mRNAs by thyroid hormones. *Endocrinology,* **116,** 873–878.
85. Gurr J. A. and Kourides I. A. (1983) Regulation of thyrotropin biosynthesis: Discordant effect of thyroid hormone on α and β subunit mRNA levels. *J. Biol. Chem.* **258,** 10208–10211.
86. Gershengorn M. C., Cohen M., and Hoffstein S. T. (1978) Cellular heterogeneity in primary monolayer cultures of mouse pituitary thyrotropic tumors. *Endocrinology* **103,** 648–651.
87. Shupnik M. A., Chin W. W., Habener J. F., and Ridgway E. C. (1985) Transcriptional regulation of the thyrotropin subunit genes by thyroid hormone. *J. Biol. Chem.,* **260,** 2900–2903.
88. Brock M. L. and Shapiro D. J. (1983) Estrogen stabilizes vitellogenin mRNA against cytoplasmic degradation. *Cell* **34,** 207–214.
89. Guyette W. A., Matusik R. J., and Rosen J. M. (1979) Prolactin-mediated transcriptional and post-transcriptional control of casein gene expression. *Cell* **17,** 1013–1023.
90. McKnight G. S. and Palmiter R. D. (1979) Transcriptional regulation of the ovalbumin and conalbumin genes by steroid hormones in chick oviduct. *J. Biol. Chem.* **254,** 9050–9058.
91. Tepper M. A., Dionne F. R., Eberwine J. H., Wilcox J. N., Roberts J. L. (1984) Regulation of beta LH gene expression during the estrus cycle and castration in the rat. 7th International Congress of Endocrinology, Quebec City, Canada. Abstract no. 2293.
93. Landefeld T. D., Kepa J., and Karsch F. J. (1983) Regulation of α subunit synthesis by gonadal steroid feedback in the sheep anterior pituitary. *J. Biol. Chem.* **258,** 2390–2393.
93. Landefeld T. and Kepa J. (1984) Regulation of LH beta subunit mRNA in the sheep pituitary gland during different feedback states of estradiol. *Biochem. Biophys. Res. Commun.* **122,** 1307–1313.
94. Corbani M., Counis R., Starzec A., and Jutisz M. (1984) Effect of gonadectomy on pituitary levels of mRNA encoding gonadotropin subunits and secretion of luteinizing hormone. *Mol. Cell. Endocrinol.* **35,** 83–87.
95. Counis R., Corbani M., and Jutisz M. (1983) Estradiol regulates mRNAs encoding precursors to rat lutropin (LH) and follitropin (FSH) subunits. *Biochem. Biophys. Res. Commun.* **114,** 65–72.
96. Chin W. W. (1985) Organization and Expression of Glycoprotein Hormone Genes, in *The Pituitary Gland* (Imura H., ed.), Raven, New York.
97. Gharib S. D., Bowers S. M., Need L. R., and Chin W. W. (1986) Regulation of rat luteinizing hormone (LH) subunit mRNAs by gonadal-steroid hormones. *J. Clin. Invest.* **77,** 582–589.

# *Chapter 8*

# Oxytocin and Vasopressin Genes

## Expression and Structure

DIETMAR RICHTER

## 1. INTRODUCTION

The nonapeptide hormone vasopressin and its structurally related counterpart oxytocin are known to be involved in the processes of water resorption and uterine contraction, respectively. The biosynthetic pathway for the two hormones vasopressin and oxytocin originates in the magnocellular neurons of the supraoptic and paraventricular nuclei of the hypothalamus (*1*). The two hormones are transported in their secretory granules in axons down the pituitary stalk to the neurohypophysis and there are released.

The first insights into the various biosynthetic steps date back to the pioneering work by Sachs and associates, who proposed that synthesis of vasopressin may occur via a larger precursor (*2*).

It is now clear that both hormones are members of a composite precursor family (*3*), which, in the case of the vasopressin precursor, consists of the hormone, its respective carrier protein neurophysin, and a glycopeptide of as-yet-unknown function. The oxytocin precursor has a similar composition, except for the glycopeptide at its C-terminus, which it lacks.

## 2. ELUCIDATION OF THE PRECURSOR STRUCTURES

The complete structural organization of the precursors became evident by a combination of cell-free translation studies and recombinant DNA techniques (*4*). The experimental concept adopted included (i) translation of hypothalamic mRNA and immunological identification of the products (*5*); (ii) tryptic peptide mapping (*6*); and (iii) cloning and sequencing of the cDNA encoding the hormones (*7,8*). The first two approaches have been outlined in detail elsewhere and will not be discussed further here (*4*).

### 2.1. Cloning Strategy

The cloned cDNAs encoding the vasopressin or oxytocin precursors were obtained from a relatively small bovine hypothalamic cDNA library (550 clones). The low number of clones was sufficient, because, first, the mRNA used for cloning was derived from membrane-bound polysomes and hence enriched for secretory precursors, and second, based on cell-free translation studies, it was calculated that 1% of the mRNA should code for the vasopressin precursor (*4*). The assumption was confirmed because, of the 550 clones analyzed, seven were positive: five specific for the vasopressin and two for the oxytocin precursor (*7,8*).

To identify the clones specific for vasopressin precursor sequences, H. Land adopted a strategy whereby cleavage sites of restriction endonucleases in an unknown DNA sequence can be predicted from a known amino acid sequence (Fig. 1). There are only a few restriction endonucleases that fulfill these requirements: e.g., *Sau*96I, which will cut the DNA at any position encoding the amino acid sequence Gly-Pro (GGNCCN). Because the neurophysin sequences contain Gly-Pro in the positions 14–15 and 23–24, it is reasonable to expect that digestion of DNA coding for neurophysin with *Sau*96I should generate a 27-bp (base pair) fragment. Clones giving rise to the expected *Sau*96I fragment were directly sequenced, and the deduced amino acid sequence was compared with the known neurophysin sequence.

According to this analysis, the bovine vasopressin preprohormone consisted of 166 amino acid residues with a molecular

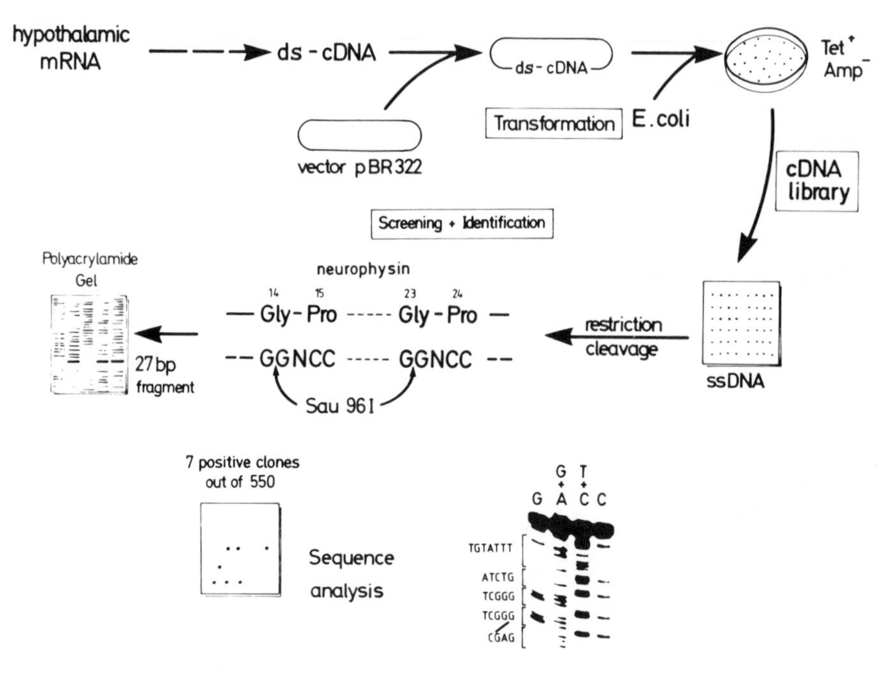


Fig. 1. Cloning strategy.

weight of 17,310; the rat counterpart, which could be predicted from its gene sequence, was almost similar in size, with 168 amino acid residues and a molecular weight of 17,826.

For identification of the cDNA encoding the bovine oxytocin precursor, about 5000 recombinants were screened with a DNA probe encoding the vasopressin precursor (*8*). Because the two neurophysins in the vasopressin or oxytocin precursors show a high homology in their amino acid sequences, considerable cross-hybridization can be expected between the two mRNA species. Clones giving a positive hybridization signal were further analyzed by restriction mapping, revealing two groups with different cDNA inserts, one specific for the vasopressin, and the other for the oxytocin precursor. The deduced bovine oxytocin precursor consisted of 125 amino acid residues with a molecular weight of 12,826.

In general, the sequence of the vasopressin- and oxytocin-associated neurophysins predicted from the cDNA agrees with the one obtained by conventional Edman sequence determination (*9*) with few exceptions. In the human neurophysin of the vasopressin precursor, the DNA sequence predicts a Gly instead of a Val in position 88 of the neurophysin (*9A*). In case of the rat neurophysin of the oxytocin precursor a Glu is replaced by a Gln residue in position 92—a difference that may be caused by microheterogeneity expressed in various rat strains.

## 2.2. Structure–Function Relations

The primary sequence of the vasopressin and oxytocin preprohormones offered a number of interesting insights into their structural organization (Fig. 2).

First, the hormones were directly adjacent to the signal peptide; hence, within the respective prohormones, vasopressin or oxytocin was N-terminally located. As with many other precursors to secretory peptides, the signal peptide and the prohormone are separated by a small neutral amino acid, in this case an Ala residue (*10*). The cleavage site and the size of the putative signal peptides are in good agreement with the results of

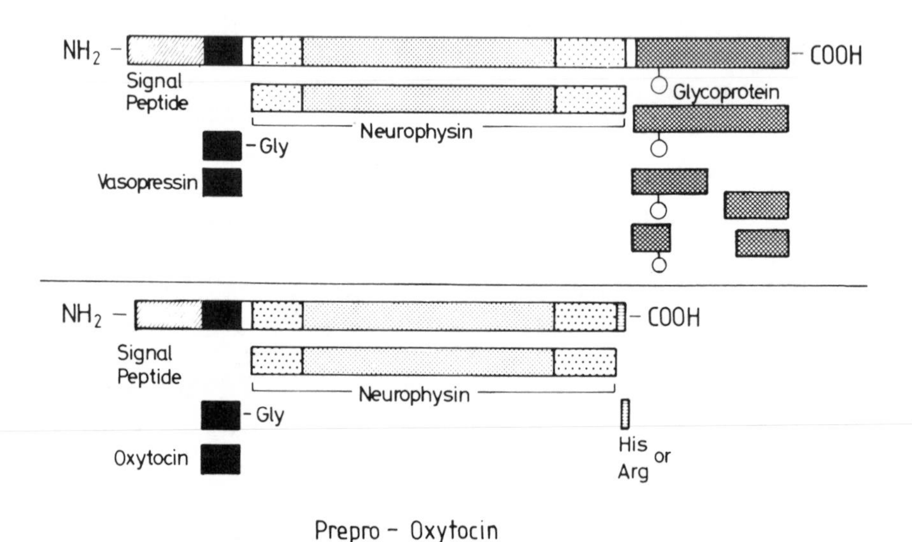


Fig. 2. Schematic comparison of the vasopressin and oxytocin precursors from calf and rat.

the membrane-supplemented translation experiments and tryptic fingerprinting of the prepro- and prohormones (*4*).

Except for the rat vasopressin precursor, the putative signal sequences of the two hypothalamic preprohormones so far analyzed consist of 19 amino acids each (Table 1). Only the gene sequence of the rat vasopressin precursor predicts a longer putative signal sequence, with 23 amino acid residues and three possible translation start sites (*11*). According to the rule that, in most cases, an A is found three bases upstream, and a G residue four bases downstream from the AUG translation start codon (*12*), it is predictable that the rat vasopressin preprohormone starts at the third Met residue, with a 19-amino-acid signal peptide. However, as indicated in Table 1, there are exceptions to this rule, inasmuch as the rat oxytocin precursor has a G and a U(T) residue in the respective positions. Direct sequence data of the preprohormones are needed to solve this problem.

Second, the two hormones and their corresponding neurophysins are separated by the three amino acid residues, Gly, Lys, and Arg. All three are involved in the maturation reaction leading to the active hormones. The nucleotide sequences encoding the three amino acids are highly conserved, pointing also to the significance of these processing signals (*10*).

Processing of the two precursors into active vasopressin and oxytocin has to be accompanied by the amidation of the hormones. The finding of an extra Gly residue adjacent to the carboxy terminus of either vasopressin or oxytocin is consistent with the sequence of other precursors to amidated oligopeptides (Table 2) (*3,4*). An enzyme catalyzing the amidation reaction has been obtained from pituitary extracts (*13,14*). In this reaction, Gly

TABLE 1
Translation Start Sites in the Vasopressin (AVP) and Oxytocin (OT) Precursors

| | | | −23 Met | | | | −19 Met | | |
|---|---|---|---|---|---|---|---|---|---|
| AVP | RAT | ACCT | ATG | CTC | GCC | **A**TG | ATG | CTC | **G**CC |
| AVP | HUMAN | | | | | **A**GG | ATG | CCT | **G**AC |
| AVP | CALF | | | | | **A**GG | ATG | CCC | **G**AC |
| OT | RAT | | | | | **G**CC | ATG | GCC | **T**GC |
| OT | CALF | | | | | **A**CC | ATG | GCA | **G**GT |

TABLE 2
Partial Amino Acid Sequences of Precursors to Oligopeptides Amidated at the Carboxy-Terminus[a]

| | |
|---|---|
| Melittin | -Lys-Arg-Gln-Gln-**Gly**-COOH |
| α-MSH | -Gly-Lys-Pro-Val-**Gly**-Lys-Lys- |
| γ-MSH | -Trp-Asp-Arg-Phe-**Gly**-Arg-Arg- |
| Vasopressin | -Cys-Pro-Arg-Gly-**Gly**-Lys-Arg- |
| Oxytocin | -Cys-Pro-Leu-Gly-**Gly**-Lys-Arg- |
| Caerulein | -Trp-Met-Asp-Phe-**Gly**-Arg-Arg- |
| Gastrin | -Trp-Met-Asp-Phe-**Gly**-Arg-Arg- |
| Calcitonin | -Val-Glu-Ala-Pro-**Gly**-Lys-Lys- |
| Human joining peptide | -Gly-Pro-Arg-Glu-**Gly**-Lys-Arg- |
| Egg-laying hormone | -Leu-Glu-Lys-Lys-**Gly**-Lys-Arg- |
| CRF | -Leu-Asp-Ile-Ala-**Gly**-Lys |
| TRF | -Arg-Gln-His-Pro-**Gly**-Lys-Arg- |

$$-X\text{-Gly}\downarrow\text{-}\ \text{Lys}\downarrow\text{-}\ \text{Arg}\downarrow\text{-} \longrightarrow -X\text{-Gly} \xrightarrow[\text{O}_2,\ \text{ascorbate}]{\text{amidating enzyme}} -X{\cdot}\text{NH}_2 + \text{glyoxylate}$$

[a]From ref. *4*.

serves as the nitrogen donor, with glyoxylate as the other product, provided that the Gly has a free carboxyl group (Table 2).

The basic amino acids C-terminal to Gly can be either lacking (*15*) (melittin); single (*16*) (CRF); or dibasic residues (e.g., vasopressin). In the last-mentioned case, the combination Lys-Arg is preferred over Arg-Arg or Lys-Lys; the combination Arg-Lys appears to be very rare. So far, the data for the amidation process suggest that the removal of the basic amino acids is a necessary precondition for the amidation reaction to take place (*13*).

Third, neurophysin and the glycoprotein within the vasopressin precursor are separated by only a single basic amino acid, an Arg residue. Interestingly enough, the basic amino acids within the region of the C-terminus of bovine neurophysin are not cleaved by the processing enzymes, suggesting that the potential cleavage site is exposed and more easily accessible to proteolytic enzyme(s) than are the two basic amino acids in the close neighborhood.

The finding that peptide units within one precursor can be separated by either pairs of or single basic amino acid residues possibly suggests the existence of different sets of processing enzymes discriminating between the two cleavage sites (see also below). Paired basic residues represent the most common form of cleavage signal, but a single residue is sometimes found, e.g., precursors to growth-hormone-release factor and somatostatin (*17*).

Fourth, inspection of the nucleotide sequence encoding the calf and rat oxytocin precursors predicts that there is a supernumerary basic amino acid at the carboxy terminus that is not found in any of the isolated neurophysins and is either Arg (rat) (*18,19*) or His (calf) (*8*) residue (Fig. 2).

In the rat oxytocin, the position of this extra Arg residue corresponds to the position occupied by the single Arg between the neurophysin and the glycopeptide in the vasopressin precursor. It is possible that the extra amino acid is removed by posttranslational processing, predicting an exopeptidase that accepts Arg or His as the signal.

The single basic amino acid residue at the C-terminus of the oxytocin precursor may represent the rudiment of a once-present peptide that extended farther downstream from the Arg (or His) residue. This hypothetical peptide could share an ancestral origin with the glycopeptide moiety of the vasopressin precursor.

Fifth, the rat and bovine vasopressin precursors include a glycopeptide of 39 amino acids at the C-terminus with the glycosylation site Asn-Ala-Thr (*see* also Figs. 3 and 4). Comparing the glycopeptide sequence from calf with sequences from rat, pig, and humans showed remarkable homologies (*20–22*). The glycosylation site and the Leu-rich center part are well conserved in all species known so far. The consecutive Leu residues may represent alternative processing signals for converting the glycopeptide into subfractions (*21*). The glycopeptide has been localized in vasopressin-producing magnocellular neurons (*23*), yet its biological function remains unknown.

## 3. MATURATION OF THE PRECURSORS

A still-unresolved problem is the processing of the precursors and the order of these steps that lead to the mature peptides. The early kinetic studies by Sachs and Takabatake (*24*) suggested that, in the dog, it required about 1.5 h from the time of synthesis to release of the physiological nonapeptide. Pulse-chase experi-

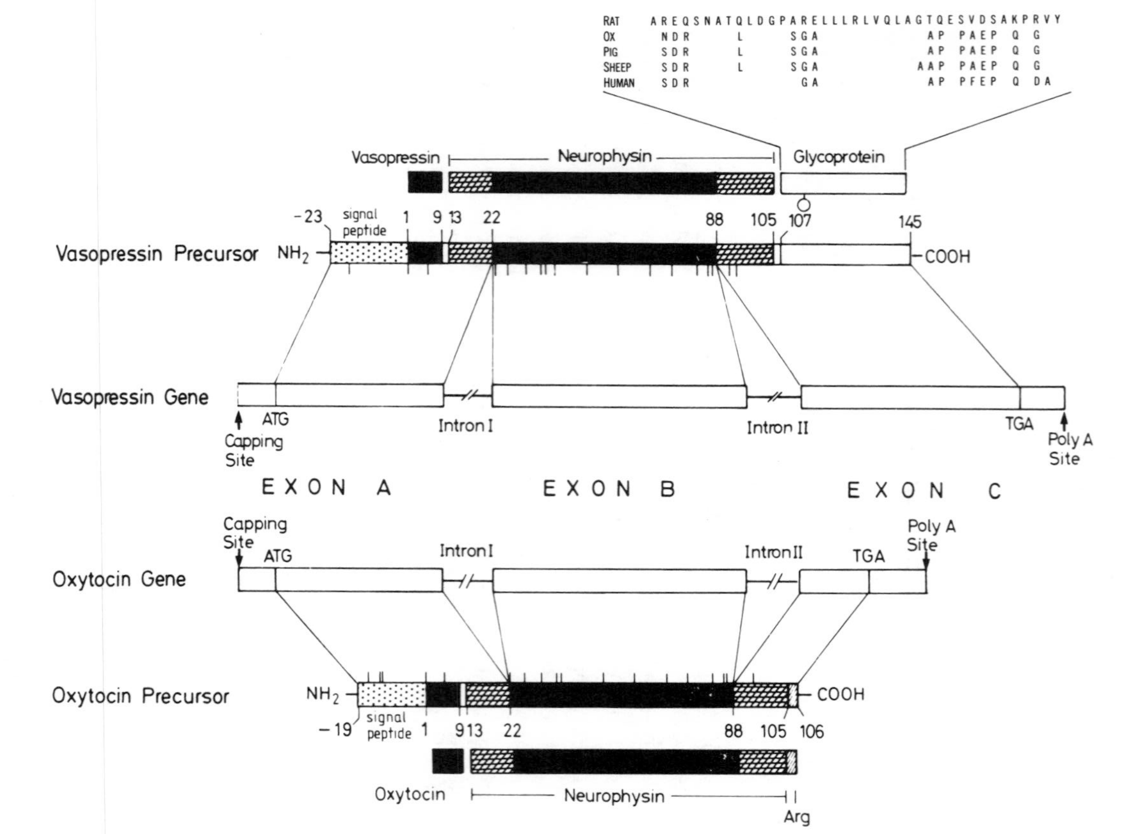


Fig. 3. Structural organization of the vasopressin and oxytocin precursors and their genes.

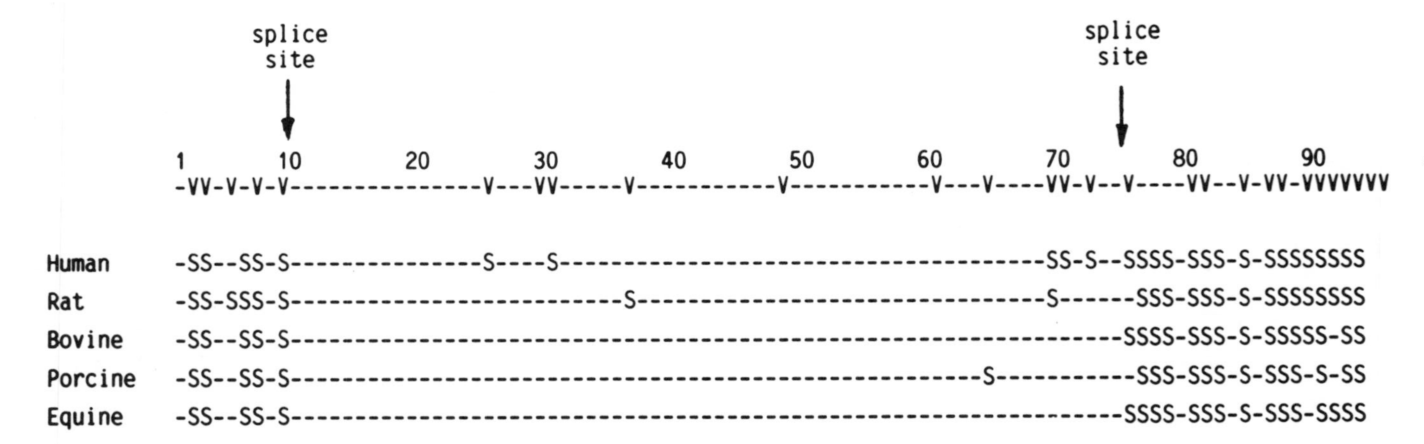


Fig. 4. Analysis of amino acid conservation in the oxytocin-associated (OT) and vasopressin-associated (AVP) neurophysins from various mammals. The upper lines show those amino acids (V) that have varied during mammalian evolution in either the OT-neurophysin or the AVP-neurophysin. Below this is listed a comparison of OT- and AVP-neurophysin within the indicated species, showing those amino acids substituted (S) between the proteins since their possible evolution by gene duplication over 400 million years ago. The noted absence of substitution in the region corresponding to the central exon of the gene, by comparison with the normal mutation rate evident at the N- or C-terminus on the variable positions in the top line, points to a genetic mechanism emphasizing the conservation of the central region (data from ref. *17*).

ments in vivo in the rat (*25*) also indicated that cleavage occurred continuously during axonal transport, with both cleaved and uncleaved radiolabeled products appearing in the neurohypophysis after a few hours.

This is in line with immunohistochemical studies using nonapeptide antisera that specifically require the C-terminal amide for recognition; substantial quantities of the free nonapeptides were located in hypothalamic cell bodies. HPLC analysis of hypothalamic nuclear extracts also shows both nonapeptides and their respective neurophysins in the proximal regions of the neurons (*26*).

Little information is yet available on intermediate forms of the vasopressin or oxytocin precursors. So far, a small ($M_r$ = 14,000) polypeptide has been isolated from mouse hypothalamus, evidently containing both vasopressin and neurophysin antigenic determinants (*27*). A glycopeptide was isolated by concanavalin-A–Sepharose chromatography from neurosecretory granules that contain a neurophysin, but not a vasopressin, antigenic moiety (*28*). A similar intermediate was detected in the posterior pituitary

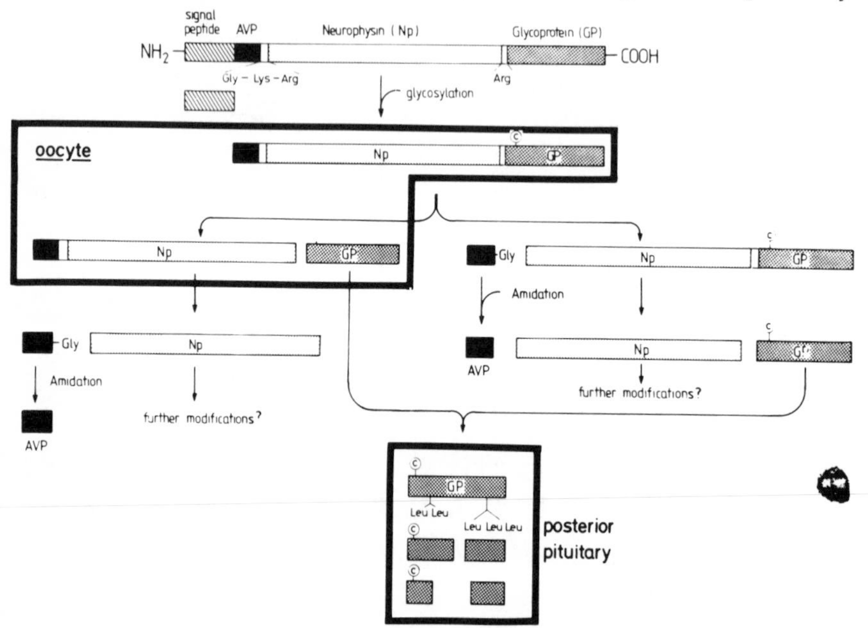


Fig. 5. Possible processing pathways of the vasopressin precursor (reproduced from ref. *10*).

of the guinea pig (*29*), implying that, in the neurohypophysis, the first processing step occurs between vasopressin and the neurophysin.

The processing model proposed in Fig. 5 is based on studies of *Xenopus laevis* oocytes programmed with hypothalamic mRNA (*17*). There, the first cleavage of the vasopressin precursor occurs between the neurophysin and the glycopeptide, which are separated by a single Arg residue. The second cleavage step between hormone and neurophysin could not be observed, which implies that the processing enzymes of the oocyte recognize a single Arg but not a pair of basic amino acids. Lack of such a processing system has also been noted for fish proinsulin, which also remains as a longer precursor when synthesized in frog oocytes (*30*). The oocyte experiments strongly suggest that there are at least two different processing systems, one acting on pairs of, and the other on single, basic amino acids.

## 4. VASOPRESSIN AND OXYTOCIN GENES

Vasopressin and oxytocin genes were isolated from calf and rat genomic libraries. According to restriction mapping and nucleotide sequence analysis, the rat and bovine vasopressin genes are approximately 2 kbp (kilobase pairs) in size (*11,31*). The oxytocin gene from rat and calf are smaller, consisting roughly of 1000 bp (*18,19,31*). As with other genes from higher eukaryotes, the two genes are organized in such a way that the principal functional domains of the hormone precursors are encoded by three distinct exons separated by two intervening sequences (Figs. 3, 6, and 7).

The first exons (A) comprise the 5′-untranslated regions, the putative signal peptides, the hormones vasopressin or oxytocin, and the N-terminal, variable parts of the respective neurophysins. The second exons (B) encode the highly conserved center parts of the neurophysins, whereas the third exons (C) contain the remaining C-termini of the neurophysins and, in the case of the vasopressin gene, the glycopeptide and the untranslated 3′ regions with the polyadenylation sites.

The structural organization of the two genes is in line with other eukaryotic split genes and includes the consensus se-

```
R   1 GATCCACGTAAGGGCTTCCTCACCCACTTGCGC

R  34 TATCCAGCCTCATTCTGAGGTATTGGATTTCTATGAAAAAACAGCTCTTGGCTAGGCTGCACCTCTACCCCCTCCAAGTCTCTTTATCCTCTTGTAGCTT
B   1 AAGCAGCA----CTC-GCTATCATCTCCCGGACCATTAGCCATTAGCCGAC---ATAACCTTGACCCGGCACAGC-CTCTGCAAATGAGGGG------GC
      *R**   *      **  ***R*  Y*YYYY ** * Y R*Y*******Y**    R*R********** R  * R  *Y**********R***     *
R 134 AGGCCTCCCCTTCTAGGCTGTGTCCCTTTTGAGCTCAGGTCATTAGCTGAGGCGGTGACCTTGACCCCAGCCCAGACCCTGCAAATGAAGGGCCTGCTTC

   86 GCGCCGGGGGCGCAGGGCCTGACCCGCAGCGGCGCTGTGACCAGTCATGCGGCTGCCCTCTTAGACACCGTTCCGCCCACGCGGCCGCCGGGCTT
        R *R* *   * * R*YY******R RR R *  ****************R*YYR***********    *YY**R**    **R*Y * *RR*
  234 TAAACAGTGTGGAACAGTTTGACCCAAGAGACCTGGCTGTGACCAGTCATGCAGTCACCCTCTTAGACTGGGCCCCACCATGGCAGTGGACAAGGC

                                                                       MetAlaGlySerSerLeuAlaCysCysLeuLeuGly
  184 AAAGGCCAG-----ACCCGAGAGACGGCCGCAGTCCCCGGCCCGGAGACCAGCGTCTGCACCATGGCAGGTTCCAGCCTCGCCTGCTGCCTGCTCGGC
      *****Y*R*     R*Y ****R** RY*R* * *  Y** Y*  R***YRR    Y  R*R******* *YY****Y*****Y***********Y***
  334 AAAGGTCGGTCTGGGCTGGAGAAACCATCACCGACGGTGGATCTCGGACTGAACACCA-ACGCCATGGCCTGCCCCAGTCTCGCTTGCTGCCTGCTTGGC
                         OT                   NP MetAlaCysProSerLeuAlaCysCysLeuLeuGly
      LeuLeuAlaLeuThrSerAlaCysTyrIleGlnAsnCysProLeuGlyGlyLysArgAlaValLeuAspLeuAspValArgThr
  279 CTCCTGGCGTTGACCTCCGCCTGCTACATTCAGAACTGCCCCCTGGGCGGCAAACGCGCGGTGCTGGACCTCGACGTGCGCACGGTGAG-AGCCCCGCCC
      ** *****  Y*****************Y************************R * ** *Y***R***** **YR****** ******   ***** *
  433 CTACTGGCTCTGACCTCCGCCTGCTACATCCAGAACTGCCCCCTGGGCGGCAAGAGGGCTGCGCTAGACCTGGATATGCGCAAGGTGAGTCTCCCCGACG
      LeuLeuAlaLeuThrSerAlaCysTyrIleGlnAsnCysProLeuGlyGlyLysArgAlaAlaLeuAspLeuAspMetArgLys

  378 TCGACCCGTGGCTCTCCGGGCTGCCCGGCCCGCTGCCACAGGGTCGCCCCGCCGCCCCCTTTCCCGCGCTGACCGCGTACCGGCCCCACCTAGCCTGGG
      Y*R *****   * Y**Y** * R    ***Y R  R*** **R   * *Y***R** Y*      * Y * ** **   Y *R**R****Y**   *
  533 CCATCCCGTCCCGTTCTGG-CAA---GGCTAAGGACCAGAGATGCTCTCCCACCTTCAGAGAGCATCCCCTCACACT-TGCCAGCCCTAC---G------

  478 AATCGAGGGAGCGGAGGAGCTTTTGACTGCCCTCCTTCCACCGCTTTTGAGCCCAAAGAGAGAGCCAGGGAGACCCGCCACCTCCCGCGCTCCTCCGGCC
          ***                        ****    **Y                          **********  *      ** YY****Y*R*
  619 ---CGA-------------------------CCTC---GCAT--------------------------GGGAGACCCG-GA------GC-TCCCTCTGAC-

                                                                                                  CysLeu
  578 GCCCTCGCCCGCCCGGCTCAGCCCCCCGCCCCACAGGGTCTCCCTCCCCGGCCGCTCCCCTCCCGCCCCCGGCTCATCCCTTCCCTCCCACCAGTGTCTC
      *** * *    ** * ***   * Y***Y*******   R    ****  Y Y** Y*     *  *****   ** *R*   **Y**Y***R ********Y
  654 GCCGT-GAAGGCAC-GCT--G-TCCCTGCCCCAC-CACAGTCCCGATATGGATGGGGGGCAG-CGCCC-ATGCGCGTG--TTTCCCCCCG-CAGTGTCTT
                                                                                                  CysLeu

      ProCysGlyProGlyGlyLysGlyArgCysPheGlyProSerIleCysCysGlyAspGluLeuGlyCysPheTrpGlyThrAlaGluAlaLeuArgCysG
  678 CCCTGCGGCCCCGGGGGCAAAGGCCGCTGCTTCGGGCCCAGCATCTGCTGCGGGGACGAGCTGGGCTGCTTCTGGGGCACGGCCGAGGCGCTGCGCTGCC
      **Y***** ***** ******** ************** ************ ***************** ****** *****************
```

```
743  CCTTGCGGACCCGGCGGCAAAGGGCGCTGCTTCGGGCCGAGCATCTGCTGCGCGGACGAGCTGGGCTGCTTCGTGGGCACCGCCGAGGCGCTGCGCTGCC
     ProCysGlyProGlyGlyLysGlyArgCysPheGlyProSerIleCysCysAlaAspGluLeuGlyCysPheValGlyThrAlaGluAlaLeuArgCysG

     lnGluGluAsnTyrLeuProSerProCysGlnSerGlyGlnLysProCysGlySerGlyGlyArgCysAlaAlaAlaGlyIleCysCysSerProA
778  AAGAGGAGAACTACCTGCCGTCGCCCTGCCAGTCCGGCCAGAAGCCCTGCGGAGCGGGGCCGCTGCGCCGCCGCCGGCATCTGCTGCAGCCCGGGTGA
     *R**************** *************Y***********Y*****R*****R************R**** ***********Y**********
843  AGGAGGAGAACTACCTGCCCTCGCCCTGCCAGTCTGGCCAGAAGCCTTGCGGAAGCGGAGGCCGCTGCGCCACCGCGGGCATCTGCTGTAGCCCGGGTGA
     lnGluGluAsnTyrLeuProSerProCysGlnSerGlyGlnLysProCysGlySerGlyGlyArgCysAlaThrAlaGlyIleCysCysSerProA

                                                                                             spGlyCysHisG
878  GTCGGGCAGGGGTCGA--GACGGGACCGGGGCTCCAGGACCAGGCGGGCTGGGCCAGGGCGGCCCTGACTCGGCGTCTCTCTGTGCAGACGGCTGCCACG
     *Y **   *****Y* *   RR*R******R     * R*** * * ****Y*   **YR      ****     * *   *R*Y***** ******Y*******R*R
943  GCAGG--AGGGGCCTAGCAGCAGGACCGACCGGCAGGGAGCCGTCGGGTT-TGCTGCTCAGGCC-ACTCACC-CATTTCTCT-TGCAGATGGCTGCCGCA
                                                                                             spGlyCysArgT

     luAspProAlaCysAspProGluAlaAlaPheSerGlnHisEnd
976  AGGACCCCGCCTGCGACCCTGAGGCCGCCTTCTCCCAGCACTGAGACCGGCCGGCCCCCGATACCGTCGGAGCGCAGCCCTCATCCCCT--CTGTAATCA
      ********************* *Y********* ***R***** ***  YY *Y   Y****** *YR*R****  **Y****Y***Y     Y* *
1037 CCGACCCCGCCTGCGACCCTGAGTCTGCCTTCTCCGAGCGCTGAGCCCGCTTTGTGA-TGATACCTTTAGGGCGCTTCCTTCATTCCCCATGGCCACTAC
     hrAspProAlaCysAspProGluSerAlaPheSerGluArgEnd

1074 TCCCCAGGAATTATGACAATGAAATAAA--------GCCGGTTTTTCCCCCTCCAA-CAAGCCTCGCATCTGAGTGTCA-GAACGGGGAGGGGAGGGCT
     Y    *RR***** R* ** R* *****        R*Y    **Y**YY*Y*** *  * ** **R***R* ***** R******RR**R*RR**
1136 CAGAAAAAAATTAAAAAAAAAACATAAATAAATAAATAAAGCAGATTTCCTTTTCAAACTTGACTGGCGTCTAATTGTCAGAAACGGGAGGGAGGAGG--

1164 TTAG
      **
1234 AAAGGCACCGGGAACGCCATGGACATTGGCAATTCAGAGGGAGAACAGCCCAGCAGGCTGCTAGGAGGACCGAGGGGTCTCTACAGGTTGTCATCTCCAG

1334 TTAGGCCCGTTTCCCACTCCCGACTGGGGTAAGCCTGCGGCAGACAGACTTAGGAGCAGCATTTGGAAGTCCAAACAGCAATGGTGAACTTCTGTGGGAA

1434 AGGGCAAAGCCAAACAGGGGCCTGGTGATGGGAAGGCAGGCGAGGCGGCAAGGGACTCATCCAGTCACTTCCATATCTGGAGGAGACCTTGGCGGCGACC

1534 TCAGAGGTTGAGGGGGTAATCCCGGCCTAGAAGAAAGGGTCTTGGCCAGGGAGGGTAGTTGCAGAACCATAAGGTTTCTACAGAGCTGGGTT
```

Fig. 6. Sequence comparison of the oxytocin genes from calf (B) and rat (R). Coding sequences are shown by bold letters: identical nucleotides by asterisks; and absence of nucleotides by dashes. Pyrimidine homologies are indiciated by Y, purine homologies by R; shaded sequence, Goldberg-Hogness box, which is strongly modified in the case of the bovine oxytocin gene (TTAAAA). Lines (above or beneath the sequences) indicate the polyadenylation sites. OT, oxytocin; Np, neurophysin.

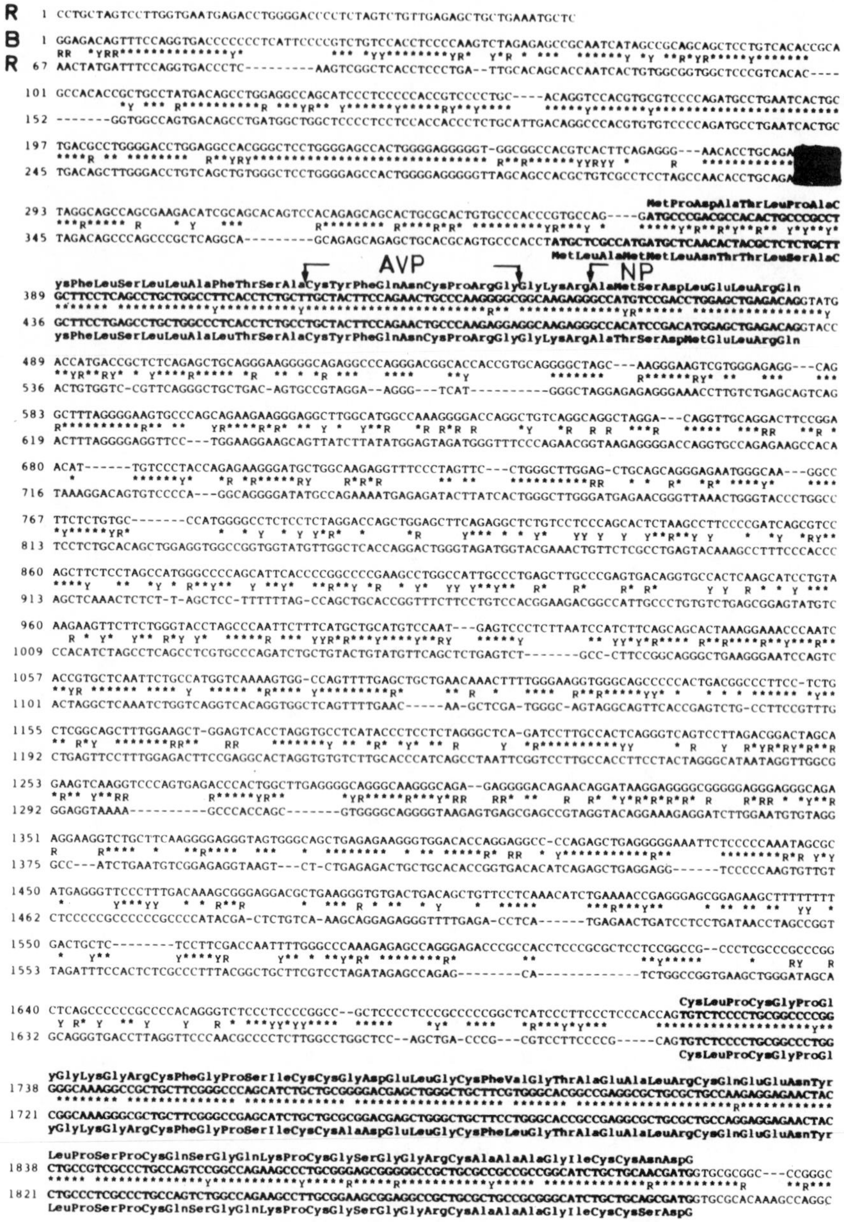

```
R    1 CCTGCTAGTCCTTGGTGAATGAGACCTGGGGACCCCTCTAGTCTGTTGAGAGCTGCTGAAATGCTC

B    1 GGAGACAGTTTCCAGGTGACCCCCCCTCATTCCCCGTCTGTCCACCTCCCCAAGTCTAGAGAGCCGCAATCATAGCCGCAGCAGCTCCTGTCACACCGCA
       RR  *YRR**************Y*          *** *YY*******YR*  Y*R * ***  ******Y *Y **R*YR*****Y*******
R   67 AACTATGATTTCCAGGTGACCCTC---------AAGTCGGCTCACCTCCCTGA--TTGCACAGCACCAATCACTGTGGCGGTGGCTCCCGTCACAC----

   101 GCCACACCGCTGCCTATGACAGCCTGGAGGCCAGCATCCCTCCCCCACCGTCCCCTGC----ACAGGTCCACGTGCGTCCCCAGATGCCTGAATCACTGC
              *Y *** R**********R ***YR** Y******Y*****RY**Y****    *****Y*******Y*************************
   152 -------GGTGGCCAGTGACAGCCTGATGGCTGGCTCCCCTCCTCCACCACCCTCTGCATTGACAGGCCCACGTGTGTCCCCAGATGCCTGAATCACTGC

   197 TGACGCCTGGGGACCTGGAGGCCACGGGCTCCTGGGGAGCCACTGGGGAGGGGGT-GGCGGCCACGTCACTTCAGAGGG---AACACCTGCAGA
       ****R ** ********  R**YRY******************************** R**R******YYRYY *     R   ***********
   245 TGACAGCTTGGGACCTGTCAGCTGTGGGCTCCTGGGGAGCCACTGGGGAGGGGGTTAGCAGCCACGCTGTCGCCTCCTAGCCAACACCTGCAGA

                                                                                  MetProAspAlaThrLeuProAlaC
   293 TAGGCAGCCAGCGAAGACATCGCAGCACAGTCCACAGAGCAGCACTGCGCACTGTGCCCACCCGTGCCAG----GATGCCCGACGCCACACTGCCCGCCT
       ***R***** R    * Y   ***          R******** R****R*R* *********YR***Y *    *****Y*R**R*Y**R** Y*Y**Y*
   345 TAGACAGCCCAGCCCGCTCAGGCA---------GCAGAGCAGAGCTGCACGCAGTGCCCACCTATGCTCGCCATGATGCTCAACACTACGCTCTCTGCTT
                                                                      MetLeuAlaMetMetLeuAsnThrThrLeuSerAlaC
                          AVP                      NP
       ysPheLeuSerLeuLeuAlaPheThrSerAlaCysTyrPheGlnAsnCysProArgGlyGlyLysArgAlaMetSerAspLeuGluLeuArgGln
   389 GCTTCCTCAGCCTGCTGGCCTTCACCTCTGCTTGCTACTTCCAGAACTGCCCAAGGGGCGGCAAGAGGGCCATGTCCGACCTGGAGCTGAGACAGGTATG
       ****** ************Y**********Y**************************R** *************YR****** ****************Y
   436 GCTTCCTGAGCCTGCTGGCCCTCACCTCTGCCTGCTACTTCCAGAACTGCCCAAGAGGAGGCAAGAGGGCCACATCCGACATGGAGCTGAGACAGGTACC
       ysPheLeuSerLeuLeuAlaLeuThrSerAlaCysTyrPheGlnAsnCysProArgGlyGlyLysArgAlaThrSerAspMetGluLeuArgGln

   489 ACCATGACCGCTCTCAGAGCTGCAGGGAAGGGGCAGAGGCCCAGGGACGGCACCACCGTGCAGGGGCTAGC---AAGGGAAGTCGTGGGAGAGG---CAG
       **YR**RY* * Y****R***** *R  ** * *R ***   ****    **Y          *******   R******RY* **   ***    ***
   536 ACTGTGGTC-CGTTCAGGGCTGCTGAC-AGTGCCGTAGGA--AGGG---TCAT----------GGGCTAGGAGAGAGGGAAACCTTGTCTGAGCAGTCAG

   583 GCTTTAGGGGAAGTGCCCAGCAGAAGAAGGGAGGCTTGGCATGGCCAAAGGGGACCAGGCTGTCAGGCAGGCTAGGA---CAGGTTGCAGGACTTCCGGA
       R**********R** **   YR****R*R* ** Y *  Y**R   *R R*R R     *Y   *R  R R  ***R   *****  ***RR   **R *
   619 ACTTTAGGGGAGGTTCC---TGGAAGGAAGCAGTTATCTTATATGGAGTAGATGGGTTTCCCAGAACGGTAAGAGGGGACCAGGTGCCAGAGAAGCCACA

   680 ACAT------TGTCCCTACCAGAGAAGGGATGCTGGCAAGAGGTTTCCCTAGTTC---CTGGGCTTGGAG-CTGCAGCAGGGAGAATGGGCAA---GGCC
        *        ******Y*  *R *R*****RY    R*R*R         ** **    **********RR   * *  R*  *R* ****Y*    ****
   716 TAAAGGACAGTGTCCCCA---GGCAGGGGATATGCCAGAAAATGAGAGATACTTATCACTGGGCTTGGGATGAGAACGGGTTAAACTGGGTACCCTGGCC

   767 TTCTCTGTGC-------CCATGGGGCCTCTCCTCTAGGACCAGCTGGAGCTTCAGAGGCTCTGTCCTCCCAGCACTCTAAGCCTTCCCCGATCAGCGTCC
       *Y*****YR*           *  * Y  * Y Y*R*  *    *R    Y*** * * Y*  YY Y  Y  Y**R**Y Y    *  *Y  *RY**
   813 TCCTCTGCACAGCTGGAGGTGGCCGGTGGTATGTTGGCTCACCAGGACTGGGTAGATGGTACGAAACTGTTCTCGCCTGAGTACAAAGCCTTTCCCACCC

   860 AGCTTCTCCTAGCCATGGGCCCCAGCATTCACCCCGGCCCCGAAGCCTGGCCATTGCCCTGAGCTTGCCCGAGTGACAGGTGCCACTCAAGCATCCTGTA
       ****Y   **  *Y * R**Y**  Y **Y*  **R**Y *R  * Y*  YY Y**Y**  R*   R*   R* R*        Y Y  R * * Y ***
   913 AGCTCAAACTCTCT-T-AGCTCC-TTTTTTAG-CCAGCTGCACCGGTTTCTTCCTGTCCACGGAAGACGGCCATTGCCCTGTGTCTGAGCGGAGTATGTC

   960 AAGAAGTTCTTCTGGGTACCTAGCCCAATTCTTTCATGCTGCATGTCCAAT---GAGTCCCTCTTAATCCATCTTCAGCAGCACTAAAGGAAACCCAATC
         R * Y*  Y** R*Y Y*  *****R *** YYR*R***Y****Y**RY   *****Y          ** YY*Y*R**** R**R****R**Y***R**
  1009 CCACATCTAGCCTCAGCCTCGTGCCCAGATCTGCTGTACTGTATGTTCAGCTCTGAGTCT-------GCC-CTTCCGGCAGGGCTGAAGGGAATCCAGTC

  1057 ACCGTGCTCAATTCTGCCATGGTCAAAAGTGG-CCAGTTTTGAGCTGCTGAACAAACTTTTGGGAAGGTGGGCAGCCCCCACTGACGGCCCTTCC-TCTG
       **YR ****** **** Y  ***** *R**** Y*********R*     ** R  *   ****  R**R*****YY* *   * * * ****** *Y**
  1101 ACTAGGCTCAAATCTGGTCAGGTCACAGGTGGCTCAGTTTTGAAC-----AA-GCTCGA-TGGGC-AGTAGGCAGTTCACCGAGTCTG-CCTTCCGTTTG

  1155 CTCGGCAGCTTTGGAAGCT-GGAGTCACCTAGGTGCCTCATACCCTCCTCTAGGGCTCA-GATCCTTGCCACTCAGGGTCAGTCCTTAGACGGACTAGCA
       ** R*Y  *******RR**   RR     *******Y ** *R* *Y* ** R     Y  *R**********YY     * R   Y  R*YR*RY*R**R
  1192 CTGAGTTCCTTTGGAGACTTCCGAGGCACTAGGTGTGTCTTGCACCCATCAGCCTAATTCGGTCCTTGCCACCTTCCTACTAGGGCATAATAGGTTGGCG

  1253 GAAGTCAAGGTCCCAGTGAGACCCACTGGCTTGAGGGGCAGGGCAAGGGCAGA--GAGGGGACAGAACAGGATAAGGAGGGGCGGGGGAGGGAGGGCAGA
       *R** Y**RR              R*****YR**         *YR*****R***Y*RR   RR* **    R   R* *Y*R*R*R* R    R*RR * *Y**R
  1292 GGAGGTAAAA----------GCCCACCAGC-------GTGGGGCAGGGGTAAGAGTGAGCGAGCCGTAGGTACAGGAAAGAGGATCTTGGAATGTGTAGG

  1351 AGGAAGGTCTGCTTCAAGGGGAGGGTAGTGGGCAGCTGAGAGAAGGGTGGACACCAGGAGGCC-CCAGAGCTGAGGGGGAAATTCTCCCCAAATAGCGC
       R      R****  *   **R****  ***  *  ********   * ** *****R* RR   * Y*********R**         *******R*R Y*Y
  1375 GCC---ATCTGAATGTCGGAGAGGTAAGT---CT-CTGAGAGACTGCTGCACACCGGTGACACATCAGAGCTGAGGAGG------TCCCCCAAGTGTTGT

  1450 ATGAGGGTTCCCTTTGACAAAGCGGGAGGACGCTGAAGGGTGTGACTGACAGCTGTTCCTCAAACATCTGAAAACCGAGGGAGCGGAGAAGCTTTTTTTT
        *        Y***YY   * * R**R          *  *** R * **  * Y   *  *****        ***R***Y**     * ** ** **  YY *
  1462 CTCCCCCGCCCCCCGCCCCATACGA-CTCTGTCA-AAGCAGGAGAGGGTTTTGAGA-CCTCA------TGAGAACTGATCCTCCTGATAACCTAGCCGGT

  1550 GACTGCTC--------TCCTTCGACCAATTTTGGGCCCAAAGAGAGCCAGGGAGACCCGCCACCTCCCGCGCTCCTCCGGCCG--CCCTCGCCCGCCCGG
        *   Y**            Y****YR       Y** * **Y*R* ********R*         **            **Y*****       *     RY   R
  1553 TAGATTTCCACTCTCGCCCTTTACGGCTGCTTCGTCCTAGATAGAGCCAGAG--------CA-------------TCTGGCCGGTGAAGCTGGGATAGCA

                                                                                       CysLeuProCysGlyProGl
  1640 CTCAGCCCCCCGCCCCACAGGGTCTCCCTCCCCGGCC--GCTCCCCTCCCGCCCCCGGCTCATCCCTTCCCTCCCACCAGTGTCTCCCCTGCGGCCCCGG
        Y R* Y   ** Y    Y   R * ***YY*YY****  *****     *Y*  ****   *R***Y*Y***        *******************Y**
  1632 GCAGGGTGACCTTAGGTTCCCAACGCCCCTCTTGGCCTGGCTCC--AGCTGA-CCCG---CGTCCTTCCCCG-----CAGTGTCTCCCCTGCGGCCCTGG
                                                                                       CysLeuProCysGlyProGl

       yGlyLysGlyArgCysPheGlyProSerIleCysCysGlyAspGluLeuGlyCysPheValGlyThrAlaGluAlaLeuArgCysGlnGluGluAsnTyr
  1738 GGGCAAAGGCCGCTGCTTCGGGCCCAGCATCTGCTGCGGGGACGAGCTGGGCTGCTTCGTGGGCACGGCCGAGGCGCTGCGCTGCCAAGAGGAGAACTAC
        ******** ************** ************* ******************* ******* *******************R************
  1721 CGGCAAAGGGCGCTGCTTCGGGCCGAGCATCTGCTGCGCGGACGAGCTGGGCTGCTTCCTGGGCACCGCCGAGGCGCTGCGCTGCCAGGAGGAGAACTAC
       yGlyLysGlyArgCysPheGlyProSerIleCysCysAlaAspGluLeuGlyCysPheLeuGlyThrAlaGluAlaLeuArgCysGlnGluGluAsnTyr

       LeuProSerProCysGlnSerGlyGlnLysProCysGlySerGlyGlyArgCysAlaAlaAlaGlyIleCysCysAsnAspG
  1838 CTGCCGTCGCCCTGCCAGTCCGGCCAGAAGCCCTGCGGGAGCGGGGGCCGCTGCGCCGCCGCCGGCATCTGCTGCAACGATGGTGCGCGGC---CCGGGC
       ***** *************Y**********Y*****R*****R***********Y***** ************R***********R       **R***
  1821 CTGCCCTCGCCCTGCCAGTCTGGCCAGAAGCCTTGCGGAAGCGGAGGCCGCTGCGCTGCCGCGGGCATCTGCTGCAGCGATGGTGCGCACAAAGCCAGGC
       LeuProSerProCysGlnSerGlyGlnLysProCysGlySerGlyGlyArgCysAlaAlaAlaGlyIleCysCysSerAspG
```

Fig. 7. Sequence comparison of the vasopressin genes from calf (B) and rat (R) (for details, *see* Fig. 6). AVP, vasopressin; GP, glycopeptide.

```
1935 GG-TGGGGCGGGGGCGGGGAGGGGGGCGAG-----GAGGGGGCGGGGCCG------GGGCCGGGGCGGGGCGGGGGCCGGGTTGATCTGGGTCCGGGTCT
      ** Y *R**R *** RR *RR ****Y*R*     *R RRR* *****YR      **R*Y*R**  RR *R **R****R   R    **       R** *
1921 GGGCTGAGCATGGGGAATGGATGGGGTGGGTGGGAGGTAAAGGGGGGCTAAGTGGGGGACTGAGG--AATCA-GGACCGGAGATGGAGGGTGAGTAGTAT

2023 GGAGGGGGTGGGAAGCGGGGCCCGGTCCCCGAGACGCGCCCGCCAGCTGCGCGCTCAGCCCGTGCTCCCCG---------------------------
     *R******* *RR**Y **R  **    *  *R * *     *  RRY**Y* *    *    *Y**Y**Y
2018 GAAGGGGGTCGAGAGTTGGA-ACGTAGCAGGGTAGGATAAAGGGGATTGTGGG-GATG---GCGCCCCTATAGGTGCGCCCACCCCAGGACGCCTGACCT

                           luSerCysValThrGluProGluCysArgGluGlyValGlyPheGlyPheProArgArgValArgAlaAsnAspArgSerA
2094 ----------------CAGAGAGCTGCGTGACCGAGCCCGAGTGCCGGGAAGGTGTCGGCTTCGGCTTCCCCCGCCGCGTTCGCGCCAACGACCGGAGCA
                     *************R*************Y**R**R*** *Y  Y*Y* **Y** *     *****Y*** *      ** *R*****
2113 CACACAGCCCTTCCTTCAGAGAGCTGCGTGGCCGAGCCCGAGTGTCGAGAGGGTTTT--TTCC-GCCTCAC---CCGCGCTCG-G-----GAGCAGAGCA
                           luSerCysValAlaGluProGluCysArgGluGlyPhe  PheA rgLeuTh   rArgAlaAr g     GluGlnSerA

     snAlaThrLeuLeuAspGlyProSerGlyAlaLeuLeuLeuArgLeuValGlnLeuAlaGlyAlaProGluProAlaGluProAlaGlnProGlyValTy
2178 ACGCGACCCTGCTGGACGGGCCGAGCGGGGCCTTGTTGCTGCGGCTGGTGCAGCTGGCGGGGGCGCCGGAGCCCGCGGAGCCCGCCCAGCCCGGCGTCTA
     **** ** * ************RR * ***  Y**Y****  *******R******** ***R*R* R***Y***Y*** Y*Y*** ***** * *****
2201 ACGCCACGCAGCTGGACGGGCCAGCCCGGGAGCTGCTGCTTAGGCTGGTACAGCTGGCTGGGACACAAGAGTCCGTGGATTCTGCCAAGCCCCGGGTCTA
     snAlaThrGlnLeuAspGlyProAlaArgGluLeuLeuLeuArgLeuValGlnLeuAlaGlyThrGlnGluSerValAspSerAlaLysProArgValTy

     rEnd
2278 CTGAG-GCGCGCCCCCC---CCTCCCCACCCCTGCCCTCGCAGCACGAAAAATAAACGTTTTAAAGGCACTGCTAGTGTGCGTCTCTGCCTCTGGGGGTG
     *****    ********    *****    *****R      *****Y*R********* *******RR *
2301 CTGAGCCATCGCCCCCCACGCCTCC---CCCCTA------CAGCATGGAAAATAAAC-TTTTAAAAA-A
     rEnd

2374 GAGAGGGGAGACGAAGGGAGGGCGGGGGATCAGTTCCTGACCCGAAGCCGCGCGCAGATC
```

Fig. 7. (*cont.*)

quences at exon–intron junctions, a modified Goldberg-Hogness sequence, CATAAAT, located 29 nucleotides upstream from the presumptive transcription start site, an A residue bounded by a pyrimidine-rich region at the transcription start site, and a polyadenylation site AATAAA at the 3′ end of the sequence (*32*).

### 4.1. Implications of the Gene Structures

A comparison of the vasopressin and oxytocin gene structures highlights several interesting points. First of all, the homologies in sequence, organization, and intron–exon structure are particularly striking for the second exon encoding the conserved part of the two neurophysins (Figs. 3, 6, and 7).

Comparison of the two rat genes shows that within exon B a stretch of 143 nucleotides is with one base difference, entirely homologous. An even higher homology of 197 nucleotides was observed for the corresponding calf sequences; there the homology extends even into the preceding intron, covering an additional 135 bp without any interruption. The nearly perfect nucleotide homology observed between the two genes from rat and calf points to a specific mechanism that has maintained this conserved region. A recent gene-conversion event seems the most likely explanation for it (*8,10,19,31*).

The term gene conversion is generally used to denote a mechanism whereby two closely homologous but nonallelic sequences on the same or different chromosomes can become similar, possibly by means of an unequal crossing-over (*33*). A comparison of the protein and nucleotide sequences for the vasopressin- and oxytocin-associated neurophysins from various mammalian species supports these observations (*17*). From a count of the silent substitutions occurring in the rat and calf genes as the result of the mutation rate, it is concluded that gene conversion took place in the calf 1 million yr ago and in the rat 10 million yr ago (Table 3).

In the top line of Fig. 4 are shown the positions of amino acids that have varied during mammalian evolution in either the oxytocin- or vasopressin-associated neurophysin. Below this line, indicative of general evolutionary variability, are listed for each species studied a comparison of the oxytocin- and vasopressin-associated neurophysins.

Significant is the noted lack of evolution in the central region, and the similarity here between the two neurophysin types, which can probably be attributed to regular mutual gene conversion. This is most clear for the ungulates. In the case of rat and human neurophysins, one or two subsequent point mutations, respectively, would account for the new amino acid substitutions. The constant residues in the C-terminal variable regions are the two Cys and the Pro, evidently of significance for the functional

TABLE 3
Nucleotide Differences Between Rat and Bovine Vasopressin and Oxytocin Precursor

| | Coding region | | | |
|---|---|---|---|---|
| | Silent substitution sites | | Replacement sites | |
| | Calf | Rat | Calf | Rat |
| Signal peptide, Gly-Lys-Arg | 56 | 53.9 | 24.4 | 29.8 |
| Hormone | 54.5 | 54.5 | 9.3 | 14 |
| N-terminal part of neurophysin | 41.7 | 64.5 | 45.5 | 38.5 |
| C-terminal part of neurophysin | 57.4 | 57.4 | 44.9 | 32.7 |
| Central part of neurophysin | 2.1 | 12.6 | 1.9 | 2.6 |

[a]From ref. *31*.

configuration of the neurophysins. That function, however, still is not defined, though specific transport functions (*34*), and/or a protective role within the neurosecretory granules have been discussed (*35*).

### 4.2. Transcriptional Control

With DNA probes specific for either hormone precursor, it is now possible to study their regulation at the transcriptional level. Two methods have been adopted to quantify the mRNA encoding the vasopressin precursor; one is based on liquid or blot hybridization assays, the other makes use of thin-layer sections of the hypothalamus that are then hybridized *in situ* to specific labeled DNA probes.

In the following, some examples are given for elevated mRNA levels under different physiological conditions. Though elevated mRNA levels found in one hypothalamic area compared to others may reflect increased transcription rate, it is premature to conclude that a gene is specifically switched on. Differences in mRNA levels under given physiological conditions, for instance, may be caused simply by different RNase activities.

#### *4.2.1. Osmotic Stress and Vasopressin-Encoding mRNA Levels*

Rats placed under osmotic stress by drinking 2% saline respond with an elevated plasma osmolality. To maintain the water balance, the stores of vasopressin in the posterior pituitary are depleted while the hormone level in the plasma rises. When studying the vasopressin levels in different hypothalamic areas, increased hormone levels are found in the supraoptic nucleus (SON) and paraventricular nucleus (PVN), not in the suprachiasmatic nucleus (SCN) (*39A*).

These data are in line with the finding that rats exposed to saline show a significant increase in mRNA encoding vasopressin in the SON and PVN, but not in the SCN (*39B*). The increase is fivefold in the SON, and twofold in the PVN. It is known that vasopressin-producing neurons of the SON and PVN project toward the posterior pituitary; those of the SCN project toward other brain areas. Hence, it has been speculated that these data point to a different regulation in the expression of the vasopressin gene in SON and PVN versus SCN, relating to peripheral and central functions of vasopressin, respectively (*39B*).

Whether the different levels of vasopressin-encoding mRNA in the three hypothalamic areas assayed reflect indeed differential regulation of the vasopressin gene must be examined more rigorously, for instance by transcription run-off experiments measuring the levels of nascent vasopressin-mRNA in the cell nucleus. Also, other physiological parameters are needed that may affect specifically the vasopressin mRNA level in the SCN but not in the other two areas.

A preliminary estimate of the vasopressin mRNA level under osmotic stress conditions has been made by using total rat hypothalami and Northern blot analysis (*39C*). In these experiments, a 20-fold increase of the vasopressin-encoding mRNA is observed.

#### 4.2.2. *In Situ* Hybridization

This method combines a number of advantages; it allows not only the cells expressing the respective gene to be identified, but also the mRNA in these cells to be quantified, as well as a study of the effect of endogenously induced factors on the anatomy and number of cells expressing the respective gene. With this technique it has been shown that thin sections obtained from rats in which osmotic stress is induced by drinking 2% saline show a significant increase in the grain numbers specific for vasopressin-encoding mRNA, whereas the oxytocin producing cells appear to show no change (*39D*). By the same technique, vasopressin-specific mRNA has also been detected in the hypothalamus of Brattleboro rats. When oxytocin-mRNA levels are assayed in female rats and compared to 15-d pregnant rats, the grain numbers increased significantly in the pregnant rat, which is in line with the increased oxytocin level in the plasma and pituitary during pregnancy.

## 5. THE VASOPRESSIN GENE IN DIABETES INSIPIDUS (BRATTLEBORO) RATS

Studies of hereditary hypothalamic diabetes insipidus (DI) in mutant (Brattleboro) rats show that these animals lack the hormone vasopressin and its corresponding neurophysin, although oxytocin and its associated neurophysin remain intact (*36,37*). Resorption of water in the distal tubules of the kidney, controlled by vasopressin, is greatly impaired. Animals with this disease therefore show a high water uptake accompanied by excretion of an excessive amount of dilute urine. The defect can be reversed by the subcutaneous application of Arg vasopressin (*38*).

Production of oxytocin and its associated neurophysin is not affected by the mutation. The observed lower levels of oxytocin in the neurohypophysis are probably caused by osmotic stress, inasmuch as normal levels can be restored when the animals are treated with vasopressin (*39*).

Studies of Brattleboro rats revealed that the neurons of the supraoptic and paraventricular nuclei are hypertrophied (*36*). Those cells that are supposed to produce vasopressin are devoid of normal neurosecretory granules, and have, instead, smaller, dense-cored granules, suggesting that the machinery for packaging secretory material is nevertheless intact. Because small granules have also been observed in normal rats, it appears unlikely that they are a source for a mutated vasopressin precursor. A candidate for this role, however, could be the peptide X, an unglycosylated protein, isolated in small quantities from Brattleboro rats, that has neurophysin- and vasopressin-like immunoreactivity (*40*).

### 5.1. Gene Sequence and the Predicted Vasopressin Precursor

The gene for the vasopressin precursor from the mutant rats was recently isolated and its sequence determined (Fig. 8) (*41*). It contains a deletion of a single G residue in exon B, which encodes the conserved part of neurophysin. This mutation gives rise to an open reading frame, predicting a hormone precursor with a different C-terminus. Theoretically, translation of the mRNA could lead to a product with a poly-Lys tail at the C-terminus of this precursor.

The single nucleotide deletion affects the predictable structure of the vasopressin precursor: Not only is the glycosylation site absent, but the C-terminus of the neurophysin is drastically changed (Fig. 9). From amino acid residue 64 of the neurophysin up to position 95, all amino acid residues have been replaced except two (*41*). These changes are particularly significant from residue 64 to 76 encoded by exon B with its highly conserved nucleotide and amino acid sequence.

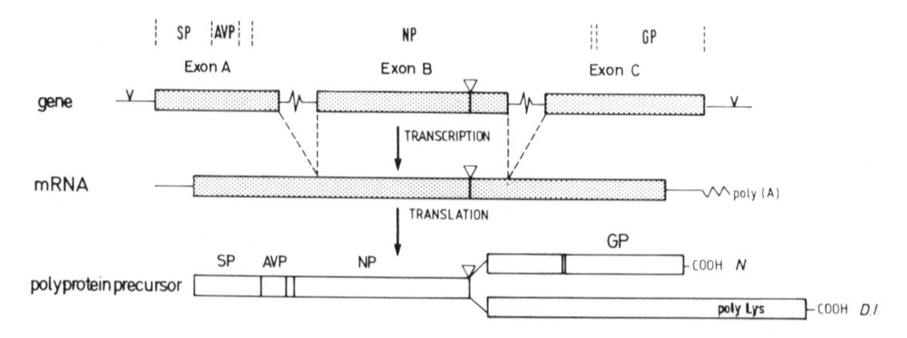


Fig. 8. Schematic representation of the vasopressin precursor from diabetes insipidus rats. The open arrow head points to the site of the deletion. Dotted bars indicate the coding, lines indicate the noncoding regions; open bars indicate the protein precursor; hatched bar shows the part of the altered sequence of the mutated vasopressin precursor. SP, signal peptide; AVP, Arg vasopressin; NP, neurophysin; GP, glycopeptide; N, normal rat; D.I., diabetes insipidus rat.

Five of fourteen Cys residues present originally in the neurophysin are replaced by other amino acids (Fig. 9); this replacement should have a severe impact on the folding of the mutated neurophysin (*35*).

The new reading frame of the mutated gene would also predict the replacement of a basic amino acid (Arg) that normally separates the neurophysin from the glycopeptide and is known to direct proteolytic cleavage. An alternative cleavage site could occur at position 110 of the mutated precursor; this would lead to a neurophysin-like protein extended for four amino acids.

### *5.2. Transcription and Translation*

A number of experiments have demonstrated that the mutated gene is correctly transcribed. The resulting mRNA also contained the single deletion (*41A*) and was otherwise identical with the mRNA from normal rats. That the two mRNAs were identically spliced could be confirmed when the vasopressin genes from normal and Brattleboro rats were injected into the nucleus of oocytes from *Xenopus laevis* or when 3T3 cells were transfected with either gene (*41A*). Also, S1 mapping of mRNAs from normal and Brattleboro rats gave identical results, thus excluding the possibility of alternative transcription start sites that could affect the secondary structure of the mRNA, and thus its translation efficiency.

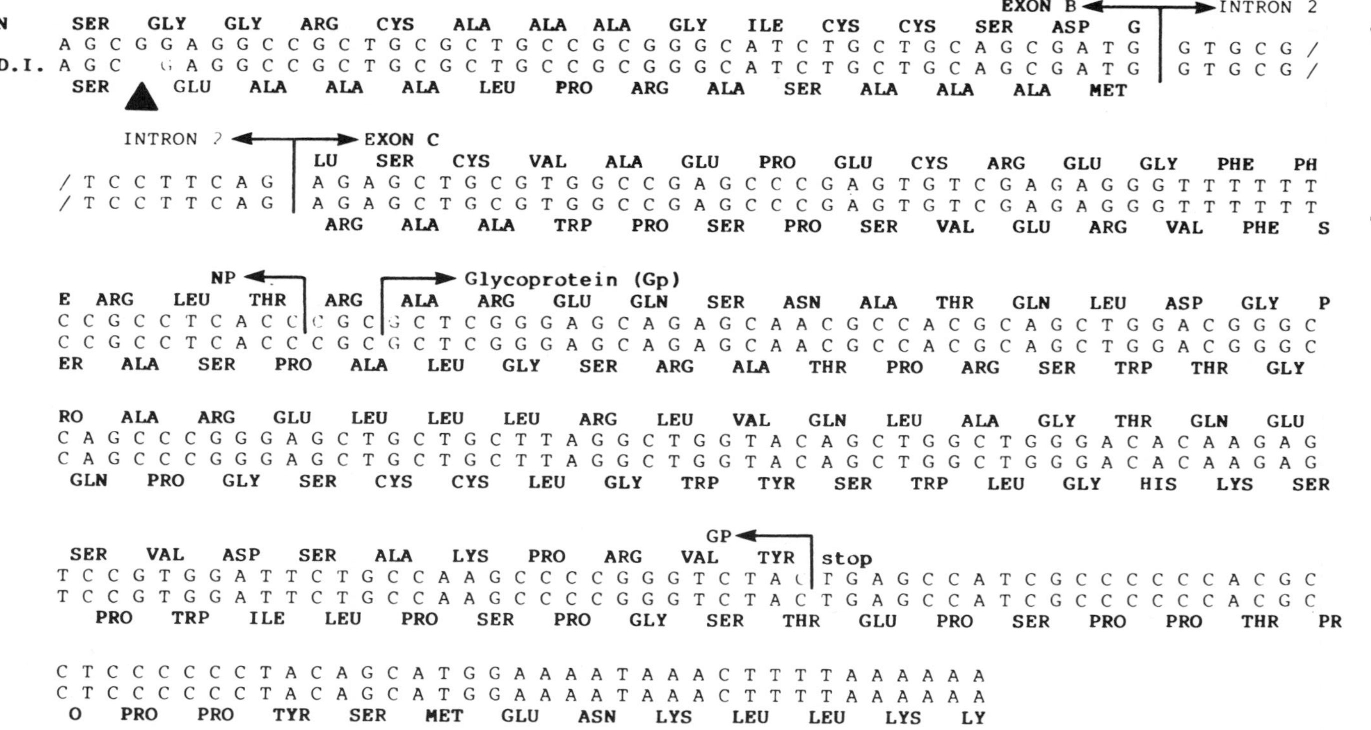


Fig. 9. Partial sequence of the vasopressin genes from normal and Brattleboro rats (downstream from the deletion site).

Comparison of Northern blots of hypothalamic poly(A)+ RNA from wild-type (Wistar) and Brattleboro rats suggested similar amounts of vasopressin and oxytocin mRNAs (*41*). Quantitative studies by dot blot analysis of hypothalamic poly(A)+ RNA from Long-Evans, Wistar, and Brattleboro rats indicated the following relative amounts of vasopressin-specific mRNA: Long-Evans, 100%; Wistar, 66%; and Brattleboro rats, 27% (*41C*). Based on the cloning experiments of hypothalamic poly(A)+ RNA from Brattleboro rats out of 5000 transformants, two contain the vasopressin precursor sequence. Taking this result into account, the vasopressin-specific mRNA amounts to 0.04% of the total poly(A)+ RNA in the Brattleboro, to 0.1% in the Wistar, and to 0.15% in the Long-Evans rat.

Recent reports claimed very low numbers of vasopressin-specific mRNA in normal (0.01% for Long-Evans rats) and Brattleboro rats (0.001%) (*41B*). From these studies it appears that saline-treated animals were used for calculation. In that case the numbers for vasopressin-specific mRNA in untreated Long-Evans rats will be even lower, because the same group reported that saline treatment increased the respective mRNA production to about 20-fold (*39C*). Taking this into consideration, only 0.0005% of the total poly(A)+ RNA should be vasopressin-specific in the untreated animal—a number that is in conflict with the cloning data outlined above. At present this discrepancy is not understood, though differences in the length of the probes used for evaluating the mRNA content may be one possible explanation.

Though vasopressin encoding mRNA can be isolated from hypothalami of Brattleboro rats, the question arose whether this mRNA is translatable. Since the deletion of a nucleotide residue in the second half of the mRNA sequence leads to a shift in the reading frame in a way that a stop codon signal is no longer read, the precise termination of the nascent peptide chain should not occur. In theory the ribosome should read through the 3′ end of the mRNA, including the poly(A) sequence, giving rise to a larger precursor product. If one calculates the maximal size of such a potential precursor, then roughly 70 amino acids have to be added to the normal-sized protein (molecular weight, 19,000) yielding a molecular weight between 26 and 28,000. The C-terminus of this precursor should consist of approximately 50 lysine residues corresponding to the 150 adenosines of the poly(A) tail.

Indeed, when Brattleboro hypothalamic mRNA was translated in a cell-free system, the normal-sized vasopressin precursor was not observed. Instead, small amounts of vasopressin-like precursors were identified with molecular weights in the range from 19,000 to 26,000 (*41C*). The heterogeneously sized translation products may reflect how far these ribosomes have been able to translate the mRNA. The products were identified by using antisera raised either against vasopressin, neurophysin, or a 14mer peptide predicted from the new reading frame and specific for the mutated Brattleboro vasopressin precursor (*41A*).

Using antibodies against the 14mer peptide (C-peptide antibodies) in immunocytochemical studies at the light microscopic level, it could be demonstrated that the enlarged cells of the supraoptic nucleus were specifically stained in Brattleboro, but not in wild-type, rat hypothalami. Typically the stain was located in the vicinity of the Nissl bodies (*42*).

Electron microscopic studies indicate that this stain is situated in the rough endoplasmic reticulum, in the inner cisternae of the Golgi apparatus, and in small lysosomes (*42A*). Since the mutated vasopressin precursor should be highly charged at its C-terminus (approximately 50–70 lysine residues!), it is not too surprising that most of the synthesized material is fixed at the membrane. Staining is restricted to the cell body and cannot be found in axons or in the pituitary. Very few cells are stained in the paraventricular nucleus, and none in the suprachiasmatic nucleus.

In summary, the Brattleboro rat represents the rare case in which a single nucleotide deletion leads to an open reading frame predicting a protein precursor that in theory should end in a poly lysine tail at its C-terminus. Another case with a frameshift mutation has been reported for hemoglobin Wayne, which is an alpha chain variant and has been explained by assuming a minus one frameshift. This leads to the extension of the hemoglobin chain by five amino acids to be followed by a stop codon (*41B*). In that example only a few nucleotide residues of the 3′ untranslated region and none of the poly(A) sequence are copied into protein.

Though the vasopressin gene is expressed in the hypothalamus of Brattleboro rats, relatively few cells in the supraoptic nucleus can be stained immunocytochemically. This could be caused by the modified precursor structure that, because of its highly charged C-terminus, might be folded in a way not easily accessible to the antibodies used. The reduced vasopressin precursor synthesis, however, may also be limited just because of the poly(A) sequence exhausting the supply of lysine, its tRNA, or the corresponding aa-tRNA synthetase.

Alternatively, translation of mRNA and packaging of the synthesized product into neurosecretory granules may be a more coordinated and feedback-controlled process than generally expected. The electron microscopic studies so far show that most of the mutated vasopressin precursor is found in the rough endoplasmic reticulum, in terminal Golgi vesicles, and in lysosomes, indicating that the "false" precursor is recognized and discarded in an unknown way by the cell. Is the modified C-peptide the wrong tag that does not allow its correct packaging in the neurosecretory granules? Is the formation of the correct precursor and neurosecretory granules a closely linked process that is halted when certain signals within the protein structure are absent? Or is the mutated precursor simply discarded because of its abnormal polylysine structure at its C-terminus? No doubt there are many questions still left to be answered before the genetic defect in diabetes insipidus rats is completely understood.

## 6. EXPRESSION OF THE OXYTOCIN GENE IN PERIPHERAL ORGANS

Recent reports have implicated the presence of vasopressin and oxytocin also in nonneural peripheral tissues such as the corpus luteum (*43,44*). Because Southern DNA analysis for oxytocin always gives only a single positively hybridizing DNA fragment, in the calf, oxytocin synthesis must result from the activity of only a single oxytocin gene.

Use of specific DNA probes encoding various parts of the oxytocin precursor showed that the oxytocin gene is not only transcribed in the bovine corpus luteum but that this transcription is phasic (*45*) and highest at the beginning of the cycle (*45A*). Our data also show that the resulting mRNA for the oxytocin precursor is similar to that of hypothalamic origin, suggesting that the luteal mRNA coding for the oxytocin precursor is not a product of an alternative splicing process (*46*). When the relative amounts per organ are calculated, the active corpus luteum altogether produces some 250 times more oxytocin mRNA than does a single hypothalamus. These data correspond well with the levels reported for oxytocin by radioimmunoassay and HPLC (*43,44*). The rather high level of oxytocin synthesis by the corpus luteum calls into question the relative importance of hypothalamic oxytocin production in peripheral reproductive functions dependent upon circulating oxytocin.

Translation studies in vitro also showed that preprooxytocin was produced just as if mRNA from the hypothalamus was used, except that in the corpus luteum the poly(A) tail of the oxytocin mRNA was shorter. Proopiomelanocortin mRNA of a different size was also found (*47*), which may also relate to the finding of alternative polyadenylation signals (*48*), as in the rat oxytocin gene (*19*). It would support the view that the 3′-untranslated region may be important for the regulation of gene expression (*47*).

Using specific vasopressin DNA probes, we found the expression of the vasopressin gene in the bovine corpus luteum to be three orders of magnitude less than that of the oxytocin gene (*41C*). No significant preprovasopressin synthesis was observed in the cell-free system using luteal mRNA (*45*).

## 7. POLYPROTEIN STRUCTURES AND THEIR IMPLICATIONS

Preprovasopressin and preprooxytocin are typical representatives of a class of precursors referred to as cellular polyproteins—a definition initially introduced for viral polyproteins. They are composed of several distinct entities whereby the structural organization of those precursors may vary (Fig. 10).

For instance, a simple form of a polyprotein can be illustrated by the precursors to the releasing factors such as corticotropin-releasing factor (CRF) (*16*), luteinizing hormone releasing factor (LHRH) (*48A*), or thyrotropin releasing factor (TRH) (*48B*). There a single peptide, the CRF LHRH is present within a precursor. The remaining sequence appears to have no other function than to serve as a spacer in the synthesizing and packaging process. The spacer, however, may be of physiological relevance, as indicated by the somatostatin spacer peptide, which is not quickly hy-

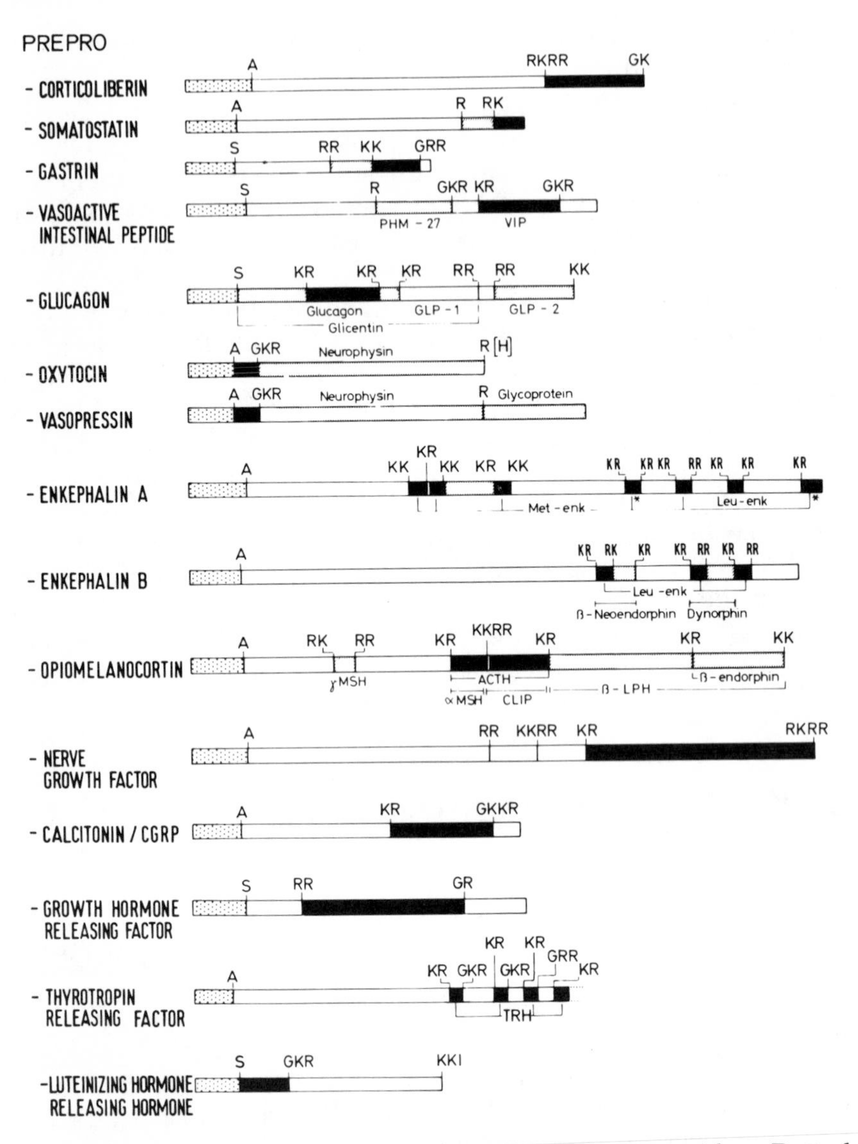


Fig. 10. Polyproteins as precursors to neuropeptides. Dotted bars, signal sequence; open bars, "spacer" sequences (bars not drawn to scale); black bars, the physiologically active hormones or neuropeptides; hatched bars, auxiliary sequences known to modify the physiology of the associated peptide hormone; ACTH, adrenocorticotropin; CGRP, calcitonin gene-related peptide; CLIP, corticotropin-like intermediate-lobe peptide; GLP-1, GLP-2, glucagon-like peptide 1,2; β-LPH, β-lipo-

drolyzed but remains *in situ*, possibly with a specific function (*49*).

More-complex polyproteins are exemplified by the somatostatin (*50,51*) and gastrin (*52,53*) precursors. In these cases, although only single hormones are included, the physiological peptides can be excised from the precursor in more than one way to give shorter or longer peptides with different functions.

Another category of polyproteins has been described for the precursors to the two enkephalins (*54,55*), cerulein (*56*), thyrotropin-releasing hormone (TRH) (*50B*), and possibly also glucagon (*57*) and vasoactive intestinal peptide (*58*). Within one precursor molecule, the physiological hormones occur in several copies, sometimes differing slightly from one another and probably resulting from gene duplications. This may have the effect of amplifying the transcriptional control.

It is interesting to note that in most cases in which the functional peptide is located at the C-terminus of the precursor it is extended by one or more basic amino acids that are absent in the biologically active peptide (e.g., CRF, NGF precursor). An extra basic amino acid is also found at the C-terminus of the neurophysin, which is part of the oxytocin precursor. These basic amino acids apparently have to be removed during posttranslational processing and most likely require specific cleavage enzyme(s). One could speculate that these amino acids are remnants of former C-terminal extensions to these precursors, which had been eliminated during evolution because of lack of function.

The class of precursors represented by the calcitonin (*46*) or substance P (*58A*) polyproteins are particularly remarkable since their genes give rise to pre-RNAs from which alternatively spliced mRNAs are generated in a tissue-specific way. For instance, translation of the mRNA in the thyroid yields the precursor to calcitonin, but in the brain, to the calcitonin gene-related protein (CGRP). It has been suggested that the two mRNAs of the

---

Fig. 10. (*cont.*) tropin; PHM-27, a 27-amino-acid peptide with an N-terminal histidine and a C-terminal methionine; Met*, C-terminally extended methionine enkephalin; MSH, melanotropin-stimulating hormone; TRH, thyrotropin-releasing hormone, VIP, vasoactive intestinal peptide; amino acids occurring as processing signals or as extension of peptides are listed as follows: A, alanine; G, glycine; H, histidine; K, lysine; I, isoleucine; R, arginine; S, serine (for futher details, *see* text).

calcitonin/CGRP gene result from a tissue-specific utilization of alternative polyadenylation sites regulated by a selective endonuclease activity, though the underlying mechanism remains to be resolved (*58B*).

In the case of substance P, a single preprotachykinin gene gives rise to two distinct mRNAs resulting from alternative splicing; one mRNA encodes substance P alone ($\alpha$-PPT), the other encodes substance P plus another neuropeptide, the tachykinin substance K ($\beta$-PPT). The relative amounts of the two mRNAs vary; $\alpha$-PPT mRNA is predominantly made in the nervous system, whereas $\beta$-PPT mRNA is found preferentially in the thyroid, intestines, and also in the nervous system (*58A*). The different mRNAs of the preprotachykinin gene may be a result of tissue-specific regulation of the splicing mechanism, e.g., by alteration in the specificity of the splicing enzymes (*58A*). Clearly this tissue-specific expression of neuropeptide genes provides a further way to increase the diversity of the neuroendocrine system.

Other complex precursors are the multifunctional polyproteins, such as proopiomelanocortin (*54,55*), provasopressin, or prooxytocin with more than one discrete functional entity. This means that different peptides with somewhat related functions are placed under the same transcriptional and translational control. Some of these peptides may mediate more than one physiological function (e.g., neurotransmitter, hormone, local regulator, and so on). Vasopressin, for instance, is known to control water resorption (*36*) and to help in ACTH release (*59*), and is considered to be involved in memory and learning processes (*60*).

Other polyproteins, such as proopiomelanocortin, provide an added sophistication, inasmuch as different hormones overlap within the precursor and are differentially expressed in different tissues (*55*). Clearly, the processing of the various polyproteins into the mature active peptides is an intriguing problem. It will be extremely interesting to study the folding and three-dimensional structure of the nascent polyproteins that together with specific amino acids as processing signals, are most probably integral parts in the proteolytic processing events.

Another feature in the neuropeptide field that has excited attention in recent years is the distribution of the peptides in the brain. Many are now found in other peripheral organs, such as the gastrointestinal tract. The expression of oxytocin in the corpus luteum is one of those examples, though its function there remains to be established. On the other hand, peptides thought to be restricted to peripheral organs turn up in the central nervous system, suggesting a wide diversification of peptide functions. Moreover, new neuropeptides could be predicted from DNA cloning experiments (*61*). Again, determination of neuropeptide functions will be a major task in the forthcoming years.

## REFERENCES


1. Brownstein M. J., Russell J. T., and Gainer H. (1980) Synthesis, transport, and release of posterior pituitary hormones. *Science* **207,** 373–378.
2. Sachs H. (1963) Vasopressin biosynthesis. II. Incorporation of [$^{35}$S]cysteine into vasopressin and protein associated with cell fractions. *J. Neurochem.* **10,** 299–311.
3. Richter D. (1983) *Trends Biochem.* **8,** 278–281.
4. Richter D. (1983) Synthesis, Processing, and Gene Structure of Vasopressin and Oxytocin, in *Progress in Nucleic Acid Research and Molecular Biology* (Moldave K. and Cohen W., eds.), Academic, New York.
5. Schmale H. and Richter D. (1981) Immunological identification of a common precursor to arginine vasopressin and neurophysin II synthesized by *in vitro* translation of bovine hypothalamic mRNA. *Proc. Natl. Acad. Sci. USA* **78,** 766–769.
6. Schmale H. and Richter D. (1981) *Neuropeptides* **2,** 47–52.
7. Land H., Schutz G., Schmale H., and Richter D. (1982) Nucleotide sequence of cloned cDNA encoding bovine arginine vasopressin–neurophysin II precursor. *Nature* **295,** 299–303.
8. Land H., Grez M., Ruppert S., Schmale H., Rehbein M., Richter D., and Schutz G. (1983) Deduced amino acid sequence from the bovine oxytocin–neurophysin I precursor cDNA. *Nature* **302,** 342–344.
9. Acher R. (1983) Principles of evolution: The neural hierarchy model, in *Brain Peptides* (Krieger D. T., Brownstein N. J., and Martin J. B., eds.), Wiley, New York.

9A. Mohr E., Hiller M., Ivell R., Haulica I. D., and Richter D. (1985) Expression of the vasopressin and oxytocin genes in human hypothalami. *FEBS Lett.* **193,** 12–16.

10. Schmale H. and Richter D. (1983) Processing Signals of the Vasopressin and Oxytocin Precursor, in *Biochemical and Clinical As-*

*pects of Neuropeptides,* (Koch G. and Richter D., eds.), Academic, New York.
11. Schmale H., Heinsohn S., and Richter D. (1983) Structural organization of the rat gene for the arginine vasopressin-neurophysin precursor. *EMBO J.* **2,** 637.
12. Kozak M. (1981) Possible role of flanking nucleotides in recognition of the AUG initiator codon by eukaryotic ribosomes. *Nucleic Acids Res.* **9,** 5233–5252.
13. Bradbury A. F., Finnie M. D. A., and Smyth D. G. (1982) Mechanism of C-terminal amide formation by pituitary enzymes. *Nature* **298,** 686–688.
14. Eipper B. A., Mains R. E., and Glembotski C. C. (1983) Identification in pituitary tissue of a peptide α-amidation activity that acts on glycine-extended peptides and requires molecular oxygen, copper, and ascorbic acid. *Proc. Natl. Acad. Sci. USA* **80,** 5144–5148.
15. Kreil G. (1981) Transfer of proteins across membranes. *Ann. Rev. Biochem.* **50,** 317–348.
16. Furutani Y., Morimoto Y., Shibahara S., Noda M., Takahashi H., Hirose T., Asai M., Inayama S., Hayashida H., Miyata T., and Numa S. (1983) Cloning and sequence analysis of cDNA for ovine corticotropin-releasing factor precursor. *Nature* **301,** 537–540.
17. Richter D. and Ivell R. (1985) Gene Organization, Biosynthesis, and Chemistry of Neurohypophyseal Hormones, in *The Pituitary Gland* (Imura H., ed.) Raven, New York, 127–148.
18. Ivell R. and Richter D. (1983) Partial Sequence of the Rat Oxytocin Gene, in *Biochemical and Clinical Aspects of Neuropeptides* (Koch G. and Richter D., eds.), Academic, New York.
19. Ivell R. and Richter D. (1984) Structure and comparison of the oxytocin and vasopressin genes from rat. *Proc. Natl. Acad. Sci. USA* **81,** 2006–2010.
20. Holwerda D. A. (1972) A glycopeptide from the posterior lobe of pig pituitaries. 2. Primary structure. *Eur. J. Biochem.* **28,** 340–346.
21. Smyth D. G. and Massey D. E. (1979) A new glycopeptide in pig, ox and sheep pituitary. *Biochem. Biophys. Res. Commun.* **87,** 1006–1010.
22. Seidah N. G., Benjannet S., Routhier R., DeSerres G., Rochemont J., Lis M., and Chretien M. (1980) Purification and characterization of the N-terminal fragment of pro-opiomelanocortin from human pituitaries: Homology to the bovine sequence. *Biochem. Biophys. Res. Comm. 95,* 1417–1424.
23. Watson S. J., Seidah N. G., and Chretien M. (1982) The carboxy

terminus of the precursor to vasopressin and neurophysin: Immunocytochemistry in rat brain. *Science,* **217,** 853–855.

24. Sachs H. and Takabatake Y. (1964) Evidence for a precursor in vasopressin biosynthesis. *Endocrinology* **75,** 943–948.
25. Gainer H., Sarne Y., and Brownstein M. (1977) Biosynthesis and axonal transport of rat neuro-hypophysial proteins and peptides. *J. Cell. Biol.* **73,** 366–381.
26. Swann R. W., Gonzalez C. B., Birkett S. D., and Pickering B. T. (1982) Precursors in the biosynthesis of vasopressin and oxytocin in the rat: Characteristics of all the components in high-performance liquid chromatography. *Biochem. J.* **208,** 339–349.
27. Masse M. J. O., Desbois-Perichon P., and Cohen P. (1982) Identification of neurophysin-related proteins in bovine neurosecretory granules. *Eur. J. Biochem.* **127,** 609–617.
28. North W. G., Mitchell T. I., and North G. M. (1983) Characteristics of a precursor to vasopressin-associated bovine neurophysin. *FEBS. Lett.* **152,** 29–34.
29. Robinson I. C. and Jones P. M. (1983) An intermediate in the biosynthesis of vasopressin and neurophysin in the guinea pig posterior pituitary. *Neurosci. Lett.* **39(3),** 273.
30. Rapoport T. A., Thiele B. J., Prehn S, Marbaix G., Cleuter Y., Hubert E., and Huez G. (1978) Synthesis of carp proinsulin in *Xenopus* oocytes. *Eur. J. Biochem.* **87,** 229–233.
31. Ruppert S., Scherer G., and Schutz G. (1984) Recent gene conversion involving bovine vasopressin and oxytocin precursor genes suggested by nucleotide sequence. *Nature* **308,** 554–557.
32. Breathnach R. and Chambon P. (1981) Organization and expression of eucaryotic split genes encoding for proteins. *Ann. Rev. Biochem.* **50,** 349–388.
33. Baltimore D. (1981) Somatic mutation gains its place among the generators of diversity. *Cell* **26,** 295–296.
34. Glasel J. A., McKelvy J. F., Hruby V. J., and Spatola A. F. (1976) Binding studies of polypeptide hormones to bovine neurophysins. *J. Biol. Chem.* **251,** 2929–2937.
35. Breslow E. (1979) Chemistry and biology of the neurophysins. *Ann. Rev. Biochem.* **48,** 251–274.
36. Valtin H., Stewart J., and Sokol H. W. (1974) Genetic control of the production of posterior pituitary principles. *Handbk. Physiol.* Section 7, **IV,** 131–171.
37. Sokol H. W. and Valtin H., eds. (1982) The Brattleboro rat. *Ann. N.Y. Acad. Sci.* **394,** 1–820.
38. Cheng S. W. T., North W. G., and Gellai M. (1982) Replacement

therapy with arginine vasopressin in homozygous Brattleboro rats. *Ann. N.Y. Acad. Sci.* **394,** 473–480.

39. Valtin H. and Schroeder H. A. (1964) Familial hypothalamic diabetes insipidus in rats (Brattleboro strain) *Amer. J. Physiol.* **206,** 425–430.

39A. Zerbe R. L. and Palkovits M. (1984) Changes in AVP content of discrete brain regions in response to stimuli for AVP secretion. *Neuroendrinology* **38,** 285–289.

39B. Burbach J. P. H., De Hoop M. J., Schmale H., Richter D., De Kloet E. R., Ten Haaf J. A., and De Wied D. J. (1984) Differential responses to osmotic stress of vasopressin-neurophysin mRNA in hypothalamic nuclei. *Neuroendocrinology* **39,** 582–584.

39C. Majzoub J. A., Rich A., v.Boom J., and Habener J. F. (1983) *J. Biol. Chem.* **258,** 14061–14064.

39D. McCabe J. T., Morrell J. I., Ivell R., Schmale H., Richter D., and Pfaff D. W. (1986) In situ hybridization technique to localize rRNA and mRNA in mammalian neurons. *J. Histochem. Cytochem.* **34,** 45–50.

40. Russell J. T., Brownstein M. J., and Gainer H. (1980) Biosynthesis of vasopressin, oxytocin, and neurophysins: Isolation and characterization of two common precursors (propressophysin and prooxyphysin). *Endocrinology* **107,** 1880–1891.

41. Schmale H. and Richter D. (1984) Single base deletion in the vasopressin gene is the cause of diabetes insipidus in Brattleboro rats. *Nature* **308,** 705–709.

41A. Schmale H., Ivell R., Breindl M., Darmer D., and Richter D. (1984) The mutant vasopressin gene from diabetes insipidus (Brattleboro) rats is transcribed but the message is not efficiently translated. *EMBO J.* **3,** 3289–3293.

41B. Majzoub J. A., Pappey A., Burg R., and Habener J. (1984) *Proc. Natl. Acad. Sci. USA* **81,** 5296–5299.

41C. Ivell R., Schmale H., Krisch B., Nahke P., and Richter D. (1986) Expression of a mutant vasopressin gene, in press.

42. Richter D., Ivell R., Schmale H., Nahke P., and Krisch B. (1985) Synthesis of Neurohypophyseal Hormones, in *Selected Topics of Neurobiochemistry,* 36th Mosbacher Kolloquium (B. Hamprecht and V. Neuhoff, eds.), Springer Verlag Berlin, Heidelberg, in the press.

42A. Krisch B., Nahke P., and Richter D. (1986) Immunocytochemical staining of supraoptic neurons from homozygous Brattleboro rats by use of antibodies against two domains of the mutated vasopressin precursor. *Cell Tiss. Res.,* in press.

43. Wathes D. C., Swann R. W., Birkett S. D., Porter D. G., and Pick-

ering B. T. (1983) Characterization of oxytocin, vasopressin, and neurophysin from the bovine corpus luteum. *Endocrinology* **113,** 693–698.

44. Flint A. P. F. and Sheldrick E. L. (1982) Ovarian secretion of oxytocin is stimulated by prostaglandin. *Nature* **297,** 587–588.
45. Ivell R. and Richter D. (1984) The gene for the hypothalamic peptide hormone oxytocin is highly expressed in the bovine corpus luteum: biosynthesis, structure and sequence analysis. *EMBO J.* **3,** 2351–2354.

45A. Ivell R., Brackett K. H., Fields M. J., and Richter D. (1985) Ovulation triggers oxytocin gene expression in the bovine ovary. *FEBS Lett.* **190,** 263–267.

46. Amara S. G., Jonas V., Rosenfeld M. G., Ong E. S., and Evans R. M. (1982) Alternative RNA processing in calcitonin gene expression generates mRNAs encoding different polypeptide products. *Nature* **298,** 240–244.
47. Civelli O., Oates E., Rosen H., Martens G., Comb M., Douglass J., and Herbert E. (1983) Expression of Opioid Genes, in *Biochemical and Clinical Aspects of Neuropeptides* (Koch G. and Richter D., eds.) Academic, New York. 45–57.
48. Heidecker G. and Messing J. (1983) Sequence analysis of zein cDNAs obtained by an efficient mRNA cloning method. *Nucleic Acids Res.* **11,** 4891–4906.

48A. Seeburg P. H. and Adelmann J. P. (1984) *Nature* **311,** 666–668.

48B. Richter K., Kawashima E., Egger R., and Kreil G. (1984) Biosynthesis of thyrotropin-releasing hormone in the skin of *Xenopus laevis:* Partial sequence of the precursor deduced from cloned cDNA. *EMBO J.* **3,** 617–621.

49. Lechan R. M., Goodman R. H., Rosenblatt M., Reichlin S., and Habener J. F. (1983) Prosomatostatin-specific antigen in rat brain: Localization by immunocytochemical staining with an antiserum to a synthetic sequence of preprosomatostatin. *Proc. Natl. Acad. Sci. USA* **80,** 2780–2784.
50. Rutter W. J., Scott J., Selby M., Crawford R. J., Shen L. U., Hobart P., Sanchez-Pescador R., and Bell G. I. (1983) Structure of Precusors Derived From the Sequences of Cloned cDNAs, in *Biochemical and Clinical Aspects of Neuropeptides: Synthesis,* (Koch G. and Richter D., eds.), Academic, New York.
51. Taylor W. L., Collier K. J., Deschenes R. J., Weith H. L., and Dixon J. E. (1981) Sequence analysis of the cDNA coding for a pancreatic precursor to somatostatin. *Proc. Natl. Acad. Sci. USA* **78,** 6694–6698.
52. Boel E., Vuust J., Norris F., Norris K., Wind, A., Rehfeld J. F.,

and Marcker K. A. (1983) Molecular cloning of human gastrin cDNA: Evidence for evolution of gastrin by gene duplication. *Proc. Natl. Acad. Sci. USA* **80,** 2866–2869.

53. Yoo O. J., Powell C. T., and Agarwal K. L. (1982) Molecular cloning and nucleotide sequence of full-length cDNA coding for porcine gastrin. *Proc. Natl. Acad. Sci. USA* **79,** 1049–1053.
54. Numa S. and Nakanishi S. (1981) Corticotropin-β-lipotropin precursor—a multihormone precursor—and its gene. *Trends Biochem. Sci.* **6,** 274–277.
55. Herbert E. (1981) Discovery of proopiomelanocortin: A cellular polyprotein. *Trends Biochem. Sci.* **6,** 184–188.
56. Hoffmann W., Richter K., Hutticher A., and Kreil G. (1983) Biosynthesis of Peptides in Amphibian Skin, in *Biochemical and Clinical Aspects of Neuropeptides* (Koch G. and Richter D., eds.), Academic, New York.
57. Bell G. I., Santerre R. F., and Mullenbach G. T. (1983) Hamster preproglucagon contains the sequence of glucagon and two related peptides. *Nature* **302,** 716–718.
58. Itoh N., Obata K., Yanaihara N., and Okamoto H. (1983) Human preprovasoactive intestinal polypeptide contains a novel PHI-27-like peptide, PHM-27. *Nature* **304,** 547–549.

58A. Nawa H., Kotani H., and Nakanishi S. (1984) *Nature* **312,** 729–734.

58B. Evans R. M., Amara S. G., and Rosenfeld M. G. (1982) *DNA* **1,** 323–328.

59. Schally A. V., Coy D. H., and Meyers C. A. (1978) Hypothalamic regulatory hormones. *Ann. Rev. Biochem.* **47,** 89–128.
60. Burbach J. P. H., Kovacs G. L., Wang X-C., and de Wied D. (1983) Metabolites of Arginine—Vasopressin and Oxytocin Are Highly Potent Neuropeptides in the Brain, in *Biochemical and Clinical Aspects of Neuropeptides* (Koch G. and Richter D., eds.), Academic, New York.
61. Sutcliffe J. G., Milner R. J., Schinnick T. M., and Bloom F. E. (1983) Identifying the protein products of brain-specific genes with antibodies to chemically synthesized peptides. *Cell* **33,** 671–682.

# Chapter 9

# The Proopiomelanocortin Genes

MICHAEL UHLER AND EDWARD HERBERT

## 1. INTRODUCTION

Proopiomelanocortin (POMC) is the name given to the protein precursor for both β-endorphin and adrenocorticotropin (ACTH). Although the proof of a common precursor for these important peptides was provided by chemical evidence (*1–3*), the elucidation of the exact structure of POMC and the prediction of other biologically important peptides in the same precursor was one of the first major demonstrations of the power of recombinant-DNA techniques (*4*). Because POMC is a member of a growing class of proteins (*5*) whose expression is regulated in a tissue-specific manner, it may serve as a model for studying the expression of other neuropeptide precursors. For these reasons, this chapter will review the determination of the structure of POMC and the studies concerning the regulation of expression of POMC.

The advantage of considering POMC a member of the growing class of polyproteins is that the striking similarity in structure in these proteins implies a similarity in the mechanism of processing. Thus, elucidation of this mechanism is facilitated by bringing together relevant findings for all members of the class. The following characteristics have been ascribed to polyproteins: (1) They are generally large (greater than 20 kdaltons) secreted proteins; (2) Within their amino acid sequence, they code for two or more peptides with specific biological activities; (3) These peptides are not biologically active in the precursor form, but are active only when they are properly proteolytically processed from the precursor; (4) Typically, processing involves proteolytic cleavage of pairs of basic amino acid residues that flank the biologically active peptide sequences; and (5) The polyprotein precursor may be processed differently in different tissues.

All of these characteristics are exemplified by POMC (*6*) and are illustrated in Fig. 1. Mouse POMC contains 235 amino acids, and has an approximate molecular mass of 28 kdaltons (*7*). The 26 amino acids at the amino-terminus of the precursor comprise the signal sequence, which is removed by a specific protease in the endoplasmic reticulum as the nascent chain penetrates the membrane during biosynthesis. In corticotrophs of the anterior lobe of the pituitary, POMC is proteolytically cleaved into three fragments: ACTH, β-lipotropin (β-LPH), and pro-γ-melanocyte-stimulating hormone (pro-γ-MSH). ACTH stimulates steroidogenesis and release of glucocorticoids from the adrenal cortex, β-LPH weakly stimulates lipolysis in adipocytes, and pro-γ-MSH has a potentiating effect on ACTH stimulation of steroidogenesis

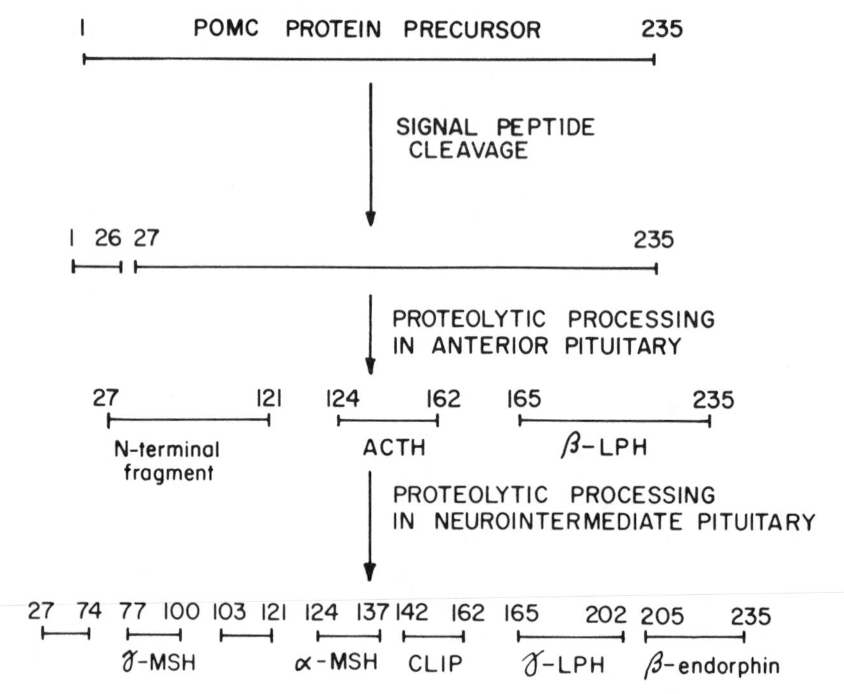


Fig. 1. Structure of the mouse POMC protein precursor and the proteolytic processing events that generate its component peptides. The numbers above the ends of a particular peptide indicate its amino acid position in the POMC precursor.

in the adrenal cortex. In the intermediate lobe of the pituitary, each of these three products is further cleaved at pairs of basic amino acid residues to yield products with entirely different biological activities (*8*). ACTH is processed to α-MSH and corticotropin-like intermediate-lobe peptide (CLIP), β-LPH is processed to β-endorphin and γ-LPH, and pro-γ-MSH is also proteolytically processed. α-MSH controls pigment production by melanocytes in lower vertebrates and plays a role in memory in mammals. β-Endorphin is a potent opiate peptide and probably serves a neuromodulatory role in synaptic transmission. However, the physiological roles of β-MSH, γ-MSH, and the other peptide products are not known. In addition to the proteolytic processing of the POMC precursor, it is also glycosylated (*9*), amidated (*10*), and phosphorylated (*11*).

The anterior and intermediate pituitary, the major sites of POMC synthesis and secretion, show tissue-specific regulation of POMC gene expression. In the anterior pituitary, POMC synthesis is inhibited by glucocorticoids (*12,13*), but stimulated by corticotropin-releasing factor (CRF) (*14*) from the hypothalamus. In the intermediate pituitary, POMC synthesis is not affected by either of these regulators, but is inhibited by dopamine (*15*). POMC is also synthesized in small amounts, in the amygdala (*16*), hypothalamus (*17*) and other areas of the brain, placenta (*18*), gut (*19*), adrenals (*20*), and some tumors, most notably oat-cell carcinoma of the lung (*21*).

## 2. CLONING OF POMC COMPLEMENTARY DNA

Complementary DNA (cDNA)—that is, DNA molecules complementary to messenger RNA (mRNA)—is generally considered the starting material for molecular biological studies since enrichment of a particular mRNA may be achieved by using an appropriate tissue source or isolation technique. The first published POMC cDNA sequence was that reported by Nakanishi et al. (*4*). POMC mRNA was purified to homogeneity by first isolating membrane-bound polysomal RNA from bovine intermediate pituitary, followed by oligo(dT)-cellulose chromatography to remove ribosomal RNA, and, finally, size-fractionation of the resulting poly-(A)-containing RNA on sucrose gradients (*22*). Using this highly purified template, cDNA was synthesized with reverse transcriptase, employing oligo(dT) as primer for the first strand reaction (*23*). The double-stranded cDNA was then treated with S1 nuclease to remove the hairpin, and the high molecular weight cDNA was isolated by electrophoresis through agarose and electroelution. Homopolymeric C tails were added to the cDNA using terminal transferase, and G tails were added to the *Pst*I-cleaved pBR322 cDNA. The cDNA and plasmid were annealed and used to transform *E. coli*. Tetracycline-resistant transformants were then screened by *in situ* hybridization for POMC cDNA inserts. One of the plasmids obtained, pSNAC20, contained 1091 base pairs (bp) of cDNA complementary to bovine POMC mRNA and contained all of the protein-coding sequence (*4*). From the predicted amino acid sequence, two major POMC structural features were discovered. First, all of the functional peptides were flanked by pairs of basic amino acid residues, suggesting that they served as processing signals for a trypsin-like enzyme. Second, a third MSH sequence, termed γ-MSH, which had been predicted, occurred in the previously "cryptic" amino-terminal region. These two discoveries showed the profound impact that recombinant-DNA techniques were to have on molecular neurobiology.

Shortly after publication of the bovine POMC cDNA sequence, a partial mouse cDNA sequence was reported (*24*); mRNA was isolated from membrane-bound polysomes, using oligo(dT)-cellulose chromatography, along with sucrose density sedimentation, to enrich the POMC mRNA. However, in this case the tissue source was a mouse anterior pituitary cell line in which POMC mRNA represents 3% of the total poly(A) mRNA. This amount is about 15-fold lower than the amount in the bovine intermediate pituitary, in which POMC mRNA represents 50% of the poly(A)-containing mRNA. Thus, Roberts (*24*) used a further enrichment step, namely, restriction of the double-stranded cDNA with *Hae*III restriction endonuclease to generate discrete fragments. Enrichment is obtained because a heterogeneous mixture of cDNA molecules of sufficiently large size will all give a single restriction fragment of a common size, distinct from other fragments produced from cDNA coding for other proteins. A single 140-bp *Hae*III fragment was found in the cDNA preparations copied from RNA samples enriched for POMC mRNA translational activity by sucrose-gradient sedimentation. This fragment was cloned by the addition of synthetic *Hin*dIII linkers, cleavage with *Hin*dIII, and insertion into the *Hin*dIII site of pBR322. Sequencing showed that this fragment corresponded to that region of POMC mRNA that codes for β-LPH.

More recently, the full-length mouse POMC cDNA was cloned using linkers to create a cDNA library of clones and hybridizing with the 140-bp mouse cDNA fragment (*7*). In addition, the porcine POMC cDNA sequence was determined from two overlapping cDNA fragments (*25*).

A comparison of the protein-coding regions of the cDNA and the protein sequences that they predict is shown in Fig. 2. Both bovine and porcine POMC contain two additional proteolytic sites relative to the mouse POMC (shown schematically in Fig. 3). One of these occurs in the pro-γ-MSH region, in which instead of the single 24-amino acid peptide found in the mouse, peptides of 12 and 13 residues for the bovine and 12 and 16 residues for the porcine, POMC are found. This additional cleavage site is also found in the human, but not in the rat, POMC gene sequence. The other additional processing site occurs in γ-LPH to generate two bovine peptides of 40 and 18 amino acids, or two porcine peptides of 40 and 16 amino acids, in contrast to the single 38 amino acids in the murine γ-LPH fragment. In addition, both the bovine and porcine cDNAs predict a 22-amino acid insertion in the large β-LPH fragments. The human POMC gene sequence also predicts this large insert and the additional processing site in human POMC, but they are not present in the rat POMC sequence. This difference in protein-coding potential of the two classes of genes—bovine, porcine, and human POMC genes on the one hand, and the rat and mouse POMC genes on the other—should be noted in view of its evolutionary importance.

Although the generation of a paired basic amino acid processing site by a point mutation may seem a simple event, it could have profound physiological consequences. For instance, one of the new peptides thus generated may evolve with time to perform a new function. Alternatively, if a large fragment with two functional domains has a glycosylation site, then a new proteolytic processing site between the domains will allow modification of biological activity via glycosylation for one of the functional domains independently of the other.

```
         1
mouse   Met pro arg phe cys tyr ser ser ser gly ala leu leu leu ala leu leu leu
        ATG CCG AGA TTC TGC TAC AGT AGC TCA GGG GCC CTG TTG CTG GCC CTC CTG CTT
bovine              leu     ser     arg
                    C G     AG      C T   G   C         C           T G
porcine             leu     gly     arg                         thr
                      G     GG      C     G             C G     A   T   C     C

             20
mouse   gln thr ser ile asp val trp ser trp cys leu glu ser ser gln cys gln asp
        GAG ACC TCC ATA GAT GTG TGG AGC TGG TGC CTG GAG AGC AGC CAG TGC CAG GAC
bovine      ala     met glu     arg gly
            G         G   A     C T G T
porcine     ala     met gly     arg gly
        C   G         G  GA     C C GG          T                     T

                     40
mouse   leu thr thr glu ser asn leu leu ala cys ile arg ala cys lys leu asp leu
        CTT ACC ACG GAG AGC AAC CTG CTG GCT TGC ATC CGG GCT TGC AAA CTC GAC CTC
bovine                                                              pro
          C           A   T               G               C       G  C
porcine     ser                                                     pro
          C T         A   T     T   T     G               C          CA   T

                             60
mouse   ser leu glu thr pro val phe pro gly asn gly asp gly gln pro leu thr glu
        TCG CTG GAG ACG CCC GTG TTT CCT GGC AAC GGA GAT GAA CAG CCC CTG ACT GAA
bovine      ala
          C GCC           G       C   C           C       G       G           G
porcine     ala                                         ala
          T GC                        C           C   C GCG   A   G       C   G

                                     80
mouse   asn pro arg lys tyr val met gly his phe arg trp asp arg phe gly pro ---
        AAC CCC CGG AAG TAC GTC ATG GGT CAC TTC CGC TGG GAC CGC TTC GGC CCC ---
bovine                                                                  arg arg
                                      C   T                              GT CGG
porcine                                                                 arg arg
                                      C                                  G  CGG

                                                                   100
mouse   --- --- arg asn ser ser ser ala --- --- --- gly ser ala ala gln arg arg
        --- --- AGG AAC AGC AGC AGT GCT --- --- --- GGC AGC GCG GCG CAG AGG CGT
bovine  asn gly ser ser         gly val --- --- ---     gly             lys
        AAT GGT   C  G          G A  T  --- --- ---   G G         C      A    C
porcine asn gly ser ser     gly gly gly gly gly gly     gly     gly     lys
              C   C  G      G   G C  G  GGC GGT GGC     G        GC      A    C

mouse   ala glu --- glu glu ala val trp gly asp gly ser pro --- --- --- --- ---
        CCG GAG --- GAA GAG GCG GTG TGG GGA GAC GGC AGT CCA --- --- --- --- ---
bovine  glu     ---     val         gly glu gly pro gly     arg gly asp asp ala
         A      ---      T          G C  A   G  CC  G G   C CGC GGC GAT GAC GCC
porcine glu     glu     val     ala gly glu gly pro gly     arg gly asp gly val
        GA      GAG   G  T       C  G C  A   G  CC  G G   C CGC GGA GAT GGC GTC
```

Fig. 2. Nucleotide sequences and predicted amino acid sequences from the mouse (*7*), bovine (*4*), and porcine (*25*) POMC cDNA sequences. The nucleotide and amino acid sequences for the bovine and porcine cDNAs are shown only when they differ from the mouse sequences. Proteolytic processing sites are underlined.

```
                            120
mouse   glu pro ser pro arg glu gly lys arg ser tyr ser met glu his phe arg trp
        GAG CCG AGT CCA CGC GAG GGC AAG CGC TCC TAC TCC ATG GAG CAC TTC CGC TGG
bovine      thr gly             asp
            A   G     G          A        T   T               A
porcine ala     gly             gln asp
         C      G C   G     C    A

                                    140
mouse   gly lys pro val gly lys lys arg arg pro val lys val tyr pro asn val ala
        GGC AAG CCG GTG GGC AAC AAA CGG CGC CCG GTG AAG GTG TAC CCC AAC GTT GCT
bovine                                                                  gly
                                  G                                      GC   C
porcine                                                                 gly
                  C           G   G                           T          GC   C

                                            160
mouse   glu asn glu ser ala glu ala phe pro leu glu phe lys arg glu leu glu gly
        GAG AAC GAG TCG GCG GAG GCC TTT CCC CTA GAG TTC AAG AGG GAG CTG GAA GGC
bovine      asp             gln                                         thr
            G             C C                   C   A                   ACC
porcine     asp     leu                                     arg         ala
            G        T    C          C          C          G             CC   G

mouse   glu arg pro --- --- --- --- --- --- --- --- --- --- --- --- --- --- ---
        GAG CGG CCA --- --- --- --- --- --- --- --- --- --- --- --- --- --- ---
bovine          leu glu gln ala arg gly pro gly ala gln ala glu ser ala ala ala
                 TC GAG CAG GCG CGC GGC CCC GAG GCC CAG GCT GAG AGT GCG GCC GCC
porcine ala pro     glu pro ala arg asp pro glu ala pro ala glu gly ala ala ala
         C   CC   C GAG CCG GCA CGG GAC CCC GAG GCC CCG GCC GAG GGC GCG GCC GCC

                                                                        180
mouse   --- --- --- --- --- --- --- leu gly leu glu his val leu glu ser asp ala
        --- --- --- --- --- --- --- TTA GGC TTG GAG CAC GTC CTG GAG TCC GAC GCG
bovine  arg ala glu leu glu tyr gly     val ala     ala glu ala     ala ala glu
        CGG GCT GAG CTG GAG TAT GGC C G  TG GC      GCG  AG GCT     G G  C   A
porcine arg ala glu leu glu tyr gly     val ala     ala glu --- --- ala ala glu
        CGG GCC GAG CTG GAG TAC GGG C G  TG GCC     GC   AG --- --- G G  C   A

                                                                            200
mouse   glu lys asp asp gly pro tyr arg val glu his phe arg trp ser asn pro pro
        GAG AAG GAC GAC GGG CCC TAC CGG GTG GAG CAC TTC CGC TGG AGC AAC CCG CCC
bovine  lys         ser                 lys met                     gly ser
        A           TCG         T AA  A     A                 G    G
porcine lys         glu                 lys met                     gly ser
        A             A  G        T AA  A                       G    G

mouse   lys asp lys arg tyr gly gly phe met thr ser glu lys ser gln thr pro leu
        AAG GAC AAG CGT TAC GGT GGC TTC ATG ACC TCC GAG AAG AGC CAG ACG CCC CTG
bovine
                      C       C   G                               A           T
porcine
                      C       C

            220
mouse   val thr leu phe lys asn ala ile val lys asn ala his lys lys gly gln
        GTG ACG CTC TTC AAG AAC GCC ATC GTC AAG AAC GCG CAG AAG AAG GGC CAG TGA
bovine
          C       G       A                           C
porcine
          C       G       A                           C   C
```

Fig. 2. *(continued)*

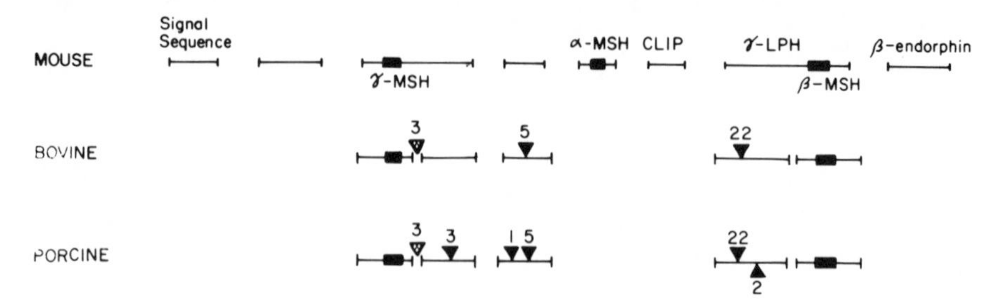


Fig. 3. Comparison of the component peptides of mouse, bovine, and porcine POMC. Differences are indicated by triangles: ▼ indicating additional amino acid residues not present in mouse POMC; ▽ indicating additional residues that generate proteolytic processing sites; and ▲ indicating deletions relative to the mouse POMC peptides.

The protein-coding regions least conserved between species are those coding for the signal sequence, the carboxy-terminal region of the mouse pro-γ-MSH and the amino-terminal region of γ-LPH. In contrast, the MSH protein sequences Met-Glu-His-Phe-Arg-Trp and their corresponding nucleotide sequences ATG-GAG-CAC-TTC-TGG-GAC are very highly conserved both between species and within the three MSH regions of a single species. This repetitious nature of POMC, first observed in the bovine cDNA, led to the hypothesis that POMC arose as a result of two tandem duplication events.

## 3. POMC GENE STRUCTURE

Two major questions concerning POMC gene structure and function could be addressed by direct sequencing of the POMC gene. One of these questions dealt with where introns (*26*) would occur in the POMC gene. Because of the functional significance of the pairs of basic amino acids as proteolytic processing sites, it was possible that introns occur at positions in the genomic protein-coding regions corresponding to these processing sites. Thus, POMC might be divided into functional exonic domains. Another possibility was that the positions of the introns would reflect the evolution of the POMC protein precursor, in that they occur between each of the three MSH sequence-coding regions.

The cloning of a gene coding for a particular protein is accomplished by constructing a library of phage-containing random genomic fragments (*27*). The library is then screened with the corresponding cDNA clone; that is, the phage plaques are transferred to nitrocellulose, denatured to expose the DNA for subsequent hybridization with radiolabeled cDNA, washed, and autoradiographed. Phage plaques that show hybridization are grown, and the DNA isolated, mapped with restriction enzymes, and sequenced. Using these techniques, the first POMC gene structure was determined; it was shown that only one intron interrupts the protein-coding sequence of the bovine POMC gene (*28*). This intron occurs in the region coding for the amino-terminal fragment of POMC. A second intron occurs in the 5′ noncoding region (*29*). These general structural features are shared by the bovine, human (*30,31*), mouse (*32,33*), and, probably, rat (*34*) POMC genes (Fig. 4). For all three, the position of the intron B is between the codons for $Leu^{44}$ and $Ala^{45}$. Thus, in contrast to what was expected, the exonic domains for the POMC gene do not correspond to the functional domains of the POMC protein precursor.

The second major question dealt with the mechanism of regulation of POMC gene expression. Direct measurement of POMC cDNA levels (*35,36*) and, more recently, of POMC gene transcription (*37*) in the pituitary showed that glucocorticoids act to inhibit POMC gene expression at a transcriptional level. Presumably, since anterior lobe corticotrophs have glucocorticoid receptors, this inhibition is mediated by interaction of the activated glucocorticoid–receptor complex with DNA sequences in the nucleus (*38*). If modulation of promoter activity parallels that in prokaryotic systems, one might expect the receptor to bind upstream from the eukaryotic promoter elements to modulate either chromatin structure or RNA polymerase-II binding, or both, to alter the rate of POMC gene transcription. Attempts to identify the DNA sequences to which the glucocorticoid receptor binds have been limited to looking for either homologies between 5′ upstream sequences of POMC genes from various species or for homologies with other glucocorticoid-regulated genes. On the basis of these homologies, several potential receptor-binding sites have been postulated, but none have been demonstrated to bind directly to the glucocorticoid receptor or to be involved in regulation

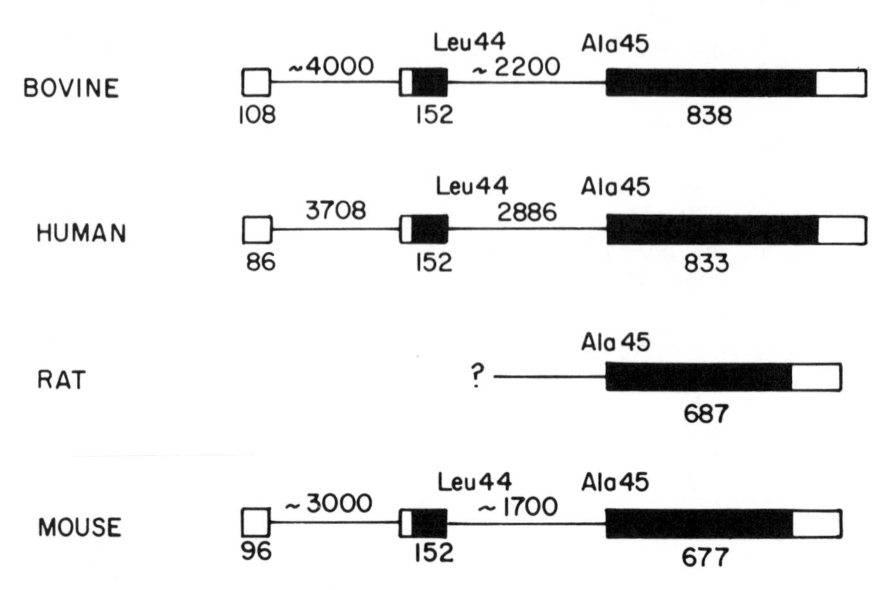


Fig. 4. Comparison of the structure of the bovine (*29*), human (*30*), mouse (*33*), and rat (*34*) POMC genes. Exons are indicated by the boxed areas. The protein-coding regions of exons are solid, whereas the untranslated regions are empty. Intronic regions are denoted by lines, and their length in base pairs appears above the lines.

of POMC gene transcription. The first of these proposed regulatory sequences was detected with the complete sequencing of the bovine POMC gene and was based on homologies with sequences of the globin α and β genes, which are transcriptionally inhibited by glucocorticoids in mouse Friend cells (*29*). The homologies detected were generally small, 5–9 bases, and no physiological significance has been attributed to them yet. Sequencing of the human POMC gene showed a homology between sequences upstream from the transcription start sites for POMC, growth hormone (GH), and mouse mammary tumor virus (MMTV) (*39*). However, subsequent studies with MMTV transcription showed that this homologous sequence does not play a significant role in glucocorticoid induction of MMTV transcription (*40*).

Finally, a recent study compared upstream sequences for the human, mouse, and bovine genes and found several conserved sequences in this region (*31*). Again, small regions of homology were noted with other glucocorticoid-controlled genes, but in this

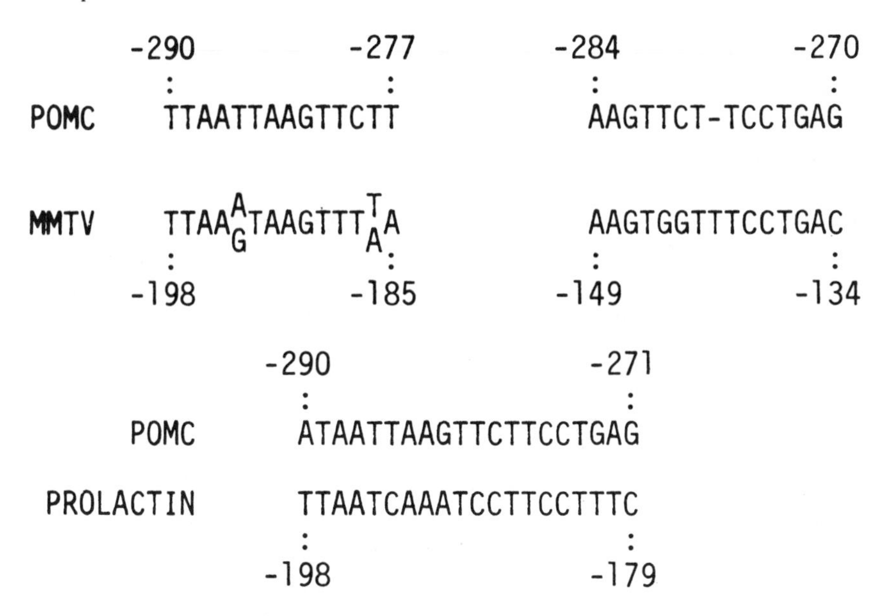


Fig. 5. Comparison of the possible regulatory sequences at the 5′ end of the POMC, MMTV, and prolactin genes (*31*). The numbers at the end of the sequences indicate the position of the sequences with respect to the start site for mRNA transcription.

case the homologies were striking; the homologies with MMTV were to regions known to be involved in MMTV induction. These homologies are shown in Fig. 5.

Although experiments identifying 5′ upstream sequences involved in glucocorticoid regulation have not been published, sequences involved in efficient and accurate transcription of the human POMC gene have been described (*41*). In these experiments, the 680 bp of the 5′ flanking region of the POMC gene and the 65 bp of its 5′ noncoding region were joined to the herpes-simplex thymidine kinase gene. The effect of deleting the POMC upstream sequences on transcription in monkey kidney cells was measured. Deletions upstream from the CAAT box had little effect, although deletion of the CAAT box itself did decrease transcription threefold. Further deletion of an imperfect palindrome between the CAAT and TATA boxes increased transcription threefold. Finally, deletions including the TATA box showed no accurate transcriptional activity. However, the human palindromic sequence that appeared to have an inhibitory effect on

POMC gene transcription is not found in this position in other species (*31*).

Although most species now appear to have one copy of the POMC gene, some organisms, such as mouse and pig, have two copies of POMC-related sequences in their genome (*32,42*). In the mouse, there is a copy of a POMC pseudogene in addition to the authentic POMC gene. The homology of the pseudogene with the authentic POMC gene begins at the codon for $Ser^{92}$ in exon III. There are no sequences corresponding to exons I or II. The 47-nucleotide differences between the pseudogene and the POMC gene result in 25 amino acid changes. Two of the most deleterious are the change of $Arg^{122}$ to a Cys, which removes a processing site, and the change of $Tyr^{205}$ to a premature ochre stop codon. Thus, the pseudogene codes for a truncated POMC precursor from which ACTH cannot be excised by trypsin-like enzymes and does not contain any β-endorphin sequence. In addition, a reiterative sequence GAGAGAGAGA . . . occurs where a poly(A) tail would appear. Thus, the mouse POMC pseudogene appears to have been generated by a reverse-transcription mechanism implicated in the origin of several pseudogenes. No transcript of the pseudogene has ever been detected. Finally, the mouse POMC gene is located on chromosome 12, whereas the pseudogene is probably located on chromosome 19. It is interesting to note that the human POMC gene was shown to be on chromosome two (*43*), closely linked to the enzyme marker acid phosphatase-1 (ACP-1), which in the mouse, is on chromosome 12. The exact structure of the two porcine POMC-related sequences has not been determined.

## 4. APPLICATIONS OF RECOMBINANT-DNA PROBES TO THE STUDY OF POMC GENE EXPRESSION

The availability of cloned DNA fragments complementary to the POMC gene sequence now makes it possible to address questions concerning POMC gene expression. The types of questions fall into three main areas: first, those dealing with which tissues express POMC; second, the extent to which POMC gene expression is regulated in various tissues and the underlying mechanisms involved; and, finally, the evolution of the POMC gene. As discussed above, POMC is expressed in the anterior and intermediate lobes of the pituitary. POMC products, such as ACTH, released from the anterior pituitary, enter the general circulation to act on their target tissues through interaction with specific receptors. POMC products in the intermediate pituitary are further processed, but their target tissues have still not been precisely identified. Some of these intermediate lobe products may be transported to the hypothalamus by retrograde axonal flow. Thus, the exact origin of POMC peptides detected in the hypothalamus and other areas of the brain is a matter of controversy. It has been argued that high concentrations of β-endorphin in certain areas of the brain may be the result of sequestration of β-endorphin transported to these tissues rather than synthesis in those regions. It is important to know which areas of the brain synthesize POMC because this may define more clearly the neuronal pathways involved and, thus, the functions of POMC in the brain.

The availability of cloned DNA probes makes it possible to detect POMC mRNA in various tissues, providing a critical test of whether a tissue synthesizes POMC. There are presently two ways to determine whether a tissue contains mRNA coding for a particular protein. First, dissection of the tissue of interest away from other tissues and use of conventional means of isolating RNA from that tissue, i.e., size fractionation of the RNA on an agarose gel, transfer to nitrocellulose, hybridization with radioactively labeled DNA fragments complementary to the RNA of interest, and autoradiography. The Northern blots generated in this way have the advantage of being very sensitive and reproducible, but have the disadvantage of requiring expert dissection techniques to avoid contamination by adjacent regions. With this technique, POMC mRNA has been detected in the hypothalamus, amygdala, and cerebral cortex, but not in the cerebellum or midbrain (*44*).

The second approach is that of *in situ* hybridization. This involves preparing slices of the particular tissue, treating them in the way that will expose the RNA molecules to allow hybridization and reduce background, hybridization with radiolabeled DNA probes, and autoradiography. This technique has the advantage of enabling identification, at the cellular level, of the location of POMC mRNA in a tissue, but has the disadvantage of not enabling easy detection of low levels of expression that can be reproduced. Nonetheless, it has been used to detect POMC mRNA in the intermediate pituitary (*45*) and in the hypothalamus (*46*). In fact, the latter study showed colocalization of POMC mRNA and ACTH immunoreactivity in the same neurons of the hypothalamus, conclusively proving that POMC is synthesized in hypothalamic neurons. However, no POMC mRNA was detected in the amygdala by *in situ* hybridization, indicating that the technique is still not as sensitive as Northern blotting. As the sensitivity and reproducibility of *in situ* hybridization improve, it will become one of the most important techniques for molecular biological analysis of gene expression in the nervous system.

Recombinant-DNA techniques have been applied to the high incidence of POMC gene expression in certain ectopic hormone-producing tumors (*21*). Numerous tumors from nonendocrine tissue produce polypeptide hormones, sometimes producing endocrinological pathologic conditions. Tumors of a particular origin and histological type often produce the same hormone or set of hormones. For example, oat-cell carcinomas of the lung generally produce large amounts of POMC. The exact cause of these tumors is a matter of keen interest because the expression of POMC may provide a means of understanding the process of oncogenesis. Two models are currently considered; one in which the cell expressed POMC before the transformation event, and the other in which the expression of POMC was the result of transformation. The proposal of a diffuse neuroendocrine system (*47*)—that is, neurosecretory cells or cells with a neurosecretory potential existing in almost all tissues—and the data that support its existence argue for the expression before transformation. However, the tumors in general lack the processing enzymes to produce authentic POMC peptide products and are therefore not neurosecretory cells in the classic sense. To date, the question has not been resolved, but the oat-cell carcinomas are aberrant in their POMC gene expression, at least to some extent. Northern blot analysis of poly(A)+ RNA showed that there are two POMC gene transcripts (*48*). One of these corresponds to the transcript found in the pituitary, whereas the other is about 200 nucleotides longer. The exact structure of the larger RNA has not been determined, but its resolution may illuminate the mechanism of POMC gene activation in these ectopic tumors.

The second major area to which POMC recombinant DNA probes have been applied is the study of regulation of POMC gene expression. As mentioned previously, protein synthesis and secretion studies have shown that these two levels of gene expression are regulated differently in the anterior and intermediate lobes of the pituitary. More recently, however, studies of mRNA abundance by nucleic hybridization techniques have shown that regulation also occurs at this level. In an early report (*36*), RNA samples extracted from either anterior or intermediate lobe were fixed to diazotized-cellulose filters and hybridized with a radiolabeled mouse cDNA probe. After washing, the filters were quantitated by liquid scintillation counting. Analysis of the results showed a 6.5-fold increase in POMC mRNA in the anterior lobe 4 d after adrenalectomy, but a 70% decrease in the amount of POMC mRNA in the intermediate lobe compared with control animals. Dexamethasone treatment of adrenalectomized animals reduced the anterior lobe POMC mRNA levels to control levels and intermediate lobe POMC mRNA levels to approximately half of control levels. It should be mentioned here that in most studies glucocorticoids have no effect on POMC protein levels in the intermediate pituitary; however, stressing of animals during experimental manipulations can have varying effects on intermediate lobe POMC levels.

In a more recent study of adrenalectomy effects on POMC gene expression, measurements of ACTH levels in the serum and anterior pituitary were correlated with POMC mRNA levels and gene transcription (*37*). After a spike in the plasma ACTH level at postadrenalectomy, there was a gradual increase in serum ACTH until, on the eighth day, serum levels were eightfold higher than controls. During this time, POMC mRNA levels rose ninefold, as determined by a very sensitive solution technique for hybridization involving generation of a radiolabeled, single-stranded cDNA probe, hybridization with an RNA sample, $S_1$-nuclease digestion to degrade unhybridized probe, and trichloroacetic acid precipitation to remove degraded probe before liquid scintillation counting. After 8 d, the animals were given dexamethasone, and, in 2 h the plasma ACTH decreased to near control levels. In 5 d, POMC mRNA levels in the anterior pituitary fell back to normal; whereas, even after 8 d the anterior pituitary ACTH levels did not reach control levels. Intermediate lobe POMC mRNA levels were not affected by adrenalectomy or dexamethasone administration. Finally, *in vitro* nuclear transcription run-off experiments were performed with anterior and intermediate pituitary nuclei isolated from control, adrenalectomized, or dexamethasone-treated adrenalectomized rats. This technique allows completion of RNA transcripts that have been initiated in the nuclei and determination of the rate of transcription by hybridization of the radiolabeled transcripts to purified cloned DNA immobilized on nitrocellulose. These experiments showed that POMC gene transcription was elevated 20-fold at 1 and 4 h after adrenalectomy; however, GH gene transcription was cut in halves after adrenalectomy. Experiments with the mouse anterior pituitary cell line, AtT-20, have been unable to reproduce the in vivo effects of adrenalectomy in tissue culture (Uhler, personal observation). Culture of the cell line with glucocorticoid antagonists, serum from adrenalectomized rats, and hypothalamic extracts from adrenalectomized rats have shown, at most, a twofold increase in POMC mRNA levels. Similar results have been obtained with primary culture experiments (M. Hinman, personal communication), making it difficult to study the glucocorticoid effect on POMC gene transcription using standard molecular biological techniques. These results are consistent with the idea that the increase in anterior pituitary POMC gene transcription is a result, to a greater extent, of increased secretion from the hypothalamus of stimulatory factors that have a secondary effect on the anterior pituitary rather than of a simple release of glucocorticoid inhibition of POMC gene transcription.

POMC mRNA levels and protein levels also were correlated in an experiment studying the dopaminergic regulation of POMC expression in the intermediate pituitary (*49*). The dopamine-receptor antagonist, haloperidol, resulted in a fivefold increase in POMC mRNA levels, as determined by dot-blot analysis; that is, fixing of RNA to nitrocellulose as described above, but quantifying by the more sensitive techniques of densitometry of an autoradiogram rather than liquid scintillation counting. Administration of ergocryptine, a dopamine agonist, caused a 70–80% decrease in POMC mRNA levels. In the intermediate pituitary, the POMC protein levels also showed negative regulation by dopaminergic compounds, but, in the anterior pituitary, the POMC mRNA or protein levels showed no change.

The third major area to be approached with molecular biological techniques is the study of the evolution of POMC and, perhaps indirectly, of the nervous system. By looking at the POMC gene structure and expression of evolutionarily diverse species, it may be possible to discern the earliest functional portions of the POMC protein. By looking to see which tissues express POMC in these species, it may be possible to determine whether the origin of POMC was a hormone or a neuropeptide. POMC-related products have been detected by radioimmunoassay in birds (*50*), fish (*51*), *Xenopus* (*52*), and *Drosophila* (*53*). Indeed, an α-MSH peptide has been sequenced in fish (*51*). However, to date, heterologous hybridization under reduced stringency has not revealed authentic POMC-encoding genes in *Drosophila* or the sea slug (*Aplysia*) (M. Uhler, personal observation). This suggests that the conservation of the peptide sequence is not accompanied by conservation of the nucleic acid sequence that is sufficient for the section of the POMC genes by heterologous hybridization in lower organisms. A POMC cDNA was recently isolated and sequenced from the salmon, *Oncorhynchus keta*, using a synthetic oligonucleotide to screen a pituitary cDNA library (*54*). The sequence of the salmon protein is quite different from the mammalian POMC proteins, with the α-MSH, β-MSH, and enkephalin-like sequences of β-endorphin being the only portions of the peptide conserved between mammals and salmon. The low degree of nucleic acid homology over short stretches of DNA sequence suggests that no hybridization to cDNAs for the mammalian POMCs would be detectable. It may be necessary to use immunological techniques in addition to molecular biological approaches to clone these POMC ancestral genes.

## 5. SUMMARY

There are several uses for POMC recombinant probes that have not been reported, but are apparent and worthwhile. For example, (a) mutation of the 5′-flanking region to identify sequences involved in glucocorticoid regulation; (b) similar experiments to identify tissue-specific enhancers involved in POMC gene expression by assay of the effect of deletions and mutations on the selective expression of the POMC gene in anterior pituitary cells compared to fibroblasts; and (c) determination of the relative importance of processing events, such as signal-sequence removal, glycosylation, proteolytic cleavage, and the like on protein maturation and secretion by mutation of the coding sequence of POMC and introduction of the mutant gene into secretory cells. The wide variety of POMC gene sequences now available will help to provide some answers to these and other questions concerning POMC gene expression.

## REFERENCES

1. Mains R. E., Eipper B. A., and Ling N. (1977) Common precursor to corticotropins and endorphins. *Proc. Natl. Acad. Sci. USA* ***74,*** 3014–3018.
2. Roberts J. L. and Herbert E. (1977) Characterization of a common precursor to corticotropin and β-lipotropin: Cell-free synthesis of the precursor and identfication of corticotropin peptides in the molecule. *Proc. Natl. Acad. Sci. USA* ***74,*** 4826–4830.
3. Roberts J. L. and Herbert E. (1977) Characterization of a common precursor to corticotropin and β-lipotropin: Identification of β-lipotropin peptides and their arrangement relative to corticotropin in the precursor synthesized in a cell-free system. *Proc. Natl. Acad. Sci. USA* ***74,*** 5300–5304.
4. Nakanishi S., Inoue A., Kita T., Nakamura M., Chang A. C. Y., Cohen S. N., and Numa S. (1979) Nucleotide sequence of cloned cDNA for bovine corticotropin-β-lipotropin precursor. *Nature* ***278,*** 423–427.
5. Herbert E. and Uhler M. (1982) Biosynthesis of polyprotein precursors to regulatory peptides. *Cell* ***30,*** 1–2.
6. Herbert E., Birnberg N., Lissitsky J. C., Civelli O., and Uhler M. (1981) Proopiomelanocortin: A model for the regulation of expression of neuropeptides in pituitary and brain. *Neurosci. Newslett.* ***12,***16–27.
7. Uhler M. and Herbert E. (1983) Complete amino acid sequence of mouse proopiomelanocortin derived from the nucleotide sequence of proopiomelanocortin cDNA. *J. Biol. Chem.* ***258,*** 257–261.
8. Rosa P., Policastro P., and Herbert E. (1980) A cellular basis for the differences in regulation of synthesis and secretion of ACTH/endorphin peptides in anterior and intermediate lobes of the pituitary. *J. Exp. Biol.* ***89,*** 215–237.
9. Phillips M. A., Budarf, M. L., and Herbert E. (1981) Glycosylation events in the processing and secretion of pro-ACTH-endorphin in mouse pituitary tumor cells. *Biochemistry* ***20,*** 1666–1675.

10. Smyth D. G. and Zakarian S. (1980) Selective processing of β-endorphin in regions of porcine pituitary. *Nature* **228,** 613–615.
11. Eipper B. A. and Mains R. E. (1982) Phosphorylation of pro-adrenocorticotropin/endorphin-derived peptides. *J. Biol. Chem.* **257,** 4907–4915.
12. Roberts J. L., Budarf M. L., Baxter J. D., and Herbert E. (1979) Selective reduction of pro-adrenocorticotropin/endorphin proteins and messenger ribonucleic acid activity in mouse pituitary tumor cells by glucocorticoids. *Biochemistry* **18,** 4907–4915.
13. Nakanishi S., Kita T., Taii S., Imura H., and Numa S. (1977) Glucocorticoid effect on the level of corticotropin messenger RNA activity in rat pituitary. *Proc. Natl. Acad. Sci. USA* **74,** 3283–3286.
14. Herbert E., Allen R. G., and Paquette T. L. (1978) Reversal of dexamethasone inhibition of adrenocorticotropin release in a mouse pituitary tumor cell line either by growing cells in the absence of dexamethasone or by addition of hypothalamic extract. *Endocrinology* **102,** 218–226.
15. Farah J. M., Jr., Malcolm D. S., and Mueller G. P. (1982) Dopaminergic inhibition of pituitary β-endorphin-like immunoreactivity in the rat. *Endocrinology* **110,** 657–659.
16. Krieger D. T., Liotta A. S., and Brownstein M. J. (1977) Presence of corticotropin in limbic system of normal and hypophysectomized rats. *Brain Res.* **128,** 575–579.
17. Krieger D. T., Liotta A. S., and Brownstein M. J. (1977) Presence of corticotropin in brain of normal hypophysectomized rats. *Proc. Natl. Acad. Sci. USA* **74,** 648–652.
18. Rees L., Burke C. W., Chard T., Evans S. W., and Letchworth A. T. (1975) Possible placental origin of ACTH in normal human pregnancy. *Nature* **254,** 620–622.
19. Larsson L. (1977) Corticotropin-like peptides in central nerves and in endocrine cells of gut and pancreas. *Lancet* **2,** 1321–1323.
20. Evans C. J., Erdelyi E., Weber E., and Barchas J. D. (1983) Identification of proopiomelanocortin-derived peptides in the human adrenal medulla. *Science* **221,** 957–960.
21. Imura H. (1980) Ectopic hormone production viewed as an abnormality in regulation of gene expression. *Adv. Cancer Res.* **33,** 39–75.
22. Kita T., Inoue A., Nakanishi S., and Numa S. (1979) Purification and characterization of the messenger RNA coding for bovine corticotropin/β-lipotropin precursor. *Eur. J. Biochem.* **93,** 213–220.
23. Nakanishi S., Inoue A., Kita T., Numa S., Chang A. C. Y., Cohen S. N., Nunberg J., and Schimke R. T. (1979) Construction of bacterial plasmids that contain the nucleotide sequence for bovine corticotropin-β-lipotropin precursor. *Proc. Natl. Acad. Sci. USA* **75,** 6021–6025.

24. Roberts J. L., Seeburg P. H., Shine J., Herbert E., Baxter J. D., and Goodman H. M. (1979) Corticotropin and β-endorphin: Construction and analysis of recombinant DNA complementary to mRNA for the common precursor. *Proc. Natl. Acad. Sci. USA* ***76,*** 2153–2157.
25. Boileau G., Barbeau C., Jeannotte L., Chretien M., and Drouin J. (1983) Complete structure of the porcine pro-opiomelanocortin mRNA derived from the nucleotide sequence of cloned cDNA. *Nucleic Acids Res.* ***11,*** 8063–8071.
26. Breathnach R. and Chambon P. (1981) Organization and expression of eucaryotic split genes coding for proteins. *Ann. Rev. Biochem.* ***50,*** 349–383.
27. Maniatis T., Hardison R. C., Lacy E., Lauer J., O'Connell C., Quon D., Sim G. K., and Efstratiadis A. (1978) The isolation of structural genes from libraries of eucaryotic DNA. *Cell* ***15,*** 687–701.
28. Nakanishi S., Teranishi Y., Noda M., Notake M., Watanabe Y., Kakidani H., Jingami H., and Numa S. (1980) The protein coding sequence of the bovine ACTH-β-LPH precursor gene is split near the signal peptide region. *Nature* ***287,*** 752–755.
29. Nakanishi S., Teranishi Y., Watanabe Y., Notake M., Noda M., Kakidani H., Jingami H., and Numa S. (1981) Isolation and characterization of the bovine corticotropin/β-lipotropin precursor gene. *Eur. J. Biochem.* ***115,*** 429–438.
30. Chang A. C. Y., Cochet M., and Cohen S. N. (1980) Structural organization of human genomic DNA encoding the pro-opiomelanocortin peptide. *Proc. Natl. Acad. Sci. USA* ***77,*** 4890–4894.
31. Takahashi H., Hakamata Y., Watanabe Y., Kikuno R., Miyata T., and Numa S. (1983) Complete nucleotide sequence of the human corticotropin-β-lipotropin precursor gene. *Nucleic Acids Res.* ***11,*** 6847–6858.
32. Uhler M., Herbert E., D'Eustachio P. D., and Ruddle F. D. (1983) The mouse genome contains two nonallelic pro-opiomelanocortin genes. *J. Biol. Chem.* ***258,*** 9444–9453.
33. Notake M., Tobimatsu T., Watanabe Y., Takahashi H., Mishina M., and Numa S. (1983) Isolation and characterization of the mouse corticotropin-β-lipotropin precursor gene and a related pseudogene. *FEBS Lett.* ***156,*** 67–71.
34. Drouin J. and Goodman H. M. (1980) Most of the coding region of rat ACTH β-LPH precursor gene lacks intervening sequences. *Nature* ***288,*** 610–613.
35. Nakamura M., Nakanishi S., Sueoka S., Imura H., and Numa S. (1978) Effects of steroid hormones on the level of corticotropin messenger RNA activity in cultured mouse-pituitary-tumor cells. *Eur. J. Biochem.* ***86,*** 61–66.

36. Schachter B. S., Johnson L. K., Baxter J. D., and Roberts J. L. (1982) Differential regulation by glucocorticoids of pro-opiomelanocortin mRNA levels in the anterior and intermediate lobes of the rat pituitary. *Endocrinology* ***110,*** 1442–1444.
37. Birnberg N., Lissitzky J. C., Hinman M., and Herbert E. (1983) Glucocorticoids regulate pro-opiomelanocortin gene expression in vivo at the levels of transcription and secretion. *Proc. Natl. Acad. Sci. USA* ***80,*** 6982–6986.
38. Yamamoto K. R. and Alberts B. M. (1976) Steroid receptors: Elements for modulation of eukaryotic transcription. *Ann. Rev. Biochem.* ***45,*** 721–746.
39. Cochet M., Chang A. C. Y., and Cohen S. N. (1982) Characterization of the structural gene and putative 5′-regulatory sequences for human pro-opiomelanocortin. *Nature* ***297,*** 335–339.
40. Payvar F., DeFranco D., Firestone G. L., Edgar B., Wrange O., Okret S., Gustafsson J., and Yamamoto K. R. (1983) Sequence-specific binding of glucocorticoid receptor to MTV DNA at sites within and upstream of the transcribed region. *Cell* ***35,*** 381–392.
41. Mishina M., Kurosaki T., Yamamoto T., Notake M., Masu M., and Numa S. (1982) DNA sequences required for transcription in vivo of the human corticotropin-β-lipotropin precursor gene. *EMBO J.* ***1,*** 1533–1538.
42. Boileau G., Gossard F., Seidah N. G., and Chretien M. (1983) Cell-free synthesis of porcine pro-opiomelanocortin: Two distant primary translation products. *Can. J. Biochem. Cell. Biol.* ***61,*** 333–339.
43. Owerbach D., Rutter W. J., Roberts J. L., Whitfield P., Shine J., Seeburg P. H., and Shows T. B. (1981) The pro-opiomelanocortin (adrenocorticotropin/β-lipotropin) gene is located on chromosome 2 in humans. *Somatic Cell Gen.* ***7,*** 359–369.
44. Civelli O., Birnberg N., and Herbert E. (1982) Detection and quantitation of pro-opiomelanocortin mRNA in pituitary and brain tissues from different species. *J. Biol. Chem.* ***257,*** 6783–6787.
45. Hudson P., Penschow J., Shine J., Ryan G., Niall H., and Coghlan J. (1981) Hybridization histochemistry: Use of recombinant DNA as a "homing probe" for tissue localization of specific mRNA populations. *Endocrinology* ***108,*** 353–356.
46. Gee C. E., Chen C. L. C., Roberts J. L., Thompson R., and Watson S. J. (1983) Identification of pro-opiomelanocortin neurons by *in situ* cDNA–mRNA hybridization. *Nature* ***306,*** 374–376.
47. Pearse A. G. E., Takor-Takor T. (1979) Embryology of the diffuse neuroendocrine system and its relationship to the common peptides. *Fed. Proc.* ***38,*** 2288–2294.
48. Tsukada T., Nakai Y., Jingami H., Imura H., Taii S., Nananishi S.,

and Numa S. (1981) Identification of the mRNA coding for the ACTH-β-lipotropin precursor in a human ectopic ACTH-producing tumor. *Biochem. Biophys. Res. Comm.* **98,** 533–540.

49. Chen. C. L. C., Dionne F. T., and Roberts J. L. (1983) Regulation of the pro-opiomelanocortin mRNA levels in rat pituitary by dopaminergic compounds. *Proc. Natl. Acad. Sci. USA* **80,** 2211–2215.
50. Iturriza F. C., Estivarez F. E., and Levitin H. P. (1980) Coexistence of α-melanocyte-stimulating hormone and adrenocorticotropin in all cells containing either of the two hormones in duck pituitary. *Gen. Comp. Endocrinology* **42,** 110–115.
51. Kawauchi H. (1983) Chemistry of pro-opiomelanocortin-related peptides in the salmon pituitary. *Arch. Biochem. Biophys.* **227,** 343–350.
52. Jegou S., Tonon M. C., Leroux P., Leboulenger F., Delarue C., Cote J., Pelletier G., and Vaudry H. (1981) Effect of dexamethasone treatment of ACTH and β-LPH concentrations in frog pituitary and hypothalamus. *Adv. Physiol. Sci. Proc. Int. Congr.* **13,** 129–135.
53. Jan Y. N. and Jan L. Y. (1983) Genetic and Immunological Studies of the Nervous System of *Drosophila melanogaster,* in *Neuropharmacology of Insects,* vol. 88, (Evered D., O'Connor M., Whelan J., ed.), Pitman, London.
54. Soma G., Kitahara N., Nishizawa T., Nanami H., Kotake C., Okazaki H., and Andoh T. (1984) Nucleotide sequence of a cloned cDNA for proopiomelanocortin precursor of chum salmon, *Onchorynchus keta. Nucleic Acids Res.* **12,** 8029–8041.

# Chapter 10

# Enkephalin Genes

Ueli Gubler

## 1. Introduction

### 1.1. Nature of the Peptides

The effects of opiate drugs on the brain in the perception of pain and pleasure have been known for a long time. The presence of an opiate receptor in the brain was realized in 1973 when it was demonstrated that opiate drugs bound stereospecifically and with high affinity at specific sites (*1–3*). This finding immediately prompted a search for the endogenous ligand(s) for these receptors because it did not seem reasonable that laboratory animals, through evolution, should have developed a receptor structure for a plant alkaloid such as morphine. The search for these endogenous ligands culminated in 1975 with the finding of two opioid peptides, termed methionine enkephalin (Met-enkephalin) and leucine enkephalin (Leu-enkephalin) (*4*). These pentapeptides were shown to have the following structures:

$NH_2$-Tyr-Gly-Gly-Phe-Met-COOH: Met-enkephalin
$NH_2$-Tyr-Gly-Gly-Phe-Leu-COOH: Leu-enkephalin

Thus, opioid peptides are endogenous (or synthetic) peptides with a spectrum of pharmacological action similar to that of morphine and other narcotic drugs. In this context, the definition of an "opiate" would be (*5*): A substance that (a) induces analgesia in vivo and whose effects are reversed or prevented by an opiate antagonist; (b) inhibits in vitro the electrically stimulated smooth-muscle contraction of the guinea pig ileum, this effect being reversed by opiate antagonists; and (c) specifically binds to the opiate receptor and whose binding is blocked by opiate antagonists.

Since the discovery of the enkephalins, the research in the field has expanded dramatically. It is realized today that (*5*):

(a) There exist numerous enkephalin-containing peptides (ECP), i.e., N- or C-terminal extensions, or both, of the pentapeptides just described.

(b) These different ECPs have other important physiological functions in addition to their opioid activity (*see* section 1.2).

(c) These peptides can be grouped into three different families of opioid peptides (derived from three distinct gene products and, thus, three different polypeptide precursors):

1. The enkephalins; the pentapeptides just discussed. Enkephalins are synthesized via a precursor called proenkephalin (*see* section 2.2).
2. The dynorphins; a group of C-terminally extended Leu-enkephalins (rimorphin, dynorphin, β-neoendorphin, leumorphin) that are synthesized via a precursor called prodynorphin (*see* section 2.6). The correct nomenclature for the complete protein precursors is preproenkephalin/preprodynorphin/preproopiomelanocortin. For simplicity, the terms proenkephalin/prodynorphin/proopiomelanocortin will be used throughout this chapter. Proenkephalin and prodynorphin have also been called proenkephalin A and B, respectively. To avoid confusion, these terms will not be used here.
3. The endorphins; a group of peptides whose name is derived from β-endorphin, an N-terminally extended Met-enkephalin of 31 amino acids. The endorphins are formed via a precursor called proopiomelanocortin (POMC; *see* chapter by Uhler and Herbert).

Molecular cloning and DNA sequencing have clearly established that these three types of peptide families are derived from three different gene products. The purpose of this chapter is to summarize the available data on gene structure of both enkephalins and dynorphins.

### 1.2. Physiologic Importance

It has become clear that ECPs have other important functions besides their opioid activity. Their physiological actions can be described as those of neuromodulators—neurotransmitters, hormones, or both—and they are mostly associated with various behavioral states and analgesia. The peptides occur both centrally and peripherally. Complete coverage of the physiological function of these peptides is impossible, inasmuch as the research has become extremely widespread and diversified. The interested reader is referred to a few recent review articles (*6–8*).

### 1.3. Rationale for Analyzing the Genes

The rationale for analyzing the gene(s) coding for the opioid peptide precursors (or, for that matter, any protein coding gene) is as follows:

(a) The mRNA for a specific precursor protein can be cloned as its complementary DNA-copy. Complementary DNAs are enzymatically generated double-stranded DNA copies of cellular poly(A)+ RNA (mRNA) that are cloned in *E. coli* plasmids. Sequencing of the DNA will allow the derivation of the amino acid sequence of the protein through the genetic code. This circumvents the often impossible task of purification and sequencing of proteins that are present in very low steady-state concentrations. In the case of proenkephalin, a number of larger biosynthetic intermediates (i.e., enkephalin-containing peptides, *see* above) had been isolated and sequenced (*5*). In order to align all these components into a single precursor structure, cDNA cloning and sequencing of the proenkephalin mRNA were undertaken in several laboratories—a task that led to the final elucidation of the human and bovine proenkephalin structure.

(b) Once cDNA clones are available, a number of new experimental approaches can be taken. Screening of genomic DNA libraries will allow the isolation and characterization of the actual gene for the protein. A genomic library consists of long [10–20 kilobase pairs (kbp)] DNA pieces derived from chromosomal DNA that are cloned into λ-phage. The complete genome of an organism under study can be cloned in this way. Such a library can be screened and rescreened for any gene of interest. The main differences between a genomic and a cDNA library are (a) the origin of the cloned DNA (double-stranded chromosomal genomic DNA versus enzymatically produced double-stranded copies of mRNA), (b) the nature of the cloned DNA (complete genome versus transcribed portion of the genetic information from a tissue or cell under study), and (c) the size of the cloned DNA (10–20 kbp in the genomic library versus usually 1–2 kbp in the cDNA library). Furthermore, using quantitative DNA–RNA hybridization techniques, the mRNA for the peptide precursors can be localized in different tissues and quantitated.

### *1.4. Screening Approaches*

In the case of both proenkephalin and prodynorphin, short synthetic oligodeoxynucleotides were used as hybridization probes to screen recombinant clones. This approach requires the knowledge of partial amino acid sequences from the protein of interest. In the case of proenkephalin and prodynorphin, this criterion was fulfilled because the amino acid sequences of enkephalins and dynorphins and other ECPs had been established (*5*).

Two other screening approaches would have been possible, involving either the monitoring of enkephalin biological activity or the immunological detection of in vitro translation products. However, both these methods would have required the translational expression of proenkephalin coding sequences and the subsequent assay of the proteins or peptides formed. In the light of these facts and the availability of quite extensive amino acid sequence information from ECPs, the reason for the use of the synthetic oligodeoxynucleotide hybridization technique for the screening is clear.

The principle of the screening approach using synthetic oligodeoxynucleotides is illustrated in Fig. 1. The mRNA sequence corresponding to a short amino acid sequence is derived through the genetic code. A synthetic oligodeoxynucleotide complementary or identical to this mRNA sequence is synthesized. These oligodeoxynucleotides are enzymatically labeled with $^{32}P$

| | | |
|---|---|---|
| 1 5 10 15 20<br>Ser - Pro - Thr - Leu - Glu - Asp - Glu - His - Lys - Glu - Leu - Gln - Lys - Arg - Tyr - Gly - Gly - Phe - Met - Arg | | |
| 25 30 35 39<br>Arg - Val - Gly - Arg - Pro - Glu - Trp - Trp - Met - Asp - Tyr - Gln - Lys - Arg - Tyr - Gly - Gly - Phe - Leu | Peptide I | |
| 5'...UGG·UGG·AUG·GA$^{U}_{C}$·UA$^{U}_{C}$·G... | mRNA | |
| 3' ACC·ACC·TAC·CTG·ATG·G 5' | DNA primer | |

Fig. 1. The amino acid sequence of peptide I, one of the bovine adrenal ECPs (top line). Amino acids number 27–32 are encoded by codons with a very low degeneracy (middle line). This mRNA sequence was used to devise a complementary synthetic probe 16 nucleotides long (bottom line) (ref *12*). By choosing G only for the third positions of the Asp- and Tyr-codons (*38*), the complexity of the probe could be reduced to one sequence (from ref. *12*, with permission).

and used as hybridization probes in the screening of recombinant clones. Probe complexity and probe length are two intimately related parameters that are technically very important. Because the genetic code is degenerate (i.e., some amino acids are coded for by as many as six different codons), an effective and useful probe must represent all the different nucleic acid sequences possible. The probe will thus consist of a pool of different sequences. The number of different sequences is termed the complexity of the probe. For practical purposes (i.e., high backgrounds and appearance of false positives during the screening), such probe pools rarely consist of more than 50–100 different sequences. The only way to decrease that complexity is to shorten the probe or to choose an amino acid sequence of low degeneracy (although such sequence stretches are rarely available). The practical upper limit on complexity will, in turn, also limit the length of the probes, which are usually 12–20 nucleotides. For DNA pieces as short as that, the stability of the double strands formed depends critically on their length and base composition. The correct conditions for hybridization have to be empirically determined in each case (*9*). Ultimately, only DNA sequencing can ascertain whether the clones isolated in this way really contain the sequences of interest.

Table 1 summarizes the important details of the synthetic probes that were used in the characterization and molecular cloning of proenkephalin and prodynorphin mRNA.

TABLE 1
Synthetic Oligonucleotides Used in the Characterization and Cloning of Proenkephalin and Prodynorphin mRNA[a]

| Protein | Partial amino acid sequence | Corresponding probe sequence | Length,[b] N | Complexity | Reference |
|---|---|---|---|---|---|
| Bovine proenkephalin | Glu Trp Trp Met Asp | 5′-TCCATCCACCA$^{T}_{C}$ TC-3′[c] | 14 | 2 | *11* |
| | Tyr Gly Gly Phe Met | 5′-CATGAAGCCCCC$^{A}_{G}$ T-3′[d] | 14 | 2 | *11* |
| | Trp Trp Met Asp Tyr Gln | 5′-GGTAGTCCATCCACCA-3′[e] | 16 | 1 | *12,13* |
| Human proenkephalin | Tyr Gly Gly Phe Met | 5′-CAT$^{A}_{G}$AAGCCGCC $^{A}_{G}$TA-3′[f] | 15 | 4 | *42,15* |
| | Met Lys Lys Met Asp | 5′-TCCAT$^{T}_{C}$TT $^{T}_{C}$TTCAT-3′[g] | 14 | 4 | *42* |
| | Trp Trp Met Asp Tyr | 5′-TAATCCATCCACC-3′[h] | 13 | 1 | *15* |
| | | *5′-TAGTCCATCCACC-3′[h] | 13 | 1 | *15* |
| Rat proenkephalin | Gly Arg Pro Glu Trp Trp Met Asp Tyr Gln Lys | 5′-TCTGGTAGTCCATCCACCACTC[j] TGCACGAC-3′ | 30 | 1 | *35* |
| Porcine prodynorphin | Lys Trp Asp Asn Gln | 5′-TG$^{A}_{G}$TT $^{A}_{G}$TCCCATTT-3′[i] | 14 | 4 | *28* |
| | | 5′-TG$^{A}_{G}$TT $^{A}_{G}$TCCCACTT-3′[i] | 14 | 4 | *28* |

[a]All the probes described were devised complementary to the mRNA sequences. For simplicity, only actual probe sequences are shown. Fig. 1 illustrates one example in more detail.
[b]N, nucleotides.

[c]One of the ECPs (*see* section 1.1a) had been shown to contain this partial nonenkephalin amino acid sequence, with three of five amino acids being coded for by only one codon. It should be noted that, generally, a sequence with such a low natural degeneracy is very rare. The probe was devised to contain two sequences complementary to all possible codons.

[d]The enkephalin sequence had, of course, been known to be part of proenkephalin for a number of years. Because of the rather high degeneracy of the codons constituting this sequence, an oligodeoxynucleotide probe was known to contain 64 different sequences. For this reason, it was not used in a primary screen. In the work referenced here, proenkephalin clones had been isolated and sequenced, using the low-degeneracy probe described in footnote *c*. The sequencing had shown the actual Met-enkephalin coding sequence. This information was in turn used to generate fully defined probes for the Met-enkephalin sequence.

[e]Using the same partial nonenkephalin amino acid sequence described in footnote *c*, a unique probe of 16 nucleotides was synthesized (*see* also Fig. 1). G was chosen for the 3rd positions of the Asp- and Tyr-codons because it had been shown that G–U basepairing is possible and does not significantly alter the stability of the duplex formed (43).

[f]For the isolation of human proenkephalin cDNA clones, one study made use of the Met-enkephalin sequence to generate a probe of 15 nucleotides, consisting of only four different sequences. The high theoretical complexity of the probe (64 different sequences) was reduced to four different sequences by choosing single codons for Gly. The basis of this choice was:

(a) observed codon usages for Gly in mammals,
(b) the coding sequence for Met-enkephalin, as determined in cloned POMC-mRNA,
(c) the fact that G–U basepairing is possible (*see* footnote *e*).

[g]In the same study, mentioned in footnote *f*, a nonenkephalin sequence occurring in one of the ECPs was used to generate a probe of 14 nucleotides, based on very-low-degeneracy codons.

[h]In another study of human proenkephalin mRNA, it was assumed that the sequence Trp-Trp-Met-Asp-Tyr (shown to occur in one of the bovine ECPs) (*see* footnotes *c* and *e*) was conserved between bovine and human proenkephalin. Two separate probes (13 nucleotides) were synthesized, taking both codons for Asp into consideration. RNA blot hybridizations were done with the two probes to establish the correct probe. The probe marked * was subsequently chosen.

[i]In the case of prodynophin, two separate probe pools (14 nucleotides) were synthesized, complementary to all possible eight sequences derived for the pentapeptide from porcine dynorphin.

[j]This sequence was based on a region in one of the ECPs that had been shown to be highly conserved between human and bovine proenkephalin mRNA.

# 2. mRNA CHARACTERIZATION AND GENE STRUCTURE

## *2.1. Characterization of Proenkephalin and Prodynorphin mRNAs*

Most studies of enkephalin biosynthesis have used the adrenal gland because it is the richest source for these peptides (*10*). Molecular cloning, however, has allowed the elucidation of the primary structure of the (rat) brain proenkephalin mRNA and protein as well and has demonstrated that both adrenomedullary and brain proenkephalin are identical (*see* section 2.4).

Before and concurrently with the molecular cloning of proenkephalin cDNA, a number of studies, using a variety of tissue sources, yielded further structural information on proenkephalin itself, as well as on its uncloned mRNA.

(a) Northern blot analysis was used to estimate both the size and abundance of proenkephalin mRNA. The strength of this technique is its ability to determine the size of a specific mRNA (Northern blot) or a DNA fragment (Southern blot) within an extensive mixture of many different sequences (i.e., the mRNA or DNA fragment(s) need not be purified). RNA or DNA is size-fractionated by electrophoresis in an agarose gel. Size standards are run in parallel. After the electrophoretic run, the nucleic acid is transferred onto a solid phase (nitrocellulose or nylon membrane). The membrane is then hybridized with a labeled probe specific for the sequence (mRNA, DNA fragment) of interest. Detection is done by autoradiography.

(b) In vitro translational assays were coupled with immunological detection of either Met-enkephalin sequences or complete proenkephalin. These experiments allowed sizing of both the mRNA and the proenkephalin protein. Table 2 briefly summarizes these data, grouped according to tissue source. In various different tissues, the proenkephalin mRNA size is consistently about 1400 nucleotides, encoding a protein product with a molecular weight of $\simeq$30,000.

TABLE 2

Summary of the Characterizations of Proenkephalin and Prodynorphin mRNAs Other Than by Molecular Cloning and cDNA Sequencing

| Tissue | Parameter determined | | | |
|---|---|---|---|---|
| | Molecular weight of proenkephalin | Size of mRNA (nucleotides) | Abundance | Reference |
| **A. Proenkephalin** | | | | |
| Bovine adrenal medulla | n.d. | 1500 | 0.1% | *12* |
| Bovine adrenal medulla | 31,000 | 4700,1500[a] | — | *44* |
| Bovine striatum | 31,000 | 1450[b] | — | *45* |
| Human pheo-chromocytoma | 36,500 | 1400 | — | *14,42* |
| Guinea pig striatum | 30,000 | — | — | *46* |
| Rat brain | 30,000 | — | — | *46* |
| Rat striatum | — | 1400[c] | — | *15* |
| **B. Prodynorphin** | | | | |
| Rat brain | — | 2400[d] | — | *41* |

[a]The mRNA of 1500 nucleotides was most abundant and probably codes for the 31,000 $M_r$ protein.

[b]Careful analysis of total RNA and specific proenkephalin mRNA content in bovine brain regions, pituitary, and adrenal medulla demonstrated that adrenal medulla had the highest concentration of proenkephalin mRNA (47). Abundancies in the 16 different brain areas analyzed varied more than 20-fold, neostriatum being highest (10× lower than adrenal medulla) and thalamus lowest (200× less than adrenal medulla). A close correlation between the levels of mRNA and immunoreactive enkephalin-peptides was noticed.

[c]Analysis of relative abundancies of rat brain proenkephalin mRNA in different brain regions showed striatum to have the highest concentration (36). The other regions analyzed could be divided into three groups: (a) 10–20% of striatal abundance–hypothalamus, pons and medulla oblongata, and spinal cord; (b) 4–10% of striatal abundance—cerebellum, midbrain, and frontal cortex; (c) 1–2% of striatal abundance—hippocampus and thalamus.

[d]Various brain and nonbrain tissues were found to be sites of synthesis for prodynorphin mRNA (41). Relative concentrations determined were striatum = hippocampus>hypothalamus>>midbrain>nucleus tractus solitarium = cortex>>>thalamus>cerebellum. Nonbrain tissues shown to contain prodynorphin mRNA were whole adrenal gland, spinal cord, testis, and anterior pituitary.

### 2.2. Molecular Cloning of Bovine and Human Proenkephalin mRNAs

The mRNA for bovine proenkephalin was cloned from bovine adrenal medulla (*11–13*); the mRNA for the human protein was cloned from human pheochromocytomas (adrenal medullary tumors) (*14,15*). cDNA clone banks were established from the respective poly(A)+ RNA pools and screened with the synthetic oligonucleotides described in Table 1. The isolated positive clones were sequenced, and the sequences were searched for the open reading frames encoding the amino acid sequences already known from the ECPs. Figure 2 shows the bovine mRNA and protein sequences; Figure 3 depicts the human sequences. The cloned and sequenced portion of bovine proenkephalin mRNA consists of 1222 nucleotides (Fig. 2). This cDNA is lacking about 65 nucleotides from the very 5′ end of the natural mRNA (*11*). Another report (*16*) described the cloning and sequencing of proenkephalin cDNA that was 48 nucleotides longer than the sequence shown in Fig. 2. When all this is considered, it seems reasonable to assign bovine proenkephalin mRNA a size of approximately 1300 nucleotides, excluding the poly(A) tail. This value agrees well with the Northern blot analysis that estimated the size of the complete mRNA to be 1400 nucleotides (*11,12*); the difference would account for the uncloned poly(A) tail. Using these assignments, proenkephalin mRNA contains a 5′-untranslated region of approximately 180 nucleotides, a coding region of 789 nucleotides, and a 3′-untranslated region of 324 nucleotides (excluding the poly(A) tail). The 3′-untranslated portion of the proenkephalin mRNA described by Gubler and coworkers (*13*) is shorter, being only 306 nucleotides. It contains only one copy of the sequence AAUAAA, located upstream from the poly(A) addition site and is believed to be involved in the polyadenylation of the mRNA (*17*). These differences could reflect variations in the poly(A) addition after transcription. The coding region of 789 nucleotides in the mRNA encodes a protein of 263 amino acids with a calculated molecular weight of 29,786. This number is in excellent agreement with the molecular weight of 30,000 found by in vitro translation studies (Table 2). At the $NH_2$-terminus of the protein, there is a signal peptide of 24 amino acids, 19 of them being nonpolar, including seven Leu residues. This signal peptide is a common feature of secreted proteins (*18*). The signal sequence usually ends in an amino acid with a small neutral side chain (*19*). Often, these general criteria have to be applied for predicting, but not determining, the actual length of the signal peptide. In the case of bovine proenkephalin, however, the assignment could be corroborated by direct amino acid sequencing data. The largest ECP sequenced (18.2 kdaltons ECP) (*5*) starts with the $NH_2$-terminal sequence Glu-Cys-Ser . . . (residue no. 25 in Fig. 2). This proves that the assignment of the signal peptide, as discussed above, was indeed correct, and that the secreted form of proenkephalin consists of 239 amino acids. This molecule contains all the adrenal ECPs previously characterized (*5*) (*see* section 1.1), thus establishing that proenkephalin is indeed their common precursor. These ECPs are 12.6 kdaltons ECP (residue nos. 25–137 in Fig. 2), 8.6 kdaltons ECP (residue nos. 25–101), 5.3 kdaltons ECP (residue nos. 140–189), peptide F (residue nos. 104–137), peptide I (residue nos. 192–230), peptide E (residue nos. 206–230), and peptide B (residue nos. 233–263). The list makes it clear that, in actuality, most of the proenkephalin protein was also sequenced (although in portions that still needed aligning). An interesting structural feature of the molecule is that it contains an $NH_2$-terminal portion of 96 amino acids that is devoid of enkephalin sequences, but contains all six Cys residues that occur in the proform (one Cys residue is removed by cleaving off the signal peptide). It can be speculated that the even number of Cys residues could have a function in the proper folding of the molecule. This in turn might be important for correct processing. Proenkephalin contains four copies of the Met-enkephalin sequence, one copy of Met-enkephalin-$Arg^6$-$Gly^7$-$Leu^8$, one copy of Met-enkephalin-$Arg^6$-$Phe^7$, and one Leu-enkephalin (Fig. 2). All these sequences are flanked by two basic amino acids, the processing sites typically found in prohormones that allow cleavage and formation of smaller active peptides (*19*). Met-enkephalin-$Arg^6$-$Phe^7$ had been isolated and characterized before and was expected to be a part of proenkephalin (*20*). The existence of the peptide Met-enkephalin-$Arg^6$-$Gly^7$-$Leu^8$ was inferred from DNA sequencing. Only later was this peptide found and characterized (*21*). The notion has been put forward (*13*) that all the ECPs might have a function of their own, possibly quite separate from the opioid activity of the basic enkephalin sequence. The extra non-enkephalin sequence might be involved in differential receptor recognition, increase of peptide stability, or transport. In the light

```
5'--------AGGACCGCGAGAGUGAGGCCCGCCCGCUUUCCUGGCUCUCCCCUCGCCGAGAGUCGCCCCGGACCGGGUUUCCACGACCGACCUGCGUGCCCCGAACAGCGGCAACCCC
                 -100                -80                -60                -40                -20               -1

1                                       10                                      20                                      30
Met Ala Arg Phe Leu Gly Leu Cys Thr Trp Leu Leu Ala Leu Gly Pro Gly Leu Leu Ala Thr Val Arg Ala Glu Cys Ser Gln Asp Cys
AUG GCG CGG UUC CUG GGA CUC UGC ACU UGG CUG CUG GCG CUC GGC CCC GGG CUC CUG GCG ACC GUC AGG GCA GAA UGC AGC CAG GAC UGC
1                       20                          40                          60                          80

                                        40                                      50                                      60
Ala Thr Cys Ser Tyr Arg Leu Ala Arg Pro Thr Asp Leu Asn Pro Leu Ala Cys Thr Leu Glu Cys Glu Gly Lys Leu Pro Ser Leu Lys
GCC ACG UGC AGC UAC CGC CUG GCG CGC CCG ACU GAC CUC AAC CCG CUG GCU UGC ACU CUG GAA UGU GAG GGG AAA CUA CCU UCU CUC AAG
            100                         120                         140                         160                         180

                                        70                                      80                                      90
Thr Trp Glu Thr Cys Lys Glu Leu Leu Gln Leu Thr Lys Leu Glu Leu Pro Pro Asp Ala Thr Ser Ala Leu Ser Lys Gln Glu Glu Ser
ACC UGG GAA ACC UGC AAG GAG CUU CUG CAG CUG ACC AAA CUA GAA CUU CCU CCA GAU GCC ACC AGU GCC CUC AGC AAA CAG GAG GAA AGC
                        200                         220                         240                         260

                                        100                                     110                                     120
His Leu Leu Ala Lys Lys [Tyr Gly Gly Phe Met] Lys Arg [Tyr Gly Gly Phe Met] Lys Lys Met Asp Glu Leu Tyr Pro Leu Glu Val Glu
CAC CUG CUU GCU AAG AAG [UAC GGG GGC UUC AUG] AAG CGG [UAU GGG GGC UUC AUG] AAG AAA AUG GAU GAG CUG UAC CCC CUG GAA GUG GAA
            280                         300                         320                         340                         360

                                        130                                     140                                     150
Glu Glu Ala Asn Gly Gly Glu Val Leu Gly Lys Arg [Tyr Gly Gly Phe Met] Lys Lys Asp Ala Glu Glu Asp Asp Gly Leu Gly Asn Ser
GAA GAG GCA AAU GGA GGU GAG GUC CUU GGC AAG AGA [UAU GGG GGC UUC AUG] AAG AAG GAU GCA GAG GAA GAU GAC GGC CUG GGC AAC UCC
                        380                         400                         420                         440

                                        160                                     170                                     180
Ser Asn Leu Leu Lys Glu Leu Leu Gly Ala Gly Asp Gln Arg Glu Gly Ser Leu His Gln Glu Gly Ser Asp Ala Glu Asp Val Ser Lys
UCC AAC CUG CUC AAG GAG CUG CUG GGA GCC GGG GAC CAG CGA GAG GGG AGC CUC CAC CAG GAG GGC AGU GAU GCU GAA GAC GUG AGC AAG
            460                         480                         500                         520                         540
```

```
                             190                                                   200                                         210
Arg Tyr Gly Gly Phe Met Arg Gly Leu Lys Arg Ser Pro His Leu Glu Asp Glu Thr Lys Glu Leu Gln Lys Arg Tyr Gly Gly Phe Met
AGA UAC GGG GGC UUC AUG AGA GGC UUA AAG AGA AGC CCC CAC CUA GAA GAU GAA ACC AAA GAG CUG CAG AAG CGA UAC GGG GGU UUC AUG
                   560                          580                          600                          620

                             220                                       230                                         240
Arg Arg Val Gly Arg Pro Glu Trp Trp Met Asp Tyr Gln Lys Arg Tyr Gly Gly Phe Leu Lys Arg Phe Ala Glu Pro Leu Pro Ser Glu
AGA AGA GUG GGU CGU CCA GAG UGG UGG AUG GAC UAC CAG AAA AGG UAC GGU GGC UUC CUC AAG CGC UUC GCC GAG CCC CUA CCC UCC GAG
            640                          660                          680                          700                  720

                             250                                       260
Glu Glu Gly Glu Ser Tyr Ser Lys Glu Val Pro Glu Met Glu Lys Arg Tyr Gly Gly Phe Met Arg Phe
GAA GAA GGC GAA AGU UAC UCC AAG GAA GUU CCU GAA AUG GAG AAA AGA UAU GGA GGA UUU AUG AGA UUU UAA UCCCCUUUCCCAUCAGUGACCUG
                   740                          760                          780                          800

AAGCCCCAGCAAGCCUUCCUCUGCCCCCAGUGAAAGACUGCUGCGCUGGUGUGUUGUAUUGUCCCGUGUCGCUUGCAUUAUAUAGUUGACUUGAGAGUCCAGAUAAUUAACUAUACAAC
  820                 840                 860                 880                 900                 920

CUGAAAGCUGUGAUCCCAGGUUCUGUGUUCUGAGAAUCUUUAAGCUUUUAAAUAUUGGUCUGUUGCAGCUGUCUUGUUUCCAUGCUCAGUUUUUGUUAUCACUUUGUCCUUUAUUUUUG
  940                 960                 980                 1000                1020                1040

ACACAAUGCCAAUAAAUGCCUACUUGUGUGUAGAUAUAAUAAACCCAUUACCCCAAGUGC--------3'
  1060                1080                1100
```

Fig. 2. The primary structure of bovine proenkephalin mRNA (*11*). The nucleotide sequence determined from cloned cDNA is shown. Nucleotides are numbered in the 5′ to 3′ direction, beginning with the first nucleotide of the initiator codon. The nucleotides 5′ to the initiator codon are given negative numbers. The cloned 5′-untranslated region does not extend to the very 5′-end of the mRNA (*see* text); the cloned 3′-untranslated region extends to the poly(A) tail, and is therefore complete. It contains two copies of the sequence AAUAAA (underlined) that is probably involved in polyadenylation. The predicted amino acid sequence of proenkephalin is shown, beginning with the initiator-methionine at position 1. The sequences of Met-enkephalin, Leu-enkephalin, Met-enkephalin-Arg$^6$-Phe$^7$, and Met-enkephalin-Arg$^6$-Gly$^7$-Le u$^8$ are boxed (from ref. *11*, with permission).

```
5' - CUGGGGCCUGGGGGCCACCAGUGGGAAAAGAUAUUAAAAUCUCAUAAAUCCUCCGUAUCUUUUUUCCAUUUCAGGAACUUCUUUGGAGUAACUUUCGCCUUCUUCG

UCGGAGGCAGAGCCCCGCAGCCUGGCCCGUGACCCCGCAGAGACGCUGAGGACCGCGACGAGUCGUGUCUGAACCCGGCUUUUCCAAUUGGCCUGCUCCAUCCGAACAGCGUCAACUCC

 1                           10                                                          20                                                          30
AUG GCG CGG UUC CUG ACA CUU UGC ACU UGG CUG CUG UUG CUC GGC CCC GGG CUC CUG GCG ACC GUG CGG GCC GAA UGC AGC CAG GAU UGC
met ala arg phe leu thr leu cys thr trp leu leu leu leu gly pro gly leu leu ala thr val arg ala glu cys ser gln asp cys

                             40                                                          50                                                          60
GCG ACG UGC AGC UAC CGC CUA GUG CGC CCG GCC GAC AUC AAC UUC CUG GCU UGC GUA AUG GAA UGU GAA GGU AAA CUG CCU UCU CUG AAA
ala thr cys ser tyr arg leu val arg pro ala asp ile asn phe leu ala cys val met glu cys glu gly lys leu pro ser leu lys

                             70                                                          80                                                          90
AUU UGG GAA ACC UGC AAG GAG CUC CUG CAG CUG UCC AAA CCA GAG CUU CCU CAA GAU GGC ACC AGC ACC CUC AGA GAA AAU AGC AAA CCG
ile trp glu thr cys lys glu leu leu gln leu ser lys pro glu leu pro gln asp gly thr ser thr leu arg glu asn ser lys pro

                             100                                                         110                                                         120
GAA GAA AGC CAU UUG CUA GCC|AAA AGG|UAU GGG GGC UUC AUG|AAA AGG|UAU GGA GGC UUC AUG|AAG AAA|AUG GAU GAG CUU UAU CCC AUG
glu glu ser his leu leu ala|lys arg|tyr gly gly phe met|lys arg|tyr gly gly phe met|lys lys|met asp glu leu tyr pro met

                             130                                                         140                                                         150
GAG CCA GAA GAA GAG GCC AAU GGA AGU GAG AUC CUC GCC|AAG CGG|UAU GGG GGC UUC AUG|AAG AAG|GAU GCA GAG GAG GAC GAC UCG CUG
glu pro glu glu glu ala asn gly ser glu ile leu ala|lys arg|tyr gly gly phe met|lys lys|asp ala glu glu asp asp ser leu
```

```
                                160                                                   170                                                   180
GCC AAU UCC UCA GAC CUG CUA AAA GAG CUU CUG GAA ACA GGG GAC AAC CGA GAG CGU AGC CAC CAC CAG GAU GGC AGU GAU AAU GAG GAA
ala asn ser ser asp leu leu lys glu leu leu glu thr gly asp asn arg glu arg ser nis his gln asp gly ser asp asn glu glu

                                190                                                   200                                                   210
GAA GUG AGC|AAG AGA|UAU GGG GGC UUC AUG AGA GGC UUA|AAG AGA|AGC CCC CAA CUG GAA GAU GAA GCC AAA GAG CUG CAG|AAG CGA|UAU
glu val ser|lys arg|tyr gly gly phe met arg gly leu|lys arg|ser pro gln leu glu asp glu ala lys glu leu gln|lys arg|tyr

                                220                                                   230                                                   240
GGG GGC UUC AUG|AGA AGA|GUA GGU CGU CCA GAG UGG UGG AUG GAC UAC CAG|AAA CGG|UAU GGA GGU UUC CUG|AAG CGC|UUU GCC GAG GCU
gly gly phe met|arg arg|val gly arg pro glu trp trp met asp tyr gln|lys arg|tyr gly gly phe leu|lys arg|phe ala glu ala

                                250                                                   260
CUG CCC UCC GAC GAA GAA GGC GAA AGU UAC UCC AAA GAA GUU CCU GAA AUG GAA|AAA AGA|UAC GGA GGA UUU AUG AGA UUU UAA
leu pro ser asp glu glu gly glu ser tyr ser lys glu val pro glu met glu|lys arg|tyr gly gly phe met arg phe end

UAUCUUUUCCCACUAGUGGCCCCCAGGCCCCAGCAAGCCUCCCUCCAUCCUCCAGUGGGAAACUGUUGAUGGUGUUUUAUUGUCAUGUGUUGCUUGCCUUGUAUAGUUGACUUCAUUGU

CUGGAUAACUAUACAACCUGAAAACUGUCAUUUCAGGUUCUGUGCUCUUUUUGGAGUCUUUAAGCUCAGUAUUAGUCUAUUGCAGCUAUCUCGUUUUUCAUGCUAAAAAUAGUUUUUG

UUAUCUUGUCUCUUAUUUUUUGACAAACAUCCAAUAAAUGCUUACUUGUAUAUAGAGAUAAUAAACCUGUUACCCCAAGUGCAUAAAAAAA - 3'
```

Fig. 3. The primary structure of human proenkephalin mRNA (*15*). The nucleotide sequence determined from cloned cDNA is shown. As was found later on, the 5′-terminal 113 nucleotides of this sequence represent an artifactual inverted repeat of the cDNA sequence starting at the second nucleotide of codon 242 (*22,23*). cDNA cloning artifacts occurring in the region corresponding to the 5′ end of mRNA have been described (*48*). The cloned 3′-untranslated region extends to the poly(A) tail and contains two copies of the sequence AAUAAA (underlined). The predicted amino acid sequence of human proenkephalin is shown, beginning with the initiator Met at position 1. The sequences of Met-enkephalin, Leu-enkephalin, Met-enkephalin-Arg$^{6}$-Phe$^{7}$ and Met-enkephalin-Arg$^{6}$-Phe$^{7}$-Leu$^{8}$ are underlined; the flanking doublebasic processing signals are boxed (from ref. *15*, with permission).

of this notion, it is interesting to think of proenkephalin as a multivalent precursor containing several copies of peptides differing slightly in their biological activity. Complete processing of the precursor yields several ECPs and free enkephalins. The isolation of large ECPs that are partial processing products indicates that the full potential of such a multivalent precursor is probably realized in vivo (*13*).

The cloned and sequenced portion of human proenkephalin mRNA was reported to be 1350 nucleotides (Fig. 3). A later report (*see* section 2.3) confirmed the presence of an artifactual inverted repeat in the 5′-untranslated portion of the cloned mRNA, shortening this region to a size of 129 nucleotides. When all these data are taken together, it seems fair to assign human proenkephalin mRNA a size of 1250 nucleotides, excluding the poly(A) tail. This value agrees well with the Northern blot analysis, which estimated the size of the mRNA to be about 1400 nucleotides. The difference would again account for the uncloned poly(A) tail. The coding region of human proenkephalin mRNA consists of 801 nucleotides, encoding 267 amino acids; the 3′-untranslated region is 320 nucleotides, containing two copies of the sequence AAUAAA (*22*).

Despite the evolutionary distance between them, both human and bovine proenkephalin are very similar (Fig. 4). The human protein is four amino acids longer than the bovine precursor, with a three amino acid insert (residue nos. 85–87) and a two-for-one insert (residue nos. 178–179). There are 35 amino acid differences overall, 19 of them being nonconservative. The positioning of the enkephalin sequences within the precursor has been completely conserved, as have the six $NH_2$-terminal Cys residues that may function in the proper folding of the molecule (Fig. 4). The double basic processing signals are all identical, except for one conservative change (position 99). The sequences Met-enkephalin-$Arg^6$-$Phe^7$ and Met-enkephalin-$Arg^6$-$Gly^7$-$Leu^8$ have been completely conserved. Peptide E (residue nos. 210–234) has been completely conserved between the two proteins, and peptide B (residue nos. 237–267) shows only one nonconservative amino acid change. Such strong conservation suggests that these features must be very important for proper functioning of the proenkephalin–enkephalin system.

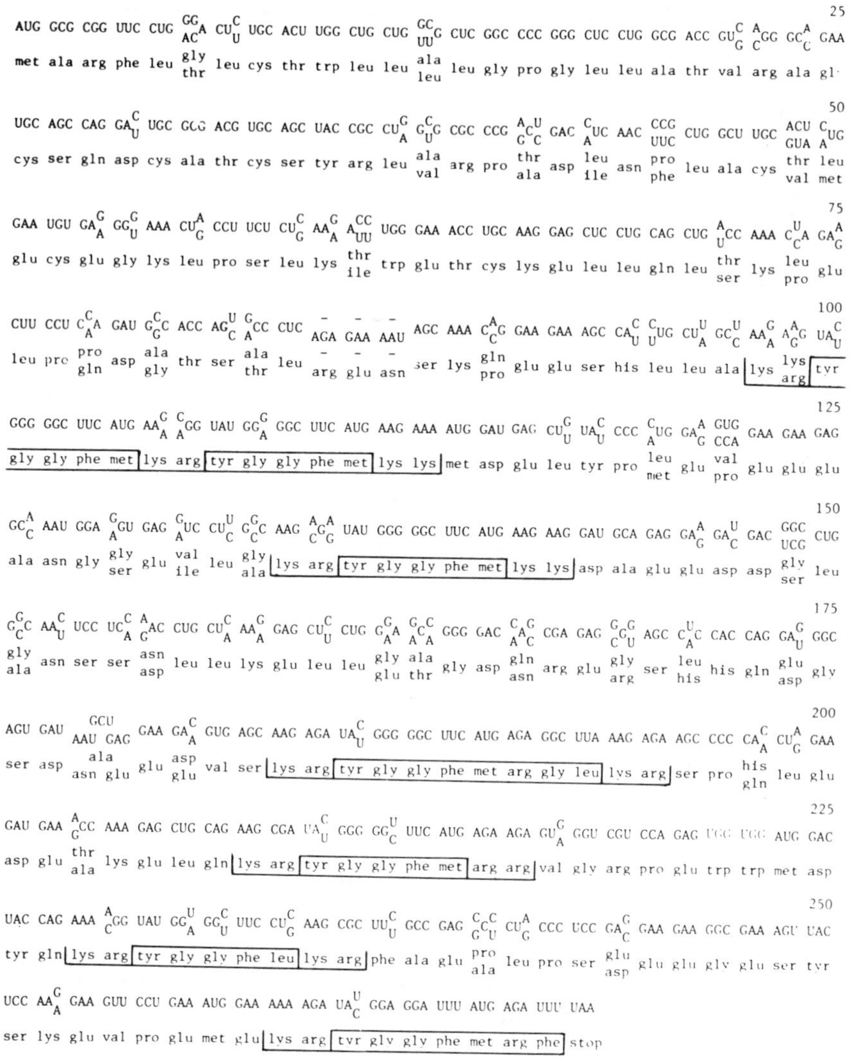

25
AUG GCG CGG UUC CUG GG AC A CU C U UGC ACU UGG CUG CUG GC UU G CUC GGC CCC GGG CUC CUG GCG ACC GU C G A C GG GC A C GAA
met ala arg phe leu gly thr leu cys thr trp leu leu ala leu leu gly pro gly leu leu ala thr val arg ala gl-

50
UGC AGC CAG GA C U UGC GCG ACG UGC AGC UAC CGC CU G A G C U G CGC CCG A C U G C GAC C A UC AAC CCG UUC CUG GCU UGC ACU GUA C A UG
cys ser gln asp cys ala thr cys ser tyr arg leu ala val arg pro thr ala asp leu ile asn pro phe leu ala cys thr val leu met

75
GAA UGU GA G A GG G U AAA CU A G CCU UCU CU C G AA G A A CC UU UGG GAA ACC UGC AAG GAG CUC CUG CAG CUG A U CC AAA C U C A GA A G
glu cys glu gly lys leu pro ser leu lys thr ile trp glu thr cys lys glu leu leu gln leu thr ser lys leu pro glu

100
CUU CCU C C A A GAU G C G C ACC AG U C G A CC CUC AGA GAA AAU AGC AAA C A C G GAA GAA AGC CA C U C U UG CU U A GC U C AA G A A G A G UA C U
leu pro pro gln asp ala gly thr ser ala thr leu - - - arg glu asn ser lys gln pro glu glu ser his leu leu ala | lys lys arg | tyr

125
GGG GGC UUC AUG AA C A C A GG UAU GG G A GGC UUC AUG AAG AAA AUG GAU GAG CU G U UA C U CCC C A UG GA A G GUG CCA GAA GAA GAG
gly gly phe met | lys arg | tyr gly gly phe met | lys lys | met asp glu leu tyr pro leu met glu val pro glu glu glu

150
GC A C AAU GGA G A GU GAG G A UC CU U C G C C AAG A C G A UAU GGG GGC UUC AUG AAG AAG GAU GCA GAG GA A G GA U C GAC GGC UCG CUG
ala asn gly gly ser glu val ile leu gly ala | lys arg | tyr gly gly phe met | lys lys | asp ala glu glu asp asp gly ser leu

175
G G C C AA C U UCC UC C A A G AC CUG CU C A AA G A GAG CU C U CUG G G A A G C A C A GGG GAC C A A G C CGA GAG G C G U AGC C U A C CAC CAG GA G U GGC
gly ala asn ser ser asn asp leu leu lys glu leu leu gly glu ala thr gly asp gln asn arg glu gly arg ser leu his his gln glu asp gly

200
AGU GAU GCU AAU GAG GAA GA C A GUG AGC AAG AGA UA C U GGG GGC UUC AUG AGA GGC UUA AAG AGA AGC CCC CA C A CU A G GAA
ser asp ala asn glu glu asp glu val ser | lys arg | tyr gly gly phe met arg gly leu | lys arg | ser pro his gln leu glu

225
GAU GAA A G CC AAA GAG CUG CAG AAG CGA UA C U GGG GG U C UUC AUG AGA AGA GU G A GGU CGU CCA GAG UGG UGG AUG GAC
asp glu thr ala lys glu leu gln | lys arg | tyr gly gly phe met | arg arg | val gly arg pro glu trp trp met asp

250
UAC CAG AAA A C GG UAU GG U A GG C U UUC CU C G AAG CGC UU C U GCC GAG C C G U CU A G CCC UCC GA G C GAA GAA GGC GAA AGU UAC
tyr gln | lys arg | tyr gly gly phe leu | lys arg | phe ala glu pro ala leu pro ser glu asp glu glu gly glu ser tyr

UCC AA G A GAA GUU CCU GAA AUG GAA AAA AGA UA U C GGA GGA UUU AUG AGA UUU UAA
ser lys glu val pro glu met glu | lys arg | tyr gly gly phe met arg phe | stop

Fig. 4. Human and bovine proenkephalin DNA and amino acid sequence. The DNA and amino acid sequences from human (*14*) and bovine proenkephalin (*11,13*) are shown. When there are differences or insertions, the human sequence is shown below, and the bovine above. The enkephalin sequences (100–104, 107–111, 146–150, 210–214, and 230–234) are boxed, as are the extended enkephalin sequences (186–193 and 261–267). The basic amino acid pairs are also shown (from ref. *49*).

### 2.3. Isolation of the Human Proenkephalin Gene

Two groups have reported the isolation of the human proenkephalin gene from an established human genomic DNA library. The human cDNA (22) or the bovine cDNA (23) were used as the screening probes. Southern blot analyses (*see* section 2.1) of both human genomic placental DNA and of the isolated genomic clones with the same probes produced identical restriction maps, supporting the conclusion that the cloned genomic fragments were derived from a single segment on the human genome. The proenkephalin gene consists of a 5.3-kbp stretch of DNA. Comparison of genomic sequences with the cDNAs allowed positioning of the introns (Fig. 5). The gene is interrupted by three introns (A,B,C). Two of them (A,B) are in the 5′-untranslated region of the mRNA. Hence, the gene consists of four exons (I–IV); exon III and IV contain the translated region of the mRNA. The translated region is interrupted between codon nos. 46 and 47. A schematic outline of the gene is shown in Table 3. Thus, the primary RNA transcript from this gene must be 5.3kb long, 3.9kb of which are subsequently removed by splicing. The 5′-termini of several independent cDNA clones and of human proenkephalin mRNA were directly mapped on the gene (22,23). Based on these experiments, the cap site, i.e., the nucleotide where RNA transcription begins, was assigned to the A residue indicated in Fig. 5. Further support for the assignment comes from two facts:

(a) A sequence TAATTATAAA, homologous to the TATA-box, is found 28 nucleotides upstream from the assigned cap site. The TATA-box is a sequence element characteristic of eukaryotic promoters (24). It is commonly found about 25–30 basepairs (bp) upstream from the transcription initiation site of a eukaryotic gene that is transcribed by RNA polymerase II. A consensus sequence 5′-TATA$^{A}_{T}$A$^{A}_{T}$-3′ has been derived for that region from a survey of a large body of sequencing data. The TATA-box is usually flanked by GC-rich regions.

(b) Most eukaryotic mRNAs start with the nucleotide A (24).

The region 5′ to the cap site in the proenkephalin gene contains other interesting features, including palindromic sequences and several other directly repeated sequence elements. A detailed study of this region has localized the functional proenkephalin

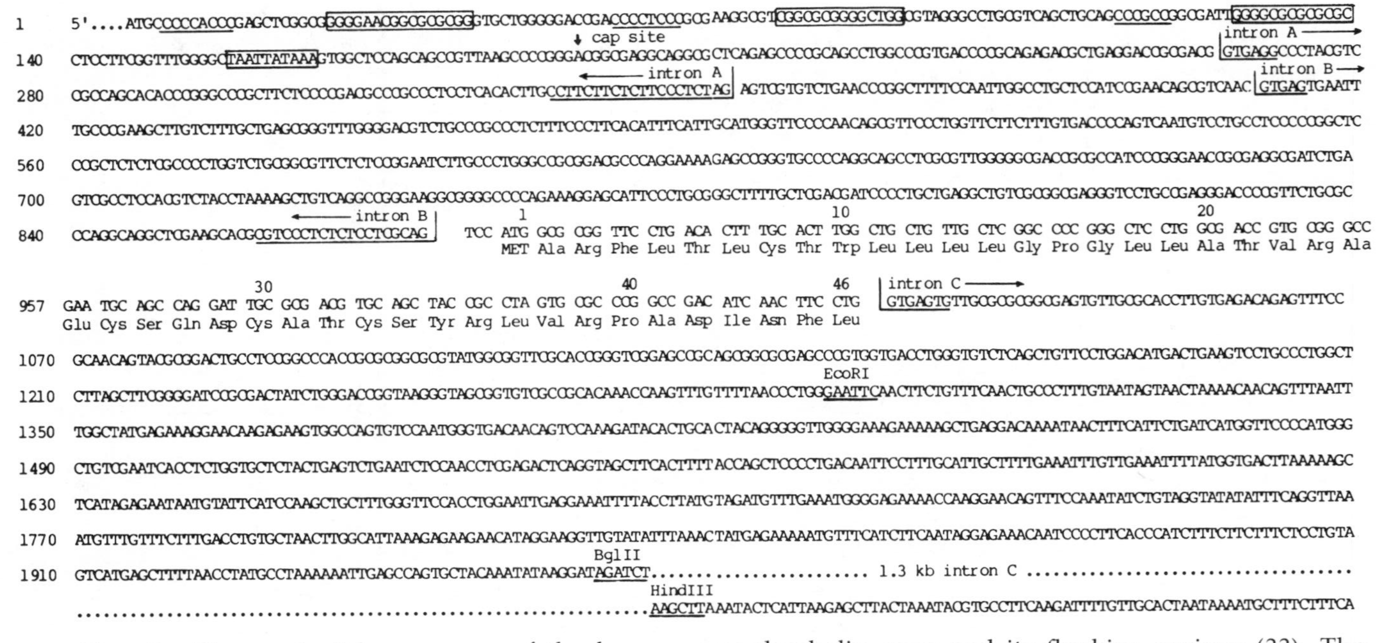


Fig. 5. The nucleotide sequence of the human proenkephalin gene and its flanking regions (*22*). The nucleotide sequence of a 1.3-kb pair region in intron C was not determined. The strand corresponding to human proenkephalin mRNA is shown, together with the deduced amino acid sequence of human proenkephalin. The positions of the exon–intron-junctions are indicated, as determined by comparison with the cloned proenkephalin cDNA sequences. Intron-exon junction sequences homologous to the splice-junction consensus sequences (*50*) are underlined. Vertical arrows indicate the assigned cap-site and the poly(A) addition sites of three independently cloned mRNAs. The TATA-sequence occurring upstream from the cap site is boxed. Other sequences

(*continued*)

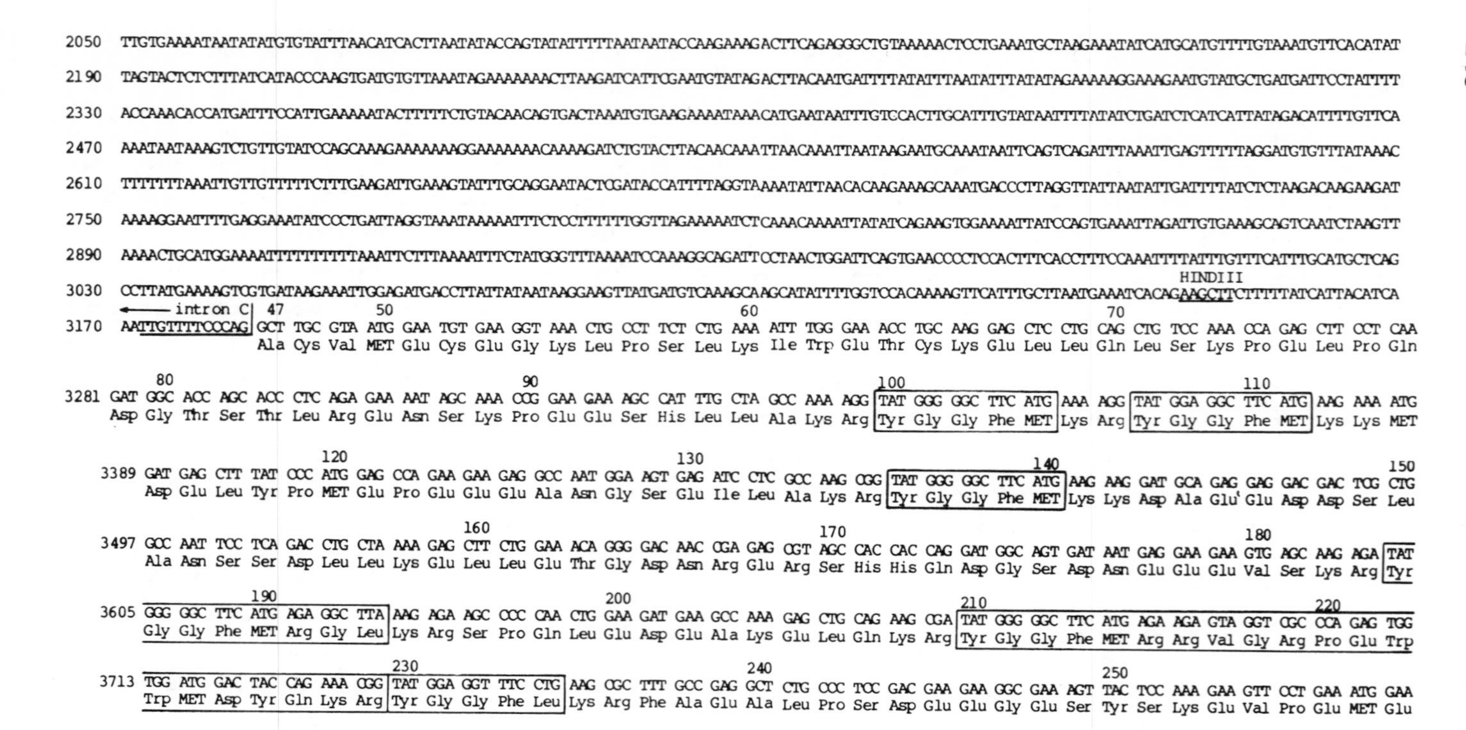

Fig. 5. (*cont.*) located farther upstream, and presumably important in transcriptional control, are boxed and underlined (*see* ref. 22 for more details). Several copies of the AAUAAA sequence occur downstream from the stop-codon (underlined). The $(GT)_{14}$ repeat element and the Alu-like repeat element are underlined. The direct 12-bp repeats flanking the Alu-like sequence are boxed (from ref. 22, with permission).

```
          260                     267
3821 AAA AGA TAC GGA GGA TTT ATG AGA TTT TAA  TATCTTTTCCCACTAGTGGCCCCAGGCCCCAGCAAGCCTCCCTCCATCCTCCAGTGGAAACTGTTGATGGTGTTTTATTGTCATGTGTTGCTTGCCTTG
     Lys Arg Tyr Gly Gly Phe MET Arg Phe END

3950 TATAGTTGACTTCATTGTCTGGATAACTATACAACCTGAAAAACTGTCATTTCAGGTTCTGTGCTCTTTTTGGAGTCTTTAAGCTCAGTATTAGTCTATTGCAGCTATCTCGTTTTCATGCTAAAATAGTTTTTGTTATC
                                                                                    ↓↓ ↓ 3' Poly(A) addition sites
4090 TTGTCTCTTATTTTTGACAAACATCAATAAATGCTTACTTGTATATAGAGATAATAAACCTATTACCCCAAGTGCATAATATCCTTGTAAGTATATTTTTCTCCAAGGCTCTCATTTACCTTTGGGTTTCCGCATCAATC

4230 CTAGACTCAGGGACAGTAGAACCATCTGCTTTTGACATAATTTTGACATCATAATCATTGTATTTCTTCTCCTTCAGCCAAGGCTCAAAAACATTTTTTATTTAATTCCTCACAGCATGAATCTTAGATTAGTAAACACG
                                                                                                        (dG-dT) repeat element
4370 TGGAGCAAATAAAATGTATTTCACTCTCATTTATCCAAATCAGATACATGAAGAAAAACCAAAGCCTAATAAAGGAGCCAGCTATGCCCCCATTAAACACACACACACACACACACACACATACTTGCATTTACATCA

4510 GATGTGTATGTCTCTTTTCTTCAGTAAAGTATTTTAGAAGTTAGTAAGAAATTAAAGATAATTTAGTCCAATTTTACAAATGAAGAAAGGGAAGCCAAGAAAGGTTGTGTAGTTGGCCCAAAGTTGCCCAGCAGTTGAAG

4650 GACAAAAATAGGATTAAAACCCCAAACTTTTTGTGCCCAGTCAATATCCTTTTCACTGCGGTTCCTTATCTCCTATCCACGTCACTAGTAAGAATTGTAAGAACGGTGGAACTACATCTTAGTCACCATTGTACACCTCT

4790 ACACTATGACTGTAGCAGATTCTCCAATATCGGATGAATGGATATATGAATGATGAATAAATTAATTTGGTGACCAGATAATGAGGAAAGATATTGTGAGTGTTTTGTAAGGTAAATCCCCAAAACACAAGGTGCTTATG

4930 ACTACTTCATATTAATGAGACAATGCTCAGATATAAGAAATGTAACTTTTCATTTTTCTTATTATTATTATTATACTTTAAGTTTTAGGGTACATGTGCACAGCATGCAGGTTTGTTACGTATGTAAACATGTGCCATTA

5070 ACTCGTCATTCTGGTTTCTCTTCAGATCGTGCAAATCTTTCGCCTTTAACTTTTCATTTTTCTAACTGAAACAACATACGGCTTAACCCCTACTCACTAAATCATTTTTCCAGGACCCCCACCAACCCTTTCTCTTGTGT

5210 TTCTCAAAACATTGGAATTTAGCAACTCTGGAAAGTGGGGACGAGTCTCTGGTCTGCAGTCTACCAAGCTGCTGCCGAGATGCTCACTGTGCAAGCCTGTGTGCCCCATCTGATAAGGCAGCTGCACTGTGCTCGGCAG

5350 TTTGTTCTCTCTGTACCAGCGCCAGGGCTCACAGGACATCCACATGATCATGCTGTTAGGAGACCCATCTCCTGGAATTAGTGCAAGGAATCACAGGGCCCCACACTCACAGAGCCCCAACTCCATCAATCTTCCTGATC

5490 CTTGGTAAACTAATTTTTTCTTCCAGCAAAGCCCGGGAGAAAAAAGATAATACAATAATTATTGATAAGATAATGCCATTATTATCTCACAATAGTCCAATATGTATACTTACATATCTCATACAATGTGTTTGAGAAAA

5630 TAACAAGTTCACCATCAGACAAATACATTTTTCAAAGCAAGCGCAGTTTCATGCGCGTCGGTGTGAAGAGACCACCAAACAGGCTTTGTGTGAGCAATAAAGCTTTTAATCATCTGGGTGCAGGCGGGCTGAGTCCGAAA

5770 AGAGAGTCAGCAAAGGGAGATAAGGGTGAGGCCATTTTATAAGATTTGGGTAGGTAAAGGAAAATTACAGTCAAAGGGGGTTGTTCTCTGGCGGGCAGGTGTGGGGGTCACAAGGTGCTCAGTAGGGGAGCTTTTGAGC

5910 CAGGACGAGCCAGGAGAAGGAATTTCACAAGATAATGTCATCACTTAAGGCAAGAACAGGCCATTTTCACTTCTTTTGTGGTAGAATGTCATCAGTTAAAGCAGGAACTGGCCATCTGGATGTGTACGTGCAGGTCACAG
                                              direct repeat                           alu-like repeat element
6050 GGGATATGATGGCTTAGCTTGGGCTCAGAGGCCTGACACACAGTTCTTTTTTTTTTTTTTAGAGATGAGGGCTCACTTGGTTGCCCAGGCTGGAGTGCAGTCGCTAGCTCACTTCAGCCTCGGCTTCCTGGCCTCAAGC
                                                                   direct repeat
6190 AATCCTCCCGCCTCAGCCTCACAAAATGCTGAGATTACAGGCGTGACACCGCGCCTGGCCAAGCACAGTTTTTTAATCTAGGAGGCTTTCTTCTCTTTTGGGGAAGCAGTGTAGACTGGGCTAAGGCACTCGTGGTAGT

6330 TAAGAGTTGTCTGCGGCTCCATTTCACCCCCACCCCCCAACACTTACCCCCCCCCAGCCCCGCCGGTCGCCTCCGCCAAGGGCCGCACGATGCTCGAGGGGTGGGACAGCGTCACACACTTGCAGTTCCTCAATTCTCTC

6470 TCATGTCATGGTGTAGAATCGGGAACCTAGAGGATGCGTTCAAAATTGCCTAGGGAAGGAAGGAACAGAGGAGAGGGGTCGGGACACGAGGAAAGGGGAAAGGGACGATCGGGGATGACTCCTGTGGAGGCGCCGGAGCG

6610 ATGCCAACCAAGCTGTGCGCCCGCCCCGGGCATTACAGGACGCAGGACCCCCTCCTCCAGAGGTCCTCAGCAGTAAGTACGGGCGGTCGTCCACAAAACATGACTCGAGGGGGTTGCGTAGGCCTCACTCACTCTCCAAA
                                                                                                                   EcoRI
6750 TTTCTCCTGAGCTGTCTGGCCTTCAATTTGAAATTGCAGCATTAGAAGTTTGCTTTCCAACTCTTAAGGGCTTTGATGTTTGATGTTTTTAAGAATTTCCTATTTGAATTC....3'
```

Fig. 5. *(continued)*

TABLE 3
Schematic Outline of the Exon–Intron Structure of the Human Proenkephalin Gene

| Segment | Length, basepairs | Remarks |
|---|---|---|
| Exon I | 70 | Comprising all but three nucleotides of the 5′-untranslated region |
| Intron A | 87 | |
| Exon II | 56 | |
| Intron B | 469 | — |
| Exon III | 141 | Contains three nucleotides of the 5′-untranslated region and codon nos. 1–46 |
| Intron C | 3500 | — |
| Exon IV | 938 | Contains codon nos. 47–267 and the 3′-untranslated region |

gene promoter to a sequence 67–171 bp upstream of the capping site (*25*). Thus, some of the short repeats and palindromes found appear to be important in promoter function. The intron–exon boundaries, as determined by sequencing all conform to the rule that an intron generally starts with GT and ends in AG (*26*). The underlined sequences at both ends of the introns are complementary to the 5′-terminal sequences of U1 small nuclear RNA (*27*). This complementarity is believed to be important in splicing. The genomic sequence shows the presence of several AAUAAA sequences downstream from the stop codon. Mapping of the actual polyadenylation site in three independent clones showed slight variations, reflecting possible variability in the transcript cleavage and subsequent addition of the poly(A) tail. Two related genes, coding for porcine prodynorphin (*28*) and human (*29*) and murine (*30*) POMC have been shown to contain multiple AAUAAA sequences as well.

Analysis of repetitive DNA elements within and surrounding the gene has shown the presence of three different classes of repeats (*22*). Repeat I is located within intron C, upstream from exon IV. Its frequency of occurrence is about $10^3$ times per haploid genome; its sequence has not yet been determined. Cross-hybridization experiments have demonstrated that repeat I is unrelated to repeats II–IV (*see* below). Repeat II is found 300 bases 3′ to the poly(A) addition site and consists of the sequence $(dGdT)_{14}$, occurring about $10^4$ times per haploid genome. Such a unit has been shown to be extremely conserved in the genomes of eukaryotes from yeast to man (*31*). This strong conservation suggests an important, but yet-unknown function. Repeat III occurs downstream from repeat II. It is approximately 72% homologous with the Alu consensus sequence. It occurs at a frequency of $\simeq 10^5$ copies per haploid genome. Repeat III is flanked by two imperfect, direct repeats, each consisting of 12 nucleotides. Repeat IV is situated 4 kb downstream from the gene terminus. It has not yet been sequenced. Hybridization studies show that it is again related to the Alu family. The overall significance of these repeats with respect to gene structure and function remains to be established—it may well be that the repeats have no specific function, but are simply a demonstration of the established presence of repetitive elements within the mammalian genome (*32*).

A careful analysis of the methylation pattern of the proenkephalin gene and its surrounding sequences was undertaken (*22*). It had been shown earlier that sequences next to an actively expressed gene were undermethylated compared with the same sequences in the vicinity of a nonexpressed gene (*33*). In mammalian DNA, methylation occurs as 5-methylcytosine. Greater than, or equal to, 90% of all those C-residues are found in the dinucleotide CpG (*34*). For the proenkephalin gene, the distribution of the dinucleotide CpG was found to be highly asymmetrical: 5′-flanking sequences, introns A and B and exons I–III, 10% CpG; intron C, 0.2% CpG; exon IV, 1.4% CpG; 3′-flanking sequences, 0.4, increasing to 8% CpG. The clustering of CpG dinucleotides in the 5′- and 3′-flanking regions is interesting in the light of a possible effect on gene expression (*see* above). Further analysis of actual methylation across the gene was undertaken (*22*). Certain restriction enzymes will not cleave the DNA if the C-residues in the recognition sites are methylated. Digestion of DNA from several tissue sources with such enyzmes, followed by Southern blot analysis, thus allows determination of whether a particular site is methylated. In the case of proenkephalin, some tissue-specific methylation was observed, although no clear correlation between methylation pattern and tissue-specific gene expression emerged.

### 2.4. Isolation and Characterization of the Rat Proenkephalin mRNA and Gene

Three reports have described the characterization of the rat proenkephalin, two of them by cloning a cDNA made from rat brain mRNA and one of them by characterizing a copy of the complete gene (*35–37*). Figure 6 shows the sequence of both the gene and the protein, as taken from reference (*37*). For the cDNA work, the screening approaches chosen were either the use of a synthetic oligonucleotide of 30 bases with a unique sequence (*see* Table 1). That sequence was based on a region coding for peptide E that had been shown to be highly conserved between human and bovine proenkephalin mRNA (*35*). The other report (*36*) describes the use of a cloned human cDNA as a probe to screen cDNA libraries established from rat brain (striatal) mRNA under conditions of reduced stringency. For the genomic cloning, a human proenkephalin cDNA probe was used to screen an established rat genomic library (*37*). The structure of the rat proenkephalin mRNA thus characterized shows the following features: total mRNA size (excluding the poly(A) tail) is 1271 nucleotides, with 155 nucleotides of 5′-untranslated region, 807 nucleotides coding region, and 312 nucleotides of a 3′-untranslated region that contains two copies of the polyadenylation signal AAUAAA. The translated portion of the mRNA codes for a protein of 269 amino acids, slightly larger than human (267 amino acids) and bovine (263 amino acids) proenkephalin. The protein shows the classical $NH_2$-terminal signal sequence, by comparison to the bovine precursor inferred to be 24 amino acids long. The mature portion of rat proenkephalin contains four copies of the Met-enkephalin sequence, one Leu-enkephalin sequence, and one copy each of Met-enkephalin $Arg^6$-$Gly^7$-$Leu^8$ and Met-enkephalin $Arg^6$-$Phe^7$. Each of these sequences is flanked by two basic amino acids. The overall structures of rat, human, and bovine proenkephalin are very similar (*see* Fig. 7). Positioning of enkephalin sequences within the precursor, the complete sequences of peptides E and F, and the positioning of the N-terminally located 6 Cys-residues have all been conserved between these species. The overall sequence homologies on the amino acid level are 82% (human/rat) and 78% (bovine/rat).

The isolation of the rat proenkephalin gene from an established rat genomic library demonstrated the following: The gene

```
5'────────>3'

1     GCTATTAGAC ATGCAAGAAA ATAGGACAGT GAGTGAAAAG TTATGACTTT CAGATAGTTG GGCAGAGGTC
71    ATCTCTGAGG TATGCCTCAA GAGGAGAGAA GGAAGGGAGC AGGTTCACCG AATCCGAGGT GTCCAACAGC
141   TCACGAAATC CAAGTCCGTC TGCCCTTTCA AACAACCCAC CCACGTGCCA CCGAAGTGAT TCATGCGGGG
211   AAAGCAGTTT CCCCTTCGTG CCCTCCCCAG AGTTTCTAAC TGATTGGTGG GGGACTCCGA CGCCCACTCG
281   CACCCGCCAC CTCGGGGGCC GCGTGCTGTG GGGACGTCCC CTCCCGCCAG CGTCGACACG GGCTGGCGTA
351   GGGCCTGCGT CAGCTGCAGC CCGATGGCGA TTGGCGCGCG CGCCTCTTCG GTTTGGGGCT AATTATAGGG
                                Cap site
421   TGGCTGTGCG GCCGCCAGAG AGGCAGGCGC ACAGAGCCCC GCAGCCCAGC GACGCTGCCG GGCGCCTCGT
491   AGAGCTCCCC GACGCGGCCA CCTCACACTT GCCTTCTTTA TTTCTCTTGC AGAGTGGCAT CTGCGATCCG
                                                 Intron A──>
561   CTCTTCCAGC TACCTGCGCC ATCTGAACAA CGGCAGCGTG AGTGACTTTG CCCAAGTCGC CGTAGCCTTG
631   CGGACCGTCT CCTGTTCCCT CGTTGCATGC GATCGTCAAC TTTCTCGGGG TTCCTCATTG TCCTGGGTGT
```

Fig. 6. Nucleotide sequence of the rat proenkephalin gene. The nucleotide sequence of the message strand, together with the deduced amino acid sequence is shown. The sequence of Met-enkephalin, Leu-enkephalin, and Met-enkephalin with carboxyl extension, together with the coding nucleotides, are boxed. The putative sites of capping and poly(A) addition are indicated by arrows. The exon-intron junctions are indicated by vertical lines and the sequences at both ends of the introns are underlined. The highly conserved sequences in the 5′-flanking region are underlined, and the TATA sequence is boxed (from ref. 37, with permission).

```
701   CCGGGTCCTA TTCCCAAGCG ACTGCT--------INTRON A ----------ACCCTGCTGA TCCTGCTCCA

771   CGACCACCCA CCCGGCAAGG TTCCCTCCTA GAGAACCTTG TCAGAGACAG AACGGGTCCC CACAGGCGCA
          <----Intron A|    1                                   10
841   TTCTTCTTTC CAACAGCCC ATG GCG CAG TTC CTG AGA CTT TGC ATC TGG CTC GTA GCG
                           MET Ala Gln Phe Leu Arg Leu Cys Ile Trp Leu Val Ala
                              20                                  30
900   CTT GGG TCC TGC CTC CTG GCT ACA GTG CAG GCA GAC TGC AGC CAG GAC TGC GCT
      Leu Gly Ser Cys Leu Leu Ala Thr Val Gln Ala Asp Cys Ser Gln Asp Cys Ala
                                      40                      46 |Intron B---->
955   AAA TGC AGC TAC CGC CTG GTA CGT CCC GGC GAC ATC AAC TTC CTG |GTAAG GTTGAATTTG
      Lys Cys Ser Tyr Arg Leu Val Arg Pro Gly Asp Ile Asn Phe Leu

1016  TGATGAGGGA-------INTRON B--------GGATCTCTAA CAAGACTCTA TTACCCTCTC CCTTCTTTGT

1067  TGCAGTGCTG CTGGACCATT GCTCATTCAG TTTTAGTGAA AGCTATGACG GAGAGGAGAT GGCTTTCTCT

1137  GCAGCGTGTG TACTAAAGCA GATAACAATA GTCCACCATT GGTTCAGAAG GGTATCTTTT AATCACTCAT
          <----Intron B| 47       50
1207  TTCCTTGACT TTGCAG| GCA TGC ACA CTC GAA TGT GAA GGG CAG CTG CCT TCT TTC
                         Ala Cys Thr Leu Glu Cys Glu Gly Gln Leu Pro Ser Phe
      60                                  70
1262  AAA ATC TGG GAG ACC TGC AAG GAT CTC CTG CAG GTG TCC AAG CCC GAG TTC CCT
      Lys Ile Trp Glu Thr Cys Lys Asp Leu Leu Gln Val Ser Lys Pro Glu Phe Pro
              80                                  90
1316  TGG GAT AAC ATC GAC ATG TAC AAA GAC AGC AGC AAA CAG GAT GAG AGC CAC TTG
      Trp Asp Asn Ile Asp MET Tyr Lys Asp Ser Ser Lys Gln Asp Glu Ser His Leu

                        100                                   110
1370  CTA GCC AAG AAG [TAT GGA GGG TTC ATG] AAA CGG [TAT GGA GGC TTC ATG] AAG AAG
      Leu Ala Lys Lys [Tyr Gly Gly Phe MET] Lys Arg [Tyr Gly Gly Phe MET] Lys Lys
```

Fig. 6. *(continued)*

```
                                     120                                                             130
1424  ATG GAT GAG CTT TAC CCC GTG GAG CCA GAA GAA GAG GCC AAT GGA GGC GAG ATC
      MET Asp Glu Leu Tyr Pro Val Glu Pro Glu Glu Glu Ala Asn Gly Gly Glu Ile
                                         140
1478  CTT GCC AAG AGG TAT GGC GGT TTC ATG AAG AAG GAT GCA GAT GAG GGA GAC ACC
      Leu Ala Lys Arg Tyr Gly Gly Phe MET Lys Lys Asp Ala Asp Glu Gly Asp Thr
      150                                         160
1532  TTG GCC AAC TCC TCC GAC CTG CTG AAA GAG CTA CTG GGA ACA GGA GAC AAC CGT
      Leu Ala Asn Ser Ser Asp Leu Leu Lys Glu Leu Leu Gly Thr Gly Asp Asn Arg
              170                                         180
1586  GCG AAA GAT AGC CAC CAA CAG GAA AGC ACC AAC AAT GAT GAA GAC AGC ACG AGC
      Ala Lys Asp Ser His Gln Gln Glu Ser Thr Asn Asn Asp Glu Asp Ser Thr Ser
                  190                                         200
1640  AAG AGG TAT GGG GGC TTC ATG AGA GGC CTC AAA AGA AGC CCC CAG CTG GAA GAC
      Lys Arg Tyr Gly Gly Phe MET Arg Gly Leu Lys Arg Ser Pro Gln Leu Glu Asp
                              210                                         220
1694  GAA GCA AAG GAG CTG CAG AAG CGC TAT GGG GGC TTC ATG AGA AGG GTC GGG CGC
      Glu Ala Lys Glu Leu Gln Lys Arg Tyr Gly Gly Phe MET Arg Arg Val Gly Arg
                                          230
1748  CCC GAG TGG TGG ATG GAC TAT CAG AAG AGA TAC GGA GGC TTC CTG AAG CGC TTT
      Pro Glu Trp Trp MET Asp Tyr Gln Lys Arg Tyr Gly Gly Phe Leu Lys Arg Phe
      240                                     250
1802  GCT GAG TCT CTA CCC TCG GAT GAA GAA GGC GAA AGT TAC TCT AAA GAA GTT CCC
      Ala Glu Ser Leu Pro Ser Asp Glu Glu Gly Glu Ser Tyr Ser Lys Glu Val Pro
              260                                         269
1856  GAG ATG GAA AAA AGA TAC GGA GGC TTT ATG CGG TTT TGA AGCCCTTCCC AGCAGTGATC
      Glu MET Glu Lys Arg Tyr Gly Gly Phe MET Arg Phe END

1915  CCGGGCCCCC ACTAGCCTGC TCCATCCCCC GTGAGCAACT GCCCCATCAG TGATGGTCTG TCATGTGCTG

1985  CTTGTGCTGT ACAGTTGCCC TCGTGGTCTG GATAACTGCT GCCTGAAAGC TGTGATTTTT AGGGGGTTGT
```

Fig. 6. *(continued)*

```
2055  CTGTATTCTT TTGAGTCTTG AAGCTCAGTA TTGGTCTCTT GTGGCTATGT TATCATGCTG AAACAGTCTG
2125  TTACCTCATC CCTTCTGACA AAACGTCAAT AAATGCTTAT TTGTATATAA AAATAATAAA CCCGTGAACC
                3'Poly (A)addition site
2195  CAACTGCACA ACGATCTTGT AAGAACTGGC TGTCATTTGT GGGTCAATGC TAGATGGAGA GACAAAGCTA
2265  TCACCTGCTA TGACAGAATA ATTGCTTGCC TTTTCCTTTC CTGACCCAGG CTGCAGAACA TTTTTCTTTA
2335  GCTCTATTAG TATGTGAACT CTTAAATAAG TAAACGCAAT AGAACAAAAT TATATGCAAA GACAATTTAA
2405  CCACATTTTA TCCATGCAGA ATGTGACAGC CAGAATGGCG GAGTACTTAG TCTAAATTTG CTCAGAAGTT
2475  GAAACATGAA TGGGATTAAG GCCCTAATCT TCTTATTCCC AGTCAGTGTA CCTTTATGCT CCATTTTCTC
2545  CCCCTCCATC CAGGAGTCTG GCAAGAACAG GAAAAAAACA TGTCACTATA TCTTGTTCAG CATTGCACAC
2615  CCTCCACCAT GGCCACAGTA GATTCTCTAA TAACTAATGA GTGAGTGAAG GAAGTGTATC GGTGAATTC
```

Fig. 6. *(continued)*

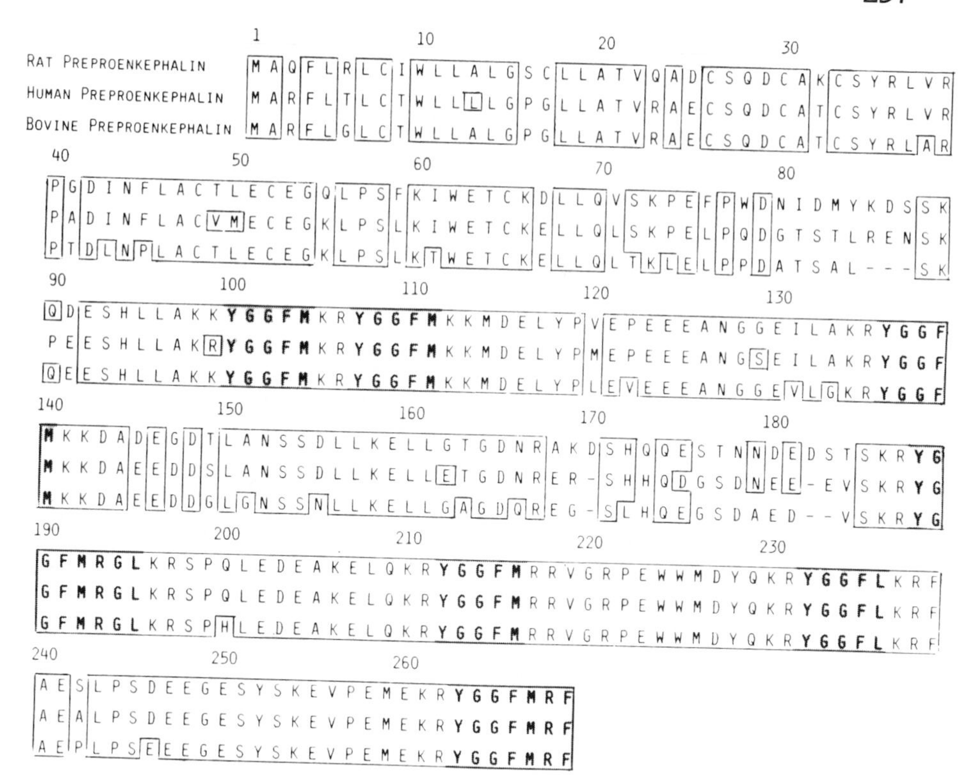


Fig. 7. Comparative amino acid sequences of rat, human, and bovine preproenkephalin (top, middle, and bottom lines, respectively, of each row). The human and bovine proteins are compared with rat preproenkephalin; regions of homology with the rat sequence are boxed. Deletions in the human and bovine sequences are denoted with a line under the corresponding rat amino acid. The sequences of (Met)enkephalin, (Leu)enkephalin, (Met)enkephalin-Arg$^6$-Gly$^7$-Leu$^8$, and (Met)enkephalin-Arg$^6$-Phe$^7$ are shaded (from ref. *35*, with permission).

stretches over approximately 5.4 kbp of DNA and was shown to be a single copy gene by genomic Southern blot analysis. Comparison of the sequences derived from the gene with the human proenkephalin gene structure and with the sequence of the rat proenkephalin cDNA allowed positioning of the intron–exon boundaries. There are three exons, two of which contain the translated portion of the mRNA and two introns, one of them in the 5′-untranslated region and one in the coding region between codon nos. 46 and 47 (Table 4).

TABLE 4
Schematic Outline of the Exon–Intron Structure of the Rat Proenkephalin Gene

| Segment | Length, basepairs | Remarks |
|---|---|---|
| Exon I | 152 | — |
| Intron A | ≃600 | — |
| Exon II | 141 | Contains three nucleotides of the 5′-untranslated region and codon nos. 1–46 |
| Intron B | ≃3500 | — |
| Exon III | 978 | Contains codon nos. 47–269 and the 3′-untranslated region |

Transcription of the whole gene should thus give rise to a precursor mRNA of ~5.5 kb. Northern blots of mRNA from striatum, a tissue rich in proenkephalin mRNA, actually show a possible precursor molecule at ~6 kb. The sites of initiation of transcription (cap-site) and polyadenylation for the rat proenkephalin mRNA were assigned based on comparisons with the sequence of the rat proenkephalin cDNA and with the studies done on the human proenkephalin gene. A comparison between human and rat genomic sequences located 5′ to the cap-site demonstrated the presence of two regions 100–150 bp long that are highly conserved between the two species. One of those regions corresponds to the sequences that in the human gene were found to represent the functional promoter region directly 5′ to the TATA box (*see* section 2.3). This finding suggests that the expression of both the human and rat gene might be controlled in a similar fashion.

### 2.5. Isolation of the Toad Proenkephalin Gene

In order to gain more insight into the evolutionary history of the proenkephalin protein precursor, the isolation of the proenkephalin gene from *Xenopus laevis* was undertaken (*38*). The human proenkephalin cDNA served as the probe to screen an established toad genomic library under conditions of reduced stringency.

Two recombinant phages containing most of the toad proenkephalin gene sequences were isolated. Comparison of the toad with the human genomic sequences allowed positioning of the intron-exon boundaries; the two main exons of the *Xenopus* gene thus isolated were characterized by DNA sequencing. The partial general structure derived for the toad proenkephalin precursor is indeed very similar to that of the human precursor (*see* Fig. 8). The toad precursor was shown to contain five copies of the Met-enkephalin sequence, and one copy each of Met-enkephalin Arg$^6$-Gly$^7$-Tyr$^8$ and Met-enkephalin Arg$^6$-Phe$^7$, but no Leu-enkephalin sequence.

This raises the interesting question of whether the Leu-enkephalin sequence was only a later evolutionary development from the (primordial) Met-enkephalin sequence. In the toad protein, all enkephalin sequences are flanked by doublebasic processing signals, analogous to the situation in higher mammals. The distribution of the different enkephalin sequences along the bovine, human, and *Xenopus* proenkephalins is very similar, with some of the spacer regions having diverged and some remaining very much the same. The highly conserved regions actually correspond to the ECP-sequences F, E, and B (*see* section 2.2) isolated from bovine adrenal medulla. This high degree of evolutionary conservation might again indicate an important, but yet unknown, physiological role for these peptides.

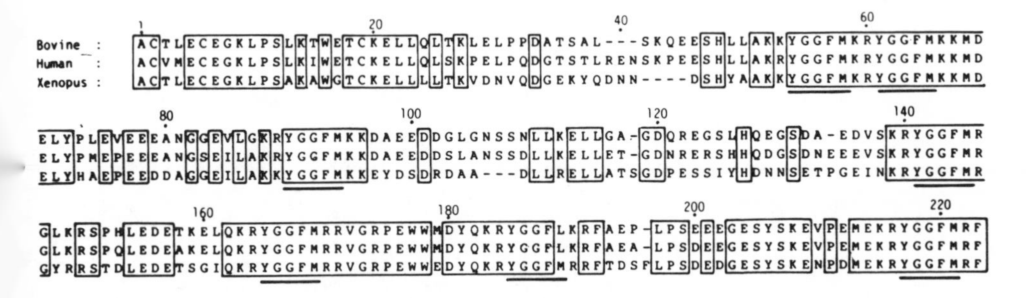


Fig. 8. Homology between *Xenopus* and mammalian proenkephalin sequences. Alignment of the amino acid sequence of part of bovine, human, and *Xenopus* proenkephalin. The one-letter amino acid notation is used. Residues identical among the three species are boxed. Gaps (−) have been introduced to achieve maximum homology. The locations of the enkephalin sequences are indicated by thick lines below the sequences (from ref. *38*, with permission).

## 2.6. Molecular Cloning of Porcine and Human Prodynorphin mRNAs

Since neither human nor bovine proenkephalin had been shown to contain the sequences for rimorphin, dynorphin, β-neoendorphin, or leumorphin, the search for yet another opioid peptide precursor continued. It led to the elucidation of the structure of prodynorphin, as cloned from porcine hypothalami (*28*). The human gene for prodynorphin was cloned and characterized (*39*) (*see* section 2.7). This study also described the partial cloning and sequencing of the human prodynorphin mRNA; the remainder of its sequence was derived from the genomic sequence. In the following discussion, the available data on the mRNAs are summarized and compared first, followed by a description of the genomic cloning.

cDNA cloning and screening with a synthetic oligonucleotide based on a partial dynorphin sequence (Table 1) were used as described for the proenkephalin just mentioned. The primary structure of porcine hypothalamic prodynorphin mRNA and the derived amino acid sequence of prodynorphin are shown in Fig. 9. The cloned portion of prodynorphin mRNA is 2333 nucleotides

```
                                          1                                   10                                      20
                                          Met Ala Trp Gln Gly Leu Leu Leu Ala Ala Cys Leu Leu Val Leu Pro Ser Thr Met Ala
5'-----------AGCACAGCAGGAAGACCCAAAACAGA AUG GCG UGG CAG GGG CUG CUG CUG GCG GCU UGC CUC CUU GUG CUC CCC UCC ACC AUG GCG
                 -20                  -1 1                   20                                   40                  60

                                    30                                      40                                      50
Asp Cys Leu Ser Gly Cys Ser Leu Cys Ala Val Lys Thr Gln Asp Gly Pro Lys Pro Ile Asn Pro Leu Ile Cys Ser Leu Glu Cys Gln
GAC UGC CUG UCC GGG UGC UCC UUG UGU GCU GUG AAG ACC CAG GAU GGG CCC AAA CCC AUC AAC CCC CUG AUU UGC UCC CUG GAA UGC CAG
                         80                         100                     120                        140

                                    60                                      70                                      80
Ala Ala Leu Gln Pro Ala Glu Glu Trp Glu Arg Cys Gln Gly Leu Leu Ser Phe Leu Ala Pro Leu Ser Leu Gly Leu Glu Gly Lys Glu
GCU GCC CUG CAG CCC GCU GAG GAG UGG GAG AGG UGC CAG GGC CUU CUG UCU UUU CUC GCU CCC UUG AGC CUC GGG CUC GAA GGC AAG GAA
            160                     180                         200                         220                         240

                                    90                                      100                                     110
Asp Leu Glu Ser Lys Ala Ala Leu Glu Glu Pro Ser Ser Glu Leu Val Lys Tyr Met Gly Pro Phe Leu Lys Glu Leu Glu Lys Asn Arg
GAC UUG GAG AGC AAG GCA GCU UUG GAA GAG CCC UCU AGU GAG CUG GUC AAG UAC AUG GGG CCC UUC UUG AAG GAG CUG GAG AAA AAC AGA
                         260                        280                     300                        320

                                    120                                     130                                     140
Phe Leu Leu Ser Thr Pro Ala Glu Glu Thr Ser Leu Ser Arg Ser Leu Val Glu Lys Leu Arg Ser Leu Pro Gly Arg Leu Gly Glu Glu
UUC CUC CUC AGC ACC CCA GCG GAG GAG ACC UCU CUG AGC AGG AGC CUG GUG GAG AAG CUC AGG AGC CUC CCU GGC AGG UUG GGG GAG GAA
            340                     360                         380                         400                         420

                        Glu         150                                     160                                     170
Thr Glu Ser Glu Leu Met Gly Asp Ala Gln Gln Asn Asp Gly Ala Met Glu Ala Ala Ala Leu Asp Ser Ser Val Glu Asp Pro Lys Glu
ACA GAG UCU GAG CUG AUG GGG GAC GCC CAG CAG AAU GAU GGU GCC AUG GAG GCU GCA GCC CUG GAU UCC AGU GUG GAG GAC CCC AAG GAG
                         A   U                     460                     480                        500
                         440
                                    180                                     190                         Gly         200
Gln Val Lys Arg [Tyr Gly Gly Phe Leu] [Arg Lys Tyr Pro] Lys Arg Ser Ser Glu Val Ala Gly Glu Gly Asp Gly Asp Arg Asp Lys Val
CAG GUC AAA CGU [UAU GGG GGC UUU CUG] [CGC AAA UAC CCC] AAA AGG AGC UCA GAA GUG GCU GGG GAG GGG GAU GGG GAC AGG GAU AAG GUG
            520                     540                         560                         580        G    C          600

                                    210                                     220                                     230
Gly His Glu Asp Leu Tyr Lys Arg [Tyr Gly Gly Phe Leu] [Arg Arg Ile Arg Pro Lys Leu Lys Trp Asp Asn Gln] Lys Arg [Tyr Gly Gly
GGU CAU GAA GAC CUG UAC AAG CGC [UAC GGG GGC UUC UUA] [CGG CGC AUU CGU CCC AAG CUC AAG UGG GAC AAC CAG] AAG CGC [UAU GGU GGU
    C                    620                        640                     660                        680

                                    240                                     250
Phe Leu] Arg Arg Gln Phe Lys Val Val Thr Arg Ser Gln Glu Asp Pro Asn Ala Tyr Tyr Glu Glu Leu Phe Asp Val
UUU CUC] CGG CGC CAG UUC AAG GUG GUU ACU CGG UCU CAG GAA GAC CCC AAU GCC UAU UAU GAA GAG CUU UUU GAU GUG UAA ACCCUUCCCCAU
            700                     720                         740                         760                         780
```

long. This cDNA is lacking about 150 nucleotides from the very 5′ end of the natural mRNA (*28*). Given these numbers, it seems reasonable to assign to porcine prodynorphin mRNA a size of ≃ 2500 nucleotides, excluding the poly(A) tail. This value agrees well with the size of the mRNA estimated by Northern blot analysis (2600 nucleotides); the difference accounts for the uncloned poly(A) tail. Prodynorphin mRNA contains a 5′-untranslated region of approximately 180 nucleotides, a coding region of 768

CCUGGAAAUGAGUCAGGAGCUUUCCCUAAGGCCCUUCCAGGUGGGAGGGCACACGUUCAUCCUCCCCUAUAGCCCUCAUUUCCACGCUCAGUUCAGCAUUGUCUAUAAAACAUCCAAACC
800 820 840 860 880 900
UCAUCUGCCUCUCUUUCCACCUGGGUGCUGUGUGUGUCUGGGUCAGUGAGAGGGAGGGUGGAGAUUCCCCUUUCCAAUAGGCUUAGUGCUUGGCUCCCACCCUACACAGCAGCAGCUCUU
920 940 960 980 1,000 1,020
GACACCAGUCCCUCCACCCCAUCCUUGUGAUGCCCCAGUUUCAGGAAUCCAGGUGGACGUCUUGACCUCUCUGGAUCUAGACUCUCCAAAGUUUACCGAUCAAUGCCCCUCUCAGAAGAU
1,040 1,060 1,080 1,100 1,120 1,140
AAAUGAGUAUACCCCCUCCCAUGCAGAACUAUACAUGUAUAUAAUAUAUACAAAUAUCAUAAAAUAUAUGAGGUAAACAUUUAAAACAAGAGUGAUAAACAAAUAAAAAUUCUAACAUUU
1,160 1,180 1,200 1,220 1,240 1,260
GAUUCUCAAAGCUCAAAAGAUCUCCCUACUGUGCAGACCAUGCUUUAGGUAAAAAGCUCAAACACUGCCUUUUAUUGGAGCAGGAAUUCCUACACUAGAAAGUUUGGUCUCCCAAGUGGG
1,280 1,300 1,320 1,340 1,360 1,380
UUAGAUAUUCAGUCAGUUACUCUGCGUUCUUCUUGUUUAGAAACACACCUUUUAAGAACGCAGUUCCUGGGCUCCAGUCCAUGUUGGUACCCUGGACAGCGCCCAGCUCCUCCGAGGAGA
1,400 1,420 1,440 1,460 1,480 1,500
GGAAGAGCGUGAAAAUCCCUUUCUGAAAUGGUUAAUGAGAGCAGCUCUCCUGAAUGCUGAAAUGAUCAAGGAGGGAGUGAGGCAAACCAAUUUGUUCUGUGCAACAGAUUCAAAAUGUGG
1,520 1,540 1,560 1,580 1,600 1,620
ACCGGUUCCCUCAGCCCUCAUUAAACUAAUUAAACUGAUCGGUAUCACACUCCAACCCCAGGAUGAACUGAAGCUGAGUUGAGCUGAUGAGGUUAAGCACAACCGUGUUCUUGAGCAGCU
1,640 1,660 1,680 1,700 1,720 1,740
GAAUCUGCCGCCAAGAGUCCAAGCCAUCUGGCCAAACAUAUGUGUUGGGCAUUGGGCGAGGCAAUCCAGAAGCAACAGCUAGAAAGAGGAGCUGGCCCUCUUUAGCCCCCAUGAUGAUUU
1,760 1,780 1,800 1,820 1,840 1,860
UUUCCUCUUAAUGUCUCAAAAUAAAACCAGAAGGAAGAAUGAAAUGAUUAAGUGCUUGAGGCCAAAUGAGUUCCCUUUAUUCAAAUAACCCAGAAACAGAGGAAGGACCUCAAUCAAAGU
1,880 1,900 1,920 1,940 1,960 1,980
CCUCUCCUCUCUCUCUGUCUCUGUCUGUCAGUCUCUUGCCUGUCCCCAGGCACUAUGGUUGGGUUUGUGGAUCCAGGAGGCUGGCCUGGAUGGGGAAGGAAAUAAAUGUAAGUUUGGUUA
2,000 2,020 2,040 2,060 2,080 2,100
UUUUGUAUAGGAUGCUGCUGAGCCCACCUCUUCAUGCAUAACCCCAGGCCUCUCAAGAUGUUCUGAAAUCUAUUGAUUGUUUUAGAGUUACUUUGUGUGCUUUUAAAAAUAUGCUUUUUU
2,120 2,140 2,160 2,180 2,200 2,220
UUUUUUUUUGCAAUUUACUUGGAAUUUGCUUAGUCCUUGUGUUAUUUCCUGCUCCUUCAGUACAAUAAAUAAAAGAAAGAUCCUG-----------3′
2,240 2,260 2,280 2,300

Fig. 9. (*opposite and above continued*) The primary structure of porcine prodynorphin mRNA (*28*). The nucleotide sequence determined from cloned cDNA is shown. Nucleotides are numbered in the 5′ to 3′ direction, beginning with the first nucleotide of the initiator codon. The nucleotides 5′ to the initiator codon are given negative numbers. The cloned 5′-untranslated region does not extend to the very 5′-end of the mRNA (*see* text). The cloned 3′-untranslated region extends to the poly(A) tail, and is therefore complete. It contains five copies of the sequence AAUAAA (underlined), two of which overlap. The predicted amino acid sequence of prodynorphin is shown, beginning with the initiator-Met at position 1. Leu-enkephalin sequences are boxed with solid lines, and their carboxy-terminal extensions with dashed lines. The nucleotide differences in the protein coding region observed between the individual clones sequenced occur at positions 440, 444, 589, 594, 606. Two of these changes, at positions 440 and 589, result in replacement of Gly at position 147 with Arg and Glu at position 197 with Gly (*see* text) (from ref. *28*, with permission).

nucleotides encoding a protein of 256 amino acids, and an unusually long 3′-untranslated region of 1539 nucleotides [excluding the poly(A) tract]. Five independently isolated cDNA clones had contained this long 3′-untranslated region; an artifact of cloning can thus be excluded. This 3′-untranslated region contains five copies of the polyadenylation sequence AAUAAA. Three of these sequences occur 200–1000 nucleotides upstream from the actual poly(A) tail. The usual distance between AAUAAA sequence and poly(A) addition site is 10–20 nucleotides. Northern blot analysis did not detect any prodynorphin mRNA substantially shorter than 2600 nucleotides, indicating that those latter polyadenylation sequences are probably not used. Sequencing of the coding region allowed the alignment of all the known remaining ECPs into one unambiguous structure (*see* subsequent discussion). There were five single nucleotide differences observed between the several clones sequenced (Fig. 9). Two of those changes result in amino acid replacements at positions 147 (Gly or Glu) and 197 (Arg and Gly); the other changes do not affect the amino acid sequence. It is possible that this reflects polymorphism, inasmuch as the hypothalami from several animals had been pooled for the analysis. Reverse-transcriptase errors cannot be completely excluded either.

The ECPs that had been isolated and sequenced align in prodynorphin in the following way: α-neoendorphin (residue nos. 175–184); β-neoendorphin (residue nos. 175–183); dynorphin 1–17 or dynorphin A (residue nos. 209–225); rimorphin or dynorphin B (residue nos. 228–240); leumorphin (residue nos. 228–256); dynorphin-32 (residue nos. 209–240). The whole precursor molecule carries an $NH_2$-terminal signal of 20 amino acids (*see* preceding discussion); the assigned signal-peptide cleavage would leave a molecule of 236 amino acids. The $NH_2$-terminal 154 amino acids of this proform do not contain any enkephalin sequences or dibasic processing signals. The similarity with proenkephalin, however, is marked. Six Cys residues are found positioned almost identical with those in proenkephalin (the spacing between Cys-residue nos. 3 and 4 is larger by one amino acid in porcine prodynorphin). Such a strict conservation of positioning makes it tempting to again speculate about certain functions of these Cys in the folding and proper processing of the precursor. The overall structural outline of prodynorphin is similar to that of proenkephalin; the processing of this multivalent precursor may again liberate a number of biologically active peptides in a coordinate fashion, thus diversifying the information contained in a single gene.

The structures of human prodynorphin and its mRNA were derived from sequencing of the cloned genomic DNA (*see* section 2.7), as well as by cloning and sequencing of partial cDNA derived from human prodynorphin mRNA (*39*). The human mRNA was assigned a very large 5′-untranslated region of about 1400 nucleotides. Using human hypothalamic poly(A)+ RNA and DNA primers, either chemically synthesized or derived from cloned DNA, partial human prodynorphin cDNAs were cloned and sequenced; 1111 nucleotides of the 5′-untranslated region, as well as the protein-coding region to amino acid residue no. 43, have been cloned and sequenced in this way. It still cannot be excluded that part of this long 5′-untranslated sequence represents an intron and is removed by splicing from the primary RNA transcripts. The coding region of the mRNA consists of 762 nucleotides, coding for a protein of 254 amino acids. The 3′-untranslated region is again very long, as in the porcine mRNA; 1566 nucleotides, excluding the poly(A) tail. The human prodynorphin protein consists of 254 amino acids. Its $NH_2$-terminus represents a putative signal peptide of 20 amino acids. Relative to the porcine precursor, there are 59 amino acid changes, 20 of them conservative, and a two-amino-acid deletion (positions 198–199, Asp-Lys) (*see* Fig. 9). The positioning of the 6 Cys residues and the ECP sequences within the precursor moiety have been completely conserved, as have the sequences for dynorphin, neoendorphin, and rimorphin. The C-terminal portion of leumorphin shows three amino acid substitutions relative to porcine leumorphin: Ser instead of Tyr (residue no. 250 in the porcine sequence) (*see* Fig. 9), Gly instead of Glu (residue no. 251); Ala instead of Val (residue no. 256). Again, the overall structural outline of the two proteins has been well conserved.

## 2.7. Isolation of the Human Prodynorphin Gene

The human prodynorphin gene was isolated from an established human genomic DNA library (*39*). The porcine cDNA to prodynorphin mRNA was used as the screening probe. Southern blot analysis of both human genomic placental DNA and the isolated genomic clones produced identical restriction maps, confirming that the DNA fragments cloned were derived from a single segment on the human genomic DNA. The prodynorphin gene thus characterized consists of a stretch of DNA of approximately 16.5 kbp. Comparison of the genomic sequences with human and porcine prodynorphin cDNA sequences allowed positioning of three introns (A, B, C) within the gene (Fig. 10). Hence, the gene consists of four exons. Two of the introns (A, B) interrupt the 5′-untranslated region of the mRNA. Assignment of exon I remains tentative; it cannot be excluded that there is another yet-unidentified intron in this region. Exons III and IV together contain the translated region of the mRNA. Intron C interrupts the protein coding region between codon nos. 43 and 44. A schematic outline of the gene is given in Table 5, and the entire gene sequence is shown in Fig. 10. Thus, the primary RNA transcript from this gene must be ≃16.4 kb, 12.8 kb of which are subsequently removed by splicing. Attempts to directly map the cap site of the mRNA on the genomic DNA by using human hypothalamic poly(A)+ RNA failed, presumably because of the low abundance of the prodynorphin mRNA in the RNA preparation used (*39*). The cap site was tentatively assigned to the A residue 25 nucleotides downstream from the TATA-box indicated in Fig. 10. Another sequence element that is often found upstream from the cap site is the CAAT-box (*40*) (*see* Fig. 10), which is possibly involved in the control of gene expression. The CAAT-box (consensus sequence 5′-GC$^{C}_{T}$CAA TCT-3′) has been found 70–80 bp upstream from the transcription initiation site. The exact interactions between TATA-box, CAAT-box, RNA-polymerase II, and other possible factors in the control of gene expression are not yet known. The presence of these two possible regulatory sequences and their distance from the cap site make its assignment plausible. The region 5′ to the cap site shows another interesting feature: There is a 68-bp sequence that is tandemly repeated three times (indicated by arrows in Fig. 7). Such a sequence arrangement might be involved in the regulation or modulation of gene expression.

A sequence comparison between the human and porcine cDNA sequences indicates that the 60-bp exon II is absent in the porcine mRNA. The exon–intron boundaries, as determined by sequencing, all conform to the rule that an intron generally starts with GT and ends in AG (*26*). The human genomic sequence shows the presence of one AAUAAA sequence, 12 nucleotides

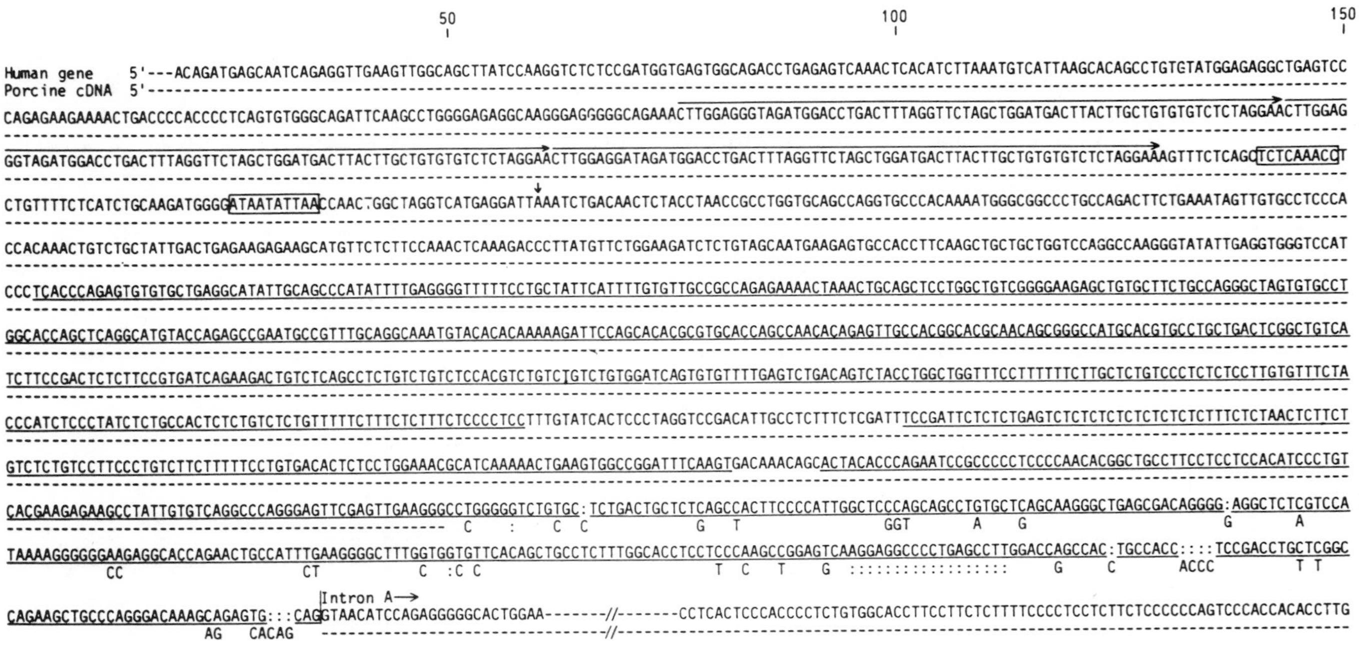

```
                                              50                                                100                                               150
                                              |                                                 |                                                 |
Human gene    5'---ACAGATGAGCAATCAGAGGTTGAAGTTGGCAGCTTATCCAAGGTCTCTCCGATGGTGAGTGGCAGACCTGAGAGTCAAACTCACATCTTAAATGTCATTAAGCACAGCCTGTGTATGGAGAGGCTGAGTCC
Porcine cDNA  5'------------------------------------------------------------------------------------------------------------------------------------------

CAGAGAAGAAAACTGACCCCACCCCTCAGTGTGGGCAGATTCAAGCCTGGGGAGAGGCAAGGGAGGGGGCAGAAACTTGGAGGGTAGATGGACCTGACTTTAGGTTCTAGCTGGATGACTTACTTGCTGTGTGTCTCTAGGAACTTGGAG
------------------------------------------------------------------------------------------------------------------------------------------------------

GGTAGATGGACCTGACTTTAGGTTCTAGCTGGATGACTTACTTGCTGTGTGTCTCTAGGAACTTGGAGGATAGATGGACCTGACTTTAGGTTCTAGCTGGATGACTTACTTGCTGTGTGTCTCTAGGAAAGTTTCTCAGCTCTCAAACCT
------------------------------------------------------------------------------------------------------------------------------------------------------

CTGTTTTCTCATCTGCAAGATGGGGATAATATTAACCAACTGGCTAGGTCATGAGGATTAAATCTGACAACTCTACCTAACCGCCTGGTGCAGCCAGGTGCCCACAAAATGGGCGGCCCTGCCAGACTTCTGAAATAGTTGTGCCTCCCA
------------------------------------------------------------------------------------------------------------------------------------------------------

CCACAAACTGTCTGCTATTGACTGAGAAGAGAAGCATGTTCTCTTCCAAACTCAAAGACCCTTATGTTCTGGAAGATCTCTGTAGCAATGAAGAGTGCCACCTTCAAGCTGCTGCTGGTCCAGGCCAAGGGTATATTGAGGTGGGTCCAT
------------------------------------------------------------------------------------------------------------------------------------------------------

CCCTCACCCAGAGTGTGTGCTGAGGCATATTGCAGCCCATATTTTGAGGGGTTTTTCCTGCTATTCATTTTGTGTTGCCGCCAGAGAAAACTAAACTGCAGCTCCTGGCTGTCGGGGAAGAGCTGTGCTTCTGCCAGGGCTAGTGTGCCT
------------------------------------------------------------------------------------------------------------------------------------------------------

GGCACCAGCTCAGGCATGTACCAGAGCCGAATGCCGTTTGCAGGCAAATGTACACACAAAAAGATTCCAGCACACGCGTGCACCAGCCAACACAGAGTTGCCACGGCACGCAACAGCGGGCCATGCACGTGCCTGCTGACTCGGCTGTCA
------------------------------------------------------------------------------------------------------------------------------------------------------

TCTTCCGACTCTCTTCCGTGATCAGAAGACTGTCTCAGCCTCTGTCTGTCTCCACGTCTGTCTGTCTGTGGATCAGTGTGTTTTGAGTCTGACAGTCTACCTGGCTGGTTTCCTTTTTTCTTGCTCTGTCCCTCTCTCCTTGTGTTTCTA
------------------------------------------------------------------------------------------------------------------------------------------------------

CCCATCTCCCTATCTCTGCCACTCTCTGTCTCTGTTTTTCTTTCTCTTTCTCCCCTCCTTTGTATCACTCCCTAGGTCCGACATTGCCTCTTTCTCGATTTCCGATTCTCTCTGAGTCTCTCTCTCTCTCTCTCTTTCTCTAACTCTTCT
------------------------------------------------------------------------------------------------------------------------------------------------------

GTCTCTGTCCTTCCCTGTCTTCTTTTTCCTGTGACACTCTCCTGGAAACGCATCAAAAACTGAAGTGGCCGGATTTCAAGTGACAAACAGCACTACACCCAGAATCCGCCCCCTCCCCAACACGGCTGCCTTCCTCCTCCACATCCCTGT
------------------------------------------------------------------------------------------------------------------------------------------------------

CACGAAGAGAAGCCTATTGTGTCAGGCCCAGGGAGTTCGAGTTGAAGGGCCTGGGGGTCTGTGC:TCTGACTGCTCTCAGCCACTTCCCCATTGGCTCCCAGCAGCCTGTGCTCAGCAAGGGCTGAGCGACAGGGG:AGGCTCTCGTCCA
------------------------------------------------------    C   :   C  C          G   T            GGT     A   G                      G     A

TAAAAGGGGGGAAGAGGCACCAGAACTGCCATTTGAAGGGGCTTTGGTGGTGTTCACAGCTGCCTCTTTGGCACCTCCTCCCAAGCCGGAGTCAAGGAGGCCCCTGAGCCTTGGACCAGCCAC:TGCCACC::::TCCGACCTGCTCGGC
           CC                      CT              C  :C C                   T  C   T   G  :::::::::::::::::::      G     C       ACCC       T T

                                    |Intron A-->
CAGAAGCTGCCCAGGGACAAAGCAGAGTG:::CAGGTAACATCCAGAGGGGGCACTGGAA-------//-------CCTCACTCCCACCCCTCTGTGGCACCTTCCTTCTCTTTTCCCCTCCTCTTCTCCCCCCAGTCCCACCACACCTTG
                           AG   CACAG  ---------------------------------//----------------------------------------------------------------------------
```

Fig. 10. The nucleotide sequence of the human prodynorphin gene and its flanking regions (*39*). The strand corresponding to human prodynorphin mRNA is shown, together with the deduced amino acid sequence for human prodynorphin. Nucleotide and amino acid sequence differences found in the porcine sequences are shown below the human sequence. For positions in which nucleotide differences are noted between individual porcine cDNA clones (Fig. 9), the residue that agrees with the human sequence has been taken; the absence of a nucleotide or an amino acid in the porcine sequence indicates that the human and porcine sequences are the same; the presence of a colon in either nucleotide sequence indicates a gap; the hyphens indicate sequences that

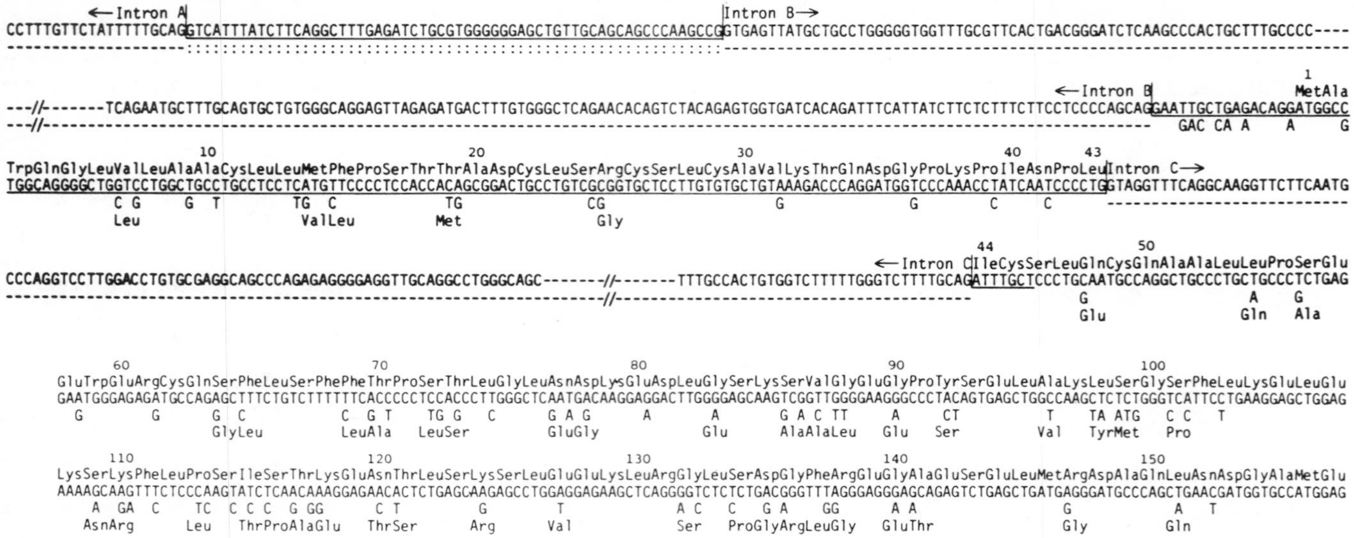

```
     ←Intron A|                                                          |Intron B→
CCTTTGTTCTATTTTTGCAGGTCATTTATCTTCAGGCTTTGAGATCTGCGTGGGGGGAGCTGTTGCAGCAGCCCAAGCCGGTGAGTTATGCTGCCTGGGGGTGGTTTGCGTTCACTGACGGGATCTCAAGCCCACTGCTTTGCCCC----
------------------::::::::::::::::::::::::::::::::::::::::::::::::::::::::::::::::----------------------------------------------------------------------

                                                                                                                                                        1
                                                                                                                                   ←Intron B|        MetAla
---//------TCAGAATGCTTTGCAGTGCTGTGGGCAGGAGTTAGAGATGACTTTGTGGGCTCAGAACACAGTCTACAGAGTGGTGATCACAGATTTCATTATCTTCTCTTTCTTCCTCCCCAGCAGGAATTGCTGAGACAGGATGGCC
---//-------------------------------------------------------------------------------------------------------------------------------------- GAC CA A   A    G

                   10                       20                        30                        40     43
TrpGlnGlyLeuValLeuAlaAlaCysLeuLeuMetPheProSerThrThrAlaAspCysLeuSerArgCysSerLeuCysAlaValLysThrGlnAspGlyProLysProIleAsnProLeu|Intron C→
TGGCAGGGGCTGGTCCTGGCTGCCTGCCTCCTCATGTTCCCCTCCACCACAGCGGACTGCCTGTCGCGGTGCTCCTTGTGTGCTGTAAAGACCCAGGATGGTCCCAAACCTATCAATCCCCTGGTAGGTTTCAGGCAAGGTTCTTCAATG
            C G     G T         TG C             TG               CG                     G              G       C     C   --------------------------
            Leu                 ValLeu          Met              Gly

                                                                                                                    44                 50
                                                                                                         ←Intron C|IleCysSerLeuGlnCysGlnAlaAlaLeuLeuProSerGlu
CCCAGGTCCTTGGACCTGTGCGAGGCAGCCCAGAGAGGGGAGGTTGCAGGCCTGGGCAGC-------//-------TTTGCCACTGTGGTCTTTTTGGGTCTTTTGCAGATTTGCTCCCTGCAATGCCAGGCTGCCCTGCTGCCCTCTGAG
-------------------------------------------------------------------//--------------------------------------           G                          A    G
                                                                                                                        Glu                        Gln  Ala

      60                        70                        80                        90                        100
GluTrpGluArgCysGlnSerPheLeuSerPhePheThrProSerThrLeuGlyLeuAsnAspLysGluAspLeuGlySerLysSerValGlyGluGlyProTyrSerGluLeuAlaLysLeuSerGlySerPheLeuLysGluLeuGlu
GAATGGGAGAGATGCCAGAGCTTTCTGTCTTTTTTCACCCCCTCCACCCTTGGGCTCAATGACAAGGAGGACTTGGGGAGCAAGTCGGTTGGGGAAGGGCCCTACAGTGAGCTGGCCAAGCTCTCTGGGTCATTCCTGAAGGAGCTGGAG
  G       G       G C             C G T   TG G   C      G A G       A       A       G A C TT    A     CT            T    TA ATG   C C    T
                  GlyLeu          LeuAla  LeuSer        GluGly              Glu       AlaAlaLeu  Glu   Ser           Val   TyrMet   Pro

      110                       120                       130                       140                       150
LysSerLysPheLeuProSerIleSerThrLysGluAsnThrLeuSerLysSerLeuGluGluLysLeuArgGlyLeuSerAspGlyPheArgGluGlyAlaGluSerGluLeuMetArgAspAlaGlnLeuAsnAspGlyAlaMetGlu
AAAAGCAAGTTTCTCCCAAGTATCTCAACAAAGGAGAACACTCTGAGCAAGAGCCTGGAGGAGAAGCTCAGGGGTCTCTCTGACGGGTTTAGGGAGGGAGCAGAGTCTGAGCTGATGAGGGATGCCCAGCTGAACGATGGTGCCATGGAG
    A  GA  C     TC  C C C  G GG      C T          G          T            A C    C  G A    GG      A A                  G             A   T
   AsnArg        Leu  ThrProAlaGlu    ThrSer        Arg        Val          Ser   ProGlyArgLeuGly    GluThr                Gly           Gln
```

Fig. 10. (*cont.*) have not been determined or cloned. The human nucleotide sequences carried by the cDNA clones are underlined. The amino acid residues in the human sequence are numbered, beginning with the initiator Met. The sequences of β-neoendorphin, dynorphin, and rimorphin are boxed with solid lines, and the carboxy-terminal 16 amino acid residues of leumorphin with a dashed line. The exon–intron junctions are indicated by vertical lines. The putative sites of capping and poly(A) addition are shown by vertical arrows. The putative TATA-box and the putative CAAT-box in the 5′-flanking region are indicated. Arrows indicate a sequence 68 bp long located 5′ to the cap site and tandemly repeated three times (from ref. *39*, with permission).

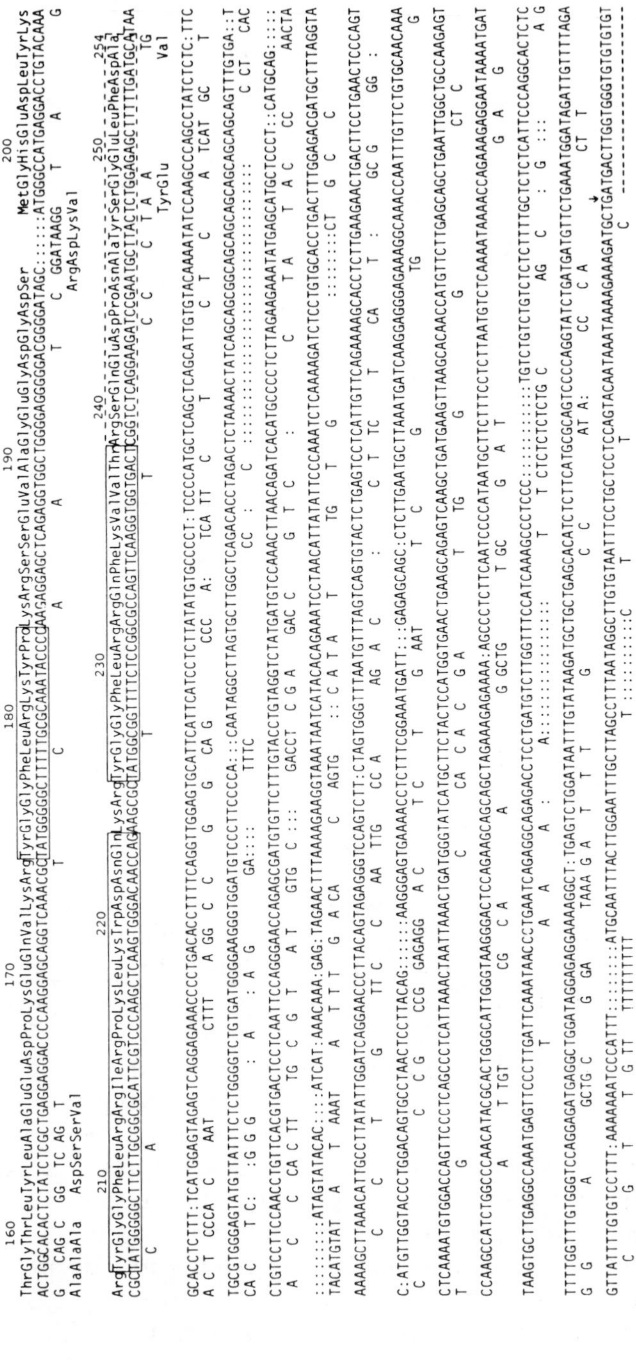

```
   160                                170                                180                               190                                 200
ThrGlyThrLeuTyrLeuAlaGluGluAspProLysGluGlnValLysArgTyrGlyGlyPheLeuArgLysTyrProLysArgSerSerGluValAlaGlyGluGlyAspGlyAspSer      MetGlyHisGluAspLeuTyrLys
ACTGGCACACTCTATCTCGCTGAGGAGGACCCCAAGGAGCAGGTCAAACGCTATGGGGGCTTTTTGCGCAAATACCCCAAGAGGAGCTCAGAGGTGGCTGGGGAGGGGGACGGGGATAGC:::::ATGGGCCATGAGGACCTGTACAAA
G   CAG C   GG  TC AG  T                               T              C              A           A                T   C  GGATAAGG    T    A          G
AlaAlaAla    AspSerSerVal                                                                                               ArgAspLysVal

          210                                220                                230                               240                          250     254
ArgTyrGlyGlyPheLeuArgArgIleArgProLysLeuLysTrpAspAsnGlnLysArgTyrGlyGlyPheLeuArgArgGlnPheLysValValThrArgSerGlnGluAspProAsnAlaTyrSerGlyGluLeuPheAspAla
CGCTATGGGGGCTTCTTGCGGCGCATTCGTCCCAAGCTCAAGTGGGACAACCAGAAGCGCTATGGCGGTTTTCTCCGGCGCCAGTTCAAGGTGGTGACTCGGTCTCAGGAAGATCCGAATGCTTACTCTGGAGAGCTTTTTGATGCATAA
    C        A                                                     T                    T                    C C      C  T A  A                  TG
                                                                                                                           TyrGlu               Val

GCACCTCTTT:TCATGGAGTAGAGTCAGGAGAAACCCCTGACACCTTTTCAGGTTGGAGTGCATTCATTCATCCTCTTATATGTGCCCCT:TCCCCATGCTCAGCTCAGCATTGTGTACAAAATATCCAAGCCCAGCCTATCTCTC:TTC
A C T  CCCA  C    AAT          CTTT   A GG  C  C     G    G   CA G        CCC   A:    TCA TT   C       T          C T   C      A  TCAT  GC     T

TGCGTGGGAGTATGTTATTTCTCTGGGGTCTGTGATGGGGAAGGGTGGATGTCCCTTCCCCA:::CAATAGGCTTAGTGCTTGGCTCAGACACCTAGACTCTAAAACTATCAGCAGCGGCAGCAGCAGCAGCAGCAGCAGTTTGTGA::T
CA C    T C:   :G G G      :  A    : A  G       GA::::      TTTC                CC  :    C  :::::::::::::::::::::::::::::::::::              C CT   CAC

CTGTCCTTCCAACCTGTTCACGTGACTCCTCAATTCCAGGGAACCAGAGCGATGTGTTCTTTGTACCTGTAGGTCTATGATGTCCAAACTTAACAGATCACATGCCCCTCTTAGAAGAAATATGAGCATGCTCCCT::CATGCAG::::
 A    C    C  CA C TT     TG  C  G  T    A T   GTG  C :::     GACCT  C G A    GAC C      G  T  C       :          C      T A      T  A C    CC        AACTA

::::::::ATAGTATACAC::::ATCAT:AAACAAA:GAG:TAGAACTTTAAAAGAAGGTAAATAATCATACACAGAAATCCTAACATTATATTCCCAAATCTCAAAAGATCTCCTGTGCACCTGACTTTGGAGACGATGCTTTAGGTA
TACATGTAT    A    T  AAAT      A    T T T   G  A CA        C    AGTG     :: C A T A     T          TG     T     G                 :::::::CT   G  C    C

AAAAGCTTAAACATTGCCTTATATTGGATCAGGAACCCTTACAGTAGAGGGTCCAGTCTT:CTAGTGGGTTTAATGTTTAGTCAGTGTACTCTGAGTCCTCATTGTTCAGAAAAGCACCTCTTGAAGAACTGACTTCCTGAACTCCCAGT
          C      C       T       G        TT C    C     AA  TTG     CC A         AG  A  C        :         C  T  TC      T      CA      T :        GC G       GG  :

C:ATGTTGGTACCCTGGACAGTGCCTAACTCCTTACAG:::::AAGGGAGTGAAAACCTCTTTCGGAAATGATT:::GAGAGCAGC:CTCTTGAATGCTTAAATGATCAAGGAGGGAGAAAGGCAAACCAATTTGTTCTGTGCAACAAA
 C                     C   C G     CCG  GAGAGG   A C       T C     T      G  AAT         T    C          G                   TG                          G

CTCAAAATGTGGACCAGTTCCCTCAGCCCTCATTAAACTAATTAAACTGATGGGTATCATGCTTCTACTCCATGGTGAACTGAAGCAGAGTCAAGCTGATGAAGTTAAGCACAACCATGTTCTTGAGCAGCTGAATTGGCTGCCAAGAGT
T              G                                        C        CA  C A  C   G A          T   TG        G              G                      CT  C

CCAAGCCATCTGGCCCAACATACGCACTGGGCATTGGGTAAGGGACTCCAGAAGCAGCAGCTAGAAAGAGAAAA:AGCCCTCTTCAATCCCCATAATGCTTCTTTCCTCTTAATGTCTCAAAATAAAACCAGAAAGAGGAATAAAATGAT
                    A      T TGT           CG    C A          A              G GCTG         T GC       G   A  T                                 G   A    G

TAAGTGCTTGAGGCCAAATGAGTTCCCTTGATTCAAATAACCCTGAATCAGAGGCAGAGACCTCCTGATGTCTTGGTTTCCATCAAAGCCCTCCC:::::::::::TGTCTGTCTGTCTCTCTTTTGCTCTCTCATTCCCAGGCACTCTC
                                      T           A    A       A  :        A::::::::::::::::          T     T CTCTCTCTCTG C            AG    C     :  G   :::            A G

TTTTGGTTTGTGGGTCCAGGAGATGAGGCTGGATAGGAGAGGAAAAGGCT:TGAGTCTGGATAATTTGTATAAGATGCTGCTGAGCACATCTCTTCATGCGCAGTCCCCAGGTATCTGATGATGTTCTGAAATGGATAGATTGTTTTAGA
G  G          A         GCTG C        G  GA        TAAA G A    T    T  T            G              C  C           AT A:          CC    C A                CT   T

                                                                                                                                                   ↓
GTTATTTTGTGTCCTTT:AAAAAAATCCCATTT::::::::ATGCAATTTACTTGGAATTTGCTTAGCCTTTAATAGGCTTGTGTAATTTCCTGCTCCTCCAGTACAATAAATAAAAGAAAGATGCTGATGACTTGGTGGGTGTGTGTGT
    C         G     T       T  G TT     TTTTTTTTT                           T ::::::::::C      T                T                         C  ----------------------

ACGTGCACGTGCACATGTGTG---3'
------------------------3'
```

Fig. 10. *(cont.)*

TABLE 5
Schematic Outline of the Intron–Exon Structure of the Human Prodynorphin Gene

| Sequence | Length, basepairs | Remarks |
|---|---|---|
| Exon I | 1316 | 5′-Untranslated region |
| Intron A | ≃1200 | |
| Exon II | 60 | 5′-Untranslated region (absent in the porcine mRNA) |
| Intron B | ≃990 | |
| Exon III | 145 | Contains 16 nucleotides of the 5′-untranslated region and codon nos. 1–43 |
| Intron C | ≃1700 | |
| Exon IV | 2196 | Contains codon nos. 44–254 and the 3′-untranslated region |

upstream from the poly(A) addition site. This is in contrast to the porcine mRNA, in which several of these sequences were found (*see* section 2.6). The overall structural outlines of the proenkephalin and prodynorphin genes are very similar (*see* below).

### 2.8. Isolation of the Rat Prodynorphin Gene

A recent report described the characterization of the main portion of the rat prodynorphin gene (*41*). A partial porcine prodynorphin cDNA had been used to screen an established rat genomic library. Comparison with the human gene sequence allowed positioning of the intron–exon boundaries. The partial structure (204 amino acids) for the rat prodynorphin thus derived turned out to be extremely similar to the human and porcine prodynorphins (Fig. 11). The rat precursor contains all the dynorphin-peptides described before, flanked by two basic amino acids, and arranged in identical manner. For the three species, the homologies in the area of the dynorphin peptides are very high (90%) on both the amino acid and nucleic acid level (Fig. 11). Rat prodynorphin mRNA is 2.4 kb long and has a very long 3′-untranslated region (1.4 kb) like the human and porcine prodynorphin mRNAs.

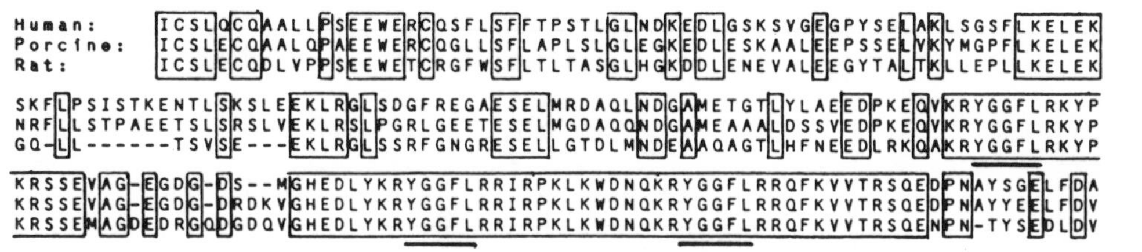


Fig. 11. Amino acid sequence homology between human, porcine, and rat prodynorphin. The single letter amino acid notation is used. Boxed areas represent residues that are identical between the three species. Gaps (–) have been introduced to achieve maximum homologies. The locations of the enkephalin sequences are indicated by a solid line below the sequences (from ref. *41*, with permission).

Northern blot analysis was used to determine sites of synthesis for the rat mRNA in both brain and nonbrain tissues. Overall, the highest concentrations were found in striatum; some size heterogeneity for the mRNA among different tissues was noticed.

## 2.9. Summary of Structural and Evolutionary Considerations

Enkephalins and enkephalin-containing polypeptides are derived from three different gene products; proenkephalin, prodynorphin, and proopiomelanocortin (POMC). As just mentioned, the overall structural outlines of the three proteins are indeed very similar (Fig. 12). They are similar in size and contain an $NH_2$-terminal signal peptide, followed by a region that is devoid of enkephalin or melanotropin sequences, but contain all the Cys residues found in the protein. In proenkephalin and prodynorphin, the positioning of these residues is almost identical (*see* section 2.6). It has been suggested that these Cys might be important for the proper folding and subsequent processing of the precursor molecules. In all three proteins, the enkephalin sequences and ECPs are clustered in the C-terminal half of the precursors. The location and spacing of the enkephalin sequences found in proenkephalin and prodynorphin are very similar (Fig. 12). An amino acid sequence comparison of proenkephalin and prodynorphin shows a high degree of homology extending over the entirety of both molecules (Fig. 13). Pairs of identical or chemically similar amino acids occupy 55% of porcine prodynorphin/human proenkephalin; 58% of porcine prodynorphin/bovine proenkephalin, 52% of human prodynorphin/human proenkephalin;

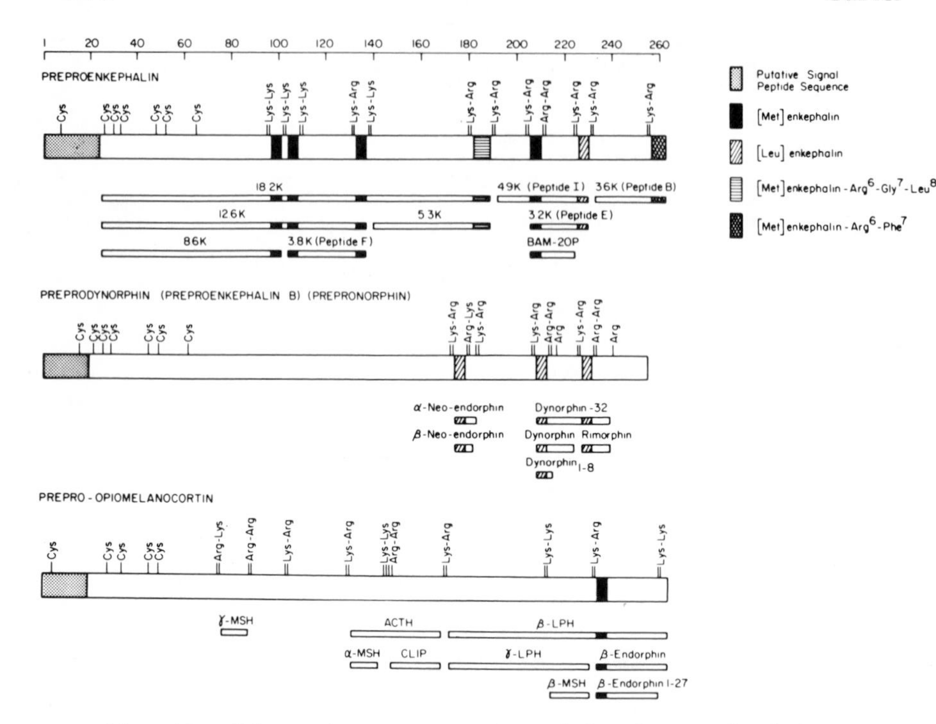


Fig. 12. Schematic representation of the three enkephalin-containing peptide gene products. The positions of Cys residues and paired and single basic amino acids serving as processing sites are indicated. Characterized products of processing are also shown (from ref. 5, with permission).

and 54% of human prodynorphin/bovine proenkephalin (*39*). Fig. 13 shows that the spacing of the enkephalin repeats has been strongly conserved, as has the region of peptide E (residue nos. 233–257 in Fig. 13). As discussed above, both enkephalin and nonenkephalin sequences could be important for the proper functioning of the peptide-receptor system. It may be hypothesized that the enkephalin sequence confers the opioid activity to the peptide, whereas the added nonenkephalin sequence confers specificity in receptor targeting or transport, increased stability, or both. The strong sequence conservation in the region of peptide E supports such a point of view.

A comparison of the structures of the genes for the three proteins is shown in Fig. 14. Positioning of exons and introns has been well conserved. Two exons contain the translated portion of the gene, one small exon contains the signal sequence and an

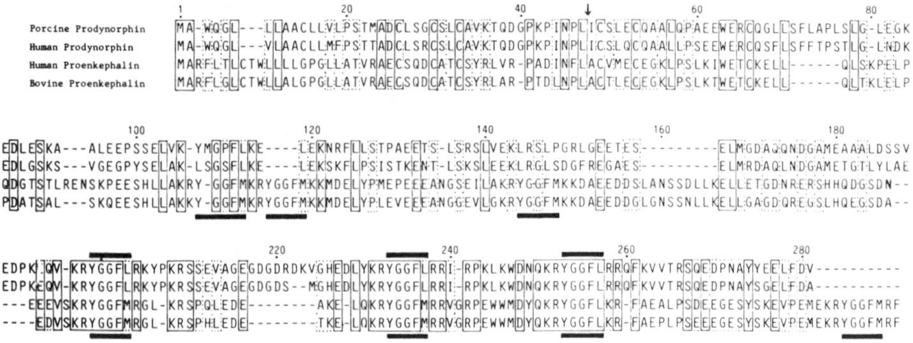

```
                       1                  20                  40     ↓            60                  80
Porcine Prodynorphin   MA-WQGL---LLAACLLVLPSTMADCLSGCSLCAVKTQDGPKPINPLICSLECQAALQPAEEWERCQGLLSFLAPLSLG-LEGK
Human Prodynorphin     MA-WQGL---VLAACLLMFPSTTADCLSRCSLCAVKTQDGPKPINPLICSLQCQAALLPSEEWERCQSFLSFFTPSTLG-LNDK
Human Proenkephalin    MARFLTLCTWLLLLGPGLLATVRAECSQDCATCSYRLVR-PADINFLACVMECEGKLPSLKIWETCKELL------QLSKPELP
Bovine Proenkephalin   MARFLGLCTWLLALGPGLLATVRAECSQDCATCSYRLAR-PTDLNPLACTLECEGKLPSLKTWETCKELL------QLTKLELP

           100                 120                 140                 160                 180
EDLESKA---ALEEPSSELVK-YMGPFLKE----LEKNRFLLSTPAEETS-LSRSLVEKLRSLPGRLGEETES---------ELMGDAQQNDGAMEAAALDSSV
EDLGSKS---VGEGPYSELAK-LSGSFLKE----LEKSKFLPSISTKENT-LSKSLEEKLRGLSDGFREGAES---------ELMRDAQLNDGAMETGTLYLAE
QDGTSTLRENSKPEESHLLAKRY-GGFMKRYGGFMKKMDELYPMEPEEEANGSEILAKRYGGFMKKDAEEDDSLANSSDLLKELLETGDNRERSHHQDGSDN--
PDATSAL---SKQEESHLLAKKY-GGFMKRYGGFMKKMDELYPLEVEEEANGGEVLGKRYGGFMKKDAEEDDGLGNSSNLLKELLGAGDQREGSLHQEGSDA--

                               220                 240                 260                 280
EDPK[illegible]QV-KRYGGFLRKYPKRSSEVAGEGDGDRDKVGHEDLYKRYGGFLRRI-RPKLKWDNQKRYGGFLRRQFKVVTRSQEDPNAYYEELFDV----------
EDPKEQV-KRYGGFLRKYPKRSSEVAGEGDGDS--MGHEDLYKRYGGFLRRI-RPKLKWDNQKRYGGFLRRQFKVVTRSQEDPNAYSGELFDA----------
---EEEVSKRYGGFMRGL-KRSPQLEDE--------AKE-LQKRYGGFMRRVGRPEWWMDYQKRYGGFLKR-FAEALPSDEEGESYSKEVPEMEKRYGGFMRF
----EDVSKRYGGFMRGL-KRSPHLEDE--------TKE-LQKRYGGFMRRVGRPEWWMDYQKRYGGFLKR-FAEPLPSEEEGESYSKEVPEMEKRYGGFMRF
```

Fig. 13. Alignment of the amino acid sequences of porcine and human prodynorphin with those of human and bovine proenkephalin. The one-letter amino acid notation is used. Sets of four identical residues are enclosed with solid lines, and sets of four residues that are identical or are considered to be favored amino acid substitutions are enclosed with dotted lines. Favored amino acid substitutions are defined as pairs of residues belonging to one of the following groups: S, T, P, A, G; N, D, E, Q; H, R, K; M, I, L, V; F, Y, W (*51*). Gaps (-) have been inserted to achieve maximum similarity. Positions in the aligned sequences, including gaps, are numbered, beginning with that of the initiator Met. The locations of the enkephalin sequences are indicated by thick lines above the prodynorphin sequences and below the proenkephalin sequences. The position at which the protein coding sequence is interrupted by an intron is shown by an arrow. The extent of similarity between the sequences was assessed statistically, as described previously (*52*). In these tests, the positions corresponding to the enkephalin and extended enkephalin sequences and the paired basic amino acid residues flanking them (positions 105–121, 142–150, 197–209, 231–239, 251–259, and 283–291) were excluded from both of the sequences compared. The alignment score (A) (*52*) and the probability (P) that the score is realized by chance, were evaluated for each pair of the aligned proenkephalin and prodynorphin sequences as follows: porcine prodynorphin/human proenkephalin, $A = 5.48$ and $P < 10^{-4}$; porcine prodynorphin/bovine proenkephalin $A = 5.45$ and $P < 10^{-4}$; human prodynorphin/human proenkephalin, $A = 2.90$ and $P = 1.9 \times 10^{-3}$; human prodynorphin/bovine proenkephalin, $A = 5.26$ and $P < 10^{-4}$ (*35*) (from ref. *39*, with permission).

$NH_2$-terminal portion of the protein, and the other exon contains the repeated enkephalin or melanotropin sequences.

For proenkephalin and prodynorphin, cDNA sequencing has shown the presence of small (≃15 nucleotides) direct duplications

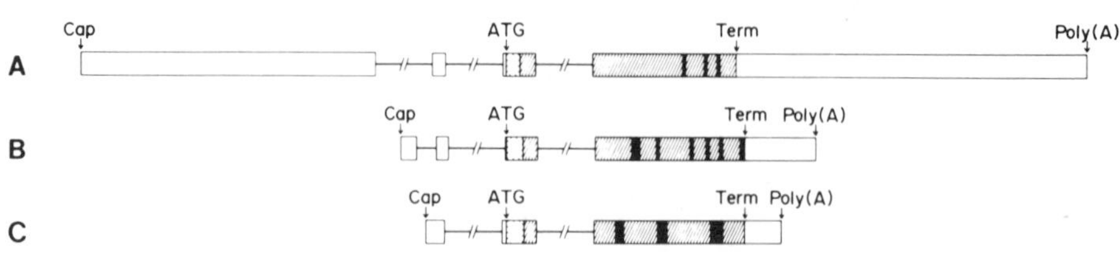


Fig. 14. Schematic representation of the structural organizations of the mammalian genes encoding prodynorphin (A), proenkephalin (B), and proopiomelanocortin (C). Exons are shown by blocks and introns by lines; the lengths of the lines do not always represent the real lengths of the introns. The sites of capping, translational initiation (ATG), translational termination (Term), and poly(A) addition are indicated; the capping site of the human prodynorphin gene has been tentatively assigned, and the possibility that the 5′-terminal region of this gene contains (an) additional intron(s) cannot be excluded (*39*). The coding regions for the repeated enkephalin or melanotropin structures are indicated by closed boxes, those for the signal peptide by stippled boxes, and the remaining protein coding regions by hatched boxes (from ref. *39*, with permission).

in addition to the enkephalin repeats (*11,28*). It is tempting to speculate that these genes evolved by a series of duplications of an ancestral opioid precursor gene. In that context, it is interesting to keep in mind the concept of the ECPs with the added nonenkephalin sequence stretch. Assuming that the enkephalin pentapeptide sequence evolved first, adding more functional sequence to it enlarged the repertoire of sequences that could be further changed by mutation. Slight changes in processing patterns for such a repetitive protein structure probably further expanded the versatility of the system. Yet, another type of mutation could have acted on the peptide–receptor genes themselves (*22*). Thus, a picture emerges in which the opioid peptide–receptor system has evolved in a number of different ways to become increasingly complex and meet the demands of more-complex organisms.

## ACKNOWLEDGMENTS

I would like to thank A. Stern for his patient help and advice during the preparation of this manuscript, M. Comb, E. Herbert, S. Numa, and S. Udenfriend for supplying original illustrations, and J. Farruggia and L. Bowen for expert secretarial help.

## REFERENCES


1. Pert C. B. and Snyder S. H. (1973) Opiate receptor: Demonstration in nervous tissue. *Science* **179,** 1011–1014.
2. Simon E. J., Hiller J. M., and Edelman I. (1973) Stereospecific binding of the potent narcotic analgesic ($^3$H) etorphine to rat-brain homogenate. *Proc. Natl. Acad. Sci. USA* **70,** 1947–1949.
3. Terenius L. (1973) Stereospecific interaction between narcotic analgesics and a synaptic plasma membrane fraction of rat cerebral cortex. *Acta Pharmacol. Toxicol.* **32,** 317–320.
4. Hughes J., Smith T. W., Kosterlitz H. W., Fothergill L. A., Morgan B. A., and Morris H. R. (1975) Identification of two related pentapeptides from the brain with potent opiate agonist activity. *Nature* (London) **258,** 577–579.
5. Udenfriend S. and Kilpatrick D. L. (1983) Biochemistry of the enkephalins and enkephalin containing peptides. *Arch. Biochem. Biophys.* **211,** 309–323.
6. Beaumont A. and Hughes J. (1979) Biology of opioid peptides. *Ann. Rev. Pharmacol. Toxicol.* **19,** 245–267.
7. Snyder S. H. and Innis R. B. (1979) Peptide neurotransmitters. *Annu. Rev. Biochem.* **8,** 755–782.
8. Childers S. R. (1980) Enkephalin and Endorphin Receptors, in *Neurotransmitter Receptors. Receptors and Recognition* Ser B Vol. 9, (Ena S. J. and Yamamura H. I., eds.), Chapman and Hall, London.
9. Wallace R. B., Shaffer J., Murphy R. I., Bonner J., Hirose T., and Itakura K. (1979) Hybridization of synthetic oligodeoxyribonucleotides to $\phi\chi$174 DNA: The effect of single base pair mismatch. *Nucleic Acids Res.* **6,** 3543–3557.
10. Lewis R. V., Stern A. S., Rossier J., Stein S., and Udenfriend S. (1979) Putative enkephalin precursors in bovine adrenal medulla. *Biochem. Biophys. Res. Comm.* **89,** 822–829.
11. Noda M., Furutani Y., Takahashi H., Toyosato M., Hirose T., Inayama S., Nakanishi S., and Numa S. (1982) Cloning and sequence analysis of cDNA for bovine adrenal preproenkephalin. *Nature* **295,** 202–206.
12. Gubler U., Kilpatrick D. L., Seeburg P. H., Gage L. P., and Udenfriend S. (1981) Detection and partial characterization of proenkephalin mRNA. *Proc. Natl. Acad. Sci. USA* **78,** 5484–5487.
13. Gubler U., Seeburg P. H., Hoffman B. J., Gage L. P., and Udenfriend S. (1982) Molecular cloning establishes proenkephalin as precursor of enkephalin containing peptides. *Nature* **295,** 206–208.
14. Comb M., Seeburg P. H., Adelman J., Eiden L., Herbert E. (1982) Primary structure of the human Met- and Leu-enkephalin precursor and its mRNA. *Nature* **295,** 663–666.

15. Legon S., Glover D. M., Hughes J., Lowry P. J., Rigby P. W. J., and Watson C. J. (1982) The structure and expression of the proenkephalin gene. *Nucleic Acids Res.* **10,** 7905–7918.
16. Gubler U. and Hoffman B. (1983) A simple and very efficient method for generating cDNA libraries. *Gene* **25,** 263–269.
17. Proudfoot N. J. and Brownlee G. G. (1976) 3′Noncoding region sequences in eukaryotic messenger RNA. *Nature* **263,** 211–214.
18. Blobel G. and Dobberstein B. (1975) Transfer of proteins across membranes: Presence of proteolytically processed and unprocessed nascent immunoglobulin light chains on membrane bound ribosomes of murine myeloma. *J. Cell Biol.* **67,** 835–851.
19. Steiner D. F., Quinn P. S., Chang S. J., Marsh J., and Tager H. S. (1980) Processing mechanisms in the biosynthesis of proteins. *Ann. NY Acad. Sci.* **343,** 1–16.
20. Stern A. S., Jones B. N., Shively J. E., Stein S., and Udenfriend S. (1981) Two adrenal opioid polypeptides: Proposed intermediates in the processing of proenkephalin. *Proc. Natl. Acad. Sci.* **78,** 1962–1966.
21. Kilpatrick D. L., Jones B. N., Kojima K., and Udenfriend S. (1981) Idientification of the octapeptide [Met] enkephalin-$Arg^6$-$Gly^7$-$Leu^8$ in extracts of bovine adrenal medulla. *Biochem. Biophys. Res. Commun.* **103,** 698–705.
22. Comb M., Rosen H., Seeburg P., Adelman J., and Herbert E. (1983) Primary structure of the human proenkephalin gene. *DNA* **2,** 213–229.
23. Noda M., Teranishi Y., Takahashi H., Toyosato M., Notake M., Nakanishi S., and Numa S. (1982) Isolation and structural organization of the human proenkephalin gene. *Nature* **297,** 431–434.
24. Breathnach R. and Chambon P. (1981) Organization and expression of eukaryotic split genes coding for proteins. *Annu. Rev. Biochem.* **50,** 349–383.
25. Terao M., Watanabe Y., Mishina M., and Numa S. (1983) Sequence requirement for transcription in vivo of the human preproenkephalin A gene. *EMBO J.* **2,** 2223–2228.
26. Breathnach R., Benoist C., O'Hare K., Gannon F., and Chambon P. (1978) Ovalbumin gene: Evidence for a leader sequence in mRNA and DNA sequences at the exon–intron boundaries. *Proc. Natl. Acad. Sci. USA* **75,** 4853–4857.
27. Lerner M. R., Boyle J. A., Mount S. M., Wolin S. L., and Steitz J. A. (1980) Are RNAs involved in splicing? *Nature* **283,** 220–224.
28. Kakidani H., Furutani Y., Takahashi H., Noda M., Morimoto Y., Hirose T., Asai M., Inayama S., Nakanishi S., Numa S. (1982) Cloning and sequence analysis of cDNA for porcine β-neoendorphin/dynorphin precursor. *Nature* (London) **298,** 245–249.
29. Whitfeld P. L., Seeburg P. H., and Shine J. (1982) The human pro-

opiomelanocortin gene: Organization, sequence and interspersion with repetitive DNA. *DNA* **1,** 133–143.

30. Uhler M., Herbert E., D'Eustachio P., and Ruddle F. D. (1983) The mouse genome contains two nonallelic pro-opiomelanocortin genes. *J. Biol. Chem.* **258,** 9444–9453.
31. Hamada H., Petriho M. G., and Kakunage T. (1982) A novel repeated element with Z-DNA forming potential is widely found in evolutionarily diverse eukaryotic genomes. *Proc. Natl. Acad. Sci. USA* **79,** 6465–6469.
32. Lewin B. (1976) *Gene Expression-2,* John Wiley and Sons, London.
33. Felsenfeld G. and McGhee J. (1982) Methylation and gene control. *Nature* **296,** 602–603.
34. Ehrlich M. and Wang R. Y. (1981) 5-Methyl-cytosine in eukaryotic DNA. *Science* **212,** 1350–1357.
35. Howells R. D., Kilpatrick D. L., Bhatt R., Monahan J. J., Poonian M., and Udenfriend S. (1984) Molecular cloning and sequence determination of rat preproenkephalin cDNA: Sensitive probe for studying transcriptional changes in rat tissues. *Proc. Natl. Acad. Sci. USA* **81,** 7651–7655.
36. Yoshikawa Y., Williams C., and Sabol S. L. (1984) Rat brain proenkephalin mRNA. *J. Biol. Chem.* **259,** 14301–14308.
37. Rosen H., Douglass J., and Herbert E. (1984) Isolation and characterization of the rat proenkephalin gene. *J. Biol. Chem.* **259,** 14309–14313.
38. Martens G. J. M. and Herbert E. (1984) Polymorphism and absence of leuenkephalin sequences in proenkephalin genes in *Xenopus laevis. Nature* **310,** 251–254.
39. Horikawa S., Takai T., Toyosato M., Takahashi H., Noda M., Kakidani H., Kubo T., Hirose T., Inayama S., Hayashide H., Miyata T., and Numa S. (1983) Isolation and structural organization of the human preproenkephalin B gene. *Nature* **306,** 611–614.
40. Benoist C., O'Hare K., Breathnach R., and Chambon P. (1980) The ovalbumin gene-sequence of putative control regions. *Nucleic Acids Res.* **8,** 127–142.
41. Civelli O., Douglass J., Goldstein A., and Herbert E. (1985) Sequence and expression of the rat prodynorphin gene. *Proc. Natl. Acad. Sci. USA* **82,** 4291–4295.
42. Comb M., Herbert E., and Crea R. (1982) Partial characterization of the mRNA that codes for enkephalins in bovine adrenal medulla and human pheochromocytoma. *Proc. Natl. Acad. Sci. USA* **79,** 360–364.
43. Agarwal K. L., Brunstedt J., and Noyes B. (1981) A general method for the detection and characterization of an mRNA using an oligonucleotide probe. *J. Biol. Chem.* **256,** 1023–1028.
44. Dandekar S. and Sabol S. L. (1982) Cell-free translation and partial

characterization of mRNA coding for enkephalin precursor protein. *Proc. Natl. Acad. Sci. USA* **79,** 1017–1021.
45. Dandekar S. and Sabol S. L. (1982) Cell-free translation and partial characterization of proenkephalin messenger RNA from bovine striatum. *Biochem. Biophys Res. Commun.* **105,** 67–74.
46. Sabol S. L., Liang C. M., Dandekar S., and Kranzler S. L. (1983) In vitro biosynthesis and processing of immunologically identified methionine-enkephalin precursor protein. *J. Biol. Chem.* **258,** 2697–2704.
47. Pittius C. W., Kley N., Loeffler J. P., and Hoellt V. (1985) Quantitation of proenkephalin A messenger RNA in bovine brain, pituitary and adrenal medulla: Correlation between mRNA and peptide levels. *EMBO J.* **4,** 1257–1260.
48. Weaver C. F., Gordon D. F., and Kemper B. (1981) Introduction by molecular cloning of artifactual inverted seqeunces at the 5′ terminus of the sense strand of bovine parathyroid hormone cDNA. *Proc. Natl. Acad. Sci. USA* **78,** 4073–4077.
49. Lewis R. V. and Stern A. S. (1983) Biosynthesis of the enkephalins and enkephalin-containing polypeptides. *Ann. Rev. Pharmacol. Toxicol.* **23,** 353–372.
50. Mount S. M. (1982) A catalogue of splice junction sequences. *Nucleic Acids Res.* **10,** 459–472.
51. Dayhoff M. O., Schwartz R. M., and Orcutt B. C. (1978) *Atlas of Protein Sequence and Structure* **5,** Suppl 3, 345–352.
52. Furutani Y., Morimoto Y., Shibahara S., Noda M., Takahashi H., Hirose T., Asai M., Inayama S., Hayashida H., Miyata T., and Numa S. (1983) Cloning and sequence analysis of cDNA for ovine corticotropin-releasing factor precursor. *Nature* **301,** 537–540.

# Chapter 11

# Calcitonin and Calcitonin-Related Peptide Genes

## Tissue-Specific RNA Processing

MICHAEL G. ROSENFELD, STUART LEFF,
SUSAN G. AMARA, AND RONALD M. EVANS

## 1. INTRODUCTION

The neuroendocrine system serves critical regulatory and communicative functions during development and in the maintenance of physiological homeostasis. To serve these functions with requisite specificity and nuance, many diverse regulatory peptides are produced by various strategies. Complex regulatory mechanisms operate to restrict the expression of genes encoding neuroendocrine peptides to precise groups in neurons in the neural tissues and in specific cell types in peripheral organs. An understanding of the mechanisms by which expression of neuroendocrine genes is regulated is critical to achieve insights into the molecular mechanisms important in developmental processes.

Regulation of gene expression can potentially occur at the level of gene transcription, RNA processing and transport, mRNA stability, and post-translational events. The regulation RNA processing has the potential to qualitatively, as well as quantitatively, alter the nature of the gene product, and the possibility of regulation at this level is implicit in the discovery of "split genes," with the definition of intervening sequences (introns) (*1–4*). Three mechanisms account for the diversity of polypeptide regulators in the neuroendocrine systems. In the case of small polypeptide hormones, proteolytic processing serves to excise a series of mature hormones from the larger protein precursors that represent the initial products of mRNA translation. An example of this type of event is provided by the generation of adrenocorticotropic hormone (ACTH), melanocyte-stimulating hormone (MSH), and β-endorphin from the primary translation product of the proopiomelanocortin mRNA (*5–7*), and such a mechanism operates in the case of many neuroendocrine genes (*8–15*). Posttranslational modifications may further increase the functional diversity of neuroendocrine gene products. A second mechanism involves the processes of gene duplication and diversification, generating families of related genes, encoding similar, but nonidentical, peptide products. A third mechanism involves the generation of multiple mRNA, each encoding discrete component neuropeptides, consequential to the alternative inclusion of exons encoding specific component polypeptide hormones. This mechanism was shown by analysis of rat calcitonin gene expression to operate in the neuroendocrine system (*16–20*).

Multiple mRNA can be generated from a single transcription unit in the case of several viral and eukaryotic genes (*21–30*). In the case of adenovirus and SV40, alternative RNA processing is used to generate a number of gene products larger than might be predicted from the length of their genomic DNA and the small number of transcription promoters; this alternative processing thus maximizes the functional utilization of the limited genetic information. In many cases, this RNA polymorphism results from the use of 3′polyadenylation sites that can be used to direct alternative splicing choices. Analysis of the expression of immunoglobulin heavy-chain genes suggests that the biological demands of the immune system are partly met by differential posttranscriptional events in addition to the developmentally determined DNA rearrangements. During B cell development, IgM antibodies are initially membrane-bound; later in development, the IgM molecules are synthesized and secreted rather than inserted in the membrane. The structural difference between the membrane-inserted and secreted form of IgM is determined by an alteration at the carboxy-terminus of the heavy chain. The carboxy-terminal change appears to be the consequence of the selective inclusion or noninclusion of exons 1850 base pairs (bp) 3′

to the end of the μ domain, in association with the use of an alternative splice site in the 3′-exon that generates mRNA encoding the membrane form ($\mu_M$) or secreted form ($\mu_S$), respectively (*21,22*). A similar mechanism is proposed in the case of delta heavy chains during their switch from $\mu_M$ to $\mu_S$ (*30*). A third case is the simultaneous synthesis of immunoglobulin heavy-chain μ by cells in which similar RNA processing events may link a particular variable region to one of two constant-region domains (*23*). It has been suggested, although not yet unambiguously proved, that similar events occur in the expression of many developmental genomic loci in Drosophila (*31–32*), and, in the case of yeast invertase gene expression, in association with putative 5′ transcriptional initiation-site alterations (*33*); parallel events clearly do occur in kininogen gene expression (*34*). Variation in the structure of several of the liver and salivary α-amylase mRNAs has been suggested to be consequential to the use of alternative transcriptional start sites (*35*); although this leads to RNA polymorphism, it does not alter the coding region. In brain, a third transcription start site has been suggested, but this transcription unit is not productive of mature mRNA (*36*). The similarity of the alternative RNA processing events in the genes of the neuroendocrine system (calcitonin gene) and of the immune system (immunoglobulin heavy-chain gene) suggests that the underlying mechanisms may operate in the expression of many eukaryotic transcription units.

## 2. CALCITONIN GENE EXPRESSION

### *2.1. Structure of the RNA Products of Calcitonin Gene Expression*

In order to initiate studies of calcitonin gene expression, it was necessary to characterize the structure of the mature RNA transcript. To this end, a cDNA complementary to total mRNA was generated, using avian myeloblastoma virus (AMV) reverse transcriptase and oligo(dT) to prime transcription of 9–13S poly(A)-rich RNA prepared from a rat calcitonin-producing medullary carcinoma of the thyroid (MTC). The cDNAs were rendered double-stranded and were S-nuclease-treated to cleave the

5′-loop. Homopolymeric (dC) tails were added, using terminal transferase, and the inserts were ligated into the oligo(dG)-tailed *Pst* I site of pBR322. A colony filter hybridization analysis of transformants was performed, using cDNA to size fractionated (9–13S) poly(A)-rich RNA from rat MTC tumors. The plasmids with inserts giving the strongest signals were identified, and hybridization selection of RNA, using several positive clones, selected an RNA that generated a 17.5-kdalton translation product in vitro by immunoprecipitation with calcitonin antisera. Plasmids were selected for subsequent analysis on the basis of restriction map and size of insert.

The preparation and sequence analysis of chimeric plasmids containing DNA complementary to calcitonin mRNA permitted prediction of the structure of the 17.5-kdalton protein precursor of the 32-amino acid calcium-regulating hormone, calcitonin (Fig. 1) (*37–39*). On the basis of these and other analyses (*17,37–40*), proteolytic processing of the precursor appears to generate an 82-amino-acid $NH_2$-terminal peptide and a 16-amino-acid COOH-terminal calcitonin cleavage product (CCP) in thyroidal C cells (*see* Fig. 1).

The complete nucleotide sequence of a 627-bp cloned cDNA fragment contains the entire coding region for the protein precursor of the hormone calcitonin. The DNA sequence presented in Fig. 1 predicts that the 32-amino-acid mature calcitonin is generated as a result of posttranslational processing events from a protein precursor to 136 amino acids. The Met codon at positions $-252$ to $-250$ appears to function as the translation initiation site. The predicted molecular weight of the precursor, based upon this initiation site, is consistent with that of the major calcitonin cell-free translation product *(13,41)*. All but one of the seven nucleotides AGGGAGG at positions $-257$ to $-263$, preceding the initiator Met codon by four nucleotides, show complementarity to a region of eukaryotic ribosomal RNA. This region of ribosomal RNA is thought to serve as a prokaryotic mRNA ribosomal binding site. This feature is often noted in eukaryotic mRNA, although eukaryotic 18S ribosomal RNA does not contain the Shine–Dalgarno sequence. The presence of this sequence in calcitonin mRNA, however, further supports the assignment of the Met codon at positions $-252$ to $-250$ as the translational initiation site.

```
                                                20
                        5'-GCACAGGAGCCGCTGCCCAGATCAAGAGTCACCG

      40                      60                      80
CCTCGCAACCACCGCCTGGCTCCATCAGGACCCCGCAGTCTCAGCTCCAAGTCATCGCTC

    100                     120                     140
ACCAG(GTGAGCCCTGAGGTTCCTGCTCAG)GGAGGCATC ATG GGC TTT CTG AAG
                                         met gly phe leu lys

                160                     180
TTC TCC CCT TTC CTG GTT GTC AGC ATC TTG CTC CTG TAC CAG GCA
phe ser pro phe leu val val ser ile leu leu leu tyr gln ala

        200                     220
TGC GGC CTC CAG GCA GTT CCT TTG AGG TCA ACC TTA GAA AGC AGC
cys gly leu gln ala val pro leu arg ser thr leu glu ser ser

240                         260                         280
CCA GGC ATG GCC ACT CTC AGT GAA GAA GAA GCT CGC CTA CTG GCT
pro gly met ala thr leu ser glu glu glu ala arg leu leu ala

                    300                     320
GCA CTG GTG CAG AAC TAT ATG CAG ATG AAA GTC AGG GAG CTG GAG
ala leu val gln asn tyr met gln met lys val arg glu leu glu

            340                     360
CAG GAG GAG GAA CAG GAG GCT GAG GGC TCT AGC TTG GAC AGC CCC
gln glu glu glu gln glu ala glu gly ser ser leu asp ser pro

        380                     400
AGA TCT AAG CGG TGT GGG AAT CTG AGT ACC TGC ATG CTG GGC ACC
arg ser lys arg cys gly asn leu ser thr cys met leu gly thr

420                     440                         460
TAC ACA CAA GAC CTC AAC AAG TTT CAC ACC TTC CCC CAA ACT TCA
tyr thr gln asp leu asn lys phe his thr phe pro gln thr ser

                480                         500
ATT GGG GTT GGA GCA CCT GGC AAG AAA AGG GAT ATG GCC AAG GAC
ile gly val gly ala pro gly lys lys arg asp met ala lys asp

            520                     540
TTG GAG ACA AAC CAC CAC CCC TAT TTT GGC AAC TAG GTCCCTCCTCT
leu glu thr asn his his pro tyr phe gly asn stop

 560                      580                     600
CCTTTCCAGTTTCCATCTTGCTTTCTTCCTATAACTTGATGCATGTAGTTCCTCTCTGGC

 620                      640                     660
TGTTCTCTGGGCTATTATGGGTTACTTTCATGAGGCAAAGGATGGTATCTGGAATCTCCA

 680                      700                     720
ATGGGTGAGGAGAAAGGGCCTACAGGCTAAAAGAGAATCACCCAGGAAGATGGCAGAGAG

 740                      760                     780
CCAGGGCAGTCATCTGGATTCCTAGTAGAGCTTCTCAGTCTAGTCTTGCTTCATAAAGTG

 800                      820
TTGGTTGTTTGGGAAATAAAGCTATTTTTCTAAAAGGA(C)(AAAAA)n
```

Fig. 1. Structure of rat calcitonin cDNA. The predicted encoded precursor peptide includes processing signals for generation of the three mature peptide products. The bracketed area in the 5′ noncoding region indicates the stochastic usage of two alternative splice sites at the end of the first exons, thereby generating two structurally different RNA products.

The calcitonin sequence is flanked by basic amino acid residues within its precursor, suggesting that posttranslational processing events occur in a manner analogous to peptide excision of other hormones, such as ACTH and insulin, from their precursor proteins. The secretion of both calcitonin and CCP has been verified in vivo. The function, if any, of the $NH_2$-terminal peptide(s) and the COOH-terminal peptide (CCP) remains unknown in both human and rats.

## 2.2. Identification of a Second Mature Transcript of the Calcitonin Gene

The production of multiple calcitonin-related mRNA was first noted during the spontaneous and permanent "switching" of serially transplanted rat MTC from states of "high" to "low" or absent calcitonin production (*16*). The unexpected explanation for the "switch" was that calcitonin gene transcription continued, but generated a series of new, structurally distinct mRNA, 50–200 nucleotides larger than calcitonin mRNA, referred to as calcitonin gene-related product mRNA (CGRP mRNA) (Fig. 2) (*18,19*). These new mRNAs encode a 16-kdalton protein product containing no immunoreactive calcitonin peptide (*18*). It was suggested on the basis of genomic DNA and RNA blotting analyses that both CGRP and calcitonin mRNA were generated by differential processing of a single gene (18,19). This model was documented by: (i) proof that the two mRNAs contain a 5′-region of identity, but completely diverge in their 3′ sequences, based upon cDNA sequence analysis (*19*) (Figs. 2 and 3); (ii) structural proof that the entire sequences of both calcitonin and CGRP mRNA are linked in a discrete genomic locus, and that a rat calcitonin genomic clone contains sequences identical to those in both mRNAs (Figs. 4 and 5); (iii) demonstration that the junction of the common and divergent sequences in calcitonin and CGRP mRNA correspond to intron–exon junctions (*4*); and (iv) evidence that these events occur physiologically, resulting in "switching" of the peptide products generated from a single gene (*20*). Based upon preparation and analysis of cDNA clones, CGRP mRNA shares sequence identity with calcitonin mRNA through nucleotide 227 of the coding region, predicting that the initial 76-*N*′-terminal amino acids are identical to those in the calcitonin mRNA-encoded precursor. The mRNA subsequently diverges entirely in

```
                                                   20
                            5'-GCACAGGAGCCGCTGCCCAGATCAAGAGTCACCG

   40                        60                        80
CCTCGCAACCACCGCCTGGCTCCATCAGGACCCCGCAGTCTCAGCTCCAAGTCATCGCTC

    100                      120                       140
ACCAG(GTGAGCCCTGAGGTTCCTGCTCAG)GGAGGCATC ATG GCC TTT CTG AAG
                                         met gly phe leu lys

                160                         180
TTC TCC CCT TTC CTG GTT GTC AGC ATC TTG CTC CTG TAC CAG GCA
phe ser pro phe leu val val ser ile leu leu leu tyr gln ala

        200                         220
TGC GGC CTC CAG GCA GTT CCT TTG AGG TCA ACC TTA GAA AGC AGC
cys gly leu gln ala val pro leu arg ser thr leu glu ser ser

 240                        260                             280
CCA GGC ATG GCC ACT CTC AGT GAA GAA GAA GCT CGC CTA CTG GCT
pro gly met ala thr leu ser glu glu glu ala arg leu leu ala

                    300                         320
GCA CTG GTG CAG AAC TAT ATG CAG ATG AAA GTC AGG GAG CTG GAG
ala leu val gln asn tyr met gln met lys val arg glu leu glu

                340                         360
CAG GAG GAG GAA CAG GAG GCT GAG GGC TCT AGT GTC ACT GCC CAG
gln glu glu glu gln glu ala glu gly ser ser val thr ala gln

        380                         400
AAG AGA TCC TGC AAC ACT GCC ACC TGC GTG ACC CAT CGG CTG GCA
lys arg ser cys asn thr ala thr cys val thr his arg leu ala

 420                        440                             460
GGC TTG CTG AGC AGG TCG GGA GGT GTG GTG AAG GAC AAC TTT GTG
gly leu leu ser arg ser gly gly val val lys asp asn phe val

                    480                         500
CCC ACC AAT GTG GGC TCT GAA GCC TTC GGC CGC CGC CGC AGG GAC
pro thr asn val gly ser glu ala phe gly arg arg arg arg asp

                520                         540             560
CTT CAG GCT TGA ACAGATAATAGCCCCAGAAAGAAGGTTACACAATAAAGATAAA
leu gln ala stop

                580                         600             620
CTCTAATTCTTTTCATGTATAATTAAAGTCATGTGTCAGAAAGGCTGATGAGAAAGGCTG

                640                         660             680
ATGGAAGACACATATATTTGCATCCTTCTTGGTACTGAAACCCTTCTCCCTATGACAGGA

                700                         720             740
AATAAAGCTAAGTGCAGAATAAGCTCACCATAGTTGGCTATTGTGCATCGTGTTGTATGT

                760                         780             800
GATTCTATATCCATAACATGACAGCATGGTTCTGGCTTATCTGGTAGCAAATCTAGTCCC

                820                         840             860
CATAAACCACCCTGTCGATGTTGATAACTCTGCTAAACCTCAAGCGGATATGAAACCACT

                880                         900             920
GCCTCTTGGTCTTCTGGGGACACATGGTAATGTTGTGACTCAATGGAACCATATGCTTAA

                940                         960             980
AGAACTGTTAATGTTGTCACTTGTGAGCTTAATCAAAATTAAAAAATATGTATTTTCGAT
                                     ------

(AAAAA)n
```

Fig. 2. Structure of rat calcitonin gene-related peptide mRNA. The RNA exhibits 5′ identity with calcitonin mRNA, but then differs in 3′ sequences. Three polypeptides are encoded by the mRNA.

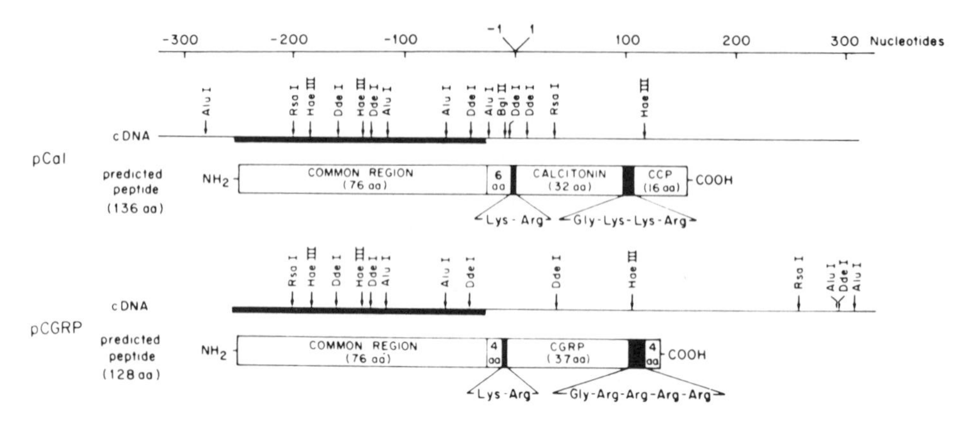


Fig. 3. Schematic diagram of rat calcitonin and CGRP mRNA and their encoded proteins.

nucleotide sequence, encoding a unique 52-amino-acid C′-terminal peptide (*19*). Protein processing signals within the C′-terminal region predict the excision of a 37-amino-acid polypeptide with a C′-terminal amidated Phe residue (*19*), (Fig. 6). Based upon calcitonin gene structure (Fig. 5), production of calcitonin mRNA involves splicing of the first three exons, present in both mRNAs, to the fourth exon, which encodes the entire calcitonin/CCP sequence. Alternatively, splicing and ligation of the first three exons to the fifth and sixth exons, which contain the entire CGRP coding sequence and the 3′-noncoding sequences, respectively, result in production of CGRP mRNA; in this case, the fourth exon is excised along with the flanking intervening sequences.

## 2.3. Calcitonin Gene Expression in the Brain

The specific hybridization of common region and CGRP exon-specific probes to poly(A)-rich RNA from rat trigeminal ganglia and hypothalamus suggested that CGRP mRNA and the predicted peptide product might be synthesized in the brain (*20*). To provide evidence for the production of the predicted peptide in the brain and to determine the precise sites of synthesis, a combined histochemical and molecular biological approach was employed.

Antibody to a synthetic peptide corresponding to the fourteen C′-terminal amino acids of CGRP was generated and used to de-

```
                                                   Exon 1
5'--ccgcggcgggaataagagcagctgcaggcgcttggaaGCACAGGACCCGCTGCCCAGATCAAGAG

                    50
TCACCGCCTCGCAACCACCGCCTGGCTCCATCAGGACCCCGCAGTCTCAGCTCCAAGTCATCGCTCACC

 100                                            Exon 2          1200
AGGTGAGCCCTGAGGTTCCTGCTCAGgt--1.06 kb--agGGAGGCATCATGGCCTTTCTGAAGTTCTC
                                                   MetGlyPheLeuLysPheSe

                                           1250
CCCTTTCCTGGTTGTCAGCATCTTGCTCCTGTACCAGGCATGCGGCCTCCAGGCAGTTCCTTTGAGgt
rProPheLeuValValSerIleLeuLeuLeuTyrGlnAlaCysGlyLeuGlnAlaValProLeuArg

          Exon 3                                          1800
--0.48 kb--ccGTCAACCTTAGAAAGCAGCCCAGGCATGGCCACTCTCAGTGAAGAAGAAGCTCGCCT
               SerThrLeuGluSerSerProGlyMetAlaThrLeuSerGluGluGluAlaArgLe

                                     1850
ACTGGCTGCACTGGTGCAGAACTATATGCAGATGAAAGTCAGGGAGCTGGAGCAGGAGGAGGAACAGGA
uLeuAlaAlaLeuValGlnAsnTyrMetGlnMetLysValArgGluLeuGluGlnGluGluGluGlnGl

           1900
GG CTGAGGGCTCTAGgtaaggtccccttacccgt--0.64kb--cgcatgtcttttccctgcagCTTGG
uAlaGluGlySerSer                                                   LeuA

Exon 4    2600                                                  2650
ACAGCCCCAGATCTAAGCGGTGTGGGAATCTGAGTACCTGCATGCTGGGCACGTACACACAAGACCTCA
spSerProArgSerLysArgCysGlyAsnLeuSerThrCysMetLeuGlyThrTyrThrGlnAspLeuA

                                                   2700
ACAAGTTTCACACCTTCCCCCAAACTTCAATTGGGGTTGGAGCACCTGGCAAGAAAAGGGATATGGCCA
snLysPheHisThrPheProGlnThrSerIleGlyValGlyAlaProGlyLysLysArgAspMetAlaL

                              2750
AGGACTTGGAGACAAACCACCACCCCTATTTTGGCAACTAGGTCCCTCCTCTCCTTTCCAGTTTCCATC
ysAspLeuGluThrAsnHisHisProTyrPheGlyAsnStop

    2800                                                  2850
TTGCTTTCTTCCTATAACTTGATGCATGTAGTTCCTCTCTGGCTGTTCTCTGGGCTATTATGGGTTACT

                                          2900
TTCATGAGGCAAAGGATGGTATCTGGAATCTCCAATGGGTGAGGAGAAACGGCCTACAGGCTAAAAGAG

                    2950
AATCACCCAGGAAGATGGCAGAGAGCCAGGGCACTCATCTGGATTCCTAGTAGAGCTTCTCAGTCTAGT

3000                                                3050
CTTGCTTCATAAAGTGTTGGTTGTTTGGGAAATAAACCTATTTTTCTAAAACGACttgagccatggtgg
                              [AATAAA boxed]
                                                          Exon 5
tcattgatgtagcccacatctcaggc--0.71kb--cactgcatcctgaatatcagTGTCACTGCCCAG
                                                            ValThrAlaGln

          3850                                                  3900
AAGAGATCCTGCAACACTGCCACCTGCGTGACCCATCGGCTGGCAGGCTTGCTGAGCAGGTCGGGAGGT
LysArgSerCysAsnThrAlaThrCysValThrHisArgLeuAlaGlyLeuLeuSerArgSerGlyGly

                                                3950
GTGGTGAAGGACAACTTTGTGCCCACCAATGTGGGCTCTGAAGCCTTCGGCCGCCGCCGCAGGACCTT
ValValLysAspAsnPheValProThrAsnValGlySerGluAlaPheGlyArgArgArgArgAspLeu

                        4000
CAGGCTTGAACAGATAATAGCCCCAGAAAGAAGgtgacttccttgtacaactgg--0.51kb--tattc
GlnAlaStop

            Exon 6                                  4600
tttttttcctcctagGTTACACAATAAAGATAAACTCTAATTCTTTTCATGTATAATTAAAGTCATGTG

                                         4650
TCAGAAACGCTGATGGAAGACACATATATTTGCATCCTTCTTGGTACTGAAACCCTTCTCCCTATGACA

          4700                                                     4750
GGAAATAAAGCTAAGTGCAGAATAAGCTCACCATAGTTGGCTATTGTGCATCGTGTTGTATGTGATTCT

                                                    4800
ATATCCATAACATGACAGCATGGTTCTGGCTTATCTGGTAGCAAATCTAGTCCCCATAAACCACCCTGT

                               4850
CGATGTTGATAACTCTGCTAAACCTCAAGCGGATATGAAACCACTGCCTCTTGGTCTTCTGGGGACACA

   4900                                              4950
TGGTAATGTTGTGACTCAATGGAACCATATGCTTAAAGAACTGTTAATGTTGTCACTTGTGAGCTTAAT

CAAAATTAAAAAATATGTATTTTCGATatcctcaagtggagtctttgctgttacttatactgtgttt--3'
    [ATTAAA boxed]
```

Fig. 4. Sequence of the rat calcitonin gene. The six component exons are labeled, and a portion of each intervening sequence is shown.

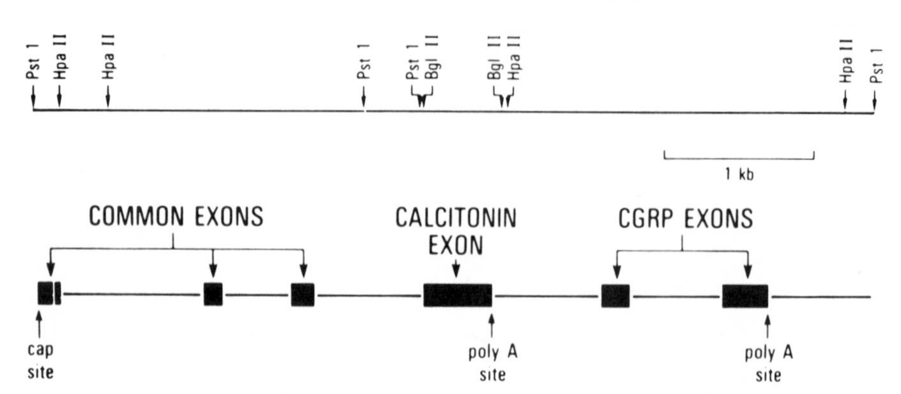


Fig. 5. Schematic diagram of the rat calcitonin/CGRP gene. The gene contains all of the coding sequences present in both calcitonin and CGRP mRNA.

termine the distribution of CGRP in neural tissues (*20*) (Fig. 7). Immunoreactive CGRP was present in a unique distribution in a large number of cell groups and pathways in the central nervous system distinct from that of any known neuropeptide. This staining was localized in discrete parts of several functional systems. First, dense terminal fields were stained throughout the substantia gelatinosa of the spinal cord and caudal part of the spinal trigeminal nucleus. These fibers arise in dorsal root and trigeminal ganglion cells and probably relay nociceptive or thermal information. Second, CGRP is found in most parts of the taste pathways, including sensory endings in taste buds and the central endings of these fibers in the rostral part of the nucleus of the solitary tract (NTS), the relay system from the parabrachial nucleus (PB) to the thalamic taste nucleus (TN), and the taste area of the cerebral cortex [posterior agranular insular area (INS)]. In addition, most motor neurons in the hypoglossal nucleus (XII), which move the tongue, were stained. Third, a small group of primary olfactory fibers (I) that end in the glomerular layer of the olfactory bulb (OB) were stained, suggesting that CGRP has a role in olfaction as well as taste. Fourth, a small number of fibers, apparently from the eighth nerve, end in the cochlear and vestibular nuclei. Fifth, CGRP is found throughout the PB, suggesting that it plays a part in the relay of visceral sensory information from the vagus (and glossopharyngeal) nerve, by way of an ascending

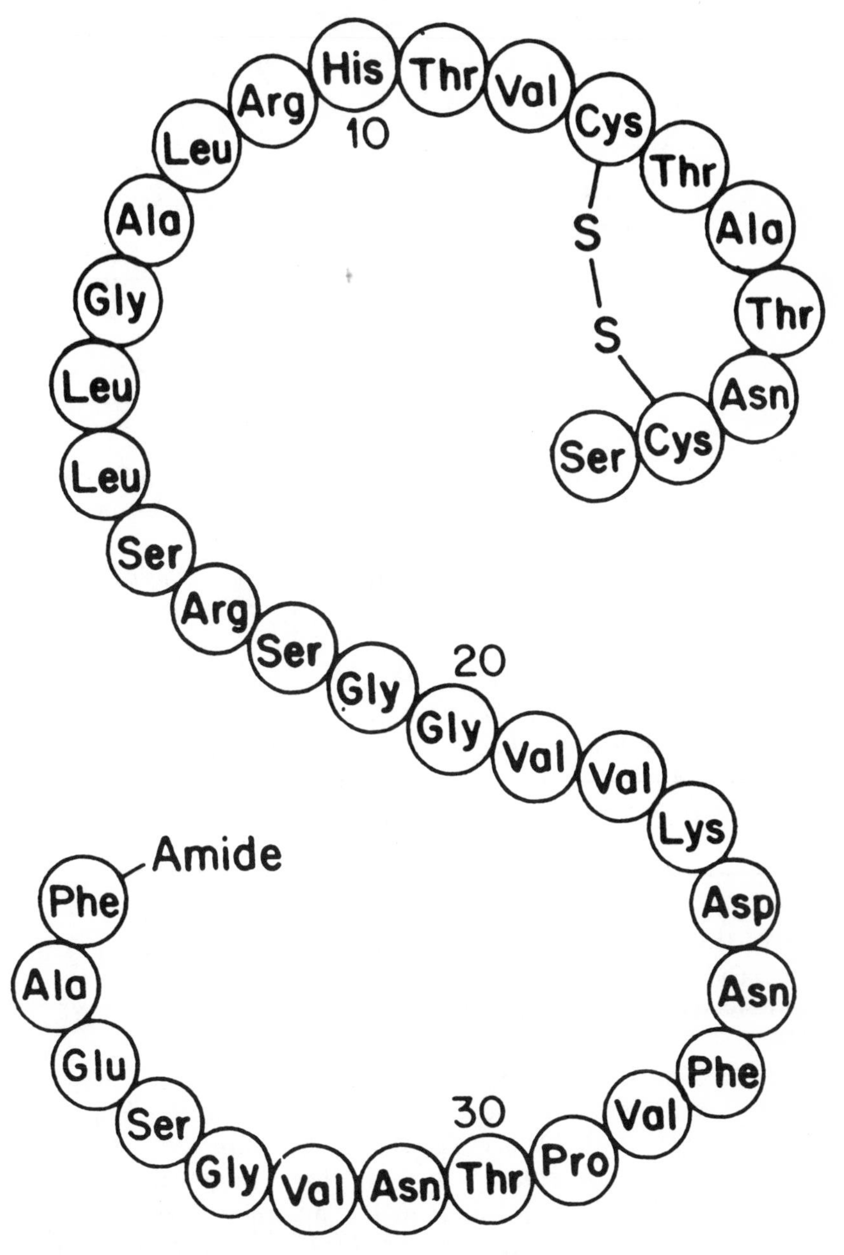


Fig. 6. The structure of the predicted, novel neuropeptide, calcitonin gene-related peptide (CGRP).

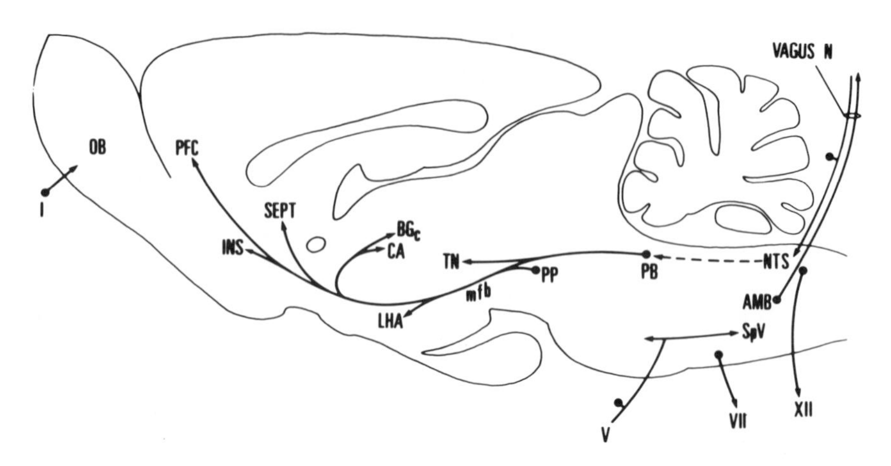


Fig. 7. Schematic diagram of the distribution of immunoreactive CGRP in the rat brain. A detailed explanation is found in the text.

pathway throughout the medial forebrain bundle (MFB). This pathway appears to arise in the PB and peripeduncular nucleus (PP) and projects to the lateral hypothalamic area (LHA), the central nucleus of the amygdala (CA), the patches in caudal parts of the caudoputamen and globus pallidus (BGc), the lateral septal nucleus and bed nucleus of the stria terminalis (SEPT), and layer III of three cortical areas: the infralimbic prefrontal area (PFC), the posterior agranular insular area, and the perirhinal area. The ascending projections in the MFB are probably modulated by a massive nonCGRP-containing pathway from the NTS to the PB (dashed line in Fig. 7). Sixth, stained motor neurons in the rostral part of the nucleus ambiguus (AMB) project through the vagus nerve and may innervate the heart, branchial muscles, or both in the pharynx that are involved, for example, in the control of swallowing.

The distribution of CGRP in specific sensory, integrative, and motor systems suggests several possible functions for the peptide. The differential distribution of CGRP immunoreactivity in the olfactory and gustatory systems, including taste buds and hypoglossal, facial, and vagal nuclei, and in the hypothalamic and limbic regions strongly suggests that it may have a functional role in ingestive behavior. Additional studies have revealed the presence of CGRP at the neuromuscular junctions in the striated muscle of upper esophagus, the first peptide identified at neuromuscular junctions in mammalian species. Administration of synthetic CGRP acts as a potent inhibitor of gastric acid secretion. CGRP is present in small trigeminal and spinal sensory ganglion cells that are known to relay thermal and nociceptive information to the brainstem and spinal cord; in the spinal ganglia, CGRP-positive cells represent 30–50% of the total population of small ganglion cells, a percentage vastly greater than that for any other neuropeptide. CGRP is present in a subset of cells in one of the vagal motor nuclei (n. ambiguus) and is extraordinarily widely distributed, but with clear patterns of tissue specificity, in fibers in vascular musculature, suggesting a role in cardiovascular homeostasis. Administration of synthetic rat CGRP produces a unique pattern of effects on blood pressure and catecholamine release in dogs and rats (*42*). CGRP is also present in the endocrine system, a subset of adrenal medullary cells, bronchiolar cells, intestinal cells, and fiber baskets that innervate the pancreatic islets (*20*).

A modified $S_1$-nuclease-protection assay was used to confirm the production of CGRP mRNA in the brain, and to identify the sites of its biosynthesis (Fig. 8). The CGRP mRNA isolated from brain regions encodes the predicted 16-kdalton CGRP precursor protein. Gel-filtration analysis of brain immunoreactive peptide suggests that this precursor is processed in brain to generate the predicted peptide product (CGRP) (*20*). Thus, a novel neuropeptide, previously unsuspected, with actions in biological systems and an unusually widespread distribution, was identified as a result of these recombinant-DNA analyses. Tissue specificity of the RNA processing events is suggested because virtually no calcitonin mRNA can be identified in the brain (*20*), whereas calcitonin and CGRP mRNA (and calcitonin and CGRP peptides) are present in a ratio of 95–98:1, respectively, in thyroidal C cells. Small amounts of CGRP peptide are present histochemically by radioimmunoassay in the thyroidal C cells, and both calcitonin and CGRP are produced within the identical cells.

### 2.4. Identification of a Second CGRP-Related mRNA

Based upon sequence analysis of clonal inserts identified during screening procedures, a second RNA species, related to CGRP

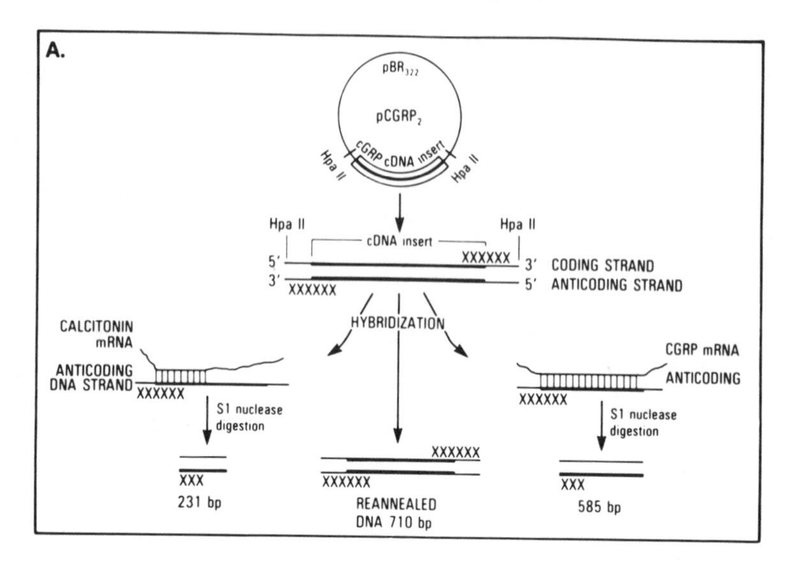
A.
pBR322
pCGRP2
cGRP cDNA insert
Hpa II
Hpa II
Hpa II
Hpa II
cDNA insert
XXXXXX
5'
3'
3'
5'
CODING STRAND
ANTICODING STRAND
XXXXXX
HYBRIDIZATION
CALCITONIN mRNA
ANTICODING DNA STRAND
XXXXXX
S1 nuclease digestion
XXX
231 bp
XXXXXX
XXXXXX
REANNEALED DNA 710 bp
CGRP mRNA
ANTICODING
XXXXXX
S1 nuclease digestion
XXX
585 bp

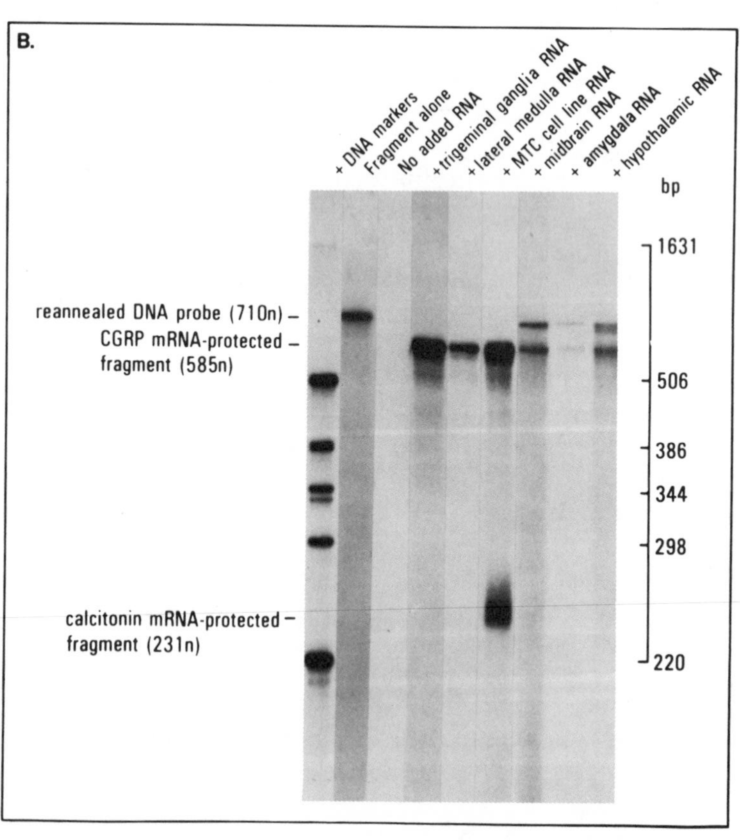
B.
+ DNA markers
Fragment alone
No added RNA
+ trigeminal ganglia RNA
+ lateral medulla RNA
+ MTC cell line RNA
+ midbrain RNA
+ amygdala RNA
+ hypothalamic RNA
bp
1631
506
386
344
298
220
reannealed DNA probe (710n)
CGRP mRNA-protected fragment (585n)
calcitonin mRNA-protected fragment (231n)

mRNA, was identified. This mRNA encodes a 134-amino-acid peptide containing the processing signals predicting the generation of five polypeptide products. One predicted product differs from CGRP by only a single amino acid (a Lys substituted for a Glu in position 35) in the entire 37-amino-acid sequence (*43*). The CGRP-related mRNA is expressed in the same brain regions (trigeminal ganglion, midbrain) as observed for the first CGRP mRNA, although at variable levels (*43*). The mRNA has considerable homology in the $NH_2$-terminal (common coding) region of CGRP and calcitonin mRNA. Therefore, a second rat gene, sharing homology with the rat calcitonin/CGRP gene, generates a CGRP-related mRNA in at least some brain regions expressing CGRP itself.

Fig. 8. (*On opposite page*) $S_1$-nuclease mapping analysis of calcitonin gene expression in the brain. The strategy used for the assay is schematically represented in panel A. The plasmid $pCGRP_2$ (585-bp CGRP) was excised, using *Hpa*II in such a way that the excised 710-bp fragment contains the 585-bp CGRP cDNA flanked by short pBR322 sequences. The fragment was labeled by 3′ exonuclease digestion and filling reactions (approximately $3x10^8$ μg DNA), hybridized with poly(A)-selected RNA from various brain regions, and subjected to electrophoresis under denaturing conditions following $S_1$ nuclease digestion. Any reannealed DNA will be 710 bases in length. A 585-nucleotide fragment will be protected if hybridized to CGRP mRNA, whereas a 231-nucleotide fragment will be protected from $S_1$-nuclease digestion if hybridized to calcitonin mRNA. In the absence of added RNA, only a trace amount of reannealed DNA would be undigested. The protection assay is shown in Panel B. Autoradiographs of: lane 1—*Hin*f-digested pBR322 standards (3-h exposure); lane 2—aliquots of undigested probe (21-h exposure); lane 3—probe hybridized to carrier RNA only (72-h exposure); lane 4—hybridization to 20 μg poly(A)-rich RNA from trigeminal ganglia (3-h exposure); lane 5—hybridization to 20 μg poly(A)-rich lateral medulla RNA (11-h exposure); lane 6—hybridization to 17 μg poly(A)-rich RNA from a rat medullary-tumor cell line producing both calcitonin and CGRP mRNa (11-h exposure); lane 7—hybridization to 20 μg poly(A)-rich RNA from midbrain (72-h exposure); lane 8—hybridization to 20 μg poly(A)-rich RNA from hypothalamus (72-h exposure). The migration of reannealed CGRP mRNA-protected probe is indicated.

### 2.5. Similar RNA-Processing Events Operate in Human Calcitonin Gene Expression

To determine whether the same events operate in the case of human calcitonin gene expression, a cDNA library was generated from a human MTC tumor and screened using rat calcitonin mRNA and CGRP mRNA-specific probes. The cDNAs were subjected to sequence analysis and, as shown in Fig. 9, established a region of identity and a 3′-region of divergence. There are two human genes containing calcitonin/CGRP-reactive sequences. One of these genes was isolated and sequenced and contains all of the coding information present in both calcitonin and CGRP mRNA. The second gene contains regions related to both common exons, CGRP and calcitonin, and encodes calcitonin-related and CGRP-related peptides. The point of divergence between the two mRNAs corresponds precisely to intervening sequence–exon boundaries. The predicted sequence of human CGRP differs by four amino acids from the rat CGRP sequence; one interesting feature is that one change (Lys for Glu in position 35) corresponds to the single amino acid alteration in the rat CGRP-related mRNA-encoded product. In fact, there is cross-hybridization between the human CGRP mRNA and a specific 3′-noncoding region clonal probe derived from rat CGRP-related cDNA. These data indicate that the alternative RNA-processing events characteristic of rat calcitonin/CGRP gene expression are also observed with regard to expression of the human calcitonin/CGRP gene (*45*).

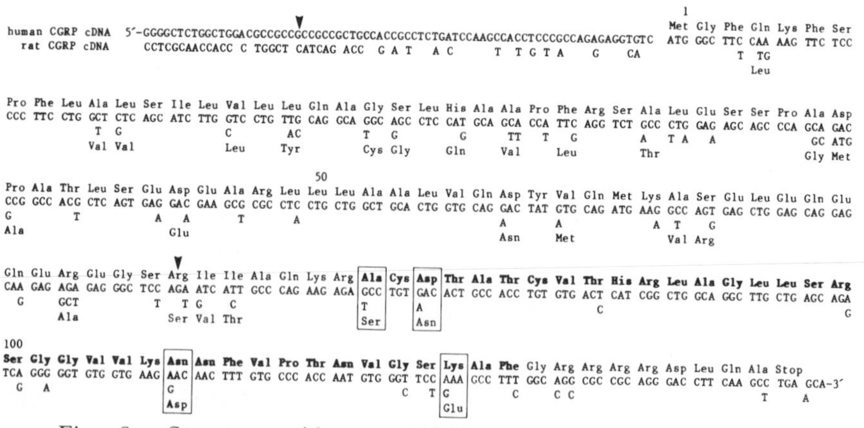


Fig. 9. Structure of human CGRP cDNA, showing sequence differences from rat CGRP.

Although the tissue-specific pattern of the human gene expression has not been fully characterized, both RNA products are found in human MTC and pituitary gland, as determined by both RNA blotting and DNA-excess hybridization analyses. Histochemical analysis performed on primate (monkey) pituitary confirms the presence of specific immunoreactive material, using calcitonin- and CGRP-specific antisera in the pars intermedia. Previous immunohistochemical reports suggested that immunoreactive calcitonin was produced in rat pituitary (*44*); however, no RNA hybridizing to calcitonin- or CGRP-specific clonal probes were detectable in rat (*46*). Using an $S_1$-nuclease mapping approach, very small amounts of authentic CGRP mRNA in rat pituitary RNA can be detected. Because both CGRP and calcitonin-immunoreactive material are easily detected in mice, there may be a variant expression of both products of the calcitonin gene in pituitaries of various mammalian species; perhaps this variance reflects regulated gene expression.

## 3. MECHANISMS OF TISSUE-SPECIFIC RNA-PROCESSING EVENTS

Structural analysis of the gene and its mRNA products has provided initial definition of the mechanisms responsible for the tissue-specific RNA-processing events. $S_1$ mapping and DNA sequence analyses have demonstrated that selective utilization of alternative poly(A) sites provides a structural basis for the production of either calcitonin- or CGRP-specific sequences (Fig. 10). Several events could account for the selective polyadenylation. First, there could be two CAP sites, and therefore, CGRP and calcitonin mRNA reflect products of separate transcription units. $S_1$-nuclease mapping, primer extension analyses, and DNA sequencing analyses have established that both mRNAs utilize an identical, unique CAP site; this defines one transcription unit in the case of the calcitonin gene (*41*). Interestingly enough, the first exon contains a second utilized splice site; however, the use of this alternative splice site in an invariant (stochastic) event, unrelated to the exon-selection events. A second possibility is that alternative sites of transcriptional termination provide the critical, regulated event. DNA-excess hybridization of nascent RNA tran-

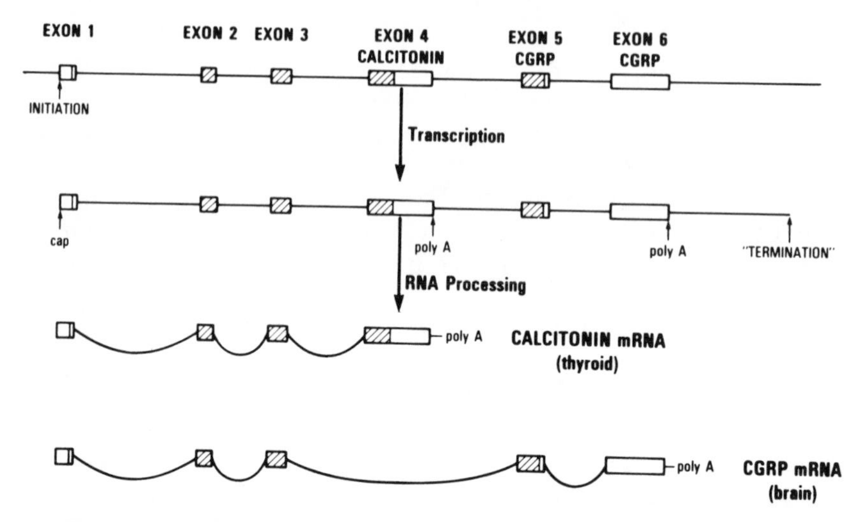


Fig. 10. Schematic diagram of the alternative RNA processing events in calcitonin/CGRP gene expression. The 5′ ends of both mRNAs were determined by use of a synthetic oligonucleotide for primer extension and DNA sequence analysis. In addition an $S_1$-nuclease mapping analysis confirmed the use of the 5′ end determined by the primer-extension analysis. Both calcitonin and CGRP mRNA appeared to use an identical cap site. $S_1$-nuclease mapping and genomic and cDNA sequence anlysis confirmed that the poly(A) sites utilized in each mRNA were present at the 3′ terminus of the calcitonin (4th) and CGRP (6th) exons, respectively.

scripts labeled by run-off elongation in isolated nuclei was performed, using subcloned, repeat-free, genomic fragments. Transcription was shown to proceed approximately 1 kilobase (kb) beyond the sixth (CGRP) exon, even in tissues producing predominantly calcitonin mRNA (Fig. 11). Genomic rearrangements provide a third potential mechanism; within the limits of restriction mapping analysis, no genomic structural alterations could be detected. Finally, the level of calcitonin gene transcription varies independently of the alternative RNA splicing events; thus, various cell lines express the calcitonin gene at transcription rates that vary over a 10-fold range in a fashion independent of the ratio of calcitonin:CGRP mRNA. Therefore, it is postulated that a cellular mechanism exists that dictates selection of a site at or near

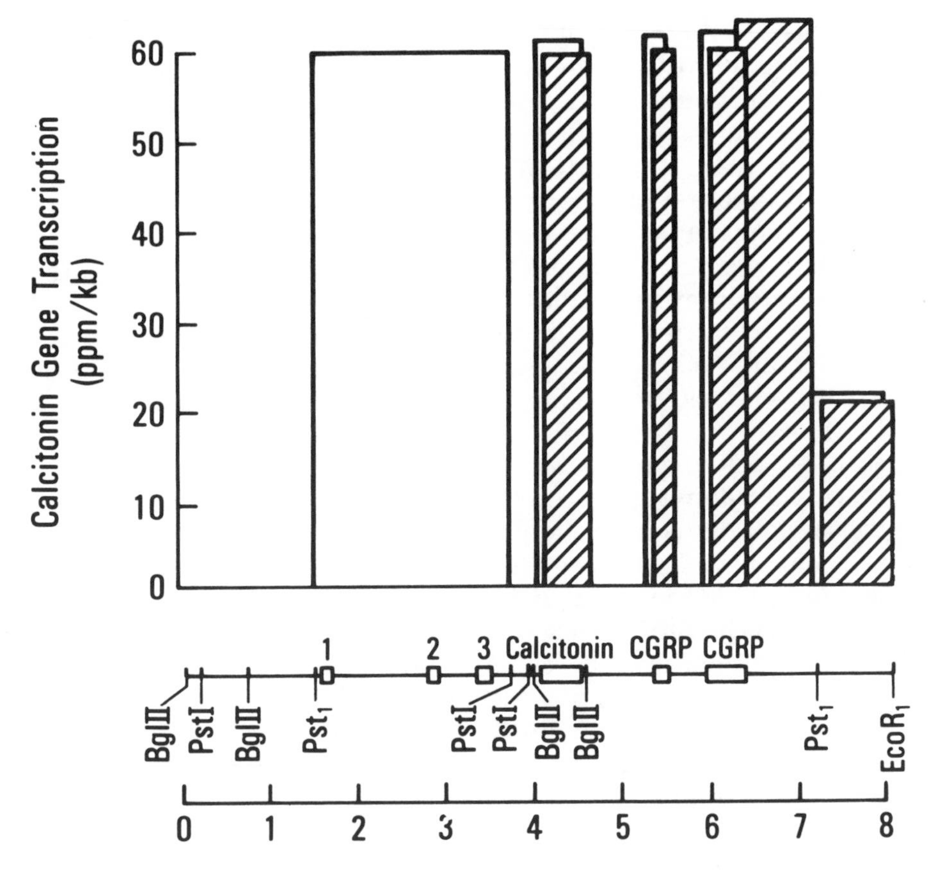


Fig. 11. The pattern of RNA transcription across the rat calcitonin gene. Nuclear run-off transcripts were quantitated under conditions of DNA excess performed in a tissue producing predominantly (>95%) calcitonin mRNA (open bars) or predominantly (>92%) CGRP mRNA (shaded bars). Transcription continues without attenuation through the CGRP exons, to a point approximately 1 kb 3′ of the CGRP poly(A) site.

one of the two alternative poly(A) sites where the critical endonucleolytic cleavage will occur.

The initial regulated event underlying the developmental association of changes in polyadenylation and use of alternative splicing pathways could be either the endonucleolytic cleavage adjacent to one of the alternative poly(A) sites, an alternative pattern of splicing of exons, or even the differential production of processing of alternatively polyadenylated primary transcripts produced in an invariant (stochastic) ratio. If direct regulation of the splicing process were to precede and dictate poly(A)-site usage in calcitonin/CGRP gene expression, production of CGRP mRNA would require that the ratio of removal of the entire sequence between the third and fifth exons, including the calcitonin exon, is more rapid than the polyadenylation reactions. The fact that we invariably observe a series of polyadenylated RNA species that appears to contain both intervening sequences flanking the calcitonin exon, including a transcript the size of the CGRP primary transcript, and that these species are present in higher concentrations in tissues producing CGRP mRNA than they are in calcitonin mRNA-producing tissues argues against initial regulation at the level of exon-splicing reactions. If any of these RNAs serve as precursors of the mature transcripts, as seems likely, then polyadenylation reactions can precede the alternative splicing reactions in calcitonin gene expression and would represent the initial, regulated step. The implication of these data is that usage of the distal poly(A) site generates a transcript in which splicing to the CGRP exon is kinetically favored over splicing to the calcitonin exon. Analysis of run-off nuclear transcripts sugggests that both calcitonin and CGRP poly(A) sites can be used in a calcitonin mRNA-producing tissue; presumably, cleavage at the first site before putative cleavage at the second site dictates a splicing pattern generating production of calcitonin mRNA. In this case, the rate-limiting reactions determining calcitonin or CGRP mRNA production would be the relative ratio of cleavage adjacent to the calcitonin poly(A) site and exon-splicing reactions joining the third and fifth (CGRP) exons.

To understand the mechanisms of alternative poly(A) site selection, it is necessary to understand the rules that dictate polyadenylation reactions. Clearly, termination of transcription occurs 3′ to both alternative poly(A) sites; therefore, a molecular mechanism other than transcriptional termination must be involved to explain the data. The sequences AAUAAA and AUUAAA appear at the appropriate distances upstream from the poly(A) sites for calcitonin and CGRP mRNA, respectively, consistent with known sequences critical for endonucleolytic cleavage at poly(A) sites; however, this hexanucleotide alone does not appear to be sufficient. In fact, a highly conserved sequence 3′ to the hexanucleotide, present at the poly(A) sites used in calcitonin gene expression, might be crucial for the regulated events.

The choice mechanisms exhibit clear tissue specificity both in vivo, and, on the basis of RNA products generated on analysis of the transfected rat calcitonin gene, in vitro. Both calcitonin and CGRP mRNA are produced when the calcitonin gene is transferred into rat fibroblast cell lines; however, only calcitonin mRNA is produced when the calcitonin gene is transfected into lymphocyte cell lines that exclusively produce the secretory form of immunoglobulin μ heavy chain. These data suggest that a *trans*-acting factor (or factors) is responsible for developmental regulation of the processing of transcripts of a large subset of transcription units. Calcitonin and CGRP mRNA exhibit similar stability, regardless of which RNA is the predominant species.

## 4. CONCLUSIONS

Many strategies are utilized in the developmental regulation of gene expression. The calcitonin/CGRP gene is an example of a complex transcription unit with a single CAP site; alternative splicing of exons at the 3′-end of the mRNA is associated with differential usage of at least two alternative poly(A) sites. These observations indicate that regulated poly(A) site selection is a physiologically significant developmental mechanism for determining neuropeptide expression and is likely to be important in establishing the phenotype of specific groups of neurons. Characterizing tissue-specific factors that determine RNA splicing and poly(A) site selection and determining whether they recognize structural elements on nascent RNA transcripts or genomic DNA will permit assessment of the potential significance of the modulation of RNA processing as a general strategy for the regulation of productive processing of neuroendocrine transcription units.

## REFERENCES


1. Darnell J. E. (1978) Transcription units for mRNA production in eukaryotic cells and their DNA viruses. *Prog. Nucleic Acid Res. Mol. Biol.* **22,** 327–353.
2. Gilbert W. (1978) Why genes in pieces. *Nature* **271,** 501.
3. Ziff E. B. (1980) Transcription and RNA processing by the DNA tumor viruses. *Nature* **287,** 491–499.

4. Abelson, J. (1979) RNA processing and the intervening sequence problem. *Ann. Rev. Biochem.* **48,** 1035–1069.
5. Eipper B. A. and Mains R. E. (1978) Analysis of the common precursor to corticotropin and endorphin. *J. Biol. Chem.* **253,** 5732–5744.
6. Roberts J. L., Phillips M., Rosa P. A., and Herbert E. (1978) Steps involved in the processing of common precursor forms of adrenocorticotropin and endorphin in cultures of mouse pituitary cells. *Biochemistry* **17,** 3609–3618.
7. Nakanishi S., Inoue A., Kita A., Nakamura M., Chang A. C. Y., Cohen S. N., and Numa S. (1979) Nucleotide sequence of cloned cDNA for bovine corticotropin-beta-lipotropin precursor. *Nature* **278,** 423–42.
8. Steiner D. F., Cho S., Oyer P. I. E., Terris S., Peterson J. D., and Rubenstein A. H. (1971) Isolation and characterization of proinsulin c-peptide from bovine pancreas. *J. Biol. Chem.* **246,** 1365–1374.
9. Noe B. D. and Bauer G. E. (1979) Evidence for glucagon biosynthesis involving a protein intermediate in islets of the anglerfish (*Lophius americanus*). *Endocrinology* **89,** 642–651.
10. Yalow R. S. and Wu N. (1973) Additional studies on the nature of big big gastrin. *Gastroenterology* **65,** 19–27.
11. Chan S. J., Keim P., and Steiner D. F. (1976) Cell-free synthesis of rat preproinsulins: Characterization and partial amino acid sequence determination. *Proc. Natl. Acad. Sci. USA* **73,** 1964–1968.
12. Lin C., Joseph-Bravo T., Sherman L., Chan L., and McKelvy J. F. (1979) Cell-free synthesis of putative neurophysin precursors from rat and mouse hypothalamic mRNA. *Biochem. Biophys. Res. Commun.* **89,** 943-950.
13. Goodman R. H., Jacobs J. W., Chin W. W., Lund P. R., Dee P. C., and Habener J. F. (1980) Nucleotide sequence of a cloned structural gene coding for a precursor of pancreatic somatostatin. *Proc. Natl. Acad. Sci. USA* **77,** 5869–5873.
14. Noyes B., Mevarech M., Stein R., and Argarwal K. (1979) Detection and partial sequence analysis of gastrin mRNA by using an oligodeoxynucleotide probe. *Proc. Natl. Acad. Sci. USA* **76,** 1770–1774.
15. Patzelt C., Tager H. S., Carrol R. J., and Steiner D. F. (1979) Identification and processing of proglucagon in pancreatic islets. *Nature* **282,** 260–266.
16. Rosenfeld M. G., Amara S. G., Roos B. A., Ong E. S., and Evans R. M. (1981) Altered expression of the calcitonin gene associated with RNA polymorphism. *Nature* **290,** 63–65.
17. Amara S. G., Jonas V., O'Neil J. A., Vale W., Rivier J., Roos B. H., Evans R. M., and Rosenfeld M. G. (1982) Calcitonin COOH-terminal cleavage peptide as a model for identification of novel

neuropeptides predicted by recombinant DNA analysis. *J. Biol. Chem.* **257,** 2129–2132.

18. Rosenfeld M. G., Lin C. R., Amara S. G., Stolarsky L. S., Ono E. S., and Evans R. M. (1982) Calcitonin mRNA polymorphism: Peptide switching associated with alternate RNA splicing events. *Proc. Natl. Acad. Sci. USA* **79,** 1717–1721.
19. Amara S. G., Jonas V., Rosenfeld M. G., Ong E. S., and Evans R. M. (1982) Alternative RNA processing in calcitonin gene expression generates mRNAs encoding different polypeptide products. *Nature* **298,** 240–244.
20. Rosenfeld M. G., Mermod J. -J., Amara S. G., Swanson L. W., Sawchenko P. B., Rivier J., Vale W. W., and Evans R. M. (1983) Production of a novel neuropeptide encoded by the calcitonin gene via tissue-specific RNA processing. *Nature* **304,** 129–135.
21. Early P., Rogers, J., Davis M., Calame K., Bond M., Wall R., and Hood L. (1980) Two mRNAs can be produced from a single immunoglobulin μ gene by alternative RNA processing pathways. *Cell* **20,** 313–319.
22. Alt F. W., Bothwell A. L., Knapp M., Siden E., Mather E. L., Koshland M. E., and Baltimore D. (1980) Synthesis of secreted and membrane-bound immunoglobulin in Mu heavy chains is directed by mRNAs that differ at their 3′ ends. *Cell* **20,** 293–301.
23. Maki R., Roeder W., Traunecker A., Sidman C., Wabe M., Raschke W., and Tonegaga S. (1981) The role of DNA rearrangement and alternative RNA processing in the expression of immunoglobulin delta genes. *Cell* **24,** 353–365.
24. Capon D. J., Seeburg P. H., McGrath J. P., Hayflick J. S., Edman U., Levinson, A. D., and Goeddel, D. V. (1983) Activation of ki-ras2 gene in human colon and lung carcinoma by two different point mutations. *Nature* **304,** 507–512.
25. Gruss P., Dhar R., and Khoury G. (1978) Simian virus 40 tandem repeated sequences as an element of the early promoter. *Proc. Natl. Acad. Sci. USA* **78,** 943–947.
26. Chow L. T. and Broker T. R. (1978) The spliced structures of adenovirus 2 fiber message and the other late mRNAs. *Cell* **15,** 497–510.
27. Nevins J. R. and Darnell J. E., Jr. (1978) Steps in the processing of Ad2 mRNA: Poly(A)$^+$ nuclear sequences are conserved and poly(A) addition precedes splicing. *Cell* **15,** 1477–1493.
28. Berget S. M. and Sharp P. A. (1979) Structure of late adenovirus 2 heterogeneous nuclear RNA. *J. Mol. Biol.* **129,** 547–565.
29. Khoury G., Gruss P., Dhar R., and Lai C.-J. (1979) Processing of expression of early SV40 mRNA: A role for RNA conformation in splicing. *Cell* **18,** 85–92.

30. Cheng H. L., Blattner F. R., Fitzmaurice L., Mushinski J. P., and Lucker P. W. (1982) Structure of genes for membrane and secreted murine IgD heavy chains. *Nature* **296,** 410–415.
31. Henikoff S., Sloan J. S., and Kelly J. D. (1983) A drosophila metabolic gene transcript is alternatively processed. *Cell* **34,** 405–414.
32. Rozek E. E. and Davidson N. (1983) Drosophila has one myosin heavy-chain gene with three developmentally regulated transcripts. *Cell* **32,** 23–34.
33. Carlson M. and Botstein D. (1982) Two differentially regulated mRNAs with different 5′ ends encode secreted and intracellular forms of yeast invertase. *Cell* **28,** 145–154.
34. Kitamura N., Takagaki Y., Furuto S., Tanaka T., Nawa H., and Nakanishi S. (1983) A single gene for bovine high molecular weight and low molecular weight kininogens. *Nature* **305,** 545–549.
35. Young R. A., Hagenbuchle D., and Schibler U. (1981) A mouse alpha-amylase gene specifies two different tissue-specific mRNAs. *Cell* **23,** 451–458.
36. Schibler U., Hagenbuchle D., Wellaver P. K., and Pittet A. C. (1983) Two promoters of different strengths control the transcription of the mouse alpha-amylase gene Amy-1[a] in the parotid gland and the liver. *Cell* **33,** 501–508.
37. Jacobs J. W., Goodman R. H., Chin W. W., Dee P. C., Habener J. F., Bell N. H., and Potts, Jr., J. T. (1981) Calcitonin messenger RNA encodes multiple polypeptides in a single precursor. *Science* **213,** 457–459.
38. Amara S. G., Rosenfeld M. G., Birnbaum R. S., and Roos B. A. (1980) Identification of the putative cell-free translation product of rat calcitonin mRNA. *J. Biol. Chem.* **255,** 2645–2648.
39. Amara S. G., Rosenfeld M. G., Roos B. A., and Evans R. M. (1980) Characterization of rat calcitonin mRNA. *Proc. Natl. Acad. Sci. USA* **77,** 4444-4448.
40. Birnbaum R. S., O'Neal J. A., Muszynski M., Aron D. C., and Roos B. A. (1982) A non-calcitonin secretory peptide derived from preprocalcitonin. *J. Biol. Chem.* **257,** 241–244.
41. Amara S. G., Evans R. M., and Rosenfeld M. G. (1984) Calcitonin/calcitonin gene-related peptide transcription unit: Tissue-specific expression involves selective use of alternative polyadenylation sites. *Mol. Cell. Biol.* **4,** 2151–2160.
42. Fisher L. A., Kikkawa D. O., Rivier J. E., Amara S. G., Evans R. M., Rosenfeld M. G., Vale W. W., and Brown M. R. (1983) Stimulation of noradrenergic sympathetic outflow by calcitonin gene-related peptide. *Nature* **305,** 534–536.
43. Amara S. G., Arriza J. L., Leff S. E., Swanson L. W., Evans R. M., and Rosenfeld M. G. (1985) Expression in brain of a messenger

RNA encoding a novel neuropeptide homologous to calcitonin gene-related peptide. *Science* **229,** 1094–1097.

44. Deftos L. S., Burton D., Bone H. G., Catherwood B. D., Parthemore J. G., Moore R. Y., Minick S., and Guillemin R. (1978) Immunoreactive calcitonin in the intermediate lobe of the pituitary gland. *Life Sci.* **23,** 743–748.
45. Jonas V., Lin C. R., Kawashima E., Semon D., Swanson L.W., Mermod J.-J., Evans R. M., and Rosenfeld M. G. (1985) Alternative RNA processing events in human calcitonin/calcitonin gene-related peptide gene expression. *Proc. Natl. Acad. Sci. USA* **82,** 1994–1998.
46. Jacobs J. W., Goltzman D., and Habener J. F. (1982) Absence of detectable calcitonin synthesis in the pituitary using cloned complementary deoxyribonucleic acid probes. *Endocrinology* **111,** 2014–2019.

# Chapter 12

# Parathyroid Hormone Genes

HENRY M. KRONENBERG

## 1. INTRODUCTION

Parathyroid hormone (PTH), a peptide of 84 amino acids, is the major hormonal regulator of the level of ionized calcium in extracellular fluid (*1*). Parathyroid glands have been found in amphibians, reptiles, birds, and mammals, but not in fish or lower vertebrates. Therefore, the appearance of parathyroid hormone in evolutionary history correlates with the movement of vertebrates from the calcium-containing sea to dry land. PTH raises the level of blood calcium by stimulating resorption of mineral from bone and by stimulating the resorption of calcium from the urine in the distal tubule of the kidney. PTH also stimulates the 1-hydroxylation of 25-hydroxyvitamin D in the kidney; 1,25-dihydroxyvitamin D, in turn, stimulates the absorption of calcium from the gut.

PTH is synthesized as a larger precursor, preproPTH (*2*). The amino-terminal "pre," or signal, sequence, directs the precursor to the endoplasmic reticulum and is removed as the protein crosses into the cisterna of the endoplasmic reticulum. The resultant intermediate precursor, proPTH, moves to the Golgi apparatus where the amino-terminal "pro" region of six residues is removed. This cleavage occurs just distal to dibasic residues and thus resembles the cleavage of "pro" sequences from precursors of other secreted proteins. PTH then moves to secretory granules and leaves the cell by exocytosis.

The secretion of PTH is regulated by the level of blood calcium in a homeostatic manner. When blood calcium falls, PTH secretion rapidly increases; when blood calcium rises, PTH secretion decreases. In less dramatic fashion, fluctuations in blood magnesium and catecholamines can also influence PTH secretion. In addition to its effects on the secretion of PTH, calcium has other effects on the parathyroid gland. Chronic lowering of blood calcium results in hyperplasia of parathyroid cells; the enlarged parathyroid gland then has the capacity to secrete substantially more PTH in response to a stimulus. The effects of calcium on the specific biosynthesis of PTH are less well defined. During a 7-h incubation in vitro, the levels of PTH mRNA in bovine parathyroid gland slices are not influenced by the exposure to varying levels of calcium (*3*). In studies of longer duration, however, the levels of PTH mRNA in dispersed parathyroid cells fall after 1–2 d in a high-calcium medium (*4*). In addition to modulating some pretranslational step, calcium also affects the fate of PTH after it is synthesized. A substantial fraction of PTH is degraded in the parathyroid gland; this fraction falls when hypocalcemia stimulates PTH secretion (*5*). Thus, calcium and perhaps other factors regulate the biosynthesis of PTH. The recent isolation and characterization of the human, bovine, and rat PTH genes should lead to a greater understanding of this regulation.

## 2. ISOLATION OF PTH GENES

Early studies of the translation of bovine PTH mRNA suggested that the PTH mRNA is a reasonably prominent mRNA, representing several percent of polyadenylated mRNA (*6*). Consequently, the isolation of bovine cDNA clones using hybrid-arrested translation to identify clones encoding PTH was straightforward (*7*). Kemper et al. (*8*) determined the sequence of a full-length representation of the bovine PTH mRNA by sequencing several independent bovine cDNA clones. These workers determined the sequence of the first few nucleotides of the mRNA using a primer-extension strategy. Hendy et al. (*9*) used bovine cDNA to screen a human cDNA library constructed by using mRNA from a parathyroid tumor removed from a patient who had primary hyperparathyroidism. The sequences of several overlapping clones provided the sequence of almost the entire human mRNA.

Human genomic DNA encoding PTH (*10*) was isolated from the Charon-4A lambda phage library constructed by Lawn et al.

(*11*). Two strategies were used. Using human cDNA as a probe, a million plaques were screened using the method of Benton and Davis (*12*), yielding one positive plaque. Southern blot analysis showed that DNA from this phage lacked the far 5′ portion of the PTH gene. Consequently, the library was rescreened using the πVX vector of Seed (*13*). By inserting 290 base pairs (bp) of PTH cDNA into the πVX vector, the PTH sequence was linked to a selectable marker, the *supF* gene encoding an amber suppressor tRNA. The phage library, carrying two amber mutations in the Charon-4A vector, was passaged through *E. coli* carrying the πVX-PTH plasmid. Recombinant phages carrying the πVX-PTH plasmid were produced by recombination between the PTH sequences on the πVX-PTH plasmid, and PTH sequences on phages carrying the human PTH gene. These phages could be identified by their ability to grow subsequently on *E. coli* that was otherwise lacking amber suppressor tRNA genes. Southern blot analysis of one phage selected in this manner indicated that that phage contained the entire transcribed PTH sequence.

Heinrich et al. (*14*) took advantage of sequence conservation between the rat and bovine PTH genes to isolate the rat PTH gene. They used a bovine cDNA as a hybridization probe to screen a rat Charon-4A lambda phage library. Two independent phages, each containing the entire gene, were analyzed.

Weaver et al. (*15*) isolated the bovine PTH gene by first constructing a Charon-30 library containing bovine DNA enriched for the PTH gene. Southern blot analysis showed that the bovine gene was contained on one 7-kb *Eco*RI fragment. Consequently, Weaver et al. digested bovine DNA to completion with *Eco*RI, isolated DNA in the appropriate size range from a sucrose gradient, and inserted the DNA into the Charon-30 vector. The library was screened by the plaque-hybridization method. Two independent positive phages were further characterized.

## 3. GENE NUMBER, LOCATION, AND POLYMORPHISMS

The parathyroid-hormone gene in humans, rats, and cows is represented only once in the haploid genome (*10,14,15*). In each species, Southern blot analysis of genomic DNA using a series of restriction enzymes revealed no heterogeneity in cleavage sites flanking the PTH gene (except for the allelic restriction enzyme–site polymorphisms discussed below). Furthermore, the intensity of the bands on Southern blots of human DNA was most consistent with the presence of only one copy of the PTH gene per haploid genome.

Naylor et al. (*16*) showed that the human PTH gene is located on the short arm of chromosome 11 by analyzing a series of mouse–human hybrid cell lines, each containing a different fraction of the human chromosome complement. They found that only cell lines containing human chromosome 11 invariably contained the human PTH gene. Lines containing only the long arm of chromosome 11 lacked the PTH gene, suggesting that the PTH gene is located on the short arm of chromosome 11.

Genetic analysis has confirmed the assignment of the PTH gene to the short arm of chromosome 11 and has allowed more precise localization of the PTH gene. Antonarakis et al. (*17*) noted that cleavage of the human PTH gene with the restriction endonuclease *Pst*I results in a pattern of cleavage sites that varies across the human population. *Pst*I always cleaves the human PTH gene in the third exon at the sequence encoding residues 28 and 29 of PTH. In 70% of human PTH genes, the next *Pst*I site is in the DNA flanking the 3′ end of the gene, and is 2800 bp away. In 30% of human PTH genes, the nearest *Pst*I site is only 2200 bp away. Thus, approximately 40% of people are heterozygotes, carrying both patterns of *Pst*I cleavage in their genomes. The differing alleles are inherited in Mendelian fashion, and thus provide a polymorphic marker for family studies.

Antonarakis et al. (*17*) analyzed families carrying the PTH gene polymorphism, as well as analogous polymorphisms in the human insulin and β-globin genes. The insulin and β-globin genes had previously been assigned to the short arm of chromosome 11. All three genes were found to be closely linked, with the gene order being centromere–PTH–β-globin–insulin. The PTH gene was 7 centimorgans (cM) from the globin gene, which in turn was 11 cM from the insulin gene.

Igarashi et al. (unpublished) recently discovered a second polymorphism in the human PTH gene. This polymorphism is determined by the presence or absence of a *Taq*I cleavage site in the second intervening sequence of the PTH gene, and is caused by a point mutation. The relative prevalence of differing alleles has not yet been systematically evaluated. Both the *Pst*I and *Taq*I polymorphisms, as well as polymorphisms in the nearby β-globin gene, are being used to define the inheritance of particular PTH alleles in families with disorders of calcium metabolism.

## 4. STRUCTURE OF MAMMALIAN PTH GENES

### *4.1. Introns*

Complete sequence analysis of the bovine PTH gene (*15*), as well as the human (*10*) and rat (*14*) PTH genes, have made possible the definition of the structure of mammalian PTH genes. These sequences are presented in Fig. 1, 2, and 3. Comparison of the sequences with the bovine and human PTH cDNA sequences reveals the presence of two intervening sequences interrupting the sequences transcribed into mRNA. The nucleotides defining the borders of the intervening sequences are similar to the consensus sequences defined by comparison of many such sequences (*18*). The intervening sequences interrupt the exon sequences at precisely the same locations in the three genes. The large first intervening sequence is located near the end of the 5′ noncoding region of the mRNA, 5 bp before the ATG encoding the initiator methionine of preproPTH. The human first intervening sequence is particularly large (3400 bp), whereas the bovine (1714 bp) and the rat (1600 bp) first intervening sequences are smaller than the human and similar in size to each other. The second intervening sequence is short and of similar size across the three species: human, 103 bp; bovine, 119 bp; and rat, 111 bp. The second intervening sequence comes between the second and third nucleotide encoding the fourth residue (Lys) of the "pro" sequence of preproPTH. Thus, the two intervening sequences separated the PTH gene into three functionally discrete domains (*19*): the 5′ noncoding region, the "prepro" sequence, and the mature hormone, along with the 3′ noncoding region. A similar pattern of intervening sequences was found in a number of genes encoding hormones, including gastrin (*20*) and insulin (*21*).

Not surprisingly, the three mammalian PTH genes resemble each other at the nucleotide level. The bovine and human proteins are identical in 85% of their amino acid residues. The rat

```
-500                                    -450                                    -400
cgaatcaggcagcaaaaggactggagattcagcattcagcatcagtccttccaatgaatattcaaggctgatttcttttagaattactaaaggaatatattatatatattatatatttataaatatataatacatatatatataatatat
-350                                    -300                                    -250
atattttataaatattttatatttatataaaatatacattttgtaaatatataatatatatttataaaatatataatatatataatatatctttttggatataggcattaatcagtcagattacaattcactatttgttagaaatcttt

gcataaacacttttccagcccacgctgttttgcttttaatatccaattatctaaaatttaagaagaatggcaccgccccatgggagtgtgtgtgctgctctatgatcctatggttaaaattcagagaattgggagtgacgtcatctgtaa
-50                                     1                                       50
caataaaaaagcttctcagtgtggaagacttatatatataaaagtcacattgaagggtctACAGCTCAATTTATCAGCCTTCTCAGGTTTACTCAACTTTGAGAAAGCATCAGCTGCTAATACATTTGAAAGAAGATTGTATCCTAAGAC
100                                     150                                     200
GTGTGgtgagtaatctttatttctcttttaagtttccatgagttctttcttacaatcaagtagtcatttaatatgtacatttcctactaatatcatgctgtttgttaaaagtaaaatgttaagggagggtttgaattatatttaatatt
250                                     300                                     350
aaaatgccacaataaaaaataaatttatgccactaaaatattgttgaaacacttatttttgtgattaaagtatttgagaggccttgctttttttttttttttaactcatctgcaattacatggaaataattaaaatttggcatgatgt
400                                     450                                     500
gctttaatgatatcttgttaaaatttgatataaagcagcctagtgagaaatgaataattctgatttaaataatcaatgtaaatttatgtctctgagttaaaattccatttccccacataatatcctcacagagagctcagtctgtgta
550                                     600                                     650
ctatagttggagaaactgttttactgctgaatatttggtcccttcacatagaacctttctagtttctatctgaaaactacacatgaatttaaggatgtttcgtgcgcttctaatgaccttcttgtgagaagcgcgcctctcagttga
700                                     750                                     800
gaataggaaagaaaggcagacgcactaggtgagtgtggcggagtggggcagactcctgtcctctgctacagtgatcttttctcatgtaaaccaaaatggaagactatccatttggtgctccttcaatagcttatattgcacagtgaaaac
850                                     900                                     950
aaatatagtaggttataatggcccttatgctacttctgatgcgcgtttcccacagccaccctatgtgtatgttaaatggagaacattaagtattaagaaaaggcacttctatcatgaaagctattgcagttgttgaaatgcagactaagg
1000                                    1050                                    1100
taagacttttctatttgagataaaaagcttaaaagcaaaagagcaaaattctaaacattatactcaaacaaaattgttaataaaatgagctgaaaagataagatacagacaagcattttatatggaatcaaagctcaaatttgtataac
```

```
     1150                                               1200                                               1250
aagtaattaattagatgatgcttaaatgctttgacacattctttttaatgctgggagaactctgtgttaagttagtaaaaaatctctaaatcagaacctataaatttggcatttatagttatcatgacattgaaaggaagtctacttatg

     1300                                               1350                                               1400
aaaaagaaatttgctaagtaaacagacgctgttgacaatttacctaattaagagaagtagctattttcaagcaaattagaagcatttagtcataattaatacaaaatgtgggaggaaattctaaacagtgaactttttgattcttcacat

     1450                                               1500                                               1550
agcttaaaattcatataacttgcacaagaactctcaaaaagatacctaaaaaagtaaaactcatctgtctcatttctaagccaattgggttatacttactatatactaattatgtgtatatatataatatacagtaattatgtgtatata

     1600                                               1650                                               1700
tataatatacagtaattatgtatatatatatataatatataaactgtaaaataaagagactgattccaaaagtctctcacaaaaaacaatccaattagtcactgtcacactctgtgcaaacttgtatggttataaatataaactactata

     1750                                               1800                    Met Met Ser Ala Lys Asp Met Val Lys Val Met Ile Val Met Leu Ala Ile Cys Phe
aatttaattgtacagtttggaatgttaaaagtattttaaatacctccatttttctttttctttttagTTAAT ATG ATG TCT GCA AAA GAC ATG GTT AAG GTA ATG ATT GTC ATG CTT GCC ATC TGT TTT

Leu Ala Arg Ser Asp Gly Lys Ser Val Ly                                   1950                                               2000
CTT GCA AGA TCA GAT GGG AAG TCT GTT AAgtaagtaccatagcctgttctgcatggtgaggtcaggggaattggattttaaggttggctttatgatttggaagaggggaactaatggagtgatcctctctgatcctgttc

          s Lys Arg Ala Val Ser Glu Ile Gln Phe Met His Asn Leu Gly Lys His Leu Ser Ser Met Gly Arg Val Glu Trp Leu Arg Lys Lys Leu Gln Asp Val His Asn
cctccag G AAG AGA GCT GTG AGT GAA ATA CAG TTT ATG CAT AAC CTG GGC AAA CAT CTG AGC TCC ATG GAA AGA GTG GAA TGG CTG CGG AAA AAG CTA CAG GAT GTG CAC AAC

Phe Val Ala Leu Gly Ala Ser Ile Ala Tyr Arg Asp Gly Ser Ser Gln Arg Pro Arg Lys Lys Glu Asp Asn Val Leu Val Glu Ser His Gln Lys Ser Leu Gly Glu Ala
TTT GTT GCC CTT GGA GCT TCT ATA GCT TAC AGA GAT GGT AGT TCC CAG AGA CCT CGA AAA AAG GAA GAC AAT GTC CTG GTT GAG AGC CAT CAG AAA AGT CTT GGA GAA GCA

Asp Lys Ala Asp Val Asp Val Leu Ile Lys Ala Lys Pro Gln End                   2300                                               2350
GAC AAA GCT GAT GTG GAT GTA TTA ATT AAA GCT AAA CCC CAG TGAaaacagatatgatcagatcactgttctagacagcatagggcaacaatattacatgctgctaatgtgttcaccttctattaagtgccagtat

                    2400                                               2450                                               2500
ttctatgaccaacctttattgctagctgtgatacctacaattttaattgagtatttgattctactttattcatctaagagctcttttaataattctatttctattgattccaataaatgaagttaagtatttctcacttgttataaaa

                    2550                                               2600
atatctttggttatgagtaacaccaatatgttaaaattgatcatgactaagaacaacaacacaaaatgtttttttaaccatatgtttcacattctgatatttgaaattatcctttttaatc
```

Fig. 1. Bovine parathyroid hormone gene. Sequences found in messenger RNA are capitalized. TATA homology sequences and AATAAA polyadenylation signal are underlined.

```
   -3840
     :         :         :         :         :         :         :         :         :         :         :         :
gattcattaatccacatagaatttttctcgatggtataattctgtatttgttaaaagtctttgcataagccccttgtcaagccaaatgctgttttcctttagtatccaattatctgaaa

   -3720
     :         :         :         :         :         :         :         :         :         :         :         :
cttaagaagagtgtgcaccgcccaatgggtgtgtgtatgtgctgctttgaacctatagttgagatccagagaattgggagtgacatcatctgtaacaataaaagagcctctcttggtaag

   -3600
     :         :         :         :         :         :         :         :         :         :         :         :
cagaagacctatatataaaagtcaccatttaaggggtctgcAGTCCAATTCATCAGTTGTCTTTAGTTTACTCAGCATCAGCTACTAACATACCTGAACGAAGATCTTGTTCTAAGACAT

   -3480
     :         :         :         :         :         :         :         :         :         :         :         :
TGTATGgtaagtaaacttaaaaattcacttctgaatctcatgagatttgataatcaagttattatttaatgtgtaccatttctacaaataccatgttgtttcttcaaggtaaaatgcta

   -3360                                          IVS 1
     :         :         :         :          IS APPROXIMATELY          :         :         :         :         :
agaagtttgagttatgtttaatataaaatgccacatacaaaaataa---  3400 BASE PAIRS LONG  ---aagcttctcgtgaaaaccaacccaattagttagtattgcattct

     -60
     :         :         :         :         :         :         :         :         :         :         :
gtgtactatagtttggaatattaaaaatatttaaaatacctccatttgcttatccttttagTGAAG ATG ATA CCT GCA AAA GAC ATG GCT AAA GTT ATG ATT GTC
                                                                   Met Ile Pro Ala Lys Asp Met Ala Lys Val Met Ile Val

40
:         :         :         :         :         :         :         :         :         :         :
ATG TTG GCA ATT TGT TTT CTT ACA AAA TCG GAT GGG AAA TCT GTT AA gtaagtactgttttgccttggaattggattttaatgttgactttatcatttcgaag
Met Leu Ala Ile Cys Phe Leu Thr Lys Ser Asp Gly Lys Ser Val Lys

     150
     :         :         :         :         :         :         :         :         :         :
tggggagctaatgggaagtggccctctctgtttctcttcttcccagG AAG AGA TCT GTG AGT GAA ATA CAG CTT ATG CAT AAC CTG GGA AAA CAT CTG AAC
                                                Lys Arg Ser Val Ser Glu Ile Gln Leu Met His Asn Leu Gly Lys His Leu Asn
```

```
250
 :                   :                   :                   :                    :                   :                   :                    :                   :
TCG ATG GAG AGA GTA GAA TGG CTG CGT AAG AAG CTG CAG GAT GTG CAC AAT TTT GTT GCC CTT GGA GCT CCT CTA GCT CCC AGA GAT GCT
Ser Met Glu Arg Val Glu Trp Leu Arg Lys Lys Leu Gln Asp Val His Asn Phe Val Ala Leu Gly Ala Pro Leu Ala Pro Arg Asp Ala

340
 :                   :                   :                   :                    :                   :                   :                    :                   :
GGT TCC CAG AGG CCC CGA AAA AAG GAA GAC AAT GTC TTG GTT GAG AGC CAT GAA AAA AGT CTT GGA GAG GCA GAC AAA GCT GAT GTG AAT
Gly Ser Gln Arg Pro Arg Lys Lys Glu Asp Asn Val Leu Val Glu Ser His Glu Lys Ser Leu Gly Glu Ala Asp Lys Ala Asp Val Asn

430
 :                   :                   :                 :               :               :               :               :               :               :               :
GTA TTA ACT AAA GCT AAA TCC CAG TGA AAATGAAAACAGATATTGTCAGAGTTCTGCTCTAGACAGTGTAGGGCAACAATACATGCTGCTAATTCAAAGCTCTATTAAGAT
Val Leu Thr Lys Ala Lys Ser Gln ***

540
 :               :               :               :               :               :               :               :               :               :               :               :
TTCCAAGTGCCAATATTTCTGATATAACAAACTACATGTAATCCATCACTAGCCATGATAACTGCAATTTTAATTGATTATTCTGATTCCACTTTTATTCATTTGAGTTATTTTAATTAT

660
 :               :               :               :               :               :               :               :               :               :               :               :
CTTTTCTATTGTTTATTCTTTTTAAAGTATGTTATTGCATAATTTATAAAAGAATAAAATTGCACTTTTAAACCTCTCTTCTACCTTAAAATGTAAAACAAAAATGTAATGATCATAAGT

780
 :               :               :               :               :               :               :               :               :               :               :               :
CTAAATAAATGAAGTATTTCTCACTCAttgcaagtatatctttttggttatcactgatacccacatgtttacattgatcatgactaggtagaacaatacaaagtatttttttagtcatgt

900
 :               :               :               :               :               :               :               :               :               :               :               :
gtttcacatttggatattttgaacatcaacgttttagtattaccaaagtattaggtttccaaatcttcactagctcaatactgttgtccttttggtttcaggaaaggaaataaaatgctc

1020
 :               :               :
agcaaaaaaagggggcataaaagtggacc
```

Fig. 2. Human parathyroid hormone gene. Sequences found in messenger RNA are capitalized. Because of uncertainty about the exact length of the first intervening sequence, the first nucleotide of the coding region is designated nucleotide 1.

```
ttccataact tccacattta caagcataag atgcttgtgt aaaaataaag atatcacatt cactgaagaa ttactgtaat aattaccaga ataccagaat taccagaata -1160
ccagattaat gtgtgcaatt aagaaatgta gttagtgttt attgaatgct attacgatac ctcaaaaacc caagttcaat gaatttaaaa ataatttctc tttggtttgg -1050
tacttcaaat ctgtatcaag cactctgaat attttcttat gtccattgtt ttcctctgtg tgcatgagtg tgagagagag agagtgtgtg tgtgtgtgtg taggcaaggc - 940
caaggccggg tatcttcctc tagttttgct ctcttccttg tttttgcata tggtacctca ctgaacctaa ggcttgctgt ttggctgaac tggctagcct gccctgggga - 830
tttgtccata tccaattcct tcccccaagc actagtctta cagtgctccc caccatggtt ggcttgtaga tggcttctag gaatggaact caggtcctca tatttgcact - 720
ttactgactg agacatctgc cagctctcca gccttatacg cattaaacat atatacaata tgtgttaaac tgatcaaagg tgataagagt tggttggaaa tggcttataa - 610
gaaaaatcat gcaacatgat tccttaagaa tacaaagtct tataatataa ggcttataat aaggctgtgc aatgcatata ctcggctaat tatactgtgc ttcataattg - 500
tcact.......80 base pairs......g aaagtgcagg cccagtggca atgtctttaa ctttgcactc atgaatagag gcaagtggat cttgagtttt gggtcagcct - 316
gatctacata ttgagttcta ggccagccag agttcatagt gagacgctat ctcaaactgt tttaaatata aaatcagtat catggattac gtcagatttt tctaagtggt - 206
tacttaagca tttgctgcaa cttcttttgc agattctttg ccagcacctt gctctttttg aatccattat ctaagtatct gaaactttag aggagtgggc accgcccgat -  96
gagggtaggt ggctgttctg attcctatga ttgagaacca gagaaccagg catgacatca tccttcccaa taaaatactc ctcttggtga gcaaaAGGCC TGCATATGAA   15
ACTCAGGCTT GAAGAACTGC AGTCCAGTTC ATCAGCTGTC TGGCTTACTC CAGCTTAATA CAGGGTCACT CCTGAAGGAT CCTCTCTGAG AGTCATTGTA TGgtaaggaa  125
tctctcaatt gccctttttaa attccgtgag atttagaaaa ttgtgctagt ttttaataca taccatttc tatcaatatg tgccattttt taaaattgga aggtaggggt  235
gaatttgcca aaatggaagg atatggcaat aaaataattt agggtaggaa attcggactc atctatgtat atggttaa .. IVS A (1600 base pairs) .. gttc 1433
tatactaaag tatctgtcct ataaagatcc cacagaccca tgaaaagtgg catcagctaa tgactgtttc cttgggtggt tatgaataac agctactggg actgtgttta 1543
                                                             -31 -30                              -20
                                                             Met Met Ser Ala Ser Thr Met Ala Lys Val Met Ile Leu Met
cacactttag agcactgaca gtgtcttaaa atatctctgt ctctccttgt agTGAAG ATG ATG TCT GCA AGC ACC ATG GCT AAG GTG ATG ATC CTC ATG  1642
                     -10        -6
Leu Ala Val Cys Leu Leu Thr Gln Ala Asp Gly Lys Pro Val Ly
CTG GCA GTT TGT CTC CTT ACC CAG GCA GAT GGG AAA CCC GTT AAgtaagtgc tgcagcccgt cgtcccaggg aagtcggaca tgaggctctg taggtttta  1742
                                                     -1 +1                                  10
                                                  s Lys Arg Ala Val Ser Glu Ile Gln Leu Met His Asn Leu Gly Lys
atgttgtggg catggggagc taatggagtg gtcctctctt tctgttctctctagG AAG AGA GCT GTC AGT GAA ATA CAG CTT ATG CAC AAC CTG GGC AAA  1842
                         20                               30                               40
His Leu Ala Ser Val Glu Arg Met Gln Trp Leu Arg Lys Lys Leu Gln Asp Val His Asn Phe Val Ser Leu Gly Val Gln Met Ala Ala
CAC CTG GCC TCT GTG GAG AGA ATG CAA TGG CTG AGA AAA AAG CTG CAA GAT GTA CAC AAT TTT GTT AGT CTT GGA GTC CAA ATG GCT GCC  1932
                         50                               60                               70
Arg Glu Gly Ser Tyr Gln Arg Pro Thr Lys Lys Glu Asp Asn Val Leu Val Asp Gly Asn Ser Lys Ser Leu Gly Glu Ala Asp Lys Ala
AGA GAA GGC AGT TAC CAG AGG CCC ACC AAG AAG GAG GAA AAT GTC CTT GTT GAT GGC AAT TCA AAA AGT CTT GGC GAG GGG GAC AAA GCT  2022
                         80          84
Asp Val Asp Val Leu Val Lys Ala Lys Ser Gln END
GAT GTG GAT GTA TTA GTT AAG GCT AAA TCT CAG TAA ATGCTGACGT ATTCTAGACC GTGCTGAGCA ATAACATCAA GTTGCTAATT CTTCTACTGT AATAA  2123

AAGTTTGAAA TTTGATTCCA CTTTTGCTCA TTTAAGGTCT CTTCCAATGA TTCCATTTCA ATATATTCTT CTTTTTAAAG TATTACACAT TTCCACTTCT CTCCTTAAAT  2233
ATAAATAAAG TTTAATGATC ATGAACCAAA taagcagtgt ttcttacttg ttaaaacttt tgtctcagtg ttggagggct ggctcagaag ttaagagtgc atactgctgc  2243
ctcaaatgac ccgagtttgc ttctcgg
```

Fig. 3. Rat parathyroid hormone gene. Sequences found in messenger RNA are capitalized. TATA homology is boxed. ATG triplets in the 5′ noncoding region of the mRNA are underlined; stop codons in the same reading frame with the ATG codons are also overlined. The dashed lines indicate sequences of alternating purines and pyrimidines, which are possible sites of formation of Z-DNA. Nucleotides are numbered on the right of the figure, beginning with +1 at the putative transcription start site. The protein-coding sequence of the gene is keyed by numbers above the assigned amino acids; +1 (alanine) begins the sequence of PTH; the minus numbers designate the signal and propeptide sequences.

protein has diverged somewhat from the other two proteins. The rat and bovine proteins have 76% of their residues in common; the rat and human proteins share 74% of their residues. This pattern is repeated at the nucleotide level with the bovine and human sequences identical over 85% of their bases; the rat/bovine identity is 78%; the rat/human identity is 77%. A similar pattern holds in the second intervening sequence (counting a gap as a one-base mismatch); bovine/human, 80%; rat/bovine, 74%; rat/human, 70%. Because only the bovine first intervening sequence has been determined in its entirety, comparisons in this region cannot be made. The 3′ noncoding regions of mRNAs are not well conserved. Even the more closely related human and bovine genes share only 48% homology. In contrast, the sequences just upstream from the gene share substantial homology. In the 108 bases adjacent to the transcription start site, the human/bovine homology is 80%; the rat/human homology is 73%; and the rat/bovine homology is 64%.

### 4.2. Promoter Sequences

The functional analysis of the PTH gene is just beginning; consequently, little is yet understood about sequences that are important in regulating transcription of the PTH gene. As noted above, comparison of the sequences flanking the 5′ end of the transcribed region demonstrates substantial conservation of sequence across the three species. Thus, this region, like the corresponding portion of other genes transcribed by RNA polymerase II, serves an important role (Fig. 4). The precise sites of initiation of transcription were determined for each gene. Weaver et al. (*8*) first used primed reverse transcription of bovine PTH mRNA to demonstrate that bovine transcripts start at two locations, 30 bp apart from each other. Subsequent sequence analysis revealed that each transcript was preceded by a consensus TATA sequence approximately 30 bp upstream. S1 nuclease digestion of DNA–mRNA hybrids (*15*) confirmed the presence of two start sites for transcription. Furthermore, high-resolution gel electrophoresis revealed that each start site contained some microheterogeneity, with transcripts starting at several different, clustered nucleotides. Sequence inspection and variation in nuclease conditions suggest that much of the heterogeneity could not be explained by the artifact caused by transient local denaturation of the ends of

```
                       20                    40
Bovine    CCTATGGTTAAAATTCAGAGAATTGGGAGTGACGTCATCTGTAACAATAAAAAAGC
          ***** *** * ** ****************** ****************** ***
Human     CCTATAGTTGAGATCCAGAGAATTGGGAGTGACATCATCTGTAACAATAAAAGAGC
          *****  ****** ********   **  **********  *  ******** * *
Rat       CCTATGATTGAGAACCAGAGAACCAGGCATGACATCATCCTTCCCAATAAAATA-C
          ****** ** * *  *******   **  **** *****  *  ******** * *
Bovine    CCTATGGTTAAAATTCAGAGAATTGGGAGTGACGTCATCTGTAACAATAAAAAAGC


             60                 80                  100
Bovine    TTCTC----AGTGTGGAAGACTTATATATATAAAAGTCAC-ATTGAAGGG-TCTAC
           ****       *  ******   **************** *** ***** *** *
Human     CTCTCTTGGTAAGCAGAAGACC--TATATATAAAAGTCACCATTTAAGGGGTCTGC
            ******** **** *** **  *  **** *** ***   ** ***    ****
Rat       TCCTCTTGGTGAGCAAAAGGCC--TGCATATGAAACTCAGGCTTGAAGAA--CTGC
          * ***     * *   *** *   *  **** *** ***   ******    ****
Bovine    TTCTC----AGTGTGGAAGACTTATATATATAAAAGTCAC-ATTGAAGGG-TCTAC
```

Fig. 4. Comparison of 5′ flanking regions of mammalian PTH genes. TATA homology regions are bracketed. Asterisks indicate conservation of assigned bases between adjacent genes. Spaces have been inserted to maximize homology.

DNA–RNA hybrids at regions of high AT content. The S1 and primer extension experiments yielded results that differed quantitatively. The S1 nuclease analysis suggested that the upstream start site was used as frequently as the downstream start site was. In contrast, the primer extension experiments suggested that only a small fraction of the transcripts started at the upstream site. Because both methods of determining the start site of transcription are indirect techniques for analyzing the 5′ end of mRNA sequences, and because each is subject to potential artifact, more-detailed quantitative analysis may require the use of other methods.

S1 nuclease analysis of rat PTH gene transcripts revealed only one start site of transcription, corresponding to the upstream start site of the bovine gene. Inspection of the DNA sequences (Fig. 4) shows that the sequence of the second TATA box of the bovine gene is changed in the rat sequence from TATATATA to TGCATATG. The alteration of the consensus sequence may be responsible for the absence of a second initiation site in the rat gene.

S1 nuclease analysis of the human PTH gene revealed a pattern like the bovine pattern. Sequence comparison shows that the human gene, like the bovine, contains two consensus TATA sequences. Recently, Igarashi et al. (submitted for publication) have used S1 nuclease and primer-extension analysis to show that both start sites are used in normal human parathyroid glands and in human parathyroid tumors. Furthermore, both sites are used when the human PTH gene is transcribed after being introduced into rat pituitary GH4 cells. The function of two different start sites for transcription is unknown; it might be to allow for greater transcription of the PTH gene. Transcription from each start site might be regulated in a different fashion. A few other genes have been shown to initiate transcription at two closely clustered locations. These genes include the chicken lysozyme gene (*22*) and the avian very-low-density apolipoprotein II gene (*23*). Although the existence of two distinct transcription start sites suggests possible regulatory consequences, the presence of only one start site in the rat PTH gene, on the other hand, suggests that the two start sites of the bovine and human PTH genes may not have great functional importance.

### *4.3. Exons*

As noted above, the coding regions of the three PTH genes are highly conserved. The 5′ noncoding regions of the human and rat sequences are unusual because they contain AUG sequences that precede the AUG used to start preproPTH translation (*see* Fig. 1 and 2). In both the human and rat sequences, an "extra" AUG is immediately followed by a UGA termination codon, then two more nucleotides, and finally the initiator AUG of preproPTH. Therefore, any translation initiated by the upstream AUG would be immediately terminated before entering the preproPTH coding region in an inappropriate reading frame. A similar pattern of upstream AUGs is found in the alpha-amylase genes of murine salivary gland and liver (*24*). In each case, the upstream AUG is followed by the codon AAA and then the termination codon, UAA. The similarity of the amylase genes to the human PTH gene is even more striking, because in both amylase genes and in the PTH gene the upstream AUG is followed immediately by an intervening sequence. Therefore the upstream AUG could conceivably initiate the translation of a completely different protein, if alternative splicing patterns associated the AUG with a different open reading frame. Similarly, a unique peptide would be encoded by completely unspliced mRNA. None of these genes, however, contain a long open reading frame in the portion of the intervening sequence immediately adjacent to the upstream AUG.

Kozak (*25,26*) compiled data that might explain why the upstream AUGs in the human and rat PTH genes do not interfere with the proper initiation of preproPTH synthesis. She first noted (*25*) that functional initiator AUGs are usually flanked by the consensus sequence ($\frac{A}{G}$XXAUGG. In contrast, upstream, apparently nonfunctional AUGs seldom contain a purine in the −3 position. Furthermore, expression of the preproinsulin gene after experimental manipulation of the sequences flanking preproinsulin's initiator AUG demonstrated the importance of the sequence context of the initiator AUG (*26*). The upstream AUGs associated with the human and rat PTH genes are found in the context of a TGTATGT sequence and thus probably do not serve as efficient initiator AUGs. In the bovine sequence, the potential initiator sequence has been changed to GUG, a much less favored initiation codon. The rat gene contains another upstream AUG sequence only 11 nucleotides from the 5′ end of the transcript. This AUG is followed by an in-frame stop codon 19 codons downstream. Like the other 5′ noncoding AUGs, this AUG is in a sequence context suggesting that it is an inefficient initiator. This upstream AUG sequence is not conserved in either the bovine or human PTH genes.

The 3′ noncoding sequences of the three PTH genes are poorly conserved and vary considerably from 222 (bovine) to 350 (human) bp. The bovine and human mRNAs both end at sites that are 14 nucleotides distal to the sequence AATAAA. In other systems, this sequence was shown to determine the site at which a longer mRNA precursor is cleaved by an endonuclease and subsequently polyadenylated (*27,28*).

## 5. CONCLUSION

The cloned human, bovine, and rat PTH genes share a common structure and conservation of sequence in both transcribed regions and flanking sequences that may have regulatory importance. The definition of the functions of these regulatory regions remains important work for the future.

## REFERENCES

1. Habener J. F. (1979) Parathyroid Hormone Biosynthesis, in *Endocrinology* vol. 2 (DeGroot, L. S., ed.), Grune and Stratton, New York.
2. Habener J. F., Rosenblatt M., Kemper B., Kronenberg H. M., Rich A., and Potts Jr., J. T. (1978) Pre-proparathyroid hormone: Amino acid sequence, chemical synthesis, and some biological studies of the precursor region. *Proc. Natl. Acad. Sci. USA* **75,** 2616–2620.
3. Heinrich G., Kronenberg H. M., Potts Jr., J. T., and Habener J. F. (1983) Parathyroid hormone messenger ribonucleic acid: Effects of calcium on cellular regulation *in vitro*. *Endocrinology* **112,** 449–458.
4. Russell J., Lettieri D., and Sherwood L. M. (1983) Direct regulation by calcium of cytoplasmic messenger ribonucleic acid coding for pre-proparathyroid hormone in isolated bovine parathyroid cells. *J. Clin. Invest.* **72,** 1851–1854.
5. Habener J. F., Kemper B., and Potts Jr., J. T. (1975) Calcium-dependent intracellular degradation of parathyroid hormone: A possible mechanism for the regulation of hormone stores. *Endocrinology* **97,** 431–441.
6. Kemper B., Habener J. F., Potts Jr., J. T., and Rich A. (1972) Proparathyroid hormone: Identification of a biosynthetic precursor to parathyroid hormone. *Proc. Natl. Acad. Sci. USA* **69,** 643–647.
7. Kronenberg H. M., McDevitt B. E., Majzoub J. A., Nathan J., Sharp P. A., Potts Jr., J. T., and Rich A. (1979) Cloning and nucleotide sequence of DNA coding for bovine preproparathyroid hormone. *Proc. Natl. Acad. Sci. USA* **76,** 4981–4985.
8. Weaver C. A., Gordon D. F., and Kemper B. (1983) Nucleotide sequence of bovine parathyroid hormone messenger RNA. *Mol. Cell. Endocr.* **28,** 411–424.
9. Hendy G. N., Kronenberg H. M., Potts Jr., J. T., and Rich A. (1981) Nucleotide sequence of cloned cDNAs encoding human preproparathyroid hormone. *Proc. Natl. Acad. Sci. USA* **78,** 7365–7369.
10. Vasicek T. J., McDevitt B. E., Freeman M. W., Fennick J. F., Hendy G. N., Potts Jr., J. T., Rich A., and Kronenberg H. M. (1983) Nucleotide sequence of the human parathyroid hormone gene. *Proc. Natl. Acad. Sci. USA* **80,** 2127–2131.
11. Lawn R. M., Fritsch E. F., Parker R. C., Blake G., and Maniatis T. (1978) The isolation and characterization of linked delta- and beta-globin genes from a cloned library of human DNA. *Cell* **15,** 1157–1174.
12. Benton W. D. and Davis R. W. (1977) Screening λgt recombinant clones by hybridization to single plaques in situ. *Science* **196,** 180–183.
13. Seed B. (1983) Purification of genomic sequences from bacterio-

phage libraries by recombination and selection *in vivo*. *Nucl. Acids Res.* **11,** 2427–2445.

14. Heinrich G., Kronenberg H. M., Potts Jr., J. T., and Habener J. F. (1984) Gene encoding parathyroid hormone. Nucleotide sequence of the rat gene and deduced amino acid sequence of rat preproparathyroid hormone. *J. Biol. Chem.* **259,** 3320–3329.
15. Weaver C. A., Gordon D. F., Kissel M. S., Mead D. A., and Kemper B. (1984) Isolation and complete nucleotide sequence of the gene for bovine parathyroid hormone. *Gene* **28,** 319–329.
16. Naylor S. L., Sakaguchi A. Y., Szoka P., Hendy G. N., Kronenberg H. M., Rich A., and Shows T. (1983) Human parathyroid hormone gene (PTH) is on the short arm of chromosome 11. *Som. Cell Gen.* **9,** 609–616.
17. Antonarakis S. E., Phillips III, J. A., Mallonee R. L., Kazazian Jr., H. H., Fearon E. R., Waher P. G., Kronenberg H. M., Ullrich A., and Meyers D. A. (1983) β-Globin is linked to the parathyroid hormone (PTH) locus and lies between the insulin and PTH loci in man. *Proc. Natl. Acad. Sci. USA* **80,** 6615–6619.
18. Mount S. M. (1982) A catalogue of splice junction sequences. *Nucl. Acids Res.* **10,** 459–472.
19. Gilbert W. (1978) Why genes in pieces? *Nature* **271,** 501.
20. Wilborg O., Berglund L., Boel E., Norris K., Rehfeld J. F., Marcker K. A., and Vuust J. (1984) Structure of a human gastrin gene. *Proc. Natl. Acad. Sci. USA* **81,** 1067–1069.
21. Bell G. I., Pictet R. L., Rutter W. J., Cordell B., Tischer E., and Goodman H. M. (1980) Sequence of the human insulin gene. *Nature* **284,** 26–32.
22. Grez M., Land H., Biesecke R., Schultz G., Jung A., and Sippel A. E. (1981) Multiple mRNAs are generated from the chicken lysozyme gene. *Cell* **25,** 743–752.
23. Hache R. J. G., Wiskocil R., Vasa M., Roy R. N., Lau P. C. K., and Deeley R. G. (1983) The 5′ noncoding and flanking regions of the avian very low density apolipoprotein II and serum albumin genes. *J. Biol. Chem.* **258,** 4556–4564.
24. Young R. A., Hagenbuchle O., and Schibler (1981) A single mouse α-amylase gene specifies two different tissue-specific mRNAs. *Cell* **23,** 451–458.
25. Kozak M. (1984) Compilation and analysis of sequences upstream from the translational start site in eukaryotic mRNAs. *Nucl. Acids Res.* **12,** 857–872.
26. Kozak M. (1984) Point mutations close to the initiator AUG codon affect the efficiency of translation of rat preproinsulin *in vivo*. *Nature* **308,** 241–246.

27. Montal C., Fisher E. F., Caruthers M. H., and Berk A. J. (1983) Inhibition of RNA cleavage but not polyadenylation by a point mutation in mRNA 3′ consensus sequence AAUAAA. *Nature* **305,** 600–605.
28. Higgs D. R., Goodbourn S. E. Y., Lamb J., Clegg J. R., Weatherall D. J., and Proudfoot N. F. (1983) α-Thalassemia caused by a polyadenylation signal mutation. *Nature* **306,** 398–400.

# Chapter 13

# The Renin Gene

François Rougeon, Jean-Jacques Panthier,
Inge Holm, Florent Soubrier,
and Pierre Corvol

## 1. INTRODUCTION

The renin angiotensin system plays a major role in the control of blood pressure and electrolyte balance. It is implicated in essential hypertension, and therefore, its inhibition constitutes the major therapeutic advance in the treatment of cardiovascular disease of these last few years. The structure and organization of the renin gene have recently been elucidated, and the advances concerning this gene in mouse and humans will be reviewed in this chapter.

## 2. THE RENIN–ANGIOTENSIN SYSTEM—ITS IMPORTANCE IN THE REGULATION OF BLOOD PRESSURE

The renin–angiotensin system consists of a cascade of proteolytic cleavages leading to the biologically active angiotensins. Renin is the enzyme that acts at the first step of the system by cleaving the decapeptide angiotensin I from the amino terminus of the renin substrate, angiotensinogen, a 55-kdalton protein synthesized by the liver. Angiotensin I has little or no significant biological activity *per se,* but it is the precursor of the potent octapeptide angiotensin II. Angiotensin II is formed when the two carboxy-terminal amino acids of angiotensin I are removed by a dipeptidyl-carboxypeptidase, also called angiotensin-converting enzyme.

All the physiological actions of angiotensin II are involved in the regulation of renal function and the maintenance of plasma volume and blood pressure. An important intrarenal effect of angiotensin II is the control of renal blood flow and glomerular filtration rate. By stimulating aldosterone secretion, it also regulates renal sodium reabsorption and plasma volume. The blood pressure levels depend on the degree of volume expansion, governed in part by angiotensin II, and also on the extremely potent direct vasopressive effect of this hormone on the vascular smooth muscle.

Renin is synthesized in the kidney, within the afferent arterioles near the vascular pole of the glomerulus. These vascular cells (juxtaglomerular cells) are differentiated into endocrine-type cells containing both myofilaments, showing their myoepithelial origin, and storage granules, as in typical endocrine cells. The secretion of renin by the kidney is regulated by beta-adrenergic nerves, transmural blood pressure, and angiotensin II.

Besides renal renin, which is the source of plasma renin and has been the most extensively studied, renin has also been reported in various other tissues. The salivary gland of mice (*1*) and the uterus of rabbits (*2*) are particularly rich sources. Renin has also been reported in human chorionic tissue (*3*), brain (*4*), adrenal glands (*5*), and blood vessel walls (*6*). The concept of a local, presumably intracellular, angiotensin synthesis was proposed and is of particular interest because of the known effects of angiotensin on most of these organs.

The contribution of the renin system to the control of blood pressure in physiological and pathological situations has been the subject of intense interest during the past several years. The synthesis of specific and competitive inhibitors of angiotensin II and converting enzyme (*7*) showed that the blockade of the renin system lowered the blood pressure in normal patients and that the blood pressure decrease was directly dependent on the sodium balance. Sodium depletion stimulates renin secretion and potentiates the blood-pressure-lowering effect of converting-enzyme inhibitors. The orally active converting-enzyme inhibitors are quite effective in the treatment of experimental and human essential hypertension, demonstrating the possible involvement of the renin system in the pathogenesis of this disease (*8*).

## 3. RENIN: A MEMBER OF THE ASPARTYL PROTEASE FAMILY?

Despite the considerable interest in the role of the renin system and its implication in hypertension, very little was known about renin biochemistry (*see* review in ref. *9*). The extremely low concentration of renin in animal and human kidney has prevented its purification in amounts sufficient to determine its entire amino acid sequence. Although the primary structure of renin was not known, several important facts led to the hypothesis that renin belonged to the class of aspartyl proteases, like pepsin, chymosin, penicillinopepsin, and cathepsin D. For example, (a) active renin has a molecular weight similar to that of pepsin (40 kdalton); (b) renin and aspartyl proteases exist in an inactive form that can be activated by acidification; (c) pepstatin, a specific and extremely potent inhibitor of all aspartyl proteases, inhibits all renin species in a competitive manner (*10,11*); and (d) studies performed by Inagami et al. (*12,13*) showed that mouse submaxillary and hog kidney renin were inactivated by diazoacetyl-norleucine methyl ester in the presence of cupric ion and by 1,2-epoxy-3-(*p*-nitrophenoxy)-propa ne. The fact that these two well-established inactivators of the aspartyl proteases were not mutually exclusive suggested that renin, like other aspartyl proteases, possessed two different carboxylic residues on its active site that were essential for its catalytic activity.

## 4. THE GENETIC SYSTEM IN THE MOUSE

The renin system is among one of the most completely characterized systems by genetic analysis in the mouse (*14,15*). In some strains of mice, such as AKR or Swiss, renin is synthesized in relatively large quantities in the submaxillary gland (SMG) of males (*14,16*). In other strains of mice, such as BALB/c, renin is 100–200 times less abundant in the SMG. In all species, renin concentration increased in estrogen-treated females and decreased after castration in males. This genetic control is specific for renin and for the SMG, inasmuch as the kidney renin content is similar in male or female of either phenotype. Analysis of heterozygotes and recombinants of inbred mouse strains led Wilson et al. (*17*) to hypothesize that the difference in renin activity between different strains was a result of the alteration of a single gene located in chromosome 1, termed the renin regulator (Rnr). Strains high in SMG renin were presumed to carry the $Rnr^s$ allele, and strains low in SMG renin, the $Rnr^b$ allele.

Kidney and submaxillary renins are similar, although not identical, by physicochemical, enzymatic, and immunological criteria (*18,19*). In addition, it was shown that precursors of SMG and kidney renin are both synthesized as 50-kdalton protein and that they are similarly immunoprecipitated by the same antibodies (*20,21*).

## 5. CLONING OF THE MOUSE SUBMAXILLARY RENIN cDNA

No data on the renin amino acid sequence were available when cloning was initiated. Because renin represents as much as 2–5% of the total protein content of the SMG of male Swiss mice, it was decided to first translate SMG poly(A) RNA to characterize the renin precursor, then to identify the renin cDNA clones by differential screening because renin mRNA was presumed to accumulate in higher quantities in the male than in the female SMG. Furthermore, results of Wilson (*15*) on strain differences in SMG renin content made an independent screening feasible for the identification of renin cDNA clones.

### *5.1. Translation of Mouse Submaxillary Renin mRNA*

RNA was extracted from the SMG of male and female mice, and poly(A)-containing mRNA was purified by oligo(dT)-cellulose affinity chromatography. Translation in vitro of mRNA was performed using the mRNA-dependent rabbit reticulocyte-lysate cell-free translation system; the renin precursor was immunoprecipitated with a rabbit serum directed against mouse renin and characterized as a 45-kdalton protein on SDS-polyacrylamide-gel electrophoresis (Fig. 1).

This 45-kdalton polypeptide was present in the translation products of male, but not female, AKR mice. The 45-kdalton protein was also absent in the translation pattern of polypeptides synthesized in response to BALB/c mRNA.

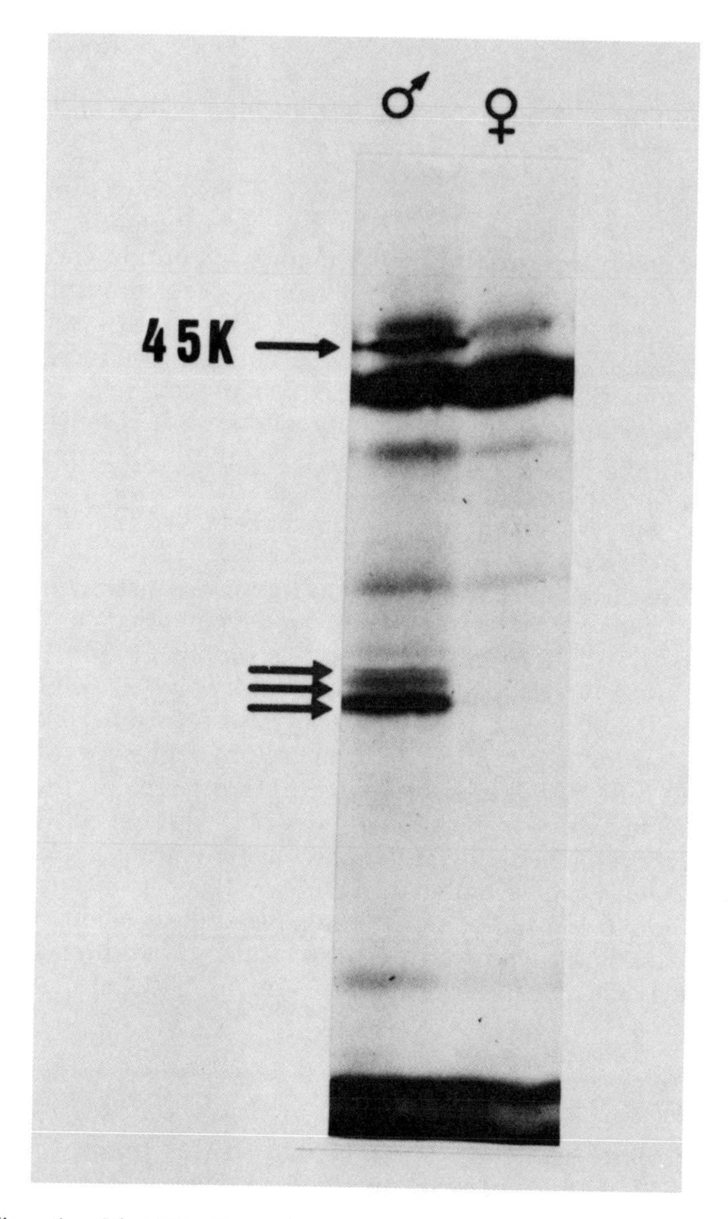


Fig. 1. Identification of preprorenin from mouse submaxillary gland. Total poly(A)-containing RNA from SMG was translated in a reticulocyte lysate system. $^{35}$S-Met labeled products were separated on SDS-polyacrylamide gel electrophoresis. On the left lane, SMG translation products from male AKR and on the right lane from female AKR. The 45 kilodalton protein was identified as preprorenin by immunoprecipitation with specific renin antibodies.

### 5.2. Cloning of Mouse Submaxillary Renin cDNA and Genetic Differential Screening

The mRNA prepared from the SMG of male Swiss mice was purified by sucrose density-gradient ultracentrifugation. The renin precursor appeared to be the main polypeptide encoded by the 16S mRNA species. Purified mRNA was transcribed into cDNA, converted into double-stranded cDNA, and inserted into the *Pst*I site of pBR322 by the dC/dG tailing procedure.

Recombinants containing renin sequences were selected by differential screening. Filters were hybridized with [$^{32}$P]-cDNA transcribed from male Swiss or male BALB/c mRNA and from female mRNA. The filters hybridizing only with male Swiss cDNA were selected. Fifteen percent of recombinant clones exhibited this pattern of hybridization.

Because of the absence of data on the primary structure of the protein, the recombinant clones were identified by hybrid-arrested translation experiments. Recombinant pRn3-5 containing a 1200-base pair (bp) insert and poly(A) mRNA were mixed and hybridized in a 1:1 molar ratio ($M_r$) relative to renin mRNA, on the assumption that this mRNA represents about 2% of total mRNA. The hybrids were translated in the rabbit reticulocyte lysate. Translation in vitro of the renin precursor was selectively abolished when total poly(A)-containing RNA was hybridized with the recombinant DNA.

Finally, Northern blot experiments performed, as described by Thomas (22), showed that the pRn3-5 clone hybridized with a 1600-nucleotide mRNA accumulated in the SMG of male Swiss mice and in the kidney of males and females, but not in the SMG of male Balb/c mice.

### 5.3. cDNA Sequence

By nucleotide sequence determination, it was found that clone pRn 3–5 (1100 nucleotides) contained a poly(A) extension, corresponding to the 3′-terminal portion of renin mRNA. Later, a clone-designed pRn 4–7 (1250 nucleotides) was isolated, containing a sequence 5′ upstream of pRn 3–5 cDNA. Because clones pRn 3–5 and pRn 4–7 shared an identical segment of about 900 nucleotides, together they represented an essentially complete transcript of renin mRNA. The nucleotide sequence of these cDNAs was determined by the method of Maxam and Gilbert

(*23*). The sequence strategy and the nucleotide sequence of mouse SMG renin mRNA, as deduced from the sequence of these two plasmids, have been reported by Panthier et al. (*24*).

# 6. STRUCTURE AND PROCESSING OF MOUSE AND HUMAN RENIN

## *6.1. Structure of Mouse Submaxillary Preprorenins—A Model for Renin Processing*

As previously described, the amino acid sequence of the mouse SMG renin precursor, preprorenin, was determined from its cDNA sequence. Panthier et al. (*24*) and Misono and Inagami (*25*) independently showed that mature active renin extracted from the SMG and purified to homogeneity consisted of two chains, A and B, linked by a disulfide bridge. The determination of the N-terminus sequence of the first 21 amino acids of chains A and B by Panthier et al. (*24*) led them to propose a model for mouse SMG renin processing (Fig. 2). This model involves at least two separate processing events.

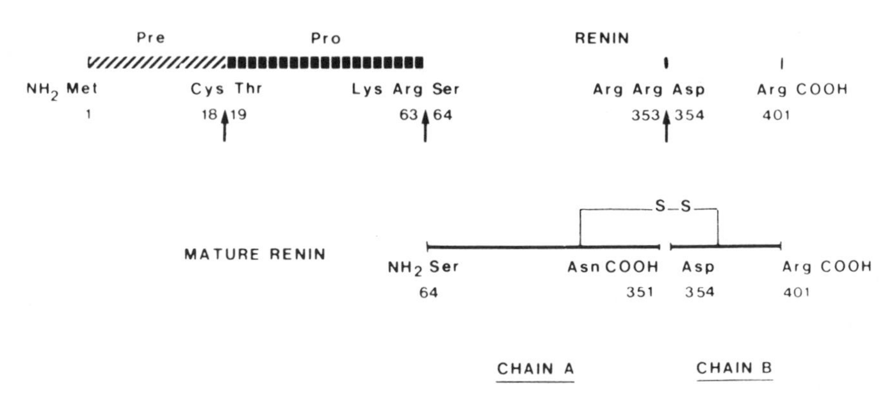


Fig. 2. Model for the mouse SMG renin precursor into mature active renin. The model involves three cleavages: first; the removal of the peptide signal ($Met_1$-$Cys_{18}$); then second; and third. The conversion of the prorenin obtained into active renin by two cleavages occurring after a dibasic peptide and giving rise to chains A and B linked by a disulphide bridge. A further processing occurs, leading to the removal of the $Arg_{352}$-$Arg_{353}$. The carboxyterminus of chain A is $Asn_{351}$ (from ref. *9*).

First, a signal peptide (presequence) is cleaved off by membrane-processing enzymes to produce prorenin. The exact length of the signal peptide is not known because the N-terminus of the prorenin (putative inactive renin, *see* below) has not been determined. Examination of the structure of the renin signal peptide revealed characteristics common to other described signal peptides—a region of charged amino acids, followed by a region rich in hydrophobic amino acids. The endopeptidase, called signal peptidase, splits the peptide bond on the carboxy-terminal side of uncharged amino acids with small side chains. In addition, the residues located in the neighborhood of the cleavage site exhibit a high potential for making beta-turns or aperiodic conformations. Therefore, the cleavage site for preprorenin might occur after Cys-18, but also after Ser-21 or Gly-25.

Second, the prorenin is subjected to two other cleavages, leading to active mature renin, containing two chains linked by an interchain disulfide bond. Each of the proteolytic cleavages occurs after a dibasic peptide; $Lys_{62}$-$Arg_{63}$ and $Arg_{352}$-$Arg_{353}$. This mechanism of peptide maturation is similar to that observed in prohormone processing in which the active hormone is released from an inactive precursor by cleavage after a pair of basic amino acid residues. After this model had been published, Misono et al. (*26*) reported the entire amino acid sequence of the mature mouse SMG renin by conventional amino acid sequencing. There was an excellent agreement between the two structures reported. These studies showed that there was no processing at the C-terminus of chain B, whereas a further proteolytic cleavage occurs in chain A after $Asn_{351}$, leading to the removal of the $Arg_{352-353}$ pair. A similar maturation event has also been reported for the vasopressin precursor in which the dibasic peptide Lys-Arg is cleaved from the C-terminus of vasopressin (*27*).

In the model proposed, chains A and B are linked by a disulfide bridge occurring between $Cys_{357}$ and, probably, $Cys_{320}$. This model is based on the conservation of the Cys residue positions within the aspartyl-protease family. On the same basis, it is likely that there are two intrachain disulfide bridges between $Cys_{114}$ and $Cys_{121}$ and between $Cys_{227}$ and $Cys_{281}$.

This model of processing has been recently confirmed by studies of renin biosynthesis in mouse SMG tissues. Renin was labeled with $[^{35}S]$-Met, and time-course experiments showed that prorenin (43 kdaltons) was synthesized first and rapidly converted to a 38-kdalton single-chain renin. Renin was then slowly hydrolyzed to produce a two-chain protein (*28,29*).

### 6.2. Structure of Mouse Kidney Renin Deduced From Genomic Sequence

As pointed out, numerous biochemical and immunological studies have shown that mouse SMG and kidney renin are closely related. However, it is now well established that the two renins are not identical. Wilson and Taylor (*30*) showed that SMG renin is thermolabile, whereas kidney renin is thermostable. Inagami et al. (*31*) showed that, in contrast to kidney renin, SMG renin is not glycosylated. Furthermore, using an SMG renin cDNA probe, Holm et al. (*32*) screened a mouse Balb/c genomic library to isolate the renin gene. Comparison of the sequence of SMG and kidney renin mRNA over 1427 nucleotides showed that there were 45 base substitutions accounting for 21 amino acid alterations. The major difference between the two proteins was the presence of three potential glycosylation sites in the kidney enzyme, located at amino acid positions 61, 139, and 319 (Fig. 3). The two basic residues involved in the maturation of the SMG renin were found at identical positions in the kidney enzyme.

### 6.3. Structure of Human Kidney Renin Deduced From cDNA Sequence

Although human renal renin has been purified to homogeneity (*33–35*), there are no data available concerning its primary structure by amino acid sequencing because of its extremely low concentration in kidney. Therefore, the general strategy was to use the mouse SMG cDNA as a molecular probe to study mRNA from the human kidney and to clone the structure of the human renin gene. An infarcted kidney synthesizing large amounts of renin was obtained at surgery, and mRNA was prepared. By the Northern blot technique, human renin mRNA was hybridized with the mouse SMG renin cDNA probe. The 1.6-kilobase pair (kb) length of this mRNA was similar to that of the mouse kidney and SMG renin mRNA.

Kidney mRNA was enriched for renin mRNA message by fractionation on a sucrose density gradient. Fractions containing renin mRNA were transcribed into cDNA and cloned, using pBr322

```
                                                               1                                   10
                                                               Met Asp Arg Arg Arg Met Pro Leu Trp Ala Leu Leu Leu
TATAAAAGAAGGCTCAGGGGGTCTGGGCTACACAGCTCTTAGAAAGCCTTGGCTGAACCAG ATG GAC AGA AGG AGG ATG CCT CTC TGG GCA CTC TTG TTG
          ------------------------------------- --- --- --G --- --- --- --- --- --- --- --- --- ---
                            20                                  30
Leu Trp Ser Pro Cys Thr Phe Ser Leu Pro Thr Arg Thr Ala Thr Phe Glu Ar                                  g Ile
CTC TGG AGT CCT TGC ACC TTC AGT CTC CCA ACA CGC ACC GCT ACC TTT GAA CG GTAACTTGGG ( ?.121 Kb) CCTGGAGCAG A ATC
--- --- --- --- --- --- --- --- --- --- --G G-- ---(  )--- --- --- --                                   - ---
                                40                                  50                                     60
Pro Leu Lys Lys Met Pro Ser Val Arg Glu Ile Leu Glu Glu Arg Gly Val Asp Met Thr Arg Leu Ser Ala Glu Trp Gly Val
CCG CTC AAG AAA ATG CCT TCT GTC CGG GAA ATC CTG GAG GAG CGG GGA GTG GAC ATG ACC AGG CTC AGT GCT GAA TGG GGC GTA
--A --- --- --- --- --C --- --- --- --- --- --- --- --- --- --- --- --- --- --- --- --- --- --- --- --- -A- ---
                                        70                                  80
Phe Thr Lys Arg Pro Ser Leu Thr Asn Leu Thr Ser Pro Val Val Leu Thr Asn Tyr Leu Asn
TTC ACA AAG AGG CCT TCC TTG ACC AAT CTT ACC TCC CCC GTG GTC CTC ACC AAC TAC CTG AAT GTGAGTCCTA (≈ 0.516 Kb) CCC
--- --- --- --- T-- --- --- --T G-- --- -T- --- --- --- --- --- --- --- --- --- ---
                                            90                                  100
        Thr Gln Tyr Tyr Gly Glu Ile Gly Ile Gly Thr Pro Pro Gln Thr Phe Lys Val Ile Phe Asp Thr Gly Ser Ala Asn
GCCACAG ACC CAG TAC TAC GGC GAG ATT GGC ATC GGT ACC CCA CCC CAG ACC TTC AAA GTC ATC TTT GAC ACG GGT TCA GCC AAC
        -G- --- --- --T --- --- --C --- --T --- --- --- --- --- --- --- --- --- --- --- --- --- --C --C --- ---
        110                                 120
Leu Trp Val Pro Ser Thr Lys Cys Ser Arg Leu Tyr Leu Ala Cys G                                   ly Ile His Ser
CTC TGG GTG CCC TCC ACC AAG TGC AGC CGC CTC TAC CTT GCT TGT G GTAAGAGTCA (≈ 0.696 Kb) CCTCTGCTAG GG ATT CAC AGC
--- --- --- --- --- --- --- --- --- --- --- --- --- --- --- -                                   -- --- --- ---
            130                                 140                                 150
Leu Tyr Glu Ser Ser Asp Ser Ser Ser Tyr Met Glu Asn Gly Ser Asp Phe Thr Ile His Tyr Gly Ser Gly Arg Val Lys Gly
CTC TAT GAG TCC TCT GAC TCC TCC AGC TAC ATG GAG AAC GGG TCC GAC TTC ACC ATC CAC TAC GGA TCA GGG AGA GTC AAA GGT
--- --- --- --- --- --- --- --- --- --- --- --- --T --A GA- --- --- --- --- --- --- --- --- --- --- --- --- ---
                    160                                                                      170
Phe Leu Ser Gln Asp Ser Val Thr                                     Val Gly Gly Ile Thr Val Thr Gln Thr Phe Gly
TTC CTC AGC CAG GAC TCG GTG ACT GTAAGTAGGA (≈ 0.713 Kb) TCTCTCACAG GTG GGT GGA ATC ACT GTG ACA CAG ACC TTT GGA
--- --- -?- --- --- --- --- ---                                     --- --- --- --- --- --- --- --- --- --- ---
                        180                                 190                                 200
Glu Val Thr Glu Leu Pro Leu Ile Pro Phe Met Leu Ala Lys Phe Asp Gly Val Leu Gly Met Gly Phe Pro Ala Gln Ala Val
GAG GTC ACC GAG CTG CCC CTG ATC CCT TTC ATG CTG GCC AAG TTT GAC GGT GTT CTA GGC ATG GGC TTT CCC GCT CAG GCC GTT
--- --- --- --- --- --- --- --- --- --- --- --- --- C-- --- --- --G --- --- --- --- --- --- --- --- --- --- --C
                                210                                 220
Gly Gly Val Thr Pro Val Phe Asp His Ile Leu Ser Gln Gly Val Leu Lys Glu Glu Val Phe Ser Val Tyr Tyr Asn Ar
GGC GGG GTT ACC CCT GTC TTT GAC CAC ATT CTC TCC CAG GGG GTG CTA AAG GAG GAA GTG TTC TCT GTC TAC TAC AAC AG GTGG
--- --- --C --- --- --- --- --- --- --- --- --- --- --- --- --G --- --- A-- --- --- --- --- --- --- --- --

                                        230                                 240
                                  g Gly Ser His Leu Leu Gly Gly Glu Val Val Leu Gly Gly Ser Asp Pro Gln His Tyr Gln
GCCTTT (≈ 1.913 Kb) TTTCCTTTAG G GGT TCC CAC CTG CTG GGG GGC GAG GTG GTG CTA GGA GGT AGC GAC CCG CAG CAT TAT CAA
                                  - --- C-- --- --- --- --- --- --- --- --- --- --C --- --- --- G-- --- --C ---
    250                                 260
Gly Asn Phe His Tyr Val Ser Ile Ser Lys Thr Asp Ser Trp Gln Ile Thr Met Lys Gl
GGC AAT TTT CAC TAT GTG AGC ATC AGC AAG ACT GAC TCC TGG CAG ATC ACG ATG AAG GG GTGGGTCAGC (≈ 0.452 Kb) GCCTCTGCAG
--- G-- --- --- --- --- --- C-- --- --- --- --T --- --- --- --- --A --- --- --
   270                                 280                                 290
y Val Ser Val Gly Ser Ser Thr Leu Leu Cys Glu Glu Gly Cys Ala Val Val Val Asp Thr Gly Ser Ser Phe Ile Ser Ala Pro
G GTG TCT GTG GGG TCT TCC ACC CTG CTA TGT GAA GAA GGC TGT GCG GTA GTG GTG GAC ACT GGT TCA TCC TTT ATC TCG GCT CCT
- --- --- --- --- --- --- --- --A --G --- --- --- --- --- -A- --- --- --- --- --- --- --- --- --- --- --- ---
            300                                 310
Thr Ser Ser Leu Lys Leu Ile Met Gln Ala Leu Gly Ala Lys Glu Lys Arg Ile Glu Glu
ACG AGC TCC CTG AAG TTG ATC ATG CAA GCC CTG GGA GCC AAG GAG AAG AGA ATA GAA GAA GTAAGAGATC (≈0.262 Kb) ATTCCCCCAG
--- --- --- --- --- --- --- --- --- --- --- --- --- --- --- --- --- C-- C-T ---
            320                                 330                                 340
Tyr Val Val Asn Cys Ser Gln Val Pro Thr Leu Pro Asp Ile Ser Phe Asp Leu Gly Gly Arg Ala Tyr Thr Leu Ser Ser Thr
TAT GTT GTG AAC TGT AGC CAG GTG CCC ACC CTC CCC GAC ATT TCC TTT GAC CTG GGA GGC AGG GCC TAC ACA CTC AGC AGT ACG
--- --- --- -G- --- --- --- --- --- --- --- --- --- --C --- --C A-- --- --- --- --- --- --- --- --- --- --- ---
                                                   350                                 360
Asp Tyr Val Leu Gln                                Tyr Pro Asn Arg Arg Asp Lys Leu Cys Thr Leu Ala Leu His Ala
GAC TAC GTG CTA CAG GTGAGGCTGG (≈ 0.567 Kb) TTCTTGCCAG TAT CCC AAC AGG AGA GAC AAG CTG TGC ACA CTG GCT CTC CAT GCC
--- --- --- --- ---                                --- --- --- --- --- --- --- --- --- --- G-- --- --- --- ---
                    370                                 380                                 390
Met Asp Ile Pro Pro Pro Thr Gly Pro Val Trp Val Leu Gly Ala Thr Phe Ile Arg Lys Phe Tyr Thr Glu Phe Asp Arg His As
ATG GAC ATC CCA CCA CCC ACT GGG CCT GTC TGG GTC CTG GGT GCC ACC TTC ATC CGC AAG TTC TAT ACA GAG TTT GAT CGG CAT AA
--- --- --- --- --- --- --- --- --- --- --- --- --- --- --- --- --- --- --- --- --- --- --- --- --- --- --- --
                        400
Asn Arg Ile Gly Phe Ala Leu Ala Arg
AAT CGC ATT GGA TTC GCC TTG GCC CGC TAAGGCCCTCTGCCACCCAGTAACCCTAGGCCAAGCCAAGCTGGCAGTCCTGGGGGCCATTTTGTCTGGCTTTGTCCC
--- --- --- --- --- --- --- --- --- ------------------------------------------C-------------------------------

CAACATAGGGACACTGGACACAGAGACCCTAACGAGTGTTTGCCCCTTCACCTGCACTCACCCTTCCCTGCTTTAAGGAAAAATCGAATAAAGATTTCATGTTTAAAGCCTGTT
--------------------------------------------------------------------------------C--------------------

TCGGATGGGTTCTTTGGAGTTTGGAGGAGGT
```

Fig. 3. Nucleotide sequence of the coding regions and of the 5′- and 3′-flanking regions of the mouse Ren 1 gene. The DNA sequence of the coding regions and the predicted amino acid sequence of the Ren 1 gene are shown. Amino acids are numbered from the $NH_2$-terminus of the preprorenin.

as a vector. Colonies were screened by hybridization with the mouse SMG renin [$^{32}$P]-cDNA. One clone, containing a 1.1-kb insert was sequenced, as previously reported (*36*). This clone corresponded to the 3′ portion of renin mRNA. Its sequence showed a high homology with mouse SMG renin. Imai et al. (*37*), using the same approach, recently reported the sequence of the entire human renin mRNA transcript. The overall homology between mouse SMG and human renal renin was 68% at the amino acid level and 76% at the level of the nucleotide sequence. When compared with mouse SMG renin and other aspartyl proteases, residues involved in the catalytic site are conserved in this structure, as discussed below. An interesting difference between mouse SMG and human renal renin is the presence of two potential glycosylation sites. This result was expected because of the ability of human renal renin to bind to concanavalin A-sepharose (*34*). A comparison of the structure of the mouse and human kidney precursors is shown in Fig. 4.

The processing of human preprorenin is not yet known because of the lack of information concerning the primary amino acid sequence of prorenin and mature renin. It is tempting to speculate,

MOUSE SUBMAXILLARY RENIN

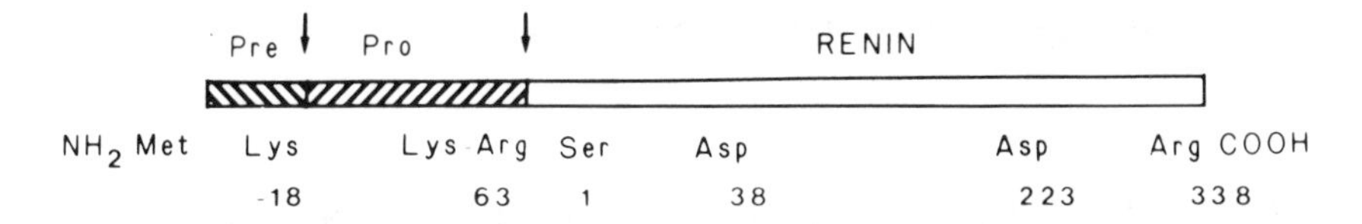


HUMAN KIDNEY RENIN

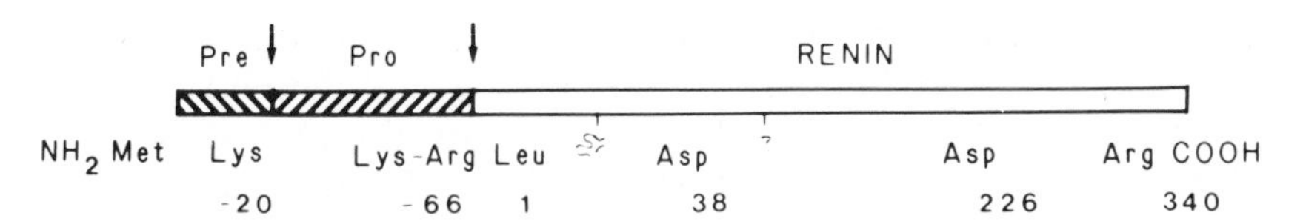


Fig. 4. Comparison of the structure of the mouse SMG and human kidney renin precursors. The putative sites of processing of the human preprorenin into prorenin and active renin are indicated by the arrows. Also indicated are the two glycosylation sites found in human kidney renin and the location of the two aspartyl residues involved in the catalytic site.

however, that the processing of human preprorenin would be similar to that of mouse SMG renin: (a) cleavage of a putative 20-amino-acid prepeptide, and (b) prorenin processing into active renin. Examination of the prorenin amino acid sequence shows the presence of two basic residues, $Lys_{65}$-$Arg_{66}$, almost at the same position as the $Lys_{62}$-$Arg_{63}$ of mouse SMG prorenin. Inasmuch as mature human renin has a molecular mass of about 40 kdalton (*33–35*), it is probable that it is processed after cleavage of a 46-amino acid profragment. Whether human renal renin is further processed into two chains is not known. The cleavage site between the two chains of SMG renin occurs after an $Arg_{353}$-$Arg_{354}$ pair. A potentially analogous cleavage site might be located at the dibasic pair $Lys_{354}$-$Lys_{355}$ in the case of human renin (*36*).

Biosynthesis experiments in a juxtaglomerular-cell tumor showed that renin is indeed synthesized as an inactive precursor of 55 kdalton that is converted into an active renin of 44 kdalton (*38*). These molecular masses are in agreement with the calculated $M_r$ for prorenin (43 kdalton) and renin (37 kdalton) deduced from the nucleotide sequence. The fact that mature renin has two glycosylation sites probably accounts for the higher $M_r$ found in cellular biosynthesis. Finally, the presence of an inactive form of renin in plasma and kidney that is activated by Ser esterases suggests that this inactive renin represents the prorenin activated by cleavage of a tryptic bond (*39*).

## 7. STRUCTURE OF THE RENIN GENE AND COMPARISON WITH OTHER ASPARTYL PROTEASES

X-ray studies of aspartyl proteases showed that these proteins have a symmetrical bilobal structure with a well-defined cleft that contains the active site. The two Asp residues of the catalytic site lie in close proximity on either side of the cleft and are localized in amino acid stretches that possess sequence homology with one another. Tang et al. (*40*) proposed that aspartyl protease genes have evolved by duplication and fusion of a primitive gene coding for a protein of about 150 residues having a fold similar to that of one lobe of pepsin.

When amino acid sequences of aspartyl proteases like pepsin, chymosin, and penicillinopepsin were compared with the renin sequence, a high similarity was found. The sequence homology was 43% between active renin and porcine pepsin and 37% between prorenin and porcine pepsinogen. Furthermore, recent studies showed that the tridimensional structure of renin may be similar to that of other aspartyl proteases (*41*). All these considerations led to the hypothesis that the aspartyl protease gene was derived from a common ancestral gene that itself originated from duplication and fusion of a primitive gene coding for a single-lobe protein.

Using the renin cDNA clone pRn 1–4 as a probe, the unique Balb/c renin gene was isolated from a library of Balb/c mouse embryo DNA fragments cloned in the vector Charon-4A (*42*). One of them yielded four *Eco*RI fragments of 8.8, 3.9, 3.2, and 0.5 kb. Southern blot analysis of this clone showed that only the 3.9-kb and the 8.8-kb *Eco*RI fragments hybridized to the renin cDNA probe (*43*). The *Eco*RI fragments of phage, which appeared to contain the entire renin gene, as well as 5′- and 3′-flanking sequences, were subcloned in the plasmid pBR322 for structural analysis. The location of the exons was established by Southern blot analysis with the pRN 1–4 probe, and these regions were sequenced by use of the Maxam and Gilbert method (*23*) (Fig. 3). The renin gene Ren 1 coding for a mRNA of about 1600 nucleotides [including the poly(A)] spans 9.6 kb and is divided into nine exons by eight intervening sequences of various lengths. The sizes of the nine exons are 91, 151, 124, 119, 197, 120, 145, 100, and 341 bp, respectively, and the sizes of the eight intervening sequences are about 3120, 510, 690, 710, 1910, 450, 260, and 570 bp, respectively. The nucleotide sequence of the nine exons of the renin gene, Ren 1, and the predicted amino acid sequence of the kidney renin precursor are shown in Fig. 3, and the general organization of the kidney renin gene is shown in Fig. 5. The renin gene is organized into two blocks of four exons separated by a 2-kb intron. The first exon encoding the peptide signal of the preprorenin was separated from the following eight by a 3-kb intron. Examination of sequence alignments of aspartyl proteases showed that each of the two lobes was roughly encoded by each of the two blocks of four exons (exons 2–5 and exons 6–9) and that the short peptide connecting the two lobes was encoded

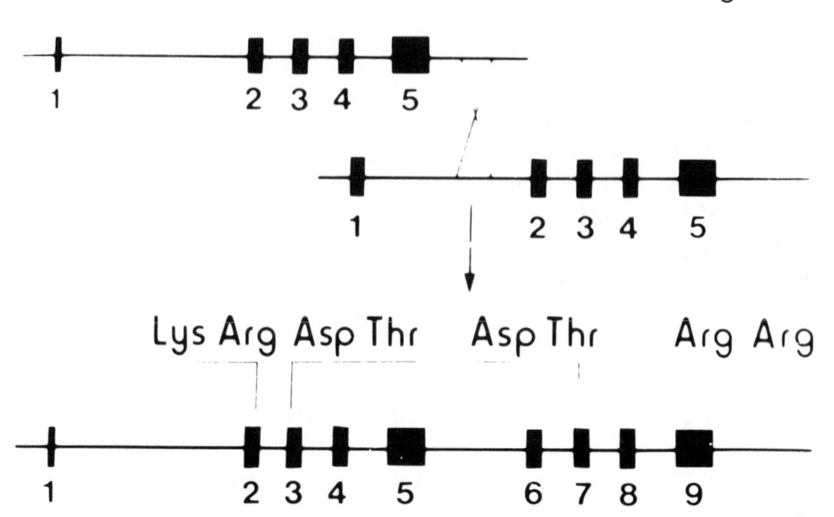


Fig. 5. A model for the evolution of the aspartyl proteases by gene duplication. The primitive gene encoding a 20-kdalton protein was divided into five exons by four intervening sequences. As a result of a duplication event by unequal meiotic crossing-over by homologous recombination between direct repeats located in the first intervening sequence and 3′ of the ancestral gene, the aspartyl protease ancestral gene is formed. Alternative models, involving first, duplication of the five exons gene and second, fusion of the two genes by deletion of the intergenic region containing the first exon, are equally valid.

This figure shows the location of the coding regions for the two dibasic residues involved in the maturation of the mouse submaxillary gland renin and the location of the two active-site aspartates.

by the first half of exon 6. Furthermore, the DNA sequences encoding the two active-site aspartates were located at equivalent positions in these clusters (*32*).

The striking 1-4-4 organization of the renin gene strongly supported the hypothesis of Tang et al. (*40*) in which aspartyl proteases have evolved by gene duplication and fusion of an ancestral gene encoding a single polypeptide chain with a size and fold similar to one lobe of pepsin. A model has been proposed in which the primitive gene was divided into five exons by four intervening sequences according to the model. The aspartyl protease ancestor gene has been generated by unequal meiotic crossover between two direct repeats of DNA located, respectively, in the first intervening sequence and 3′ end of the ancestral gene.

This model could explain the relatively large size of the fifth intervening sequence that separates the two clusters. The duplication fusion model suggests that the ancestral gene encoded a polypeptide having a fold similar to that of one lobe of pepsin (*40*). This polypeptide was probably active as a dimer in which the two identical subunits were symmetrically related. After gene duplication and fusion, the two subunits became connected by the short peptide encoded in exon 6, while retaining the same twofold relation and permitting the correct association between the two subunits.

Blundell et al. (*41*) showed that the three-dimensional structure of renin is probably similar to that of other aspartyl proteases. This suggests that every aspartyl protease gene is derived from a common ancestor and has the 1-4-4 basic organization of the renin gene (Fig. 6). The 9.6-kb human pepsinogen gene is also di-

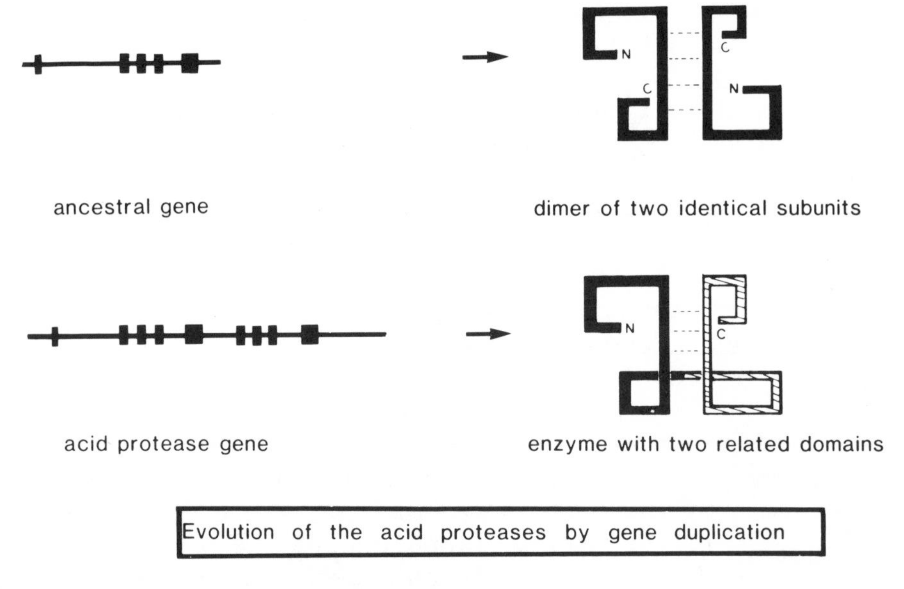


Fig. 6. Evolution of the acid proteases by gene duplication. The duplication model suggests that the ancestral gene encoded a polypeptide chain that was probably active as a dimer in which the two identical units were related by a twofold symmetry axis. After gene duplication and fusion, the two subunits are connected. This gene fusion could have been selectively important by permitting divergent evolution between the two lobes (symbolized by black and dashed lines).

vided into nine exons separated by eight intervening sequences (*44*). Except for the second exon corresponding to the activation segment that is located near the first exon, the pepsinogen and the renin genes have the same organization: The DNA sequences at the exon–intron junctions are relatively well conserved, and the same splice junctions are used in both cases.

## 8. TISSUE-SPECIFIC EXPRESSION OF THE RENIN GENE

The juxtaglomerular cells of the afferent arteriole in the kidney constitute the main source of renin. Nevertheless, other sources of renin or renin-like enzymes have been described (*see* above). Furthermore, in the mouse, the renin is also synthesized in very large quantities (2% of the total protein) in the SMG under both genetic and hormonal control. Wilson et al. (*14*) showed that differences in SMG renin activity between different mouse strains are under the control of a single gene located on chromosome 1 and termed the renin regulatory gene. Thus, the genetic situation in the mouse offers a unique opportunity to analyze mechanisms involved in tissue-specific expression of the renin gene at the molecular level.

When these experiments were undertaken, the problem of whether SMG and kidney renins were products of the same gene was not clear. There were two alternative hypotheses. The first was that SMG and kidney renins were the products of the same gene that was under different control mechanisms in different tissues. The second was that both renins are encoded by highly homologous nonallelic genes. To examine whether differences could be seen at the DNA level between the two distinct groups of inbred mice, restriction fragments of chromosomal DNA that bear renin genes were characterized by the Southern DNA blotting technique, using cloned cDNA probes derived from the complete transcript of SMG mRNA. It was shown that strains carrying the b allele, such as Balb/c, C3H, or C57B1/q, which have low basal and induced SMG renin activity, have only one renin gene per haploid genome, and strains carrying the s allele, such as Swiss or AKR mice, which have high basal and induced level of SMG renin activity, have two renin genes per haploid genome (*43*). At the same time, Piccini et al. (*45*) reported results on the DNA polymorphism at the renin locus consistent with a gene duplication.

This clear correlation between a high level of SMG renin activity and the renin gene duplication strongly suggested that SMG and kidney renin were not the products of the same gene. That hypothesis was supported by the observations of Wilson and Taylor (*30*), showing that the SMG renin activity from animals carrying the $Rnr^b$ allele is thermostable, whereas SMG renin activity from animals carrying the $Rnr^s$ allele is thermolabile. Furthermore, kidney renin activity from mice carrying either the $Rnr^s$ or the $Rnr^b$ allele is thermostable.

To demonstrate that SMG and kidney renin mRNA are not transcribed from the same gene, cDNA copies of Swiss mouse kidney renin mRNA were cloned. The comparison of the sequence of the kidney renin mRNA with the previously reported sequence of the SMG mRNA demonstrated without ambiguity that the Swiss mice bearing the s allele express two nonallelic renin genes (*46*) (Table 1).

Thus, strains of mice with a very low level of SMG renin possess one renin gene (Ren 1) per haploid genome. That gene encodes the glycosylated kidney renin. Strains of mice with a high level of SMG renin have two renin genes, Ren 1 and Ren 2. In these animals, Ren 2 is expressed in the SMG, but not in the kidney.

Basal and induced levels of Ren 2 gene transcription in the SMG are, at most, 100 times higher than those of Ren 1. Clearly,

TABLE 1
Partial Nucleotide Sequences of Mouse Kidney and Submaxillary Gland cDNA[a]

| | 69<br>Asn | 70<br>Leu | 71<br>Thr...... | 139<br>Asn | 140<br>Gly | 141<br>Ser...... | 319<br>Asn | 320<br>Cys | 321<br>Ser |
|---|---|---|---|---|---|---|---|---|---|
| Kidney | AAT | CCT | ACT..... | AAC | GGG | TTC..... | AAC | TGT | AGC |
| SMG | G- | — | -T- | —C | —A | GA- | -G- | — | — |
| | Asp | Leu | Ile | Asn | Gly | Asp | Ser | Cys | Ser |

[a]The three potential glycosylation sites corresponding to the canonical sequence Asn X Ser/Thr are shown. Residues are numbered from the $NH_2$-terminus of the preprorenin (adapted from ref. *32*).

these differences cannot be explained by a gene dosage effect. It is very likely that the Ren 2 gene is the result of a recent duplication event occurring 10 million years ago (*32*). Nevertheless, it is difficult to explain how a simple duplication event can modify so dramatically both tissue-specific expression and rate of transcription of the extra copy of the ancestral gene.

To analyze the differential expression of renin genes at the molecular level, the 5′-flanking regions of Ren 1 and Ren 2 genes, which are presumed to contain transcriptional control elements, have been sequenced and compared (*47*). The results showed that, although promoter sequences are highly homologous up to a point 214 nucleotides upstream from the 5′ end from the AUG initiation codon, the sequences diverge profoundly in the region beyond this nucleotide. Thus, the differences between the 5′-flanking sequences downstream from the promoter region suggest that tissue-specific expression of the renin gene is associated with sequences located farther upstream. These results are in agreement with those of Walker et al., showing that sequences located 200–300 bp upstream from the transcription start site are necessary for cell-specific expression of insulin and chymotrypsin gene.

The Ren 2 copy of renin genes was mapped on mouse chromosome 1 near the Pep-3 locus (*17*). Because of the lack of a known allelic form of Ren 1 gene, it is not known whether the renin genes are closely linked. The fact that both genes exhibit high similarity (*32*) suggested that they may result from a recent duplication (*42*). Nevertheless, the 3′- and 5′-flanking sequences of Ren 1 and Ren 2 are nonhomologous. This could be the result of either the duplication of Ren 1 followed by the transposition of a DNA fragment containing regulatory sequences near the extra copy, or the transposition of Ren 2 itself in a different environment. This complex DNA rearrangement has changed the transcriptional control operating on Ren 2.

## 9. CONCLUSION

In this review, we have presented results of the cloning of the mouse cDNA encoding SMG renin. The structure of mouse submaxillary preprorenin was deduced from the cDNA sequence, and a model for renin processing in the SMG is proposed. The mouse submaxillary renin cDNA clone was then used to screen an mRNA library for mouse and human kidney renin.

Partial structures of mouse kidney and human kidney renins were reported. Studies performed on tissue-specific expression of the renin gene revealed that mice producing high levels of renin in SMG have two renin genes, whereas mice producing low levels have one renin structural gene per haploid genome. In humans, evidence is provided for a single renin gene expressed in the kidney and in an extrarenal tissue, the chorion (*36*). The organization of the mouse renin gene shows that aspartyl proteases have evolved by duplication and fusion of an ancestral gene containing five exons. Finally, an overall comparison of renin with other aspartyl proteases can be made at several levels: at the nucleotide and amino acid sequences, active site, and the tertiary structure.

## REFERENCES


1. Werle E., Vogel R., and Goldel L. F. (1957) Uber ein Blutsrucksteigerndes prinzip in extrakten aus der glandula submaxillaris der weisen maus. *Arch. Exp. Pathol. Pharmacol.* **230,** 236–244.
2. Gross F., Schaechtelin G., Ziegler M., and Berger M. (1964) A renin-like substance in the placenta and uterus of the rabbit. *Lancet* **1,** 914–916.
3. Symonds E. M., Stanley M. A., and Skinner L. (1968) Production of renin by in vitro cultures of human chorion and uterine muscle. *Nature* **217,** 1152–1153.
4. Ganten D., Minnich J. L., Granger P., Hayduk K., Brecht H. M., Barbeau A., Boucher R., and Genest J. (1971) Angiotensin enzyme in brain tissue. *Science* **173,** 64–65.
5. Ryan J. W. (1972) Distribution of a renin-like enzyme in the bovine adrenal gland. *Experientia* **29,** 407–408.
6. Gould A. B., Skeggs L. T., and Kahn J. R. (1964) Presence of renin activity in blood vessel walls. *J. Exp. Med.* **119,** 389–399.
7. Ondetti M. A., Rubin B., and Cushman D. W. (1977) Design of specific inhibitors of angiotensin-converting enzyme: New class of orally antihypertensive agents. *Science* **196,** 441–444.
8. Heel R. C., Brogden R. N., Speight T. M., and Avery G. S. (1980) Captopril: A preliminary review of its pharmacological properties and therapeutic efficacy. *Drugs* **20,** 409–452.
9. Corvol P., Galen F. X., Devaux C., and Menard J. (1983) Biochemi-

cal Characteristics of Human and Animal Renin, in *Hypertension,* (J. Genest ed.) Mc Graw Hill, New York.

10. Umezawa H., Aoyagi T., Morishima H., Matsuaki M., Hameda M., and Tadeuchi T. (1970) Pepstatin, a new inhibitor produced by actinomyces. *J. Antibiot.*(Tokyo) **23,** 259–263.
11. Corvol P., Devaux C., and Menard J. (1973) Pepstatin, an inhibitor for renin purification by affinity chromatography. *FEBS Lett.* **34,** 180–192.
12. Inagami T., Misono K., and Michelakis A. M. (1974) Definitive evidence for similarity in the active site of renin and acidic proteases. *Biochem. Biophys. Res. Commun.* **56,** 503–509.
13. Misono K. and Inagami T. (1980) Characterization of the active site of mouse submaxillary gland renin. *Biochemistry* **19,** 2616–2622.
14. Wilson C. M., Erdos E. G., Dusin J. F., and Wilson J. D. (1977) Genetic control of renin-activity in the submaxillary gland of the mouse. *Proc. Natl. Acad. Sci. USA* **74,** 1185–1189.
15. Wilson C. M., Cherry M., Taylor B. A., and Wilson J. D. (1981) Genetic and endocrine control of renin activity in the submaxillary gland of the mouse. *Biochem. Genet.* **19,** 509–523.
16. Oliver W. J. and Gross F. (1967) Effect of testosterone and duct ligation on submaxillary renin-like principle. *Am. J. Physiol.* **213,** 341–346.
17. Wilson C. M., Erdos E. G., Wilson J. D., and Taylor B. A. (1978) Location on chromosome 1 of Rnr, a gene that regulates renin in the submaxillary gland of the mouse. *Proc. Natl. Acad. Sci. USA* **75,** 5623–5626.
18. Cohen S., Taylor J. M., Murakami K., Michelakis A. M., and Inagami T. (1972) Isolation and characterization of renin-like enzymes from mouse submaxillary glands. *Biochemistry* **11,** 4286–4292.
19. Michelakis A. M., Yoshida H., Menzie J. W., Murakami K., and Inagami T. (1974) A radioimmunoassay for the direct measurement of renin in mice and its application to submaxillary gland and kidney studies. *Endocrinology* **94,** 1101–1110.
20. Poulsen K., Vuust J., Lukkegaard S., Nielsen A. H, and Lund T. (1979) Renin is synthesized as a 50,000 dalton single chain polypeptide in cell-free translation system. *FEBS Lett.* **98,** 135–138.
21. Poulsen K., Vuust J., and Lund T. (1980) Renin precursor from mouse kidney identified by cell free translation of messenger RNA. *Clin. Sci.* **59,** 297–299.
22. Thomas P. S. (1980) Hybridization of denatured RNA and small DNA fragments transferred to nitrocellulose. *Proc. Natl. Acad. Sci. USA* **77,** 5201–5205.
23. Maxam A. M. and Gilbert W. (1977) A new method for sequencing DNA. *Proc. Natl. Acad. Sci. USA* **74,** 560–564.

24. Panthier J. J., Foote S., Chambraud B., Strosberg A. D., Corvol P., and Rougeon F. (1982) Complete amino acid sequence and maturation of the mouse submaxillary gland renin precursor. *Nature* **298,** 90–92.
25. Misono K. S. and Inagami T. (1982) Structure of mouse submaxillary gland renin. *J. Biol. Chem.* **257,** 7536–7540.
26. Misono K. S., Chang J. J., and Inagami T. (1982) Amino acid sequence of mouse submaxillary gland renin. *Proc. Natl. Acad. Sci. USA* **79,** 4858–4862.
27. Land H., Schutcz G., Schmale H., and Richter D. (1981) Nucleotide sequence of cloned cDNA encoding bovine arginine vasopressin-neurophysin II precursor. *Nature* **295,** 299–301.
28. Catanzaro O. F., Mullins J. J., and Morris B. J. (1983) The biosynthetic pathway in mouse submaxillary gland. *J. Biol. Chem.* **258,** 7364–7368.
29. Pratt R. T., Onelette A. J., and Dzau V. J. (1983) Biosynthesis of renin: Multiplicity of active and intermediate forms. *Proc. Natl. Acad. Sci. USA* **80,** 6809–6813.
30. Wilson C. M. and Taylor B. A (1982) Genetic regulation of thermostability of mouse submaxillary gland renin. *J. Biol. Chem.* **257,** 217–223.
31. Inagami T., Murakami K., Yokosawa H., Matoba T., Yokosawa N., Takii Y., and Misono K. (1980) Purification of Renin and Precursors, in *Enzymatic Release of Vasoactive Peptides,* (Gross F. and Vogel H. F., eds.) Raven, New York.
32. Holm I., Ollo R., Panthier J. J., and Rougeon F. (1984) Evolution of aspartyl proteases by gene duplication: The mouse renin gene is organized in two homologous clusters of four exons. *EMBO J.* **3,** 557–562.
33. Galen F. X., Devaux C., Guyenne T., Menard J., and Corvol P. (1979) Multiple forms of human renin: Purification and characterization. *J. Biol. Chem.* **254,** 4848–4855.
34. Yokosawa H., Holliday L. A., Inagami T., Haas E., and Murakami K. (1980) Human renal renin: Complete purification and characterization. *J. Biol. Chem.* **255,** 3498–3502.
35. Slater E. E. and Strout H. V., Jr. (1981) Pure human renin. *J. Biol. Chem.* **256,** 8164–8171.
36. Soubrier F., Panthier J. J., Corvol P., and Rougeon F. (1983) Molecular cloning and nucleotide sequence of a human renin cDNA. *Nucleic Acids Res.* **11,** 7181–7190.
37. Imai T., Miyazaki H., Hirose S., Hori H., Hayashi T., Kageyama R., Ohkubo H., Nakanishi S., and Murakami K. (1983) Cloning and sequence analysis of cDNA for human renin precursor. *Proc. Natl. Acad. Sci. USA* **80,** 7405–7409.

38. Galen F. X., Corvol M. T., Devaux C., Gubler M. C., Mounier F., Camilleri J. P., Houot A. M., Menard J., and Corvol P. (1984) Renin biosynthesis by human tumoral juxtaglomerular cells. Evidences for a renin precursor. *J. Clin. Invest.* **73,** 1144–1155.
39. Leckie B. J. (1981) Active renin: An attempt at a perspective. *Clin. Sci.* **60,** 119–130.
40. Tang J., James M. N. G., Hsu I. N., Jenkins J. A., and Blundell T. L. (1978) Structural evidence for gene duplication in the evolution of acid proteases. *Nature* **271,** 618–621.
41. Blundell T., Sibanda B. L., and Pearl L. (1983) Three dimensional structure, specificity and catalytic mechanisms of renin. *Nature* **304,** 273–275.
42. Ollo R., Auffray C., Sikorav J. L., and Rougeon F. (1981) Mouse heavy chain variable regions: Nucleotide sequence of a germ line gene segment. *Nucleic Acids Res.* **9,** 4099–4109.
43. Panthier J. J., Holm I., and Rougeon F. (1982) The mouse Rn locus: S allele of the renin regulator gene results from a single structural gene duplication. *EMBO. J.* **1,** 1417–1421.
44. Sogawa K., Fujii-Kuriyama Y., Mizumami Y., Ichihara T., Takahashi K. (1983) Primary structure of human pepsinogene gene. *J. Biol. Chem.* **258,** 5306–5311.
45. Piccini N., Knopf J. L., and Gross K. W. (1982) A DNA polymorphism, consistent with gene duplication, correlates with high renin levels in the mouse submaxillary gland. *Cell* **30,** 205–213.
46. Panthier J. J. and Rougeon F. (1983) Kidney and submaxillary gland renins are encoded by two non-allelic genes in Swiss mice. *EMBO. J.* **2,** 675–678.
47. Panthier J. J., Dreyfus M., Tronik-Lerous D., and Rougeon F. (1984) Mouse kidney and submaxillary gland renin genes differ in their 5′-putative regulatory sequences. *Proc. Natl. Acad. Sci. USA* **81,** 5489–5493.

# Chapter 14

# Nerve Growth Factor

## mRNA and Genes That Encode the β-Subunit

Gerhard Heinrich

## 1. Introduction

Beta-nerve growth factor (β-NGF) is a polypeptide of 118 amino acids, and is a member of the group of polypeptide growth factors (PGF) that includes epidermal growth factor (EGF), platelet-derived growth factor (PDGF), fibroblast growth factor (FGF), and several other growth factors (*see* Table 1). PGFs are polypeptides that regulate cellular growth, proliferation, and differentiation (*1*).

β-NGF is required for the development and maintenance of the sympathetic nervous system. In newborn rodents, neutralization of β-NGF with antisera completely abolishes, and β-NGF administration markedly increases, the number and size of sympathetic neurons. In vitro, β-NGF enhances the survival of primary cultures of sympathetic and sensory ganglia and is capable of inducing morphologic, biochemical, and electrochemical differentiation of a rat pheochromocytoma cell line into predominantly adrenergic neurons (*2,3*).

The biologic actions of β-NGF are initiated through interaction of β-NGF with specific receptors (*1*). Both plasma membrane and nuclear receptors have been identified. Binding of β-NGF to receptor initiates, through unknown mechanisms (possibly involving cAMP), a multitude of effects, including stimulation of synthesis of macromolecules, ion and nutrient transport, and, ultimately, expression and maintenance of a specific neuronal phenotype both in vivo and in vitro (*2,3*).

TABLE 1
Representative Polypeptide Growth Factors and Related Substances

| | |
|---|---|
| A. *Detailed characterization* | C. *Initial characterization* |
| Epidermal growth factor (EGF) | B-cell growth factor |
| Insulinlike growth factors I & II (IGF I & II) | Bone-derived growth factor |
| Interleukin-2 (T-cell growth factor) (IL-2) | Chondrocyte growth factor |
| Nerve growth factor (NGF) | Endothelial-derived growth factors |
| Platelet-derived growth factor (PDGF) | Macrophage-derived growth factor |
| Transforming growth factor (Type I or α) (TGF) | Neurotrophic growth factors |
| | Transforming growth factors (γ) |
| B. *Partial characterization* | D. *Growth-regulating agents* |
| Cartilage-derived growth factor | Growth hormone |
| Colony-stimulating factors (CSFs) | Insulin |
| Endothelial-cell growth factors (ECGFs) | Placental lactogen |
| Erythropoietin | Plasminogen activators |
| Eye-derived growth factor (EDGF) | Prolactin |
| Fibroblast-derived growth factor (FDGF) | Relaxin |
| Fibroblast growth factors (FGFs) | Thrombin |
| Glial growth factor (GGF) | Transferrin |
| Osteosarcoma-derived growth factor (ODGF) | Vasopressin |
| Thymosin | |
| Transforming growth factor (Type II or β) (TGF) | |

β-NGF activity was first discovered in a sarcoma (*4*); subsequent attempts to purify the activity led to the discovery and purification of NGF from the male submaxillary gland (*5*). NGF is also present in bovine and guinea pig prostate (*6*) and snake venom (*7*). To date, β-NGF production has not been conclusively demonstrated in any other tissue. In particular, target tissue of β-NGF-dependent sympathetic and sensory neurons, i.e., heart, eye (iris), muscle spindle, and the like, contain β-NGF-like bioactivity, but definitive identification of the responsible factor (or factors) has not yet been accomplished.

In the submaxillary gland, β-NGF forms a tight complex with two additional polypeptides, α- and γ-NGF (*2*). The 7S complex consists of two α, two β, and two γ subunits. The γ subunit is an Arg-specific kallikrein. The α subunit is also a kallikrein, but enzymatically inactive. The 7S complex has no β-NGF activity. Production of β-NGF by the submaxillary gland is androgen-dependent and is stimulated by thyroxine, pituitary hormones, and glucocorticoids (*2,3*). β-NGF secretion is under α-adrenergic control (*2,3*).

Although much has been learned about NGF since its discovery over 20 years ago, many important questions remain: Precisely which tissues and which cells produce extra-submaxillary NGF and under what circumstances? Are the structures of submaxillary and extrasubmaxillary NGF in fact identical? If not, and several NGF exist, how are they related? Are the NGFs encoded by separate genes? If they are not, is the diversity of the NGFs created by differential splicing of the primary transcript of a unique NGF gene, or alternatively by tissue-specific processing of a common protein precursor? Is β-NGF associated in all tissues with α and γ subunits? What is the developmental regulation of β-NGF production? Are β-NGF synthesis and secretion regulated by the same factors in all tissues, i.e., by androgens, thyroxine, and α-adrenergic stimulation? What are the mechanisms that regulate NGF gene expression in the submaxillary gland? How is NGF synthesized, and secreted? What abnormalities of β-NGF gene expression exist? What human diseases might be related to such defects, and how could they be corrected?

As a consequence of recent advances in recombinant DNA and molecular-cloning technology and improved capabilities for the analysis and synthesis of DNA and protein sequences, many of these questions can now be examined with new molecular tools. Thus, both a DNA complementary to the mRNA that encodes mouse salivary β-NGF (β-NGF cDNA) and a partial human β-NGF gene have recently been cloned. With these cloned DNAs, much new information on β-NGF biology has already been obtained.

A description follows of the molecular cloning of β-NGF mRNA and genes and their structures. A discussion is given of questions posed earlier in the light of recently gained knowledge.

## 2. CLONING OF DNA COMPLEMENTARY TO THE mRNA THAT ENCODES THE β-SUBUNIT OF NERVE GROWTH FACTOR

The cloning of cDNAs that encode β-NGF sequences was achieved independently and simultaneously by two groups. Each group took a different approach. Scott et al. (*8*) used the so-called

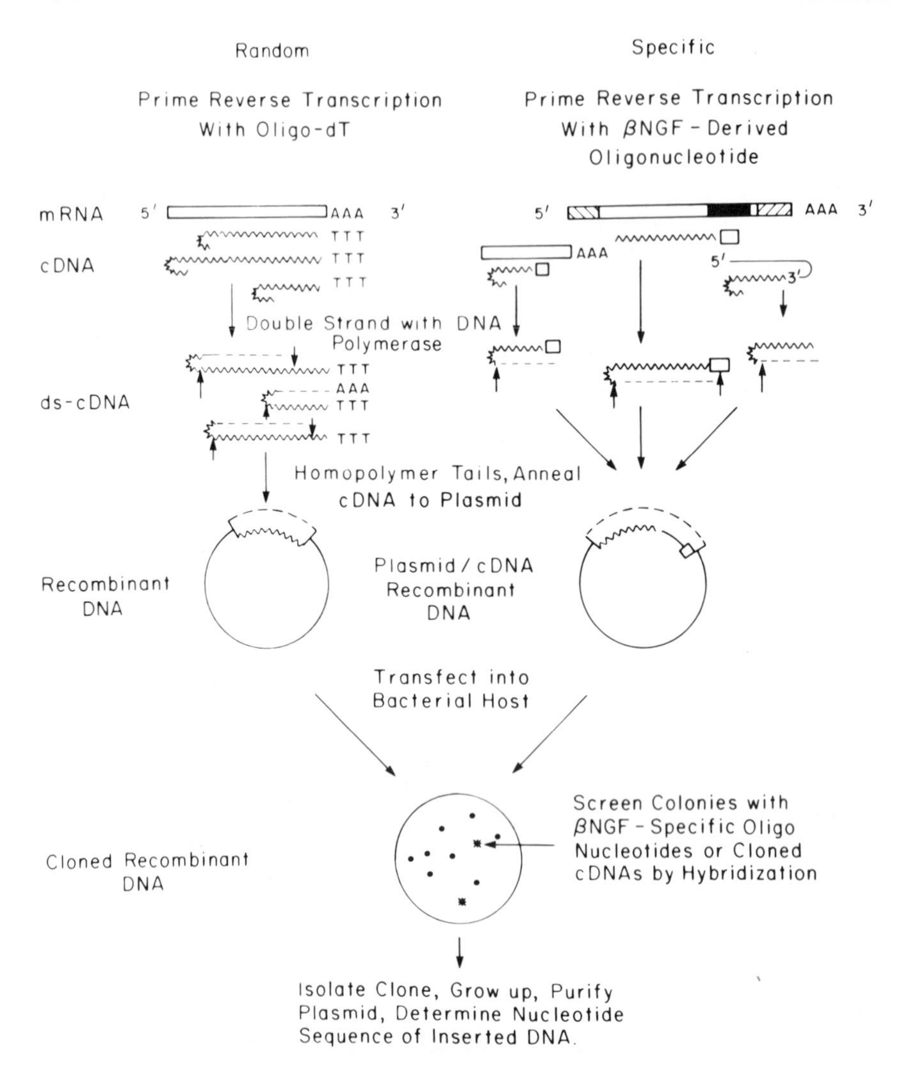


Fig. 1. cDNA strategies employed for the cloning of mouse salivary gland β-NGF mRNA. The random or "shotgun cloning" strategy is outlined in the left column, and the primer-specific strategy in the right column. The two strategies differ primarily in the initial step of preparing DNA complementary in sequence to mRNA (cDNA, wiggly lines). In the shotgun strategy, cDNA synthesis is primed by oligo(dT) (TTT) that is hybridized to the poly(A)-tail (AAA) at the 3′ end of mRNA. The primer-specific strategy exploits knowledge of the amino acid sequence encoded by the desired mRNA. cDNA synthesis is then primed with oligonucleotides that are complementary to the nucleotide sequence of the mRNA, as deduced from "reverse translation" of the

"shot-gun" cloning method, whereas Ullrich et al. (*9*) attempted to clone β-NGF mRNA specifically. The two approaches are outlined in Fig. 1 and will be discussed in more detail below. The cloning by Scott et al. was straightforward. First, a cDNA library was constructed. For cDNA construction, poly(A)+ mRNA was prepared from male mouse submaxillary gland, the richest source of NGF (*5*). Reverse transcription of the mRNA was primed with oligodeoxythymidilate, oligo(dT) 11–18, which hybridizes to the 3′-poly(A) extension of mRNA. Thus, priming of reverse transcription with oligo(dT) 12–18 yields a random collection of cDNAs. This collection of cDNAs was made double-stranded and was inserted into a plasmid cloning vector by complementary homopolymer extensions added to vector and cDNA (*10*).

For detection in the library of cloned bacteria that contain recombinant cDNA sequences, advantage was taken of the known amino acid sequence of β-NGF (*2*). A mixture of oligonucleotides with all possible nucleotide sequences derived from the genetic code was synthesized. A six-amino acid region of β-NGF having the least-degenerate codons, i.e., Gln-Tyr-Phe-

---

Fig. 1. (*continued*) protein according to the genetic code. However, as indicated, nonspecific priming and self-priming are important side reactions in this strategy that will depend on primer length, sequence, conditions of hybridization of primer to mRNA, and reverse transcription.

Subsequent steps are similar in the two strategies, and consist of synthesis of a second strand of DNA (ds-cDNA), tailing of the ds-cDNA with homopolymer extensions, and annealing of the homopolymer extensions to complementary extensions on a cloning vector. The collection of recombinant DNA is next introduced into a host in which it can replicate and that can be cloned.

Host clones containing individual recombinant cDNA are then detected by hybridization. Note that in both strategies, cloned β-NGF sequences were detected by NGF-derived oligonucleotides. In the outline, stippled boxes denote mRNA and small open boxes are NGF-derived oligonucleotides.

Abbreviations: A, deoxyadenosinemonophosphate; T, thymidinemonophosphate. Wavy lines denote DNA-complementary to mRNA. Broken lines (--------) denote second strand of synthesized DNA. Solid lines are vector DNA. The bottom circle represents a filter on which host clones have formed colonies (dots). Clones that contain recombinant β-NGF sequences and therefore hybridize to NGF-derived oligonucleotides are shown as stars.

Phe-Glu-Thr, was chosen (*see* Fig. 2,3). All 32 possible 17-base oligonucleotides were then prepared by the phosphotriester method (*11*), labeled with $^{32}P$ at the 5′ ends, and used to screen the cDNA library by the colony hybridization method (*12*). Because NGF is a relatively abundant protein in the male mouse submandibular gland (≈1% of protein), direct screening with the oligonucleotide probes of the library was feasible, and no specific measures to enrich the library in β-NGF sequence was required.

This approach yielded several overlapping cDNA clones from which >90% of the nucleotide sequence of β-NGF mRNA (≈1300 bases) and all of the amino acid sequence of a precursor of β-NGF could be deduced. To determine additional sequences at the 5′ end of β-NGF mRNA that were not represented in the cDNA clones, oligonucleotides complementary to cloned regions near the 5′ end of β-NGF mRNA were synthesized and used to prime the reverse transcription of 5′ regions of the mRNA. From the resulting reverse transcribed cDNA, additional nucleotide sequences to within a few bases of the 5′ cap of β-NGF mRNA were determined. The nucleotide sequence of mouse β-NGF mRNA and the deduced amino-acid sequences of mouse β-NGF are shown in Fig. 2.

A more-targeted approach to the cloning of β-NGF mRNA was taken by Ullrich and coworkers (*9*) (*see* Fig. 1). An attempt was made to enrich a cDNA library prepared from male mouse SMG in β-NGF cDNA sequences. To accomplish this, reverse transcription of poly(A)+ mRNA was primed with a pool of β-NGF-specific, 14-base oligonucleotides complementary to all possible codons of β-NGF amino acids 93–97, rather than with oligo(dT) 12–18, which hybridizes to the 3′-poly(A) extension of mRNA and is, therefore, a nonspecific primer of cDNA synthesis. Double-stranded cDNA was prepared and cloned by the same standard methods used by Scott et al. The resulting cDNA library was then screened separately with the oligonucleotide pool used for prim-

Fig. 2. (*on opposite page*) Nucleotide sequence of mouse submaxillary-gland NGF mRNA and the deduced amino acid sequence of the NGF precursor (*8*). The number of nucleotides is indicated in the right-hand margin. The amino acids of preproNGF are numbered 1 to 307. The sequence of NGF is underlined (from ref. *8*).

GAGCGCCUGGAGCCGGA[illegible]GGGAGCGCAUCGAGUGA[illegible]UUUGGAGCUGGCCUUAUA[illegible]UUGG 5[illegible]

1
met leu cys leu lys
AUCUCCCGGGCAGCUUUUUGGAAACUCCUAGUGAAC AUG CUG UGC CUC AAG 110

10 20
pro val lys leu gly ser leu glu val gly his gly gln his gly
CCA GUG AAA UUA GGC UCC CUG GAG GUG GGA CAC GGG CAG CAU GGU 155

30
gly val leu ala cys gly arg ala val gln gly ala gly trp his
GGA GUU UUG GCC UGU GGU CGU GCA GUC CAG GGG GCU GGA UGG CAU 200

40 50
ala gly pro lys leu thr ser val ser gly pro asn lys gly phe
GCU GGA CCC AAG CUC ACC UCA GUG UCU GGG CCC AAU AAA GGU UUU 245

60
ala lys asp ala ala phe tyr thr gly arg ser glu val his ser
GCC AAG GAC GCA GCU UUC UAU ACU GGC CGC AGU GAG GUG CAU AGC 290

70 80
val met ser met leu phe tyr thr leu ile thr ala phe leu ile
GUA AUG UCC AUG UUG UUC UAC ACU CUG AUC ACU GCG UUU UUG AUC 335

90
gly val gln ala glu pro tyr thr asp ser asn val pro glu gly
GGC GUA CAG GCA GAA CCG UAC ACA GAU AGC AAU GUC CCA GAA GGA 380

100 110
asp ser val pro glu ala his trp thr lys leu gln his ser leu
GAC UCU GUC CCU GAA GCC CAC UGG ACU AAA CUU CAG CAU UCC CUU 425

120
asp thr ala leu arg arg ala arg ser ala pro thr ala pro ile
GAC ACA GCC CUC CGC AGA GCC CGC AGU GCC CCU ACU GCA CCA AUA 470

130 140
ala ala arg val thr gly gln thr arg asn ile thr val asp pro
GCU GCC CGA GUG ACA GGG CAG ACC CGC AAC AUC ACU GUA GAC CCC 515

150
arg leu phe lys lys arg arg leu his ser pro arg val leu phe
AGA CUG UUU AAG AAA CGG AGA CUC CAC UCA CCC CGU GUG CUG UUC 560

160 170
ser thr gln pro pro pro thr ser ser asp thr leu asp leu asp
AGC ACC CAG CCU CCA CCC ACC UCU UCA GAC ACU CUG GAU CUA GAC 605

180
phe gln ala his gly thr ile pro phe asn arg thr his arg ser
UUC CAG GCC CAU GGU ACA AUC CCU UUC AAC AGG ACU CAC CGG AGC 650

Nerve growth factor 200
lys arg ser ser thr his pro val phe his met gly glu phe ser
AAG CGC UCA UCC ACC CAC CCA GUC UUC CAC AUG GGG GAG UUC UCA 695

210
val cys asp ser val ser val trp val gly asp lys thr thr ala
GUG UGU GAC AGU GUC AGU GUG UGG GUU GGA GAU AAG ACC ACA GCC 740

220 230
thr asp ile lys gly lys glu val thr val leu ala glu val asn
ACA GAC AUC AAG GGC AAG GAG GUG ACA GUG CUG GCC GAG GUG AAC 785

240
ile asn asn ser val phe arg gln tyr phe phe glu thr lys cys
AUU AAC AAC AGU GUA UUC AGA CAG UAC UUU UUU GAG ACC AAG UGC 830

250 260
arg ala ser asn pro val glu ser gly cys arg gly ile asp ser
CGA GCC UCC AAU CCU GUU GAG AGU GGG UGC CGG GGC AUC GAC UCC 875

270
lys his trp asn ser tyr cys thr thr thr his thr phe val lys
AAA CAC UGG AAC UCA UAC UGC ACC ACG ACU CAC ACC UUC GUC AAG 920

280 290
ala leu thr thr asp glu lys gln ala ala trp arg phe ile arg
GCG UUG ACA ACA GAU GAG AAG CAG GCU GCC UGG AGG UUC AUC CGG 965

300
ile asp thr ala cys val cys val leu ser arg lys ala thr arg
AUA GAC ACA GCC UGU GUG UGU GUG CUC AGC AGG AAG GCU ACA AGA 1,010

307
arg gly OP
AGA GGC UGA CUUGCCUGCAGCCCCCUUCCCCACCUGCCCCCUCCACACUCUCUUGGG 1,067

CCCCUCCCUACCUCAGCCUGUAAAUUAUUUUAAAUUAUAAGGACUGCAUGAUAAUUUAUC 1,127

GUUUAUACAAUUUUAAAGACAUUAUUUAUUAAAUUUUCAAAGCAUCCUGAAAAAAAA

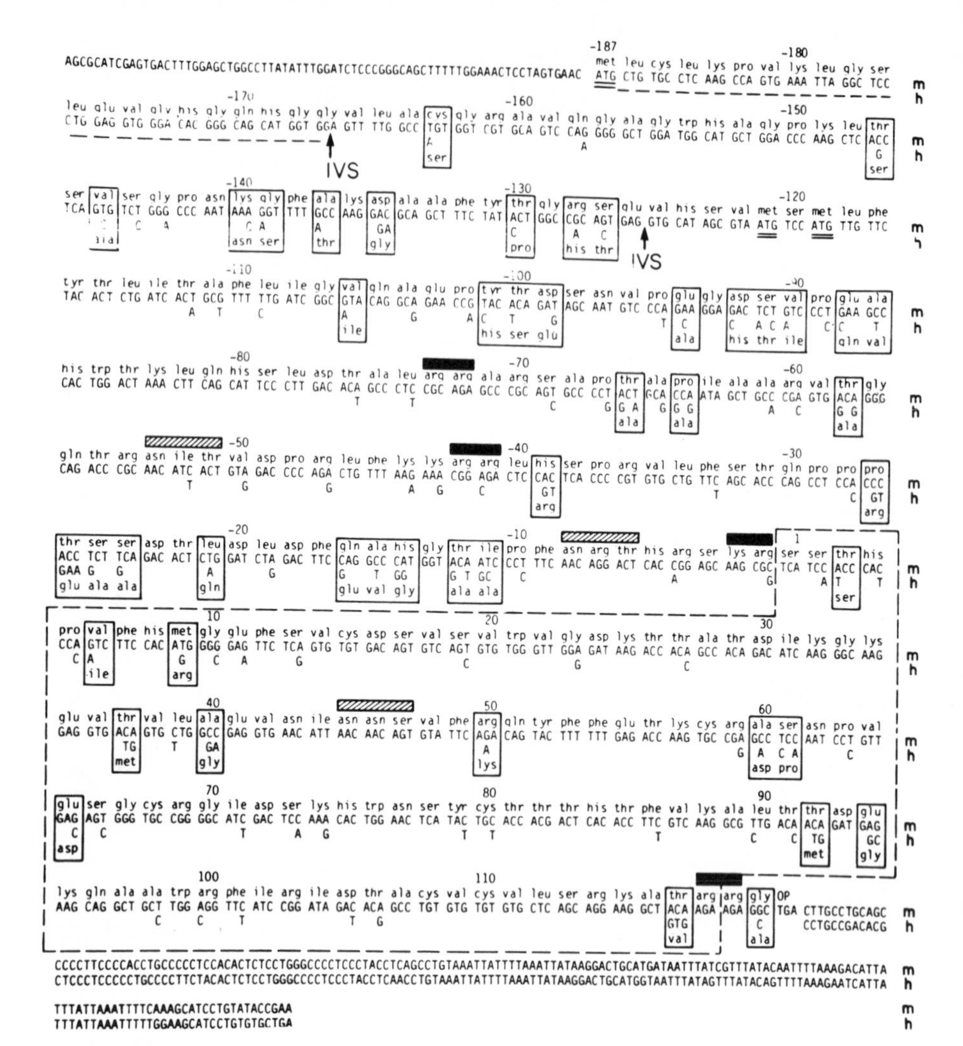

Fig. 3. Nucleotide sequence of cloned mouse submaxillary-gland β-NGF mRNA and comparison with human sequences (*9*). The mouse β-NGF cDNA sequence is given (m) and the deduced amino acid sequence indicated above the nucleotide sequence. Human gene sequences (h) that differ from mouse are written below, together with deduced amino acids that also differ from mouse (boxes). Missing human sequences are underlined with a dashed line. Positions of intervening sequences in the human β-NGF-gene are indicated by arrows and are designated IVS. Note that the missing human sequences are situated upstream from intron I (IVS I). Additional human β-NGF gene sequences are shown in Fig. 5. Solid bars indicate pairs of basic amino acids that might be sites of proteolytic processing of proβ-NGF. Hatched bars denote potential sites of N-linked glycosylation of proβ-NGF (from ref. *9*).

ing cDNA synthesis, and two additional pools of oligonucleotides, each complementary to a separate region of β-NGF mRNA lying farther upstream, i.e., complementary to all possible codons of amino acids 74–77, and 52–56 of β-NGF, respectively. Of 10,000 bacterial colonies screened in this fashion, six (<0.1%) hybridized with all three pools of oligonucleotide probes, strongly suggesting that they contained β-NGF mRNA sequences. The colony with the recombinant plasmid with the longest cDNA insert (pmβN-9G1) was completely sequenced and shown to contain 75% of the β-NGF mRNA sequences.

To obtain the missing, more-5′-situated sequences of β-NGF mRNA, restriction fragments of pmβN-9G1 were used to prime the synthesis of cDNA from poly(A)+ mRNA. The cDNA was then separated by electrophoresis on polyacrylamide gels. A cDNA band specifically synthesized by primer-dependent transcription was eluted from the gel. The resulting collection of cloned cDNAs was screened with a fragment of the longest original cDNA, pmβN-9G1. Two hybridization-positive cDNAs were obtained, and yielded the missing 5′ sequences of β-NGF mRNA.

However, 3′ sequences were still incomplete. Inasmuch as reverse transcriptase transcribes only from 3′ → 5′ on the mRNA template, and 3′-untranslated sequences cannot be inferred from a known amino acid sequence to obtain specific primers for cDNA synthesis, missing 3′ sequences of β-NGF mRNA were obtained as follows. The size of β-NGF mRNA was estimated by agarose gel electrophoresis of male mouse SMG mRNA and blotting of the size-fractionated mRNA onto a membrane. The membrane-bound β-NGF mRNA was then visualized by hybridization to the membrane of a radioactively labeled β-NGF cDNA, followed by washing and autoradiography. Using this technique (Northern blotting), salivary β-NGF mRNA was estimated to be 1300 ± 50 bases. Consequently, male SMG mRNA was size-fractionated on a urea-agarose gel; the fractions ranging from 800 to 1500 bases were eluted and cloned. To enhance the chances of cloning the complete 3′end of β-NGF mRNA, cDNA synthesis was primed at the 3′-poly(A) extension of the size-fractionated mRNA with oligo(dT) 12–18, i.e., nonspecifically, but at the extreme 3′ end of the mRNA. Two of the 4400 cloned cDNAs contained 3′-directed sequences, but none of them included the poly(A) extension. Therefore, some, albeit little, of the 3′ end of β-NGF mRNA was still missing. However, the complete set of cDNA clones obtained in this three-step, targeted approach contained 95% of the β-NGF mRNA sequence, from which the complete amino acid sequence of a β-NGF precursor could be deduced (*see* Fig. 3).

It is instructive to compare the two approaches to the cloning of β-NGF mRNA (*see* Fig. 1). Scott et al. did not enrich for β-NGF mRNA and did not prime cDNA synthesis specifically. Their random library contained 0.1% β-NGF-related clones. Thus, screening 40,000 cDNA clones produced sufficient β-NGF-related sequences so that 90% of the β-NGF mRNA nucleotide sequence could be determined, and missing 5′ sequences were easily obtained by cDNA "read out." In contrast, Ullrich et al. attempted to enrich the β-NGF cDNA sequences by specific priming of cDNA synthesis. However, despite specific priming, only 0.1% of the cDNA clones was β-NGF-related, a frequency similar to that obtained by random priming with oligo(dT) 12–18. It is not clear why specific priming should not result in cDNA enrichment. Such failure may be related to the low frequency of β-NGF mRNA, i.e., 0.1% of poly(A)+ mRNA, the degeneracy of the primer mixtures, the small size in nucleotides of the individual primers (14-mer) (*13*), and the possibility of self-priming of cDNA synthesis by RNA.

Although specific priming of cDNA synthesis did not enrich the β-NGF sequences, it appears to have biased the β-NGF related cDNA against sequences 3′ to the specific primer. Therefore, to obtain these missing 3′-sequences, additional nonspecifically primed cDNA had to be cloned. For this cloning, another attempt was made to enrich the β-NGF sequences, this time by enriching the mRNA template in β-NGF mRNA by size selection. However, again the apparent frequency of β-NGF cDNA clones was 0.1%. Thus, neither enrichment by size selection of mRNA nor by specific priming appears to have significantly increased the frequency of β-NGF sequences over the frequency observed in the nonspecific priming of unselected mouse salivary poly(A) RNA with oligo(dT) 12–18. Note that each cloning approach required special attempts to obtain 5′ sequences near the mRNA cap. In each case, these sequences were derived from specifically primed cDNA, although Scott et al. did not clone the cDNA. It should be mentioned that methods for the cloning of full-length cDNA have been reported (*14*), and such methods should be applicable to mRNA of 0.1% abundance.

## 3. HUMAN β-NGF GENE

A partial human β-NGF gene was isolated by Ullrich et al. from a human gene library (*12*). The library was obtained from Lawn et al. (*15*), who had cloned random fragments of human DNA in bacteriophage vector Charon-4A. The same library was successfully screened for a number of human genes.

To isolate the human β-NGF gene, the library was screened by the plaque hybridization method of Benton and Davis (*16*); the hybridization probe was a restriction fragment of pmβN-9G1. Thus, a mouse cDNA was used to screen the human library. The restriction fragment spans most of the β-NGF sequence, where greatest conservation between the mouse and human sequences is expected. With this hybridization probe, 27 hybridizing recombinant phages were isolated from the library. Analysis of all 27 phages with digestion by restriction enzymes revealed that they shared restriction fragments, suggesting that they were derived from the same genomic region. The largest clone, λhβN8, was further analyzed. The human DNA fragment, carried by λhβN8, contained ≃17,000 base pairs (bp), 15,000 of which were sequenced by the method of Maxam and Gilbert (*17*). To facilitate sequencing, *Eco*RI and *Hin*dIII restriction fragments were subcloned into a plasmid pBR322.

To localize possible β-NGF sequences within it, the cloned human DNA fragment was screened for regions of homology with the mouse cDNA by blot hybridization analysis of restriction-enzyme digests. Its nucleotide sequence was translated into all possible encoded amino acid sequences, and its nucleotide sequence was compared directly with that of the mouse cDNA. Using these approaches, two regions of extensive homology with the mouse cDNA were detected. On the basis of these homologies, two intervening sequences could be identified in the human β-NGF gene, separating three exons (Figs. 3 and 4). Exon II is 90% homologous with the mouse cDNA. Attempts to definitively identify exon I in any of the 27 cloned human β-NGF-related DNA fragments failed. Several explanations are possible. First, mouse and human sequences are not sufficiently homologous in this exon to permit cross-hybridization of nucleic acids or recognition of the deduced human amino acid sequences as β-NGF-related. Second, the cloned and analyzed human gene is a

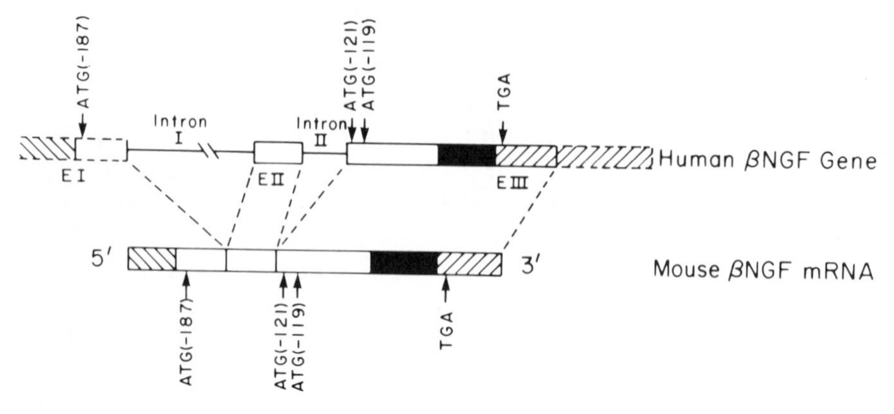


Fig. 4. Outline of mouse salivary-gland β-NGF mRNA and human β-NGF gene. The partial human β-NGF gene is shown on top, and mouse β-NGF mRNA at the bottom. The three exons are boxes designated EI, II, and III. Introns I and II are shown as lines. Intron II contains 6.6 kb; the length of intron I is unknown. Exons I–III are connected to the corresponding regions of the mRNA with dashed lines. Portions of exons enclosed by broken lines indicate that the length or chromosomal location, or both, of the exon are still incompletely known. These unknown portions of Exons I and III may also be interrupted by additional introns.

The open and solid boxes indicate segments of gene and mRNA that are known or potential coding regions. Arrows point to sites where translation of coding sequences may be initiated; the locations of the ATG initiator codons are given in the parentheses and correspond to the numbering of amino acids in Fig. 3. The single TGA stop codon is shown.

pseudogene, or another gene more distantly related to NGF. Third, human and mouse 5′-exons are closely homologous, but the human 5′-exon is not present in any of the 27 recombinant clones. The first explanation is less probable, but difficult to evaluate, at present, and ultimately will require sequence analysis of the human β-NGF mRNA. The second explanation is less probable because analysis of human DNA by Southern hybridization under high stringency revealed only a single human β-NGF gene.

Therefore, the cloned and sequenced gene is not a pseudogene, but must represent the expressed human β-NGF gene. The third explanation is the most probable and would make intron I at least 8.5 kilobases (kb). This possibility, i.e., that the 5′-exon of the NGF gene is not present in any of the cloned DNA fragments analyzed, is supported by the finding that a mouse cDNA containing this exon did hybridize in Southern hybridization experiments to DNA restriction fragments not present in λhβN8, albeit only under loosened stringency of hybridization. Moreover, if it is assumed that translational initiation occurs at the Met initiator codon at position −187 (Fig. 4), an assumption most consistent with data derived from studies of β-NGF biosynthesis (*18*), then exon I will contain another segment of amino acid coding sequence, and not solely 5′-untranslated region. Therefore, the 90% homology seen in the exon II should at least extend through this coding region, and may then drop off in the 5′-untranslated tract (*see* Figs. 3–5). Thus, the human and mouse sequences are expected to cross-hybridize. Because no sequences hybridizable to this region of the mouse cDNA were present in any of the recombinant human clones, but could be seen on Southern analysis of human DNA, it is most probable that these clones did not contain the most 5′-exons, the promoter, or the 5′-flanking regions of the human β-NGF gene. Therefore, at present, only a partial human β-NGF gene appears to have been cloned. A condensed nucleotide sequence of the partial human NGF gene is shown in Fig. 5. To summarize the foregoing discussion, current knowledge of β-NGF gene structure may be stated as follows: (a) intron I is at least 8 kb long, (b) exon II contains 123 bp, (c) intron II spans 6.6 kb, (d) exon III contains at least 900 bp, and (e) the size of exon I is unknown.

It should be noted that even if the human and mouse 5′-exons were insufficiently homologous to detect the human exon in the human gene library by using mouse cDNA as hybridization probes for further screenings, and if the human β-NGF cDNA were difficult to obtain, as it may well be, it would still be possible to find the human exon. A similar situation was encountered during cloning and sequencing of the rat parathyroid hormone gene (*19*). A bovine cDNA was used to obtain the rat gene from a rat liver library. During gene analysis it became apparent that the 5′-untranslated region of the PTH mRNA is encoded by a sepa-

A

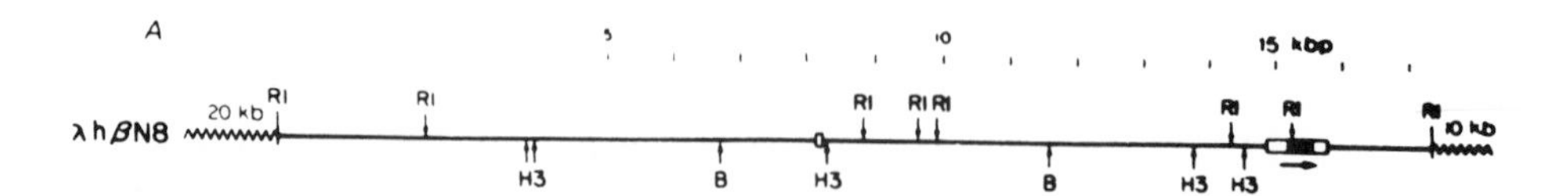


B

```
GAAGAGAGAATACGGACAGGAAAGTTAAGATGTCATTCTAGAACTTTATTGGGAGGGCATCTCCACCCTACAACAAATTCTGTGATGGACATAATCATTC        100
ATTCATTTATCCGTAAATATCACCCTCTTGTTCAAAGCCCTCCACTGCCTTCCTAATATCCTGAGGATAAAACCATAGCTCCTTGCTGTGTCTCTGTAGA        200
CCTGGCTCTTCCTGGCTCTCCAGCTCATTTTCTAGGTCTCGTTACTTCATGCTCAGAACCTTTGTCTTGTTTCTAGCTCAGGGCCTTTGCACTTGTTCTT        300
GCTGCCTAGAATGTTCTCTCCCTCATTCCTTCTCATCCTCCAGATCTCAACTTGAAGGCCATCTCCTCAGAGCTCCTCGCTGAGCGTCCTGTCTACAGTG        400
GCCCCTCGATACATCCTGCAGTTGCTCTCTATCATCAGACCCTGTAATTGCCTTCATGGCATATAAAGAATCTGGAGTATCTTGCTTATTTACACAACAC        500
TGTAAGCTCCATGAGAGCAGAGGCCTTGTTTGTCTTGTTTACTGCTGCTCAGCACCAAAAACAGTGCCTGGCACATAGTCGGTGCCCAGAAAATATTGTG        600
AATGAATGAAGTGCCTACATAGATTACATTATAGAAGTGAGAGGAGAATAGAAAACTTCCATTGTTTCTAGAAACTACAGCCTAAAATTGATTTTTTAAA        700
ATTGTATCAGCTCCATAGCTTCCAATCCTAAAATCTGCCTTTCAGTGTGGTACTCTGAGATTCCTGTCTGATTCTGTGAGAGCTCCACATTCTCTCTCAA        800
ATGGTCAGTCTGTCTTATTTGTCACCATTACTCATCTGCATTTTTATCAAAGCACCAACTTGCTCTGAATTGTCAGGGATTTTGCGTCTGTATAAGGTAT        900
TTTAGGCTGGTTCAGAGTTGGATCTGTTATGTCTGCATGTGTAATGTACTGAACAATTTCTATTTTGATGCCAGATTAGGGATCTGCTGGGGCAAGACTT       1,000
TGGCATGTGTCTAGAAACACCTGCACTAGGTGCAAGATCAGCCATGGACTGTGTCCAGGCTGAAACCAAAAGGTATGGCGCAAGAGTGAGAGGCAGGTGC       1,100
CACCACAGGACCATGAGAGGCCAAGCTCCGGTATAATTTTGGTAGACCAAATTCTAGCTCCTTCCTGGGCCTTGATGCTGGTAAAATCCCAGAACTCAAG       1,200

                                                                                                  Va
GAAATGGAATTTGTCCTATTGGCACATGCCTCCCCACTGTGTAGGGCACAGGGAATGTGGTGAGGTACAGTCTAATGCCAGCTCTCCCCCTCCACAGAGT       1,300
lLeuAlaSerGlyArgAlaValGlnGlyAlaGlyTrpHisAlaGlyProLysLeuSerSerAlaSerGlyProAsnAsnSerPheThrLysGlyAlaAla
TTTGGCCAGTGGTCGTGCAGTCCAAGGGGCTGGATGGCATGCTGGACCCAAGCTCAGCTCAGCGTCCGGACCCAATAACAGTTTTACCAAGGGAGCAGCT       1,400
PheTyrProGlyHisThrGlu                                     ValHisSerValMetSerMetLeuPheTyrThrLeuIle
TTCTATCCTGGCCACACTGAGGTAAGTGCCT ----6.6kb--- TTCATTCCAGGTGCATAGCGTAATGTCCATGTTGTTCTACACTCTGATC       1,480
ThrAlaPheLeuIleGlyIleGlnAlaGluProHisSerGluSerAsnValProAlaGlyHisThrIleProGlnValHisTrpThrLysLeuGlnHisS
ACAGCTTTTCTGATCGGCATACAGGCGGAACCACACTCAGAGAGCAATGTCCCTGCAGGACACACCATCCCCCAAGTCCACTGGACTAAACTTCAGCATT       1,580
erLeuAspThrAlaLeuArgArgAlaArgSerAlaProAlaAlaAlaIleAlaAlaArgValAlaGlyGlnThrArgAsnIleThrValAspProArgLe
CCCTTGACACTGCCCTTCGCAGAGCCCGCAGCGCCCCGGCAGCGGCGATAGCTGCACGCGTGGCGGGGCAGACCCGCAACATTACTGTGGACCCCAGGCT       1,680
uPheLysLysArgArgLeuArgSerProArgValLeuPheSerThrGlnProProArgGluAlaAlaAspThrGlnAspLeuAspPheGluValGlyGly
GTTTAAAAAGCGGCGACTCCGTTCACCCCGTGTGCTGTTTAGCACCCAGCCTCCCCGTGAAGCTGCAGACACTCAGGATCTGGACTTCGAGGTCGGTGGT       1,780
AlaAlaProPheAsnArgThrHisArgSerLysArgSerSerSerHisProIlePheHisArgGlyGluPheSerValCysAspSerValSerValTrpV
GCTGCCCCCTTCAACAGGACTCACAGGAGCAAGCGGTCATCATCCCATCCCATCTTCCACAGGGGCGAATTCTCGGTGTGTGACAGTGTCAGCGTGTGGG       1,880
alGlyAspLysThrThrAlaThrAspIleLysGlyLysGluValMetValLeuGlyGluValAsnIleAsnAsnSerValPheLysGlnTyrPhePheGl
TTGGGGATAAGACCACCGCCACAGACATCAAGGGCAAGGAGGTGATGGTGTTGGGAGAGGTGAACATTAACAACAGTGTATTCAAACAGTACTTTTTTGA       1,980
uThrLysCysArgAspProAsnProValAspSerGlyCysArgGlyIleAspSerLysHisTrpAsnSerTyrCysThrThrThrHisThrPheValLys
GACCAAGTGCCGGGACCCAAATCCCGTTGACAGCGGGTGCCGGGGCATTGACTCAAAGCACTGGAACTCATATTGTACCACGACTCACACCTTTGTCAAG       2,080
AlaLeuThrMetAspGlyLysGlnAlaAlaTrpArgPheIleArgIleAspThrAlaCysValCysValLeuSerArgLysAlaValArgArgAlaEnd
GCGCTGACCATGGATGGCAAGCAGGCTGCCTGGCGGTTTATCCGGATAGATACGGCCTGTGTGTGTGTGCTCAGCAGGAAGGCTGTGAGAAGAGCCTGAC       2,180
CTGCCGACACGCTCCCTCCCCCTGCCCCTTCTACACTCTCCTGGGCCCCTCCCTACCTCAACCTGTAAATTATTTTAAATTATAAGGACTGCATGGTAAT       2,280
TTATAGTTTATACAGTTTTAAAGAATCATTATTTATTAAATTTTTGGAAGCATCCTGTGTGCTGATGCTGGTTATTTTTTTTGAGTAAAATCATCTGCAA       2,380
GTCTGAGGAAGATGCAGGGGGAATTGTCTGAAGCACCCCCTGGCTCCTTCTAAGGCCCACCTGTAACCCTTACCCCACCCCCAGCCAGTGCTGCAACTTC       2,480
AGGAAAGGCTAACTGGTTTCAGATATTAGCCTAAGCCAGGCATGATTAGCAAAGGAAGCCTGCTGGATTGCAACTTTGTTCTACATTCCAGAGCCAAAGC       2,580
TACCTAATCATTGGATTACCTGCCCACGGCCTGGAAGCATGGGCCCTGGCTCCTCCCTTGGAGAAGTTAGTGGAGCCTCATCAGGAGGCAGATCACCACT       2,680
GGGCATGGGGCTGCCCTGGCCTCAGAGGCACAGTCTCAGTCCTGGGGGCACTCGATAGACAGGAAGAGGCTTGTACCAAAGATCTGTGAGCTGCTCATTT       2,780
GTAGAGGGAGTGTGCTGTCTGGAGAGTTGTAACAGAGTAACTCACGAAGCCTGCCAAAGTAAGATCTAGGGTTTCATGTTCAGCTGGCCTTAGTACTCGT       2,880
TTTATCCCTCAGCAGCCTCCTGCAGAAAGACTTTCATTACCCCAAGAAAACATTAGTGCCTCCGACACTCTTGCAACCCACTAACAGCCACAGTAGCAAC       2,980
TTCCAGGGGCCCCACCATCAGCCCACAGCCAGGGAAGCCAGAGACGGGAGAGAACTTCTGAAAATATTTTTATTCTTAATTGGATGATTTCCATAAGTGA       3,080
CCAAAGGTGACTGGGGCCGTGGGACCTGCAGAGAAGAGGGTGGGGCAGAAAGACTAAGTAAATGGAGTAAGAACCAAATAAACCTGGAGATGTGGAGGAC       3,180
AGGTGGGCCGGCAGGAACCACGGCTTTCTGCTTCTCCCAAACAGGGGAAATGAAACAAGGATGACAAAGAGAAGGCCAGAGGTGGATGTGAAGAAGGGGA       3,280
GGGGAGGACAGAGCAGGAGGAGGAGGGAGGTGTGATCTCTCATCTTTACCAGCACTCTCCAGCCTCCAGGAAAGTTAGCAACCTGGGTGGTTCTTGTGGG       3,380
GCTTTTTGGATGTGCTAATAAACACTGACTCCATCCAGAGCATCTTAGGAGTGAAGGTGACAGAGGGACTCCAGGGTTCAGCCTGCAGCAGGGCTCCACC       3,480
CAGCCTCACACGACCCCCTTCCCCTGCAACTTTCATTTTGTATTTTCCTTGAAGCCACAATATTCCCCCAGACAGAGCACTTATATTAAGACAGAAAATT       3,580
TCCCCTCCTATCCATCCTGTCTCCCCCCGAGGATGAAATGCCTACACTGGTTTTCAAGAGAACAGGCCCAAATTGTCCATCCGAGAAAAGAGCAAATGGT       3,680
CTTATTCATTCCCTCTGTAATCCTCGCTGTCCGGCCCCACCTCCAACGCCATCCAAATAGGGCAAATTAGAAACGTAGGGAGCTTTGGCATGAGAGCCAT       3,780
TTGATTAGGGTAGAAGTATGATGAAATGTGCCCTGTTTTTTCTATCAGAGCCACTGCAATTGGAATAAAGGTTTATGCTCAAGAAACACCAGCTAGAACA       3,880
TCTGTGCTGAGATTCTAAGTGAGATAA
```

rate exon, as it is in the human PTH gene (see chapter by Henry M. Kronenberg in this book), and that bovine, human, and rat 5′-exons did not cross-hybridize. Therefore, to determine whether the 5′ rat exon was present in the cloned rat DNA, a large restriction fragment of the cloned rat PTH gene was split into smaller fragments, each fragment labeled with $^{32}P$ then individually hybridized to rat PTH mRNA bound to a membrane. A single fragment that hybridized to PTH mRNA was sequenced, and the sequence was searched for homology with the human PTH gene by computer analysis. A region with 60% homology with the human PTH sequences was found that contained all the hallmarks of a 5′-untranslated region with adjacent promoter and enhancer sequences. The authenticity of this region so discovered was confirmed by S1 nuclease mapping and PTH mRNA readout experiments. A similar strategy could be used for mapping of the 5′-exon and promoter of the human β-NGF gene. Clearly, these regions are of great interest. Of course, an important prerequisite for this strategy is a source of human mRNA that contains hybridizable amounts of β-NGF sequences.

Fig. 5. (*on opposite page*) Physical map and partial nucleotide sequence of human β-NGF gene (*12*). (A). Diagrammatic representation of a fragment of human DNA cloned in bacteriophage Charon-4A. Wavy lines indicate vector Charon-4A DNA. In the human DNA insert, exons are shown as boxes, and introns and flanking DNA as lines. Restriction-endonuclease cleavage sites are marked by arrows and labeled R1 (*Eco*RI), H3 (*Hin*d III), and B (*Bam*HI). A scale in 1000 bp (1000 bp = 1 kb) is shown above. (B). Partial nucleotide sequence of the human β-NGF gene. The sequence is given from 5′ (left upper boxes) to 3′ (right lower corner). Nucleotides are numbered in the right margin. Regions homologous with mouse β-NGF mRNA are translated into the deduced amino acid sequence indicated above the nucleotide sequences. The segment corresponding to the secreted form of β-NGF is boxed. Only 20 of the 6.6 kb of nucleotides of intron II are shown. A typical TATAAA transcriptional initiation site, or TATA box, is doubly underlined. Three potential sites of translational initiation, or ATG initiator codons, are underlined with dashes. Three AATAAA polyadenylation sites in the 3′ region of the gene are underlined. Note that 15 kb of the 17-kb insert of λhβN8 were sequenced, and the complete nucleotide sequence is available upon request from Ullrich et al. (*9*).

## 4. STRUCTURE OF β-NGF mRNA AND GENE

The nucleotide sequences of the mouse salivary β-NGF mRNA and of the corresponding human gene are compared in Fig. 3. The β-NGF mRNA is 1300 ± 100 bp, as determined by sizing on agarose gels (*12*). The size of the gene is presently unknown.

At the nucleotide level, mRNA and gene contain no unusual features; however, there is still uncertainty about the precise length of the 5′-and 3′-untranslated regions of the mRNA and the assignment of intervening and 5′-flanking sequences for the gene. Inasmuch as the length of the 5′-untranslated region of β-NGF mRNA is directly related to the question of where translation of the mRNA begins, and therefore what the earliest precursor of β-NGF might be, this question will be dealt with later.

The exact length of the 3′-untranslated tract of β-NGF mRNA is uncertain. The cDNA sequence reported by Scott et al. contains a poly(A) tail, suggesting that it covers the entire 3′-untranslated region, in this case 156 bases, of β-NGF mRNA. However, the polyadenylation signal preceding the poly(A) tail is a relatively rare variant, i.e., ATTAAA rather than AATAAA. The cDNA sequence reported by Ullrich et al. continues beyond the poly(A) tail of the Scott sequence, but does not itself contain a poly(A) tail. It is most reasonable to assume that the extended cDNA sequence reported by Ullrich is the result of some infidelity of polyadenylation that creates heterogeneity at the 3′ end, possibly because the variant polyadenylation signal is less strong. This interpretation is supported by the observation that an oligonucleotide (20 bases) that includes the last nine bases of the cDNA sequence reported by Ullrich did not hybridize to male mouse submaxillary poly(A)+ mRNA (unpublished observations by G. Heinrich, 1984). Moreover, the human gene is highly homologous with the mouse in the 3′-untranslated region and contains a similar variant polyadenylation signal in a position identical to that of the mouse. However, another explanation for the differences between the two reported mRNA sequences is that there may be a more classic AATAAA sequence in the mouse gene downstream from the variant ATTAAA sequence, as is the case in the human gene (*see* Fig. 5). If this were so, the possibility that two different polyadenylation sites are alternatively used in mouse β-NGF mRNA would arise. It is interesting in this regard that the homology between mouse cDNA and human gene sequences appears to end rather abruptly where the poly(A) tail of the cDNA sequence reported by Scott begins. Moreover, this breakpoint has a classic "GT" splice donor site. Thus, is it possible that the human gene may contain an intron in this location? Clearly, to evaluate these speculations, the sequences of the mouse β-NGF gene and the human β-NGF mRNA must be completely determined.

The human β-NGF gene exhibits no unusual structural features, except for the possibility just discussed that an intron may be located in the 3′-untranslated region of the encoded mRNA and the further possibility that two, rather than one, introns are located in the 5′-untranslated region. The presence of two introns in a 5′-untranslated region would be unique. This question will be discussed in more detail in the next section.

## 5. STRUCTURE OF THE β-NGF PRECURSOR ENCODED BY β-NGF mRNA AND GENE

β-NGF mRNA encodes a protein larger than β-NGF. This was expected on two grounds: (a) most secreted proteins are synthesized as larger precursors, and (b) biosynthesis studies using radioactive amino acids and pulse-labeling techniques strongly suggested the presence of larger β-NGF precursors (*18*). The largest of these was 29 kdaltons. Indeed, an intracellular form of ≃32 kdalton was reported. Such large precursors suggest initiation of translation at the ATG codon at position −187 with or without removal of a signal sequence in case of the 29 and 32 kdalton precursors, respectively (all codon positions refer to Fig. 3). The difficulty is that the amino acid sequence between amino acids −187 and −160 is not that of a typical signal sequence, i.e., there are several charged amino acids in the central portion of this putative signal sequence, whereas most signal sequences have a hydrophobic core. In view of this, Ullrich et al. favor initiation of translation at initiation codons −121 or −119. Initiation at these codons would produce a more-classic signal sequence. However, neither ATG-121 nor ATG-119 is an optimal initiator codon according to observations of Kozak (*20*), whereas ATG-187 is (*19*).

Thus, biosynthesis studies and studies of ATG-initiator strength both suggest that translation of β-NGF mRNA is initiated at ATG-187.

Additional considerations support this view. First, if initiation occurred at $ATG_{-127}$, the 5′-untranslated region of β-NGF mRNA would be unusually long, i.e., >200 nucleotides. Moreover, the human 5′-untranslated region would be unusually homologous with the mouse 5′-untranslated region. However, an example of unusually homologous 5′-untranslated regions exists for rat and human somatostatin mRNA (*21,22*). Finally, initiation of translation at codon −121 would imply that two introns are located in the 5′-untranslated region of a β-NGF mRNA. This situation would be unprecedented. Thus, it is likely, but not proved, that translational initiation occurs at ATG-187.

It should be noted that the mode of secretion of β-NGF from the salivary gland is unknown. It is therefore possible that it may not occur along the classic storage-type of secretory pathway, but instead may occur along an alternative pathway. For example, it is conceivable that processing of preproβ-NGF is not cotranslational, but rather takes place posttranslationally. Such a mechanism of secretion would resemble more closely the "secretion" of mitochondrial proteins. The signal sequences of mitochondrial proteins are similar to the presequence of preproβ-NGF in that they contain charged amino acids in their cores (*23*).

## 6. KNOWLEDGE GAINED—AND THE FUTURE

The cloning and structural analyses of β-NGF mRNA and gene provided the following new insights into NGF biology.

(a) Mouse and human β-NGF are encoded by unique and distinct genes.
(b) β-NGF is derived from a larger precursor. However, the structure and presecretory processing of this precursor are incompletely understood.
(c) The amino acid sequences of β-NGF and the β-NGF precursors are highly conserved from mouse to human (>90%).
(d) The β subunit of 7S NGF is encoded by a separate mRNA, as are the γ and α subunits, i.e., β, γ, and α

subunits do not arise from a common precursor. Indeed, the mouse β-NGF gene was localized to chromosome 3 (*24*), whereas the α- and γ-subunit genes are found in a cluster of kallikrein genes on chromosome 7 (*25*).

Further work will be necessary to complete the structural analysis of the β-NGF mRNA and gene; once these structural studies are completed, it will then be possible to identify and delineate regulatory sequences within and around the β-NGF gene and to learn how the expression of the β-NGF gene is regulated during development, maturity, and aging.

## ACKNOWLEDGMENTS

This work was supported by Howard Hughes Medical Institute. I wish to thank Janice Canniff for excellent secretarial help.

## REFERENCES

1. James R. and Bradshaw R. A. (1984) Polypeptide growth factors. *Annu. Rev. Biochem.* **53,** 259–292.
2. Bradshaw R. A. (1978) Nerve growth factor. *Annu. Rev. Biochem.* **47,** 191–216.
3. Harper G. P. and Thoener H. (1980) Nerve growth factor: Biologic significance, measurement, and distribution. *J. Neurochem.* **34,** 5–16.
4. Levi-Montalcini R., and Hamburger V. (1951) Selective growth stimulating effects of mouse sarcoma on the sensory and sympathetic nervous system of the chick embryo. *J. Exp. Zool.* **116,** 323–361.
5. Cohen S., Levi-Montalcini R., and Hamburger V. (1954) A nerve growth-stimulating factor isolated from sarcomas 37 and 180. *Proc. Natl. Acad. Sci. USA* **40,** 1014–1018.
6. Rubin J. S. and Bradshaw R. A. (1981) Isolation and partial amino acid sequence analysis of nerve-growth factor from guinea pig prostate. *J. Neurosci. Res.* **6,** 451–64.
7. Hogue-Angeletti R. A., Frazier W. A., Jacobs J. W., Niall H. D., and Bradshaw R. A. (1976) Purification, characterization, and partial amino acid sequence of nerve growth factor from cobra venom. *Biochemistry* **15,** 26–34.
8. Scott J., Selby M., Urdea M., Quiraga M., Bell G. I., and Rutter W.

J. (1983) Isolation and nucleotide sequence of cDNA encoding the precursor of mouse nerve growth factor. *Nature* **302,** 538–541.

9. Ullrich A., Gray A., Berman C., and Dull T. J. (1983) Human β-nerve growth factor gene sequence highly homologous to that of mouse. *Nature* **303,** 821–825.
10. Roychoudhury R. and Wu R. (1980) Terminal transferase-catalyzed addition of nucleotides to the 3′ termini of DNA. *Methods Enzymol.* **65,** 43–62.
11. Hirose T., Crea R., and Itakura K. (1978) Rapid synthesis of trideoxyribonucleotide blocks. *Tetrahedron Letters* **28,** 2449–2452.
12. Grunstein M. and Hogness D. S. (1975) Colony hybridization: A method for the isolation of cloned cDNA that contains a specific gene. *Proc. Natl. Acad. Sci. USA* **72,** 3961–3965.
13. Smith M. (1983) Synthetic Oligodeoxyribonucleotides as Probes for Nucleic Acids and as Primers in Sequence Determination, in *Methods of DNA and RNA Sequencing* (Sherman M. and Weissman M. D., eds.), Praeger, New York.
14. Land H., Grez M., Hanser H., Lindenmeier W., and Schutz G. (1981) 5′-Terminal sequences of eukaryotic mRNA can be cloned with high efficiency. *Nucleic Acids Res.* **9,** 2251–2266.
15. Lawn R. M., Fritsch E. F., Parker R. C., Blake G., and Maniatis T. (1978) The isolation and characterization of linked γ- and β-globin genes from a cloned library of human DNA. *Cell* **15,** 1157–1174.
16. Benton W. D. and Davis R. W. (1977) Screening λgt recombinant clones by hybridization to single plaques *in situ*. *Science* **196,** 180–182.
17. Maxam A. and Gilbert W. (1980) Sequencing end-labeled DNA with base-specific chemical cleavages. *Methods Enzymol.* **65,** 499–560.
18. Darling T. L. J., Petrides P. E., Begium P., Frey P., Shooter E. M., Selby M., and Rutter W. J. (1983) The biosynthesis and processing of proteins in the mouse 7S nerve growth factor complex. *Cold Springs Harbor Symp. Quant. Biol.* **48,** 427–434.
19. Heinrich G., Kronenberg H. M., Potts J. T., Jr., and Habener J. F. (1984) Gene encoding parathyroid hormone: Nucleotide sequence of the rat gene and deduced amino acid sequence of rat preproparathyroid hormone. *J. Biol. Chem.* **259,** 3320–3329.
20. Kozak M. (1981) Possible role of flanking nucleotides in recognition of the AUG initiator codon by eukaryotic ribosomes. *Nucleic Acid Res.* **9,** 5233–5252.
21. Montminy M. R., Goodman R. H., Horovitch S. J., and Habener J. F. (1984) Primary structure of the gene encoding rat preprosomatostatin. *Proc. Natl. Acad. Sci. USA* **81,** 3337–3340.
22. Shen L. P. and Rutter W. J. (1984) Sequence of the human somatostatin I gene. *Science* **224,** 168–171.

23. Jadler I., Suda K., Schatz G., Kandewitz F., and Haid A. (1984) Sequencing of the nuclear gene for the yeast cytochrome $C_1$ precursor reveals an unusually complex amino-terminal pre-sequence. *EMBO J.* **3,** 2137–2143.
24. Cox D. R., Mason A. J., Evans B. A., Shine J., and Richards R. I. (1983) Assignment of the Murine Glandular Kallikrein Gene Family to Mouse Chromosome >7, in *Human Gene Mapping 7, Seventh International Workshop on Human Gene Mapping*. Basel, Karger, Los Angeles.
25. Zabel B. U., Naylor S. L., and Sakaguchi T. B. (1983) Gene Mapping by *In Situ* Hybridization, *Human Gene Mapping 7, Seventh International Workshop on Human Gene Mapping*. Basel, Karger., Los Angeles.

# *Chapter 15*

# Bombesin-Like Peptide Genes

ELIOT R. SPINDEL

## 1. INTRODUCTION

Bombesin, a tetradecapeptide, was first isolated from the skin of the frog *Bombina bombina* by Anastasi and coworkers in 1971 (*1*). Subsequently, related peptides were found in the skin of other amphibian species (*2–4*). After it was isolated, bombesin was found to have marked biological activity despite the fact that no mammalian counterpart was known. Synthetic amphibian bombesin released gastrointestinal (GI) hormones (*2, 5, 6*), stimulated gastric motility (*6*), and, when injected into the brain, produced hypothermia (*7*).

In 1976 and 1977, radioimmunoassays of synthetic amphibian bombesin were established, and bombesin immunoreactivity was found to be extensively distributed throughout the brain, lung, and GI tract of mammals and birds (*8–10*). The exact structure of mammalian bombesin remained unknown until 1979, when McDonald and coworkers isolated a 27-amino-acid, bombesin-like peptide from porcine gastric tissue (*11*). This peptide was isolated on the basis of its potent release of gastrin (*12*) and hence, was named gastrin-releasing peptide (GRP), a name that has stuck. Mammalian bombesin is also sometimes referred to as bombesin-27, and amphibian bombesin as bombesin-14. Here, mammalian bombesin shall be referred to as either mammalian bombesin or GRP, and amphibian bombesin as either amphibian bombesin or bombesin.

Bombesin and porcine GRP have the same carboxy-terminal heptapeptide sequences and share nine out of ten amino acids in their carboxy-terminal decapeptides. The amino terminus of GRP is unrelated to bombesin. When other GRPs were isolated from chicken proventriculus (*13*) and canine small intestine (*14*), the canine, porcine, and avian GRP were all quite homologous—especially in their carboxytermini, where they shared highest homology with bombesin. Not surprisingly, only the carboxy-terminus of GRP is required for biological activity (*15,16*). All biological actions of bombesin appear to be duplicated by GRP (*17–21*).

Mammalian bombesin immunoreactivity is present in normal lung (*10*), small-cell carcinoma of the lung, and carcinoid tumors of the lung (*22, 23*). Investigation of bombesin as a potential tumor marker and growth factor has been aided by the existence of tumor cell lines that actively synthesize bombesin (*24*).

Perhaps because mammalian bombesin is a relatively new peptide and its major physiologic roles are only now being defined, exploration of its molecular biology has just begun. In 1984, a human GRP-encoding mRNA was cloned from a pulmonary carcinoid tumor (*25*), and now the human GRP gene is undergoing its first analysis.

## 2. STRUCTURE OF BOMBESIN-LIKE PEPTIDES

### *2.1. Amphibian Bombesin-Like Peptides*

Bombesin-like peptides are divided into three families based on the peptide's carboxy-terminal amino acids (Fig. 1). The prototype is the bombesin family, which includes bombesin and alytesin, and is characterized by a Leu as the amino acid second from the carboxy-terminus. The ranatensin family is characterized by Phe as the amino acid second from the carboxy-terminus. The phyllolitorin family has a Ser as the third-to-last amino acid (Fig. 1) (*2–4*).

Each frog genus is characterized by its own family of bombesin-like peptides. For example, different *Rana* species have different ranatensins, and each of the Bombina species, *B. bombina*, *B. variegata*, and *B. orientalis*, contains bombesin (*1,2*). As the skin of more species of frogs is evaluated, it is likely that more bombesin-like peptides will be discovered (perhaps foreshadowing the discovery of more mammalian bombesin-like peptides). The concen-

A. Bombesin family

| | 1 ... 10 ... 14 |
|---|---|
| Bombesin: | pGlu-Gln-Arg-Leu-Gly-Asn-Gln-Trp-Ala-Val-Gly-His-Leu-Met.NH2 |
| Alytesin: | pGlu-Gly-Arg-Leu-Gly-Thr-Gln-Trp-Ala-Val-Gly-His-Leu-Met.NH2 |

B. Ranatensin family

| | 1 ... 10 |
|---|---|
| Ranatensin: | pGlu-Val-Pro-Gln-Trp-Ala-Val-Gly-His-Phe-Met.NH2 |
| Ranatensin-C: | pGlu-Thr-Pro-Gln-Trp-Ala-Thr-Gly-His-Phe-Met.NH2 |
| Ranatensin-R: | Ser-Asn-Thr-Ala-Leu-Arg-Arg-Tyr-Asn-Gln-Trp-Ala-Thr-Gly-His-Phe-Met.NH2 |
| Litorin: | pGlu-Gln-Trp-Ala-Val-Gly-His-Phe-Met.NH2 |
| Glu(OMe)-litorin: | pGlu-Glu(OMe)-Trp-Ala-Val-Gly-His-Phe-Met.NH2 |
| Glu(OEt)-litorin: | pGlu-Glu(OEt)-Trp-Ala-Val-Gly-His-Phe-Met.NH2 |

C. Phyllolitorin family

| | 1 ... 9 |
|---|---|
| $Leu^8$-phyllolitorin: | pGlu-Leu-Trp-Ala-Val-Gly-Ser-Leu-Met.NH2 |
| $Phe^8$-phyllolitorin: | pGlu-Leu-Trp-Ala-Val-Gly-Ser-Phe-Met.NH2 |

Fig. 1. Structures of amphibian bombesin-like peptides (for reviews, *see* references 2–4). pGlu represents pyroglutamate; $.NH_2$ indicates an amidated carboxy-terminal residue.

trations of peptides in frog skin are astronomical; levels greater than 2000 μg/g of skin are not unusual (*1–4*).

Within frog skin, the peptides are localized in cutaneous granular glands, which are sometimes referred to as poison glands (*26*). These glands are myoepithelial with alpha-adrenergic innervation (*27–30*). Injection of norepinephrine or alpha agonists into the dorsal lymph sac of the frog stimulates peptide secretion and can cause up to 95% depletion of skin peptide levels (*26,28*). High peptide concentration and regulation by neurotransmitters make frog skin a valuable model for the study of neuropeptide gene expression.

The bombesin family consists of bombesin and alytesin, isolated, respectively, from the European frogs *B. bombina* and *Alytes obstetricans* (*1*). Both bombesin and alytesin are tetradecapeptides with Leu in position 13 (Fig. 1). As is characteristic for all the bombesin-like peptides (mammalian and amphibian), the carboxy-terminal amino acid of bombesin and alytesin is an amidated Met. The carboxy-terminal hexapeptide, Trp-Ala-Val-Gly-His-Leu-Met.$NH_2$, with minor substitution, characterizes all bombesin-like peptides. Alytesin and bombesin are immunologically cross-reactive and share biological actions (*3,16*).

The ranatensin subfamily has six identified members (Fig. 1). The prototype member, ranatensin, isolated from the common North American leopard frog, *Rana pipiens*, is an undecapeptide with Phe as the amino acid second from the carboxy-terminus (*31*). Levels of ranatensin in the skin of *R. pipiens* are approximately 100 μg/g of skin (*32*). Ranatensin is also found in the skin of the American bullfrog, *R. catesbiana*, but at lower levels. Ranatensin is roughly as potent as bombesin in stimulating smooth muscle contractions (*2, 33*), but is less potent in stimulating GI hormone release (*2*). Litorin and its derivatives have a similar spectrum of biological activities. Ranatensin C and R have a Thr substituted for Val at position 5 from the carboxy-terminus and have less biological activity than ranatensin (*2–4*).

The phyllolitorins, isolated from the South American frog, *Phyllomedusa sauvagei*, have a Ser instead of His as the amino acid third from the carboxy-terminus. They occur in both a bombesin-like "Leu" form and in a ranatensin-like "Phe" form (*3, 4*).

The structure of amphibian bombesin-like peptides is predictive of the structure of mammalian bombesin-like peptides. The continued characterization of new amphibian peptides is probably limited only by the enthusiasm of comparative endocrinologists. The techniques of molecular biology will undoubtedly increase the rate of frog skin peptide characterization.

## 2.2. Mammalian Bombesin-Like Peptides

At present, only two families of mammalian bombesin-like peptides have been characterized; the bombesin-like family composed of gastrin-releasing peptides (GRP), and the ranatensin-like family composed of neuromedin B. Although there have been extensive studies on GRP, little is known about neuromedin B.

### 2.2.1. Gastrin-Releasing Peptide

Chromatographic analysis of mammalian tissue reveals two forms of bombesin immunoreactivity, one larger than amphibian bombesin and one of roughly equivalent size. The larger form was identified as GRP by McDonald et al. in porcine nonantral tissue (Fig. 2). GRP was also isolated from avian proventriculus (*13*) and canine small intestine (*14*) (Fig 2). The structure of human GRP was determined by molecular cloning (*25*) and conventional isolation (*34*) (Fig. 2).

All four GRPs are highly homologous (Fig. 2). The three mammalian GRPs share an identical carboxy-terminal 15 amino acids, and avian GRP differs from them by only two amino acids. All four GRPs have the same carboxy-terminal heptapeptide that bombesin has. The four GRPs also show homology in their amino-terminus. This suggests a role for the amino-terminus, though a significant difference between the bioactivity of bombesin and GRP has yet to be demonstrated.

GRP-10 (Fig. 2) was isolated and sequenced from porcine spinal cord (*35*), canine small intestine (*14*), and human pulmonary carcinoid tumor (*34*). Such consistent isolation suggests that GRP-10 is not an extraction artifact, but an authentic peptide. GRP-10 is probably the small form of mammalian bombesin immunoreactivity that elutes close to amphibian bombesin. In human tissue, this small peak of immunoreactivity was demonstrated by reverse phase high-pressure liquid chromatography to elute exactly with GRP-10 (*36*).

### 2.2.2. Mammalian Ranatensin

At present, the only ranatensin-like peptide sequenced from mammalian tissue is neuromedin B, which was isolated from porcine spinal cord by Minamino et al. (*24*) (Fig. 2). Neuromedin B

A. Gastrin-releasing Peptides

| | 1 | | | | | | | | | 10 | | | 13 |
|---|---|---|---|---|---|---|---|---|---|---|---|---|---|
| Human GRP: | Val- | Pro- | Leu- | Pro- | Ala- | Gly- | Gly- | Gly- | Thr- | Val- | Leu- | Thr- | Lys- |
| Porcine GRP: | Ala- | Pro- | Val- | Ser- | Val- | Gly- | Gly- | Gly- | Thr- | Val- | Leu- | Ala- | Lys- |
| Canine GRP: | Ala- | Pro- | Val- | Pro- | Gly- | Gly- | Gln- | Gly- | Thr- | Val- | Leu- | Asp- | Lys- |
| Avian GRP: | Ala- | Pro- | Leu- | Gln- | Pro- | Gly- | Gly- | Ser- | Pro- | Ala- | Leu- | Thr- | Lys- |

| | | 14 | | | 17 | | | 20 | | | | | | | 27 |
|---|---|---|---|---|---|---|---|---|---|---|---|---|---|---|---|
| Human GRP: | | Met- | Tyr- | Pro- | Arg- | Gly- | Asn- | His- | Trp- | Ala- | Val- | Gly- | His- | Leu- | Met.NH2 |
| Porcine GRP: | | Met- | Tyr- | Pro- | Arg- | Gly- | Asn- | His- | Trp- | Ala- | Val- | Gly- | His- | Leu- | Met.NH2 |
| Canine GRP: | | Met- | Tyr- | Pro- | Arg- | Gly- | Asn- | His- | Trp- | Ala- | Val- | Gly- | His- | Leu- | Met.NH2 |
| Avian GRP: | | Ile- | Tyr- | Pro- | Arg- | Gly- | Ser- | His- | Trp- | Ala- | Val- | Gly- | His- | Leu- | Met.NH2 |
| GRP-10: | | | | | | Gly- | Asn- | His- | Trp- | Ala- | Val- | Gly- | His- | Leu- | Met.NH2 |
| Bombesin: | pGlu- | Gln- | Arg- | Leu- | | Gly- | Asn- | Gln- | Trp- | Ala- | Val- | Gly- | His- | Leu- | Met.NH2 |

B. Ranatensin-like Peptides

| | 1 | | | | | | | | | 10 |
|---|---|---|---|---|---|---|---|---|---|---|
| Porcine Neuromedin B: | Gly- | Asn- | Leu- | Trp- | Ala- | Thr- | Gly- | His- | Phe- | Met.NH2 |
| Ranatensin C: | pGlu- | Thr- | Pro- | Gln- | Trp- | Ala- | Thr- | Gly- | His- | Phe-Met.NH2 |

Fig. 2. Structures of mammalian bombesin-like peptides. GRP-10 was isolated from canine (*14*), porcine (*35*), and human tissue (*34*). Neuromedin B was isolated from porcine spinal cord. The structures of amphibian bombesin and ranatensin are given for reference. $.NH_2$ indicates an amidated carboxy-terminal residue.

has the same carboxy-terminal heptapeptide as ranatensin R and C. In rat brain, ranatensin-like immunoreactivity distinct from bombesin-like immunoreactivity can be detected by immunohistochemistry and radioimmunoassay (*37, 38*). Ranatensin can also be detected in the brain of several frog species, including *Xenopus laevis*, a species that has no bombesin or ranatensin in its skin (*32*). A big form of neuromedin B has also recently been characterized in porcine brain and spinal cord (*88*).

### 2.3. Conclusions

GRP, the mammalian equivalent of bombesin, has been isolated from several species and shows high sequence conservation.

Only one ranatensin-like mammalian peptide has so far been characterized, but radioimmunoassay studies suggest more to come. Although sequence homology suggests that mammalian GRP evolved from amphibian bombesin, the finding in frog stomachs of a large form of bombesin immunoreactivity that is consistent in size with GRP (*32,39,40*) suggests that GRP may have evolved from bombesin before the vertebrate radiation. If so, the presence of high levels of bombesin in the skin of some frog species is unrelated to the presence of GRP as a neuromodulator in the stomach of frogs. Conversely, a species of frog might be found that has GRP as its skin peptide. The exact evolution of bombesin into GRP will be clarified when the genes encoding amphibian bombesins and amphibian GRPs are cloned.

## 3. DISTRIBUTION AND PHYSIOLOGY

Gastrin-releasing peptide is found in the GI tract, the pulmonary tract, and the nervous system. Ranatensin-like peptides have been localized only in the CNS. Although GRP has many effects, its major roles are only now being defined. Nothing yet is known about the function of mammalian ranatensin.

Whereas the structure of amphibian peptides is helpful in predicting the structure of mammalian bombesin-like peptides, the physiology of frog skin is not germane to mammals. The role of peptides in frog skin is unknown, but, possibly, they serve as defensive toxins or as regulators of electrolyte and oxygen balance.

### *3.1. Gastrin-Releasing Peptide*

#### *3.1.1. GRP in the Pulmonary Tract*

GRP is localized in the pulmonary neuroendocrine cells (Kultschitzky, or K, cells) of the lung (*10, 41*) and its distribution follows the distribution of those cells. Levels are highest in bronchioles and neuroepithelial bodies and are lowest in the peripheral lung parenchyma (*41–43*). Levels of GRP in lung correlate with the number of pulmonary neuroendocrine cells; therefore, levels of GRP are high at birth, highest just after birth (*42,44*), and slowly decline thereafter. GRP is elevated in conditions associated with K-cell hyperplasia, including adult-respiratory-distress syndrome (*43*) and bronchopulmonary dysplasia stemming from hyaline-membrane disease (*42*). Chromatography of lung extracts shows the presence of both GRP and GRP-10 (*45,46*).

The presence of GRP in the lung during development suggests that GRP may be a pulmonary-growth factor. In tissue culture, GRP stimulates clonal growth of pulmonary fibroblasts and colony formation in soft agar (*47*). A monoclonal antibody to the bombesin receptor inhibits clonal growth of lung small-cell carcinoma cell lines in vitro and causes tumor regression in nude mice in vivo (*48*).

Much of what is known about the molecular biology of GRP has come from the study of GRP expression in pulmonary neoplasms. Pulmonary carcinoids and small-cell carcinomas contain the highest levels of GRP (*22,23,49,50*), but squamous, and large-cell carcinomas, and adeno-carcinomas often have low but detectable levels of GRP (*49,50*). The expression of GRP in these neoplasms is not truly ectopic because the cells of origin of some of these tumors normally make GRP. As is consistent with a neural crest origin of GRP-producing cells (*10,43*), thyromedullary carcinomas often make GRP as well as calcitonin (*51*).

### *3.1.2. GRP in the GI Tract*

Within the GI tract, levels of GRP are highest in the gastric fundus (*9,52*), reaching, in humans, levels of approximately 15 pg/mg tissue (*52*); levels in the antrum are about 50% lower. Levels in the muscular layers of the stomach are about twice as high as in the mucosa (*9,52*). GRP levels in the pancreas and small intestine are roughly the same as in the antrum; levels in the large intestine are somewhat lower (*9,52,53*). Most of the GRP in the GI tract appears to be located in neuronal elements (*54*), although there is some controversy over whether GRP also occurs in GI endocrine cells.

Within the stomach, bombesin-like immunoreactivity occurs primarily in nerve fibers in the myenteric and submucosal plexus. In the large intestine, the immunoreactivity is mostly in fibers of the myenteric plexus. In the rat, some of the GRP-containing neurons of the GI tract project to the celiac-superior and inferior mesenteric ganglia (*54–56*).

In the GI tract, GRP is a pan-releasor of GI hormones and a stimulant of GI functions. In both humans (*19*) and dogs (*17,18*), infusion of nanogram amounts of GRP increases serum levels of gastrin, insulin, glucagon, pancreatic polypeptide, and gastric inhibitory peptide. Stomach gastric acid increases 15 min after the gastrin increase. These effects are independent of vagal integrity (*57*) and can be duplicated by either GRP or bombesin (*58, 59*). Gastric motility is also stimulated by GRP (*60*).

GRP stimulates the release of pancreatic enzymes, increasing secretion of pancreatic lipase and trypsin, as well as bicarbonate, into the small intestine (*61*). Bombesin stimulates anion secretion in the ileum, and gut motility in the colon (*21*). The diversity and potency of the actions of GRP, coupled with its distribution in GI neurons, make it an important peptidergic modulator of GI function. GRP probably acts as an autocrine hormone at its site of release; the half-life of circulating GRP is 2–3 min (*19*).

#### *3.1.3. GRP in the Nervous System*

GRP is found in both the peripheral and the central nervous system. It is distributed throughout the brain with highest levels in the ventral medial hypothalamus and high levels throughout the forebrain, hippocampus, midbrain, and brain stem (*62–65*). High levels also occur in the nucleus of the solitary tract and the trigeminal tract (*63–65*). GRP is present in paravertebral sympathetic ganglia (*66,67*) and also in dorsal-root ganglia (*68,69*), from which it projects to the superficial layers of the posterior horns of the spinal cord (*68,70*). In some dorsal-root ganglion cells, GRP may coexist with substance P (*69*).

GRP has multiple actions within the CNS. Bombesin is a potent hypothermic agent when injected into the cerebral ventricles (*7*). Intraventricular bombesin also causes hyperglycemia and increases blood pressure through sympathoadrenal mediation (*71,72*). Paradoxically, intraventricular bombesin inhibits gastric acid secretion, in opposition to its peripheral effect (*73*). GRP stimulates the pituitary release of prolactin and growth hormone. As little as 30 ng of bombesin, injected iv, is sufficient to increase serum prolactin in the rat (*16*). GRP may also act to potentiate CRH-induced ACTH release (*74*). Ranatensin either does not have these central effects or has them at reduced potency (*16*).

### *3.2. Mammalian Ranatensin*

At present, all that is known about mammalian ranatensin is the sequence of one peptide, neuromedin B (*24*), and data on the distribution of ranatensin-like immunoreactivity in rat brain

(*37,38,75,89*). In rat brain, ranatensin immunoreactivity is detected in the hypothalamus, amygdala, septum, pineal, olfactory tract, and the neurointermediate lobe of the pituitary (*38,89*). In the dorsal pontine tegmentum, a high concentration of ranatensin-containing cell bodies projects to the hippocampus (*75*). The role of mammalian ranatensin is not known, but, by analogy with other neuropeptides, ranatensin is undoubtedly a neurotransmitter or neuromodulator. Synthetic oligonucleotides complementary to porcine neuromedin B and amphibian ranatensin will help to identify and map other mammalian ranatensin-like peptides.

## 4. MOLECULAR BIOLOGY

### 4.1. Gastrin-Releasing Peptide

At the time of the writing of this chapter, human GRP (hGRP) is the only mammalian bombesin-like peptide whose cloning has been reported (*25*). The hGRP-encoding mRNA was cloned and sequenced from a pulmonary carcinoid tumor (*76*), rich in bombesin-like immunoreactivity, through the use of mixed synthetic oligonucleotides complementary to amphibian bombesin (Fig. 3).

The deduced amino acid sequence showed that hGRP is part of a 148-amino-acid prohormone, beginning with an AUG start codon and terminating with three successive stop codons. The amino-terminal portion of the precursor contains a cluster of hydrophobic amino acids consistent with a signal sequence. GRP itself immediately follows the signal sequence. Cleavage of the signal peptide occurs at the consensus signal-peptidase recognition site (*77*), as confirmed by the partial sequence of human GRP published by Orloff et al. (*34*).

Human GRP is flanked by two basic amino acids at its carboxy-terminus. A Gly follows the terminal Met as the source for the carboxy-terminal amide (*78*). The protein-coding region is followed by 289 bases before the poly(A) tail. There is a typical AAATAA polyadenylation signal (*79*). Within the carboxy-terminal extension peptide, there are no obvious cryptic neuropeptides set off by dibasic amino acids. A search of a computer data base did not reveal any homologies with other known peptides.

```
    5' ....  AGTCTCTGCTCTTCCCAGCCTCTCCGGCGCGCTCCAAGGGCTTCCCGTCGGGACC ATG CGC GGC  64
                                                                   Met Arg Gly
                                                                   -23

CGT GAG CTC CCG CTG GTC CTG CTG GCG CTG GTC CTC TGC CTG GCG CCC CGG GGG CGA GCG ↓ 124
Arg Glu Leu Pro Leu Val Leu Leu Ala Leu Val Leu Cys Leu Ala Pro Arg Gly Arg Ala
                                                                            -1

GTC CCG CTG CCT GCG GGC GGA GGG ACC GTG CTG ACC AAG ATG TAC CCG CGC GGC AAC CAC  184
Val Pro Leu Pro Ala Gly Gly Gly Thr Val Leu Thr Lys Met Tyr Pro Arg Gly Asn His
+1                     Gastrin-Releasing Peptide

TGG GCG GTG GGG CAC TTA ATG GGG AAA AAG AGC ACA GGG GAG TCT TCT TCT GTT TCT GAG  244
Trp Ala Val Gly His Leu Met Gly Lys Lys Ser Thr Gly Glu Ser Ser Ser Val Ser Glu
                        +27

AGA GGG AGC CTG AAG CAG CAG CTG AGA GAG TAC ATC AGG TGG GAA GAA GCT GCA AGG AAT  304
Arg Gly Ser Leu Lys Gln Gln Leu Arg Glu Tyr Ile Arg Trp Glu Glu Ala Ala Arg Asn

TTG CTG GGT CTC ATA GAA GCA AAG GAG AAC AGA AAC CAC CAG CCA CCT CAA CCC AAG GCC  364
Leu Leu Gly Leu Ile Glu Ala Lys Glu Asn Arg Asn His Gln Pro Pro Gln Pro Lys Ala

CTG GGC AAT CAG CAG CCT TCG TGG GAT TCA GAG GAT AGC AGC AAC TTC AAA GAT GTA GGT  424
Leu Gly Asn Gln Gln Pro Ser Trp Asp Ser Glu Asp Ser Ser Asn Phe Lys Asp Val Gly

TCA AAA GGC AAA GTT GGT AGA CTC TCT GCT CCA GGT TCT CAA CGT GAA GGA AGG AAC CCC  484
Ser Lys Gly Lys Val Gly Arg Leu Ser Ala Pro Gly Ser Gln Arg Glu Gly Arg Asn Pro

CAG CTG AAC CAG CAA  TGA  TAA  TGA  TGGCCTCTCTCAAAAGAGAAAAACAAAACCCCTAAGAGACTGC  551
Gln Leu Asn Gln Gln  Stop Stop Stop

GTTCTGCAAGCATCAGTTCTACGGATCATCAACAAGATTTCCTTGTGCAAAATATTTGACTATTCTGTATCTTTCATCC  630

TTGACTAAATTCGTGATTTTCAAGCAGCATCTTCTGGTTTAAACTTGTTTGCTGTGAACAATTGTCGAAAAGAGTCTTC  709

CAATTAATGCTTTTTTATATCTAGGCTACCTGTTGGTTAGATTCAAGGCCCCGAGCTGTTACCATTCACAATAAAAGCT  788

TAAACACAT .... Poly(A)        797
```

Fig. 3. Nucleotide and amino acid sequence of the mRNA encoding preproGRP deduced from cDNA sequences. The signal peptide is numbered −23 to −1. The signal peptide cleavage point is marked by an arrow. Glycine immediately follows GRP and is the source for the carboxy-terminal amide. Two Lys indicate +6 cleavage site. There are three consecutive stop codons at bases 500–508. The polyadenylation signal at bases 779–784 is underlined. The sequence shown here is corrected from the previously published (*25*) at bases 65, 106, 456, 551, and 640.

Human GRP contains two internal tryptic cleavage points that could generate hGRP-(*14–27*) or hGRP-(*18–27*). Protein isolation and HPLC show that hGRP-(*18–27*) is formed in vivo and is the peptide designated GRP-10 (*14,34–36*) (Fig. 2). Probably, GRP-10 is formed by alternate proteolytic processing of the same prohormone that forms GRP-27, similar to the mechanism of formation of somatostatin-28 and -14. The smaller GRP-like peptide could be encoded in a second mRNA, but there are no data to support that hypothesis.

Southern blot analysis of human genomic DNA, using both 5′- and 3′-directed probes, shows a simple pattern consistent with the presence of only a single GRP-encoding gene (Fig. 4). The probability of a single gene and the lack of major differences in the restriction analyses of other recombinant clones suggests that GRP-10 is not encoded in a second mRNA. Cytogenetic analysis localizes the GRP gene to human chromosome 18 (*90*).

Blot analysis of mRNA from the pulmonary carcinoid from which GRP was cloned shows a major mRNA of approximately 950 bases and a smaller mRNA of approximately 875 bases (Fig. 5). Under higher resolution, each band appears to be composed of several bands. The smaller GRP-encoding mRNA (Fig. 4) lacks part of the 3′-untranslated region, as determined by the use of synthetic oligonucleotide probes. Studies are now underway to

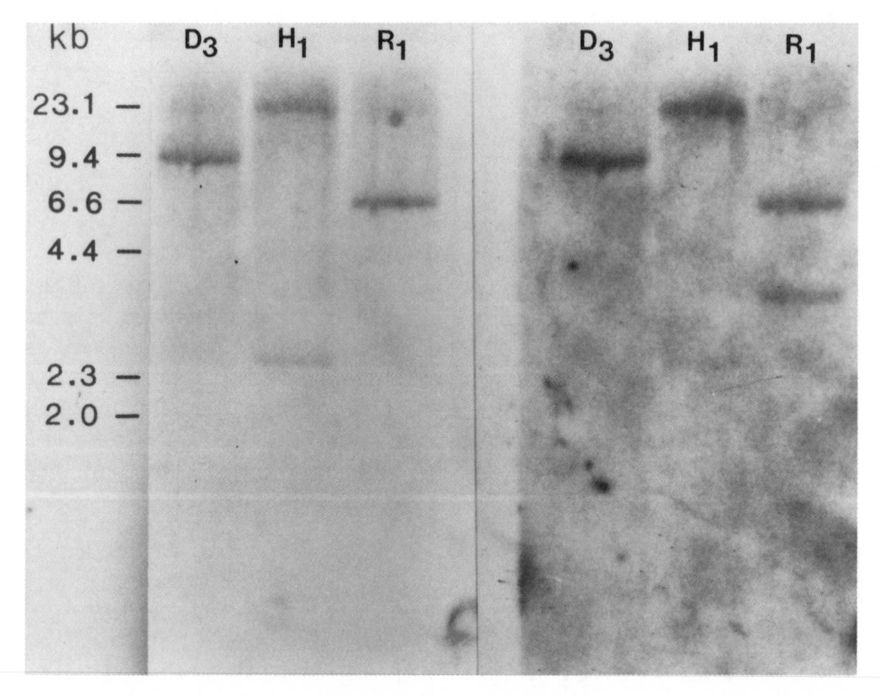


Fig. 4. Human genomic DNA digested with *Hin*dIII (D3), *Bam*HI (H1), or *Eco*RI (R1). The first three lanes were probed with a GRP-encoding cDNA extending from the 5′-untranslated region to base 400 of the GRP mRNA. The second three lanes were probed with a GRP-encoding cDNA extending from base 180 to the poly(A) tail of the GRP mRNA.

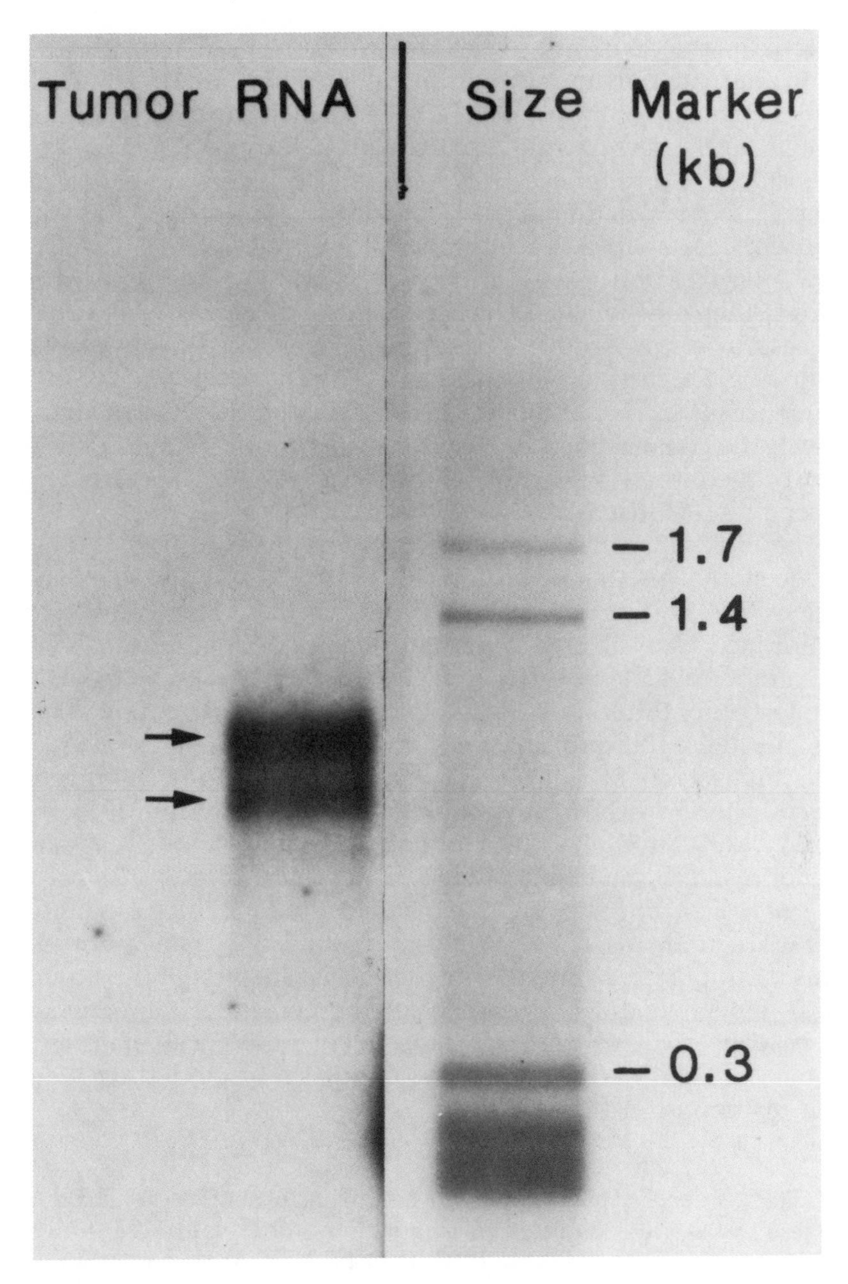


Fig. 5. RNA hybridization analysis. Five micrograms of poly(A)+ RNA from a pulmonary carcinoid tumor was fractionated on 1.5% agarose and probed with a near-full-length GRP-encoding cDNA.

determine the exact difference. The variation could arise from use of an alternate polyadenylation signal, but there are no other consensus polyadenylation signals upstream. Another possibility would be the use of an alternate exon splice.

Besides the difference in the 3′-untranslated region, there is another, subtler variation in the GRP-encoding mRNA (*80*). Restriction enzyme analysis of multiple GRP-encoding cDNA reveals two internal *Pvu*II pieces of different sizes (Fig. 6). Sequencing shows that the two *Pvu*II fragments are identical, except for a 19-base insert in the larger piece (Fig. 7A). The 19-base insert is bound by GT and AG, the mandatory intron donor, and acceptor sequences (*81*). Five of the first six bases of the 5′ end of the insert match the consensus sequence for intron donor sites (*81*); the 3′ end of the insert ends with AG, but otherwise deviates from the acceptor-site consensus.

The 19-base insert, because it is not a multiple of three, causes a frame shift. This frame shift causes a stop codon to be bypassed. The mRNA without the 19-base insert reaches a stop codon after amino acid 138. The mRNA with the insert does not terminate until amino acid 148. This creates two forms of the GRP prohormone, differing by 1 kdalton (Fig. 7B). In RNA blot analyses, all detectable precursor RNA forms contain the insert (*80*). This suggests that the insert is spliced out as the message is processed. The 19-base insert could be an unspliced mini-intron or it could be the product of an alternate intron splice. Such a splice could result from using either an alternate donor (*82*) or an alternate acceptor site (*83, 84*). The strong donor-site consensus sequence in the insert suggests the use of the alternate donor site. This is supported by preliminary sequence analysis of the human GRP gene, which shows an intron immediately following the 19-base insert. Thus the insert most likely arises by an alternative donor site splice with the second donor site provided by the 5′ end of the intron (*91*).

Fig. 6. (*on opposite page*) Restriction digest of seven GRP-encoding clones. DNAs were digested to completion with *Pvu*II and *Pst*I. The cDNAs were inserted into the *Pst*I site of pBR322. Arrows point to the two different sizes of internal *Pvu*II fragments present in the cDNA inserts; the other low-molecular-weight bands are the *Pst*–*Pvu*II fragments of each insert. The higher-molecular-weight bands are from pBR322.

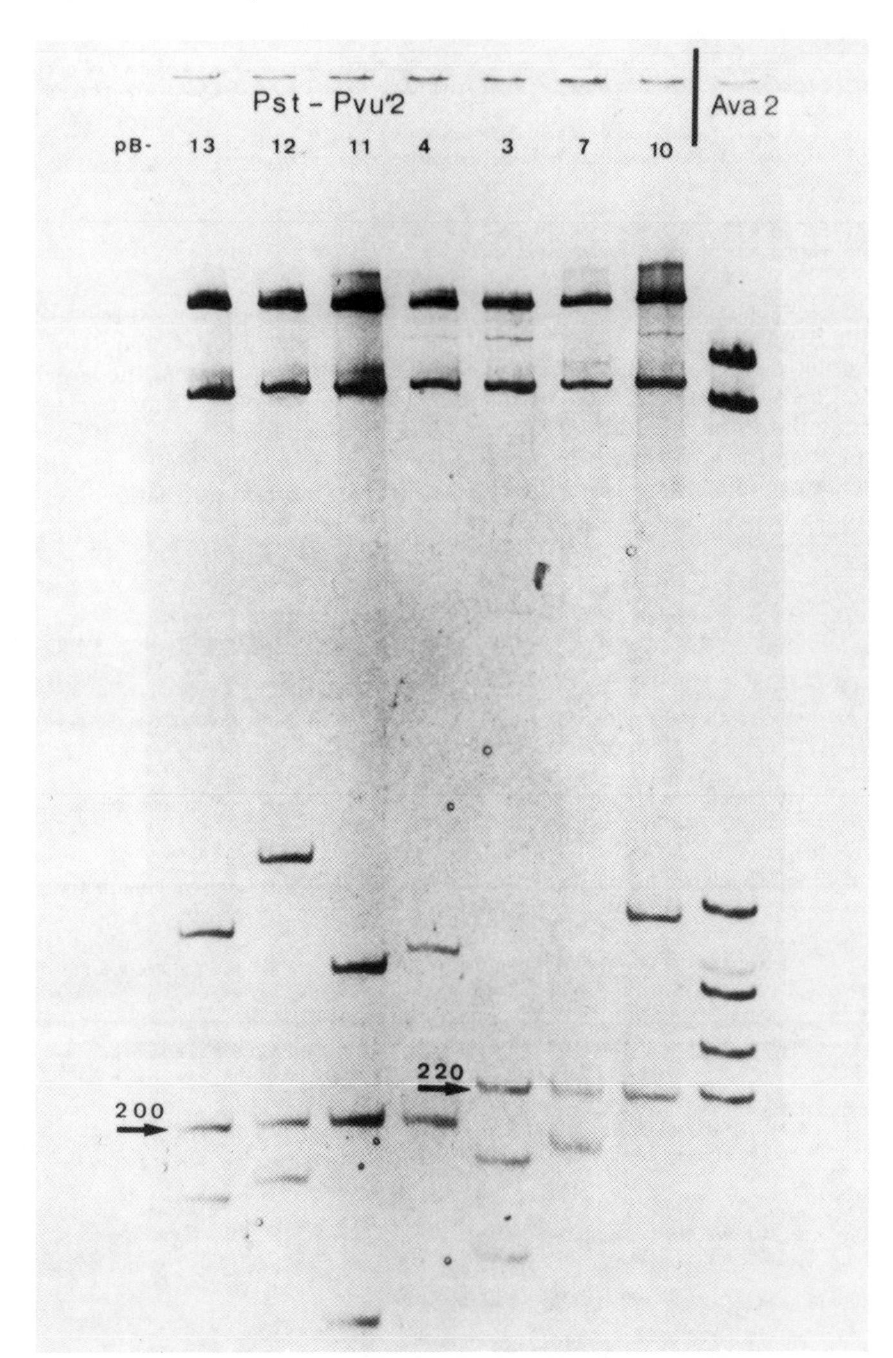

Pst - Pvu 2
Ava 2
pB-
13
12
11
4
3
7
10
220
200

```
     263
 I:  CAGCTGAGAG AGTACATCAG GTGGGAAGAA GCTGCAAGGA ATTTGCTGGG TCTCATAGAA GCAAAGGAGA ACAGAAACCA CCAGCCACCT
II:  CAGCTGAGAG AGTACATCAG GTGGGAAGAA GCTGCAAGGA ATTTGCTGGG TCTCATAGAA GCAAAGGAGA ACAGAAACCA CCAGCCACCT
     353
 I:  CAACCCAAGG CCTTGGGCAA TCAGCAGCCT TCGTGGGATT CAGAGGATAG CAGCAACTTC AAAGAT.....................TTGGT
II:  CAACCCAAGG CCTTGGGCAA TCAGCAGCCT TCGTGGGATT CAGAGGATAG CAGCAACTTC AAAGATGTAG GTTCAAAAGG CAAAGTTGGT
                                                                         |---- 19 bases -----|
     443                                      485
 I:  AGACTCTCTG CTCCAGGTTC TCAACGTGAA GGAAGGAACC CCC
II:  AGACTCTCTG CTCCAGGTTC TCAACGTGAA GGAAGGAACC CCC
```

Fig. 7. Origin of two forms of preproGRP. (A) DNA sequence of the two internal *Pvu*II pieces of different sizes indicated in Fig. 6. Numbering corresponds to Fig. 3, which includes the sequence of the larger fragment. (B) Translation of the two forms of preproGRP that result from the 19-base insertion. PreproGRP II includes the longer *Pvu*II fragment and is also shown in Fig. 3. PreproGRP I has the *Pvu*II fragment without the 19-base insert; this causes a frame shift and translational termination 10 amino acids before preproGRP II.

```
       1
 I:    Met-Arg-Gly-Arg-Glu-Leu-Pro-Leu-Val-Leu-Leu-Ala-Leu-Val-Leu-Cys-Leu-Ala-Pro-Arg-
II:     "   "   "   "   "   "   "   "   "   "   "   "   "   "   "   "   "   "   "   "

       21
 I:    Gly-Arg-Ala-Val-Pro-Leu-Pro-Ala-Gly-Gly-Gly-Thr-Val-Leu-Thr-Lys-Met-Tyr-Pro-Arg-
II:     "   "   "   "   "   "   "   "   "   "   "   "   "   "   "   "   "   "   "   "

       41
 I:    Gly-Asn-His-Trp-Ala-Val-Gly-His-Leu-Met-Gly-Lys-Lys-Ser-Thr-Gly-Glu-Ser-Ser-Ser-
II:     "   "   "   "   "   "   "   "   "   "   "   "   "   "   "   "   "   "   "   "

       61
 I:    Val-Ser-Glu-Arg-Gly-Ser-Leu-Lys-Gln-Gln-Leu-Arg-Glu-Tyr-Ile-Arg-Trp-Glu-Glu-Ala-
II:     "   "   "   "   "   "   "   "   "   "   "   "   "   "   "   "   "   "   "   "

       81
 I:    Ala-Arg-Asn-Leu-Leu-Gly-Leu-Ile-Glu-Ala-Lys-Glu-Asn-Arg-Asn-His-Gln-Pro-Pro-Gln-
II:     "   "   "   "   "   "   "   "   "   "   "   "   "   "   "   "   "   "   "   "

       101
 I:    Pro-Lys-Ala-Leu-Gly-Asn-Gln-Gln-Pro-Ser-Trp-Asp-Ser-Glu-Asp-Ser-Ser-Asn-Phe-Lys-
II:     "   "   "   "   "   "   "   "   "   "   "   "   "   "   "   "   "   "   "   "

       121                                                                 138
 I:    Asp-Leu-Val-Asp-Ser-Leu-Leu-Gln-Val-Leu-Asn-Val-Lys-Glu-Gly-Thr-Pro-Ser-STOP
II:     "  -Val-Gly-Ser-Lys-Gly-Lys-Val-Gly-Arg-Leu-Ser-Ala-Pro-Gly-Ser-Gln-Arg-Glu-Gly-

       141                         148
II:    Arg-Asn-Pro-Gln-Leu-Asn-Gln-Gln STOP             Mr(I)  = 15 kd
                                                        Mr(II) = 16 kd
```

Fig. 7B.

The extent to which these GRP prohormones are expressed in normal tissue remains to be established. Small-cell-carcinoma lines appear to express both forms, as determined by RNA blot analysis, using probes specific for the presence or absence of the 19-base insert (*80*). There are, as yet, little data on GRP message expression in normal tissue. RNA blot analyses of normal neonatal lung, normal adult lung, small-cell-carcinoma cell lines, and GI tissues appear to show similar patterns of GRP mRNAs. In human fetal lung, levels of GRP mRNAs are markedly elevated from 15 to 30 wk of gestation. S1 nuclease mapping shows that fetal lung contains the same alternately spliced GRP mRNAs as does neoplastic lung (*92*).

### 4.2. Amphibian Bombesin-Like Peptides

Little work has yet been done on the molecular biology of frog skin peptides, despite the high message levels and abundant tissue. Kreil and coworkers cloned mRNA encoding the *Xenopus laevis* skin peptides TRH (*85*) and cerulein (*86*), but little has been done with the *Bombina* and *Rana* species. A partial cDNA encoding the mRNA for ranatensin was cloned from *Rana pipiens* and used to study the noradrenergic regulation of peptide-encoding mRNA in frog skin (*87*). Unpublished studies by Spindel show that ranatensin is encoded in a typical precursor; the peptide sequence is set off by dibasic amino acids, a Glu codes for the amino-terminal pyroglutamate, and a Gly provides the amide for the carboxy-terminal methioninamide. RNA blot analysis shows that the message is approximately 600 bases. A full-length cDNA has not yet been sequenced, nor have genomic blot analyses been done.

The finding of GRP in frogs, as well as mammals (*32,39,40*), suggests that GRP developed before the vertebrate radiation, which would make the bombesin family quite old. Consistent with this theory, dot-matrix analysis shows little homology between the frog ranatensin mRNA and the human GRP mRNA. Because both peptides belong to the bombesin family, the only regions of strong homology are those in which the amino acids are the same.

## 5. SUMMARY AND PERSPECTIVES

The physiologic roles of mammalian bombesin-like peptides are just now being established, as is the molecular biology of these peptides.

Evidence suggests that GRP may be an important pulmonary growth factor. Further knowledge of the molecular biology of GRP may lead to advances in the understanding of lung cancer. GRP or its receptor may well be related to as-yet-undiscovered oncogenes. The potency of GRP as a releasor of GI hormones has been established for several years, although there is no overall framework for GRP's GI functions to date. Detailing the regulation of GRP gene expression in GI tissue under varioius physiologic stimuli will help to clarify the functions of GRP. The extent to which alterations in GRP expression underlies GI disorders, such as peptic ulcer disease, remains to be established. If GRP is a pulmonary-growth factor, it might also be a GI-growth factor.

Perhaps the biggest challenge lies in determining the role of GRP and ranatensin-like peptides in brain. The potent effects of GRP in brain suggest that these roles will be important. Correlation of message levels with physiologic states will be informative. As the sophistication of molecular biology advances, studies that turn GRP expression on and off within specific brain regions will be of great interest.

## ACKNOWLEDGMENTS

This work was supported by the Howard Hughes Medical Institute and by USPHS grant 1R23-CA-39237-01 from the National Cancer Institute.

## REFERENCES

1. Anastasi A., Erspamer V., and Bucci M. (1971) Isolation and structure of bombesin and alytesin, two analogous active peptides from the skin of the European amphibians. *Bombina* and *Alytes*. *Experientia* **27,** 166–167.
2. Erspamer V., Melchiorri P., Falconieri-Erspamer G., and Negril L. (1978) Polypeptides of the amphibian skin active on the gut and their mammalian counterparts. *Adv. Exp. Med. Biol.* **106,** 51–64.

3. Erspamer V., Falconieri-Erspamer G., Massanti G., and Endean R. (1984) Active peptides in the skins of one hundred amphibian species from Australia and Papua New Guinea. *Comp. Biochem. Physiol.* **77C,** 99–108.
4. Nakajima T. (1981) Active peptides in amphibian skin. *Trends Pharmac. Sci.* **2,** 202–206.
5. Bertaccini G., Erspamer V., Melchiorri P., and Sopranzi N. (1974) Gastrin release by bombesin in the dog. *Br. J. Pharmacol.* **52,** 219–225.
6. Erspamer V. and Melchiorri P. (1975) Actions of Bombesin on Secretions and Motility of the Gastrointestinal Tract, in *Gastrointestinal Hormones: A Symposium,* (Thompson J. C., ed.) University of Texas, Austin.
7. Brown M., Rivier J., and Vale W. (1977) Bombesin: Potent effects on thermoregulation in the rat. *Science* **196,** 998–1000.
8. Brown M., Allen R., Villarreal J., Rivier J., and Vale W. (1978) Bombesin-like activity: Radioimmunologic assessment in biological tissues. *Life Sci.* **23,** 2721–2728.
9. Walsh J. H., Wong H. C., and Dockray G. J. (1979) Bombesin-like peptides in mammals. *Fed. Proc.* **38,** 2315–2319.
10. Wharton J., Polak J. M., Bloom S. R., Ghatei M. A., Solcia E., Brown M. R., and Pearse A. G. E. (1978) Bombesin-like immunoreactivity in the lung. *Nature* **273,** 769–770.
11. McDonald T. J., Jornvall H., Nilsson G., Vagne M., Ghatei M., Bloom S. R., and Mutt V. (1979) Characterization of a gastrin releasing peptide from porcine non-antral gastric tissue. *Biochem. Biophys. Res. Commun.* **90,** 227–233.
12. McDonald T. J., Nilsson G., Vagne M., Ghatei M., Bloom S. R., and Mutt V. (1978) A gastrin releasing peptide from the porcine nonantral gastric tissue. *Gut* **19,** 767–774.
13. McDonald T. J., Jornvall H., Ghatei M., Bloom S. R., and Mutt V. (1980) Characterization of an avian gastric (proventricular) peptide having sequence homology with the porcine gastrin-releasing peptide and the amphibian peptides bombesin and alytensin. *FEBS Lett* **122,** 45–48.
14. Reeve J. R., Jr., Walsh J. H., Chew P., Clark B., Hawke D., and Shively J. E. (1983) Amino acid sequences of three bombesin-like peptides from canine intestine extracts. *J. Biol. Chem.* **258,** 5582–5588.
15. Broccardo M., Erspamer G. F., Melchiorri P., Negri L., and De Castiglione R., (1975) Relative potency of bombesin-like peptides. *Biochemistry* **55,** 221–227.
16. Rivier C., Rivier J., and Vale W. (1978) The effect of bombesin and related peptides on prolactin and growth hormone secretion in the rat. *Endocrinology* **102,** 519–523.

17. McDonald T. J., Ghatei M. A., Bloom S. R., Adrian T. E., Mochizuki T., Yanaihara C., and Yanaihara N. (1983) Dose-response comparisons of canine plasma gastroenteropancreatic hormone responses to bombesin and the porcine gastrin-releasing peptide. *Regul. Pept.* **5,** 125–137.
18. McDonald T. J., Ghatei M. A., Bloom S. R., Track N. S., Radziuk J., Dupre J., and Mutt V. (1981) A qualitative comparison of canine plasma gastroenteropancreatic hormone response to bombesin and the porcine gastrin-releasing peptide (GRP). *Regul. Pept.* **2,** 293–304.
19. Knigge U., Holst J. J., Knuhtsen S., Petersen B., Krarup T., Holst-Pedersen J., and Christiansen P. M. (1984) Gastrin-releasing peptide: Pharmacokinetics and effects on gastro-enteropancreatic hormones and gastric secretion in normal men. *J. Clin. Endocrinol. Metab.* **59,** 310–314.
20. Wood S. M., Jung R. T., Webster J. D., Ghatei M. A., Adrian T. E., Yanaihara N., Yanaihara C., and Bloom S. R. (1983) The effect of the mammalian neuropeptide, gastrin-releasing peptide (GRP), on gastrointestinal and pancreatic hormone secretion in man. *Clin. Sci.* **65,** 365–371.
21. Fox J. E. and McDonald T. J. (1984) Motor effects of gastrin-releasing peptide (GRP) and bombesin in the canine stomach and small intestine. *Life Sci* **35,** 1667–1673.
22. Erisman M. D., Linno R. I., Hernandez O., DiAngustine R. P., and Lazarus L. H. (1982) Human lung small-cell carcinoma contains bombesin. *Proc. Natl. Acad. Sci. USA* **79,** 2379–2383.
23. Moody T. W., Pert C. B., Gazdar A. F., Carney D. N., and Minna J. D. (1981) High levels of intracellular bombesin characterize human small-cell lung carcinoma. *Science* **214,** 1246–1248.
24. Minamino N., Kangawa K., and Matsuo H. (1983) Neuromedin B: A novel bombesin-like peptide identified in porcine spinal cord. *Biochem. Biophys. Res. Commun.* **114,** 541–548.
25. Spindel E. R., Chin W. W., Price J., Rees L. H., Besser G. M., and Habener J. F. (1984) Cloning and characterization of cDNAs encoding human gastrin-releasing peptide. *Proc. Natl. Acad. Sci. USA* **81,** 5699–5703.
26. Bennet G. W., Balls M., Clothier R. H., Marsden C. A., Robinson G., and Wemyss-Holden G. D. (1981) Location and release of TRH and 5-HT from amphibian skin. *Cell. Biol. Int. Rep.* **5,** 151–158.
27. Benson B. J. and Hadley M. E. (1969) In vitro characterization of adrenergic receptors controlling skin gland secretion in two anurans *Rana pipiens* and *Xenopus laevis*. *Comp. Biochem. Physiol.* **30,** 857–864.
28. Dockray G. J. and Hopkins C. R. (1975) Caerulein secretion by dermal glands. *J. Cell. Biol.* **64,** 724–733.

29. Sjoberg E. and Flock A. (1976) Innervation of skin glands in the frog. *Cell Tissue Res.* ***172,*** 81–91.
30. Holmes C. and Balls M. (1978) In vitro studies on the control of myoepithelial cell contraction in the granular glands of *Xenopus laevis* skin. *Gen. Comp. Endocrinol.* ***36,*** 255–263.
31. Nakajima T., Tanimura T., and Pisano J. J. (1970) Isolation and structure of a new vasoactive polypeptide. *Fed. Proc.* ***29,*** 282.
32. Walsh J. H., Lechago J., Wong H. C., and Rosenquist G. L. (1982) Presence of ranatensin-like and bombesin-like peptides in amphibian brains. *Regul. Pept.* ***3,*** 1–13.
33. Geller R. G., Govier W. C., Louis W. J. and Clineschmidt B. V. (1970) Pharmacology of ranatensin. *Fed. Proc.* ***29,*** 282.
34. Orloff M. S., Reeve J. R., Jr., Ben-Avram C. M., Shively J. E., and Walsh J. H. (1984) Isolation and sequence analysis of human bombesin-like peptides. *Peptides* ***5,*** 865–870.
35. Minamino N., Kangawa K., and Matsuo H. (1984) Neuromedin C: A bombesin-like peptide identified in porcine spinal cord. *Biochem. Biophys. Res. Commun.* ***119,*** 14–20.
36. Roth K. A., Evans C. J., Weber E., Barchas J. D., Bostwick D. G., and Bensch K. G. (1983) Gastrin-releasing peptide-related peptides in a human malignant lung carcinoid tumor. *Cancer Res.* ***43,*** 5411–5415.
37. Moody T. W., O'Donohue T. L., and Jacobwitz D. M. (1981) Biochemical localization and characterization of bombesin-like peptides in discrete regions of rat brain. *Peptides* ***2,*** 75–95.
38. Chronwall B. M., Pisano J. J., Bishop J. F., Moody T. W., and O'Donohue T. L. (1985) Biochemical and histochemical characterization of ranatensin immunoreactive peptides in rat brain: Lack of coexistence with bombesin/GRP. *Brain Res.* ***338,*** 97–113.
39. Lechago J., Holmquist A. L., Rosenquist G. L., and Walsh J. M. (1978) Localization of bombesin-like peptides in frog gastric mucosa. *Gen. Comp. Endocrinol.* ***36,*** 553–558.
40. Lechago J., Crawford B. G., and Walsh J. H. (1981) Bombesin (like)-containing cells in bullfrog gastric mucosa: Immunoelectronmicroscopic characterization. *Gen. Comp. Endocrinol.* ***45,*** 1–6.
41. Cutz E., Chan W., and Track N. S. (1981) Bombesin, calcitonin and leuenkephalin immunoreactivity in endocrine cells of the human lung. *Experientia* ***37,*** 765–767.
42. Johnson D. E., Lock J. E., Elde R. P., and Thompson T. R. (1982) Pulmonary neuroendocrine cells in hyaline membrane disease and bronchopulmonary dysplasia. *Pediatr. Res.* ***16,*** 446–454.
43. McDonald T. J., Ghatei M. A., Bloom S. R., Track N. S., Radziuk J.,

Dupre J., and Mutt V. (1981) A qualitative comparison of canine plasma gastroenteropancreatic hormone response to bombesin and the porcine gastrin-releasing peptide (GRP). *Regul. Pept.* **2,** 293–304.

44. Track N. S. and Cutz E. (1982) Bombesin-like immunoreactivity in developing human lung. *Life Sci.* **30,** 1553–1556.
45. Price J., Penman E., Bourne G. L., and Rees L. H. (1983) Characterization of bombesin-like immunoreactivity in human fetal lung. *Regul. Pept.* **7,** 315–322.
46. Yoshizaki K., de Bock V., and Solomon S. (1984) Origin of bombesin-like peptides in human fetal lung. *Life Sci.* **34,** 835–843.
47. Willey J. C., Lechner J. F., and Harris C. C. (1984) Bombesin and the C-terminal tetradecapeptide of gastrin-releasing peptide are growth factors for normal human bronchial epithelial cells. *Exp. Cell Res.* **153,** 245–248.
48. Cuttitta F., Carney D. N., Mulshine J., Moody T. W., Fedorko J., Fischler A., and Minna J. D. (1985) Bombesin-like peptides can function as autocrine growth factors in human small cell lung cancer. *Nature* **316,** 823–826.
49. Wood S. M., Wood J. R., Ghatei M. A., Lee Y. C., O'Shaughnessy D., and Bloom S. R. (1981) Bombesin, somatostatin and neurotensin-like immunoreactivity in bronchial carcinoma. *Endorcrinol. Metab.* **53,** 1310–1312.
50. Bostwick D. G., Roth K. A., Evans C. J., Barchas J. D., and Bensch K. G. (1984) Gastrin-releasing peptide, a mammalian analog of bombesin is present in human neuroendocrine lung tumors. *Am. J. Pathol.* **117,** 195–200.
51. Yamaguchi K., Abe K., Adachi I., Suzuki M., Kimura S., Kameya T., and Yanaihara N. (1984) Concomitant production of immunoreactive gastrin-releasing peptide and calcitonin in medullary carcinoma of the thyroid. *Metabolism* **33,** 724–727.
52. Price J., Penman E., Wass J. A. H., and Rees L. H. (1984) Bombesin-like immunoreactivity in human gastrointestinal tract. *Regul. Pept* **9,** 1–10.
53. Ghatei M. A., George S. K., Major J. H., Carlei F., Polak J. M., and Bloom S. R. (1984) Bombesin-like immunoreactivity in the pancreas of man and other mammalian species. *Experientia* **40,** 884–886.
54. Dockray G. J., Vaillant C., and Walsh J. H. (1979) The neuronal origin of bombesin-like immunoreactivity in the rat gastrointestinal tract. *Neuroendocrinology* **4,** 1561–1568.
55. Dalsgaard C-J., Hokfelt T., Schultzberg M., Lundberg J. M., Terenius L., Dockray G. J., and Goldstein M. (1983) Origin of peptide-containing fibers in the inferior mesenteric ganglion of the guinea-pig: Immunohistochemical studies with antisera to sub-

stance P, enkephalin, vasoactive intestinal polypeptide, cholecystokinin and bombesin. *Neuroscience* **9,** 191–211.

56. Schultzberg M. and Dalsgaard C-J. (1983) Enteric origin of bombesin immunoreactive fibres in the rat coeliac-superior mesenteric ganglion. *Brain Res.* **269,** 190–195.
57. Modlin I. M., Lamers C., Walsh J. H., and Jaffe B. M. (1981) Bombesin: A vagally independent stimulator of gastrin release. *Am. J. Surg.* **141,** 98–104.
58. Bruzzone R., Tamburrano G., Lala A., Mauceri M., Annibale B., Severi C., de Magistris L., and Delle-Fave G. (1983) Effect of bombesin on plasma insulin, pancreatic glucagon, and gut glucagon in man. *J. Clin. Endocrinol. Metab.* **56,** 643–647.
59. Modlin I. M., Lamers C. B., and Walsh J. H. (1981) Stimulation of canine pancreatic polypeptide, gastrin, and gastric acid secretion by ranatensin, litorin, bombesin nonapeptide and substance P. *Regul. Pept.* **1,** 279–288.
60. McDonald T. J. and Fox J. E. (1984) Effects of porcine gastrin releasing peptide (GRP) on canine antral motility and gastrin release in vivo. *Life Sci.* **35,** 1415–1422.
61. Labo G., Vezzadini P., Gullo L., Sternini C., and Bonora G. (1983) Effect of bombesin on serum immunoreactive trypsin in healthy subjects and in patients with chronic pancreatitis. *Gastroenterology* **85,** 323–327.
62. Moody T. W. and Pert C. B. (1979) Bombesin-like peptides in rat brain: Quantitation and biochemical characterization. *Biochem. Biophys. Res. Commun.* **90,** 7–14.
63. Roth K. A., Weber E., and Barchas J. D. (1982) Distribution of gastrin releasing peptide-bombesin-like immunostaining in rat brain. *Brain Res.* **251,** 277–282.
64. Panula P., Yang H-Y. T., and Costa E. (1982) Neuronal location of the bombesin-like immunoreactivity in the central nervous system of the rat. *Regul. Pept.* **4,** 275–283.
65. Ghatei M. A., Bloom S. R., Langevin H., McGregor G. P., Lee Y. C., Adrian T. E., O'Shaugnessey D., and Uttenthal L. O. (1984) Regional distribution of bombesin and seven other regulatory peptides in the human brain. *Brain Res.* **293,** 101–109.
66. Schultzberg M. (1983) Bombesin-like immunoreactivity in sympathetic ganglia. *Neuroendocrinlogy* **2,** 363–374.
67. Helen P., Panula P., Yang H-Y. T., Hervonen A., and Rapoport S. I. (1984) Location of substance P, bombesin-gastrin-releasing peptide, [Met5]enkephalin- and [Met5]enkephalin-Arg6-Phe-like immunoreactivities in adult human sympathetic ganglia. *Neuroscience* **12,** 907–916.

68. Panula P., Hadjiconstantinou H., Yang -Y. T., and Costa E., (1983) Immunohistochemical localization of bombesin/gastrin-releasing peptide and substance P in primary sensory neurons. *J. Neurosci.* ***3,*** 2021–2029.
69. Fuxe K., Agnati L. F., McDonald T., Locatelli V., Hokfelt T., Dalsgaard C-J., Battistini N., and Cuello A. C. (1983) Immunohistochemical indications of gastrin releasing peptide-bombesin-like immunoreactivity in the nervous system of the rat. Codistribution with substance P-like immunoreactive nerve terminal systems and coexistence with substance P-like immunoreactivity in dorsal root ganglian cell bodies. *Neurosci Lett.* ***37,*** 17–22.
70. Moody T. W., Thoa N. B., O'Donohue T. L., and Jacobowitz D. M. (1981) Bombesin-like peptides in rat spinal cord: Biochemical characterization, localization and mechanism of release. *Life Sci.* ***29,*** 2273–2279.
71. Brown M., Rivier J., and Vale W. (1977) Bombesin affects the central nervous system to produce hyperglycemia in rats. *Life Sci.* ***21,*** 1729–1734.
72. Brown M., Tache Y., and Fisher D. (1979) Central nervous system action of bombesin's mechanism to induce hyperglycemia. *Endocrinology* ***105,*** 660–665.
73. Tache Y., Marki W., River J., Vale W., and Brown M. (1981) Central nervous system inhibition of gastric secretion in the rat by gastrin-releasing peptide, a mammalian bombesin. *Gastroenterology* ***81,*** 298–302.
74. Hale A. C., Price J., Ackland J. F., Doniach I., Ratter S., Besser G. M., and Rees L. H. (1984) Corticotrophin-releasing factor-mediated adrenocorticotrophin release from rat anterior pituitary cells is potentiated by C-terminal gastrin-releasing peptide. *J. Endo.* ***102,*** R1–R3.
75. Chronwell B. M., Skirboll L. R., O'Donohue T. L. (1985) Demonstration of a pontine-hippocampal projection containing a ranatensin-like peptide. *Neurosci Lett.* ***53,*** 109–114.
76. Price J., Nieuwenhuijzen-Kruseman A. C., Doniach I., Howlett T. A., Besser G. M., and Rees L. H. (1985) Bombesin-like peptides in human endocrine tumors: Quantitation, biochemical characterization and secretion. JCEM ***60,*** 1097–1103.
77. Perlman D. and Halvorson H. (1983) A putative signal peptidase recognition site and sequence in eukaryotic and prokaryotic signal peptides. *Mol. Biol.* ***167,*** 391–409.
78. Bradbury A. F., Finnie M. D. A., and Symth D. G. (1982) Mecha-

nism of C-terminal amide formation by pituitary enzymes. *Nature* ***298,*** 686–688.
79. Proudfoot N. J. and Brownlee G. G. (1976) 3′ Non-coding region sequences in eukaryotic messenger RNA. *Nature* ***263,*** 211–214.
80. Spindel E. R., Zilberberg M. Z., Habener J. F., and Chin W. W. (1986) Two prohormones for gastrin-releasing peptide prohormones are encoded by two mRNAs differing by 19 bases. *Proc. Natl. Acad. Sci., USA.* ***83,*** 19–23.
81. Mount S. M. (1982) A catalogue of splice junction sequences. *Nucleic Acids Res.* ***10,*** 459–472.
82. Early P., Rogers J., Davis M., Calame K., Bond M., Wall R., and Hood L. (1980) Two mRNAs can be produced from a single immunoglobulin μ gene by alternative RNA processing pathways. *Cell* ***20,*** 313–319.
83. Kress M., Glaros D., Khoury G., and Jay G. (1983) Alternative RNA splicing in expression of the H-2K gene. *Nature* ***306,*** 602–604.
84. Oates E. and Herbert E. (1984) 5′ sequence of porcine and rat proopiomelanocortin mRNA. *J. Biol. Chem.* ***259,*** 7421–7425.
85. Richter K., Kawashima E., Egger R., and Kreil G. (1984) Biosynthesis of thyrotropin releasing hormone in the skin of *Xenopus laevis:* Partial sequence of the precursor deduced from cloned cDNA. *EMBO J.* ***3,*** 617–621.
86. Hoffman W., Bach T. C., and Kreil G. (1983) Synthesis of caerulein in the skin of *Xenopus laevis:* Partial sequences of precursors as deduced from cDNA clones. *EMBO J.* ***2,*** 111–114.
87. Spindel E. R., Chin W. W., and Habener J. F. (1984) Noradrenergic stimulation of TRH, ranatensin and caerulein mRNA levels in frog skin. *Society for Neuroscience, 14th Annual Meeting Abst.* ***201,*** 1.
88. Minamino N., Sudoh T., Kangawa K., and Matsuo H. (1985) Neuromedin B-32 and B-30: Two "big" neuromedin B identified in porcine brain and spinal cord. *Biochem. Biophys. Res. Commun.* ***130,*** 685–691.
89. Minamino N., Kangawa K., and Matsuo H. (1984) Neuromedin B is a major bombesin-like peptide in rat brain: Regional distribution of neuromedin C in rat brain, pituitary and spinal cord. *Biochem. Biophys. Res. Commun.* ***124,*** 925–932.
90. Naylor S. L., Spindel E. R., Chin W. W., and Sakaguchi A. Y. (1985) Gastrin-releasing peptide gene is located on human chromosome 18. *Eighth International Human Gene Mapping Meeting.*
91. Sausville E. A., Lebacq-Verheyden A., Spindel E. R., Cuttitta F., Gazdar A. F., and Battey J. F. (1986) Expression of the gastrin-releasing peptide (GRP) gene in human small cell lung cancer: Evi-

dence for alternative processing resulting in three distinct mRNA. *J. Biol. Chem.* (in press).

92. Spindel E. R., Sunday M. E., Hofler H., Dobrzanksa E., Wolfe H. J., Habener J. F., and Chin W. W. (1986) Transient elevation of mRNAs encoding gastrin-releasing peptide (GRP), a putative pulmonary growth factor, in human fetal lung (abstract). *Clin. Res.* (in press).

# Chapter 16

# Biosynthesis of Peptides in the Skin of *Xenopus laevis*

## cDNAs Encoding Precursors of Caerulein, PGL[a], Xenopsin, and Thyrotropin-Releasing Hormone

K. Richter, W. Hoffmann, R. Egger, and G. Kreil

## 1. INTRODUCTION

Amphibian skin is a rich source for a variety of peptides with diverse physiological actions (*1–4*). It is surprising that these skin peptides are virtually always similar or identical to peptide hormones found in mammalian brain or other parts of the nervous system or in the gastrointestinal (GI) tract. This "brain–gut–skin triangle" (*5*) has intrigued pharmacologists and biochemists for many years. Table 1 shows a list of a few prototypes of skin peptides and their mammalian counterparts. Even after two decades of research, new peptides are still discovered in amphibian skin, as exemplified by dermorphin, a potent opioid peptide with the unusual feature of having one D-amino acid in its sequence (*6*).

In certain amphibian species, up to several milligrams of a given peptide may be present per gram of skin, and therefore, it is to be expected that its biosynthesis should be amenable to experimental analysis. Using recombinant DNA techniques, the precursors of three different skin peptides have now been sequenced (*7–9*), and an additional peptide was discovered during this work (*10*). In all cases, skin of *Xenopus laevis* was used as starting mate-

TABLE 1
Homologous Peptides From Amphibian Skin, Brain, and Gastrointestinal Tract of Mammals[a]

| Amphibian skin | Mammalian tissues |
|---|---|
| Caerulein | Cholecystokinin |
| | Gastrin |
| Xenopsin | Neurotensin |
| TRH | TRH |
| Tachykinins | Substance P |
| Bombesins | Bombesin-like peptides |
| Sauvagine | Corticotropin-releasing hormone |
| Dermorphins | Enkephalins |
| Bradykinins | Bradykinin |

[a]Modified from ref. 2.

rial for the preparation of total mRNA, which was transcribed into cDNA and used for the construction of a cDNA library. Screening with different oligonucleotides then led to the identification of clones containing inserts derived from mRNA for the precursors of these peptides. The structures of these precursors suggest the existence of complex activation mechanisms for the liberation of the final products. Moreover, different peptides originate from homologous precursors.

## 2. PRECURSOR SEQUENCES DEDUCED FROM CLONED cDNAs

### *2.1. Precursors of Caerulein*

Caerulein was discovered about 15 years ago in the skin of the Australian frog *Hyla caerulea* (*11*) and was subsequently found in several other species, including *X. laevis* (*12*). This decapeptide, with a Tyr-*O*-sulfate residue, has the same carboxy-terminal sequence as the mammalian hormones cholecystokinin and gastrin. It also resembles these hormones in its physiological action inasmuch as it stimulates the musculature of the gall bladder and the small intestine, as well as secretion of the pancreas and the gastric mucosa.

Total mRNA from skin of X. *laevis* was transcribed into double-stranded cDNA and inserted into the unique *Pst*I site of plasmid pUC8 via poly(dG)-poly(dC) homopolymeric extensions. From ampicillin-resistant transformants, 300 were selected at random and screened with radioactive cDNA primed with the decamer d(AGTCCATCCA), which is complementary to the codons of the sequence Trp-Met-Asp-Phe of caerulein (*7*). Among the clones strongly hybridizing with this radioactive cDNA, three were found to contain inserts coding for parts of caerulein precursors. Clone pUF37 had an open reading frame for 97 amino acids, with two copies of the end product caerulein in its sequence; a second clone had a deletion of 45 base pairs (bp), including one caerulein sequence.

Several more clones derived from mRNAs for caerulein precursors have since been analyzed. We now have evidence for four different types that contain between one and four copies of caerulein in a single polypeptide chain. One type with three copies is shown in Fig. 1. This sequence is derived from the inserts of two clones that overlap over a long segment. The different caerulein precursors all show an internal homology, indicating a duplication of a primary segment, which was then further modified through insertions, deletions, and additional duplications of parts of these structures. Our current knowledge is insufficient to give a coherent picture of the possible evolution of these diverse sequences. However, it is quite clear that a small family of mRNAs and, probably genes, for caerulein precursors exists that codes for polypeptides of different lengths, with a varying number of caerulein copies. Moreover, the amino-terminal region of these precursors, including the signal peptide, shows sequence homology to the precursor for PGL$^a$ (P = peptide; G = glycine, the first amino acid; L = leucine, the last amino acid; $^a$ = amide) and xenopsin (*see* section 2.5 of this review).

In the predicted amino acid sequence of the caerulein precursors, pairs of Lys/Arg residues are present close to the sequence of the final product. Cleavage at, and excision of, these basic amino acids yields an intermediate form of caerulein with an extension of two or four amino acids at the $NH_2$-terminus and an additional Gly at the other end. The latter is oxidized to yield the terminal amide (*13,14*), whereas the further processing at the amino end is probably catalyzed by a dipeptidylaminopeptidase.

```
                    1
                    ATAACTGTGAGGAATCACAAAGTGCACTCTTTACTTAGTACCTTTCAAAATCTCTTGCAATTACCTTCTGAAAGC

 ATG TTT AAA GGG ATA TTA CTT TGT GTG TTA TTT GCT GTG CTC TCT GCA AAC CCA TTG TCA CAG CCA GAA GGC
 Met-Phe-Lys-Gly-Ile-Leu-Leu-Cys-Val-Leu-Phe-Ala-Val-Leu-Ser-Ala-Asn-Pro-Leu-Ser-Gln-Pro-Glu-Gly-
  1

 TTT GCA GAT GAA GAA CGA GAT GTC CGA GGG CTT GCA TCT TTC CTA GGT AAA GCT TTA AAG GCT GGT TTA AAA
-Phe-Ala-Asp-Glu-Glu-Arg-Asp-Val-Arg-Gly-Leu-Ala-Ser-Phe-Leu-Gly-Lys-Ala-Leu-Lys-Ala-Gly-Leu-Lys-

 ATT GGT GCA CAT TTT CTG GGA GGA GCA CCT CAA CAA CGG GAA GCC AAT GAC GAA CGT CGC TTT GCT GAT GGA
-Ile-Gly-Ala-His-Phe-Leu-Gly-Gly-Ala-Pro-Gln-Gln-Arg-Glu-Ala-Asn-Asp-Glu-Arg-Arg-Phe-Ala-Asp-Gly-

 CAA CAA GAC TAC ACA GGT TGG ATG GAT TTT GGC CGC CGC GAC GGA CAA CAA GAC TAC ACA GGT TGG ATG GAT
-Gln-Gln-Asp-Tyr-Thr-Gly-Trp-Met-Asp-Phe-Gly-Arg-Arg-Asp-Gly-Gln-Gln-Asp-Tyr-Thr-Gly-Trp-Met-Asp-

 TTT GGC CGC CGT GAT GAT GAA GAT GAT GTA AAT GAA CGA GAT CTC CGA GGA TTT GGC TCT TTC CTA GGT AAA
-Phe-Gly-Arg-Arg-Asp-Asp-Glu-Asp-Asp-Val-Asn-Glu-Arg-Asp-Leu-Arg-Gly-Phe-Gly-Ser-Phe-Leu-Gly-Lys-

 GCT TTA AAG GCT GCT TTA AAA ATT GGT GCA AAT GCG CTG GGA GGA GCA CCT CAA CAA CGA GAA GCC AAT GAC
-Ala-Leu-Lys-Ala-Ala-Leu-Lys-Ile-Gly-Ala-Asn-Ala-Leu-Gly-Gly-Ala-Pro-Gln-Gln-Arg-Glu-Ala-Asn-Asp-

 GAA CGT CGC TTT GCT GAT GGA CAA CAA GAC TAC ACA GGT TGG ATG GAT TTT GGC CGC CGC AAT GGT GAA GAT
-Glu-Arg-Arg-Phe-Ala-Asp-Gly-Gln-Gln-Asp-Tyr-Thr-Gly-Trp-Met-Asp-Phe-Gly-Arg-Arg-Asn-Gly-Glu-Asp-

 GAT TAATATTCTTCTTGAAAACCTCAAATGTATAAAACTACATCTGTTTCTGTACAGAGGAAATAAAGCATTTACTGAAGpolyA
-Asp ///                                                                          659
 169
```

Fig. 1. Nucleotide sequence of the cDNA inserts of clones 11/9 and 11/63 and the amino acid sequence of preprocaerulein III. Only the coding strands with the poly(dA) tail are shown. Pairs of Arg residues, terminal Gly, and caerulein copies are underlined.

Activation of precursors by stepwise cleavage of dipeptides has previously been observed for honeybee promelittin (*15*) and yeast proalpha-mating factor (*16*). High amounts of a dipeptidylaminopeptidase have been detected in skin secretions of *X. laevis,* and this enzyme appears to have the substrate specificity required for the cleavage of the amino-terminal extensions present in the processing intermediates (*17*). The same mechanism should operate in the liberation of xenopsin from its precursor (see subsequent sections).

### 2.2. Precursors for PGL[a], a New Peptide From the Skin of X. laevis

During the search for the caerulein precursor, several clones were detected that contained parts of the mRNA coding for an unknown polypeptide (*10*). At least two slightly different mRNAs of this type are present in skin. The largest clone found in these initial studies had an insert starting at the 5′ end with an ATG codon followed by an open reading frame for 63 amino acids. More recently, a clone was found with an insert containing 64 nucleotides preceding this initiation codon (*see* Fig. 2). In this extension, two stop codons are in phase with the ATG codon, which demonstrated that the predicted polypeptide represents the total coding capacity of these mRNAs.

In the predicted amino acid sequence, there is an amino-terminal segment of mostly hydrophobic residues, as is typical for signal peptides. The polypeptide contains a stretch of 24 amino acids that is flanked by a Lys-Arg sequence at the amino-terminus and by a Gly-Arg-Arg sequence at the other end. Typical prohormone processing at these pairs of basic amino acids and use of the terminal Gly exposed during processing for the formation of an amide would then yield a peptide composed of 24 amino acids, starting with Tyr and ending with leucineamide. Following the nomenclature suggested by Tatemoto and Mutt (*18*), this hypothetical peptide was called PYL[a]. The precursor of this peptide contains only 64 amino acids, which makes this the smallest prepropeptide yet found in nature.

These clones were also used for hybrid selected translation. It could be shown that mRNA coding for one of the major translation products of total skin mRNA is selected by these clones (*10*).

```
                               1                                                              64
                               AGCACAACAATTGTACGGAGCACTTTGCTACTTCTAGTTTTGAAGAGCTATTACATTTGGAAGG

ATG TAC AAA CAA ATT TTC CTC TGT CTG ATC ATT GCA GCA CTC TGT GCA ACC ATA ATG GCA GAG GCT
Met-Tyr-Lys-Gln-Ile-Phe-Leu-Cys-Leu-Ile-Ile-Ala-Ala-Leu-Cys-Ala-Thr-Ile-Met-Ala-Glu-Ala-
 1                                                                                ↑

TCA GCA TTA GCA GAT GCA GAT GAT GAC GAT GAC AAG CGT TAC GTC CGA GGA ATG GCA TCT AAA GCT
-Ser-Ala-Leu-Ala-Asp-Ala-Asp-Asp-Asp-Asp-Asp-Lys-Arg-Tyr-Val-Arg-Gly-Met-Ala-Ser-Lys-Ala-
                                             -------         ---

GGA GCA ATT GCG GGA AAA ATT GCT AAA GTT GCT CTA AAG GCT CTT GGA CGT CGT GAC TCG TAGGACT
-Gly-Ala-Ile-Ala-Gly-Lys-Ile-Ala-Lys-Val-Ala-Leu-Lys-Ala-Leu-Gly-Arg-Arg-Asp-Ser ///
                                                             -----------      64

TCAGCGGTCTCAATGGAATGTAAATGACAACACTTGGCTGGACAGATTTTGAACATTGATGTATTGAAGAATAACATAAACCCTCCA

CAATCCCTCAAACCTATAAATAAATCTGTTGGACAGGAAATAAAATGACCTATATGCATATpolyA
                                                            411
```

Fig. 2. Nucleotide sequence of the insert of clone 12/128 and amino acid sequence of preproPGL[a]. The sequence is identical to the one for clone pUF81 (*see* ref. *10*) after base 109. Processing sites are underlined, the arrow marks the possible end of the signal peptide.

The predicted peptide PYL[a] has been synthesized and used as a reference substance to search for the natural counterpart. A peptide was found in the skin secretion of *X. laevis* that differs from PYL[a] in the lack of the first three amino acids (*19*). Additional processing of PYL[a] after the single Arg in position three yields PGL[a], a 21-amino-acid peptide with amino terminal glycine and carboxyl-terminal leucineamide. Recently, several peptides liberated from the caerulein precursors by processing at single arginine residues have also been detected in skin secretion (*20,21*).

## 2.3. Precursor of Xenopsin

More than ten years ago, the octapeptide xenopsin was isolated from skin secretion of *X. laevis* (*22*) and was found to be homologous to neurotensin isolated from mammalian brain (*23*). As judged by immunological criteria, *X. laevis* also contains a neurotensin-like peptide, but xenopsin could not be detected in mammalian tissues (*24*). It is currently not known whether peptides identical or similar to xenopsin are also present in other frog species.

Using the known amino acid sequence of xenopsin, two different oligodeoxynucleotide mixtures were used to screen a cDNA library constructed from total skin mRNA of *X. laevis*. One clone, designated pXP, was found that contained an insert of 426 nucleotides (*8*). The nucleotide sequence of this insert had an open reading frame coding for a polypeptide composed of 81 amino acids, including the initiating Met and a putative signal peptide region (*see* Fig. 3). A single copy of the sequence of xenopsin was found at the carboxy end of this precursor. The nucleotide and amino acid sequence of the xenopsin precursor shows homology with those of caerulein and, particularly, PGL[a] (*see* section 2.5).

Processing of this precursor at the single pair of basic amino acids would yield an intermediate form of xenopsin, with the extra sequence Glu-Ala-Met-Leu-Arg-Ser-Ala-Glu-Ala at the amino end. In view of our results with the precursors for caerulein and PGL[a] found in skin secretion, the following mechanism can be postulated for the liberation of xenopsin from its precursor: Cleavage also occurs after the single Arg residue of this amino-terminal extension, and the remaining tetrapeptide Ser-Ala-Glu-Ala is removed through the action of dipeptidylaminopeptidase.

```
                              1                                                     62
                              ACAATCATCTGTCTTTTGTACCTTGCTACTTCTGTTACTGGAAAACACTACATTTGGAAAGG

ATG TAC AAA GGG ATA TTT CTT TGT GTT TTA CTT GCT GTG ATC TGT GCA AAC TCA CTG GCA ACG CCA
Met-Tyr-Lys-Gly-Ile-Phe-Leu-Cys-Val-Leu-Leu-Ala-Val-Ile-Cys-Ala-Asn-Ser-Leu-Ala-Thr-Pro-
 1

AGT AGC GAT GCA GAT GAA GAT AAT GAT GAA GTC GAA CGC TAT GTA CGA GGC TGG GCA TCT AAA ATA
-Ser-Ser-Asp-Ala-Asp-Glu-Asp-Asn-Asp-Glu-Val-Glu-Arg-Tyr-Val-Arg-Gly-Trp-Ala-Ser-Lys-Ile-

GGA CAA ACT TTG GGA AAG ATA GCT AAA GTT GGA TTA AAG GAA TTA ATT CAA CCA AAA CGA GAG GCA
-Gly-Gln-Thr-Leu-Gly-Lys-Ile-Ala-Lys-Val-Gly-Leu-Lys-Glu-Leu-Ile-Gln-Pro-Lys-Arg-Glu-Ala-

ATG CTA CGC AGC GCT GAG GCC CAA GGC AAG AGA CCA TGG ATA CTC TAAATGAACAGAAAAACTGCTTTGCTG
-Met-Leu-Arg-Ser-Ala-Glu-Ala-Gln-Gly-Lys-Arg-Pro-Trp-Ile-Leu ///
                                                           81

ACAAAGCATTTACCCCGGTCTGAAGAATAAACACAGCCCTCAGATAAACTCGAGACCTTTAAAAATACATGATTCCAATGCTAATAA

AATAATCpolyA
    426
```

Fig. 3. Nucleotide sequence of the insert of clone pXP and amino acid sequence of preproxenopsin (from ref *8*). Possible processing sites and the sequence of xenopsin are underlined.

## *2.4. Precursor of Thyrotropin-Releasing Hormone (TRH)*

Relatively large amounts of TRH were first detected in skin of the frogs *Bombina orientalis* (*25*) and *Rana pipiens* (*26*). In the latter case, it was assumed that the TRH found in frog blood originated from skin. For *X. laevis,* however, it was shown that at least part of the TRH was stored, together with 5-hydroxytryptamine, in the dermal granular glands, from where it is discharged to the outside by adrenergic agonists (*27*). Current evidence therefore suggests that TRH is made in large quantities in frog skin, part of which may be released into the blood stream, whereas another part is a constituent of the skin secretion.

Using a mixture of two synthetic oligonucleotides, one clone was recently found with an insert coding for part of the TRH precursor in *X. laevis* skin (*9*). The insert corresponds to a cloned mRNA fragment that contains a 5′-untranslated region and an open reading frame for 123 amino acids ending abruptly at the poly(C)-tail (*see* Fig 4). In this predicted polypeptide, the sequence Lys-Arg-Gln-His-Pro-Gly-$^{\mathrm{Lys}}/_{\mathrm{Arg}}$-Arg is found three times, and a fourth copy terminating at the Pro residue is present at the carboxyl end. Typical prohormone processing at these sequences

```
1
AGCACAGAGCAGCACAAGGACACACTCTGCATATTGTGCTGCCGGACAAGGAGGTGACAGCCAGTCAGGCTGAGACAAAGGA
                           109
ACTTCCAGACCTCTGACAGCAGGAAAG ATG GTG TCT GTC TGG TGG TTG CTG CTT CTC GGT ACA ACC GTA TCT
                            Met-Val-Ser-Val-Trp-Trp-Leu-Leu-Leu-Leu-Gly-Thr-Thr-Val-Ser-
                             1                                                        15
CAC ATG GTG CAC ACA CAA GAG CAG CCT TTA CTG GAG GAG GAC ACA GCA CCA TTA GAT GAC TCG GAT
-His-Met-Val-His-Thr-Gln-Glu-Gln-Pro-Leu-Leu-Glu-Glu-Asp-Thr-Ala-Pro-Leu-Asp-Asp-Ser-Asp-

GTT CTT GAG AAA GCC AAA GGT ATC CTG ATC CGC AGT ATC CTG GAG GGA TTT CAA GAA GGG CAA CAA
-Val-Leu-Glu-Lys-Ala-Lys-Gly-Ile-Leu-Ile-Arg-Ser-Ile-Leu-Glu-Gly-Phe-Gln-Glu-Gly-Gln-Gln-

AAC AAT AGA GAT CTA CCA GAT GCA ATG GAA ATT ATA TCT AAG CGC CAG CAC CCA GGG AAA CGA TTC
-Asn-Asn-Arg-Asp-Leu-Pro-Asp-Ala-Met-Glu-Ile-Ile-Ser-Lys-Arg-Gln-His-Pro-Gly-Lys-Arg-Phe-

CAG GAG GAG ATA GAA AAG AGA CAA CAC CCT GGA AAG AGG GAT CTG GAA GAT CTG AAT CTA GAG CTT
-Gln-Glu-Glu-Ile-Glu-Lys-Arg-Gln-His-Pro-Gly-Lys-Arg-Asp-Leu-Glu-Asp-Leu-Asn-Leu-Glu-Leu-
                                                                                  478
TCC AAA AGG CAA CAC CCC GGA AGA AGA TTT GTG GAT GAT GTA GAG AAG AGG CAA CAT CCA CCCCC...
-Ser-Lys-Arg-Gln-His-Pro-Gly-Arg-Arg-Phe-Val-Asp-Asp-Val-Glu-Lys-Arg-Gln-His-Pro .....
                                                                             123
```

Fig. 4. Nucleotide sequence of the insert of clone 8/136 and partial amino acid sequence of preproTRH. Processing sites and the three amino acids found in the end product are underlined.

should yield the hypothetical intermediate Gln-His-Pro-Gly, and, through cyclization of the amino-terminal Glu and formation of a terminal amide by oxidation of the Gly residue, pGlu-His-Pro. $NH_2$, i.e., TRH would be generated. The segment of prepro-TRH shows an internal homology, and the TRH copies can be arranged in pairs separated by the homologous hexapeptides Phe-Gln-Glu-Glu-Ile-Glu and Phe-Val-Asp-Asp-Val-Glu, respectively.

The precursors for caerulein, PGL[a], and xenopsin show sequence homology. This is not true, however, for the part of the TRH precursor analyzed so far. The size of this latter precursor is presently not known, and to search for clones containing the carboxy-terminal regions of this polypeptide will be interesting.

### *2.5. Comparison of the Precursors for Caerulein, PGL[a], and Xenopsin*

It is quite surprising that the three skin peptides, caerulein, PGL[a], and xenopsin, which have different structures and, as far as one knows, different physiological actions, are derived from precursors with striking sequence homology. For the 62 nucleotides preceding the initiation codon, this similarity ranges from 37% between the mRNA for preprocaerulein and preproPGL$^a$ to 55% between those for preproxenopsin and preproPGL$^a$, and these percentages increase in the coding region. As is shown in Fig. 5, the sequence homology is also readily discernible for the corresponding polypeptides. First, it is clear that the signal peptides of these precursors are very similar. Among the first 20 positions in the sequence, there are only two in which all three polypeptides differ from each other. Beyond this region, the homology is again greatest for the precursors of PGL$^a$ and xenopsin, but only up to residue 59, after which no relation between the amino acid or nucleotide sequences is discernible. The amino acid sequence of preprocaerulein also shows about 50% homology with the two other precursors, if a gap of five amino acids is introduced (*see* Fig. 5). Again, the similarity ends after about 60 amino acids. However, the mRNA for this precursor shows homology with the mRNA for preproPGL$^a$ also in the 3′-untranslated region (*10*).

This comparison indicates that the amino-terminal regions of these three precursors have evolved from a common origin. Caerulein has been considered the ancestor of cholecystokinin and gastrin, and homologous sequences have been detected in invertebrates by immunological techniques (*28–30*). If one assumes that a primordial gene for preprocaerulein without an internal duplication existed, the gene for the PGL$^a$ precursor could have arisen through a deletion of the coding region between about residue 60 and the amino-terminus of caerulein. This would explain the similarities found at both ends of the mRNAs for these two

```
a) MET-phe-LYS-GLY-ILE-leu-LEU-CYS-VAL-LEU-phe-ALA-VAL-LEU-ser-ALA-ASN-pro-LEU-ser-gln-PRO-
b) MET-TYR-LYS-GLY-ILE-PHE-LEU-CYS-VAL-LEU-leu-ALA-VAL-ile-CYS-ALA-ASN-ser-LEU-ALA-thr-PRO-
c) MET-TYR-LYS-gln-ILE-PHE-LEU-CYS-leu-ile-ile-ALA-ala-LEU-CYS-ALA-thr-ile-met-ALA-glu-ala-

a)-glu-gly-phe-ALA-ASP-GLU-                             -GLU-ARG-asp-VAL-ARG-GLY-leu-ALA-SER-phe-leu-
b)-SER-ser-asp-ALA-ASP-GLU-ASP-asn-ASP-glu-val-GLU-ARG-TYR-VAL-ARG-GLY-trp-ALA-SER-LYS-ile-
c)-SER-ala-leu-ALA-ASP-ala-ASP-asp-ASP-asp-asp-lys-ARG-TYR-VAL-ARG-GLY-met-ALA-SER-LYS-ala-

a)-GLY-lys-ala-LEU-lys-ala-gly-leu-LYS-ile-GLY-ala-his-phe-LEU-GLY-gly-ala-pro-gln-gln....
b)-GLY-gln-thr-LEU-GLY-LYS-ILE-ALA-LYS-VAL-GLY-LEU-LYS-glu-LEU-ile-gln-pro-lys-arg-glu....
c)-GLY-ala-ile-ala-GLY-LYS-ILE-ALA-LYS-VAL-ala-LEU-LYS-ala-LEU-GLY-arg-arg-asp-ser
```

Fig. 5. Comparison of the amino-terminal sequence of the precursors for (a) caerulein, (b) xenopsin, and (c) PGL$^a$. Amino acids present in at least two of the three sequences are written in capital letters.

precursors. A more recent event would then have been the duplication of a segment derived from the 5′ half of this PGL[a] gene and the insertion of one copy next to the genetic information for xenopsin. Finally, the caerulein gene would have undergone a number of duplications and additional internal changes, yielding several variants without changing the homology of corresponding segments to any extent.

In view of the similarity between these precursor polypeptides, it is also tempting to speculate that the processing giving rise to PYL[a] and its truncated form may also take place with the other proforms. In case of preproxenopsin, processing not only at the Lys-Arg pair, but also at single Arg residues, would yield a peptide very similar to PGL[a], with the amino-terminal sequence Gly-Trp . . . and the sequence . . . Pro-Lys at the carboxyl end (*see* Figs. 3 and 5). In preprocaerulein, three single Arg are present in the segments preceding the caerulein sequences, and thus, several processing alternatives are possible. In our search for PGL[a] in skin secretion of *X. laevis,* a few peptides with a basic net charge have been detected and it will be interesting to check whether any of these correspond to segments of these precursors.

## 3. CONCLUSIONS

Using cDNA libraries from the skin of *X. laevis,* the mRNAs for the precursors of four peptides have been analyzed. Among these, the caerulein precursors appear to be unusually complex inasmuch as evidence for at least four different mRNAs coding for polypeptides containing between one and four copies of the end product indicate. The origin of these variants, through duplication of genes or gene segments, insertions, and deletions is at present only partly understood. In the case of PGL[a], at least two different mRNAs have been observed as well.

It is quite unexpected that the mRNAs coding for the precursor of caerulein, PGL[a], and xenopsin show sequence homology in their 5′-untranslated parts as well as in the region coding for the first 60 amino acids of the respective polypeptides. This homology indicates that the corresponding segment in the genome, possibly a particular exon, has been duplicated several times and also inserted into other regions of the chromosomes. Available evidence suggests, moreover, that this common segment may itself contain the genetic information for basic peptides found in skin secretion. It will be most interesting to analyze the exon–intron organization of the genes for these precursors, which may yield some information about their evolutionary history.

The precursors for caerulein and TRH are polyproteins containing more than one copy of the final product in one polypeptide chain. Liberation of TRH, the smallest peptide hormone, from its precursor appears to be quite simple, whereas the processing of the other precursors is more complex, also involving hydrolysis at single Arg residues and, in particular, stepwise cleavage of dipeptides by a dipeptidylaminopeptidase.

The few studies carried out so far demonstrated that the biosynthesis of amphibian skin peptides yields some interesting and unexpected results. The most surprising finding, apparently without precedence, is the fact that totally different end products originate from precursors with similar pre- and proregions. It is our hope that the experience gained in these studies will facilitate similar experiments on the formation of hormones and neuropeptides in mammals and invertebrates.

## ACKNOWLEDGMENTS

This work was supported by grants from the "Politzer Stiftung" of the Austrian Academy of Sciences and from the Austrian Fonds zur Foerderung der wissenschaftlichen Forschung (grant S29 T4). We thank Drs. Irmi Sures and Marco Crippa for communicating their results on the xenopsin precursor before publication.

## REFERENCES

1. Erspamer V. and Melchiorri P. (1973) Active polypeptides of the amphibian skin and their synthetic analogs. *Pure Appl. Chem.* **35,** 463–494.
2. Erspamer V. and Melchiorri P. (1980) Active polypeptides: From amphibian skin to gastrointestinal tract and brain of mammals. *Trends Pharmacol. Sci.* **1,** 391–395.

3. Nakajima T. (1980) Evolutionary Aspects of Hormone-Like Peptides, in *Hormones, Adaptation and Evolution* (S. Ishii, T. Hirano, and M. Wada, eds.), Springer-Verlag, Berlin.
4. Erspamer V., Falconieri-Erspamer G., Mazzanti G., and Endean R. (1984) Active peptides in the skin of one hundred amphibian species from Australia and Papua New Guinea. *Comp. Biochem. Physiol.* **77C,** 99–108.
5. Erspamer V., Melchiorri P., Broccardo M., Erspamer G. F., Falaschi P., Improta G., Negri L., and Renda T. (1981) The brain-gut-skin triangle: New peptides. *Peptides* **2,** (suppl. 2), 7–16.
6. Montecucchi P. C., de Castiglione R., Piani S., Gozzini L., and Erspamer V. (1981) Amino acid composition and sequence of dermorphin, a novel opiate-like peptide from the skin of *Phyllomedusa sauvagei. Int. J. Peptide Protein Res.* **17,** 275–288.
7. Hoffmann, W., Bach T. C., Seliger H., and Kreil G. (1983) Biosynthesis of caerulein in the skin of *Xenopus laevis:* Partial sequences of precursors as deduced from cDNA clones. *EMBO J.* **2,** 111–114.
8. Sures I. and Crippa M. (1984) Xenopsin: The neurotensin-like octapeptide from *Xenopus* skin at the carboxyl terminus of its precursor. *Proc. Natl. Acad. Sci. USA* **81,** 380–384.
9. Richter K., Kawashima E., Egger R., and Kreil G. (1984) Biosynthesis of thyrotropin releasing hormone in the skin of *Xenopus laevis:* Partial sequence of the precursor deduced from cloned cDNA. *Embo J.* **3,** 617–621.
10. Hoffmann W., Richter K., and Kreil G. (1983) A novel peptide designated PYL[a] and its precursor as predicted from cloned mRNA of *Xenopus laevis* skin. *EMBO J.* **2,** 711–714.
11. Anastasi A., Erspamer V., and Endean R. (1968) Isolation and amino acid sequence of caerulein, the active decapeptide of the skin of *Hyla caerulea. Arch. Biochem. Biophys.* **125,** 57–68.
12. Anastasi A., Bertaccini G., Cei J. M., de Caro G., Erspamer V., Impicciatore M., and Roseghini M. (1970) Presence of caerulein in extracts of the skin of *Leptodactylus pentadactylus labyrinthicus* and of *Xenopus laevis. Brit. J. Pharmacol.* **38,** 221–228.
13. Bradbury A. F., Finnie, M. D. A., and Smyth D. G. (1982) Mechanism of C-terminal amide formation by pituitary enzymes. *Nature* **298,** 686–688.
14. Eipper B. A., Mains R. E., and Glembotski C. C. (1983) Identification in pituitary tissue of a peptide α-amidation activity that acts on glycine-extended peptides and requires molecular oxygen, copper, and ascorbic acid. *Proc. Natl. Acad. Sci. USA* **80,** 5144–5148.
15. Kreil G., Haiml L., and Suchanek G. (1980) Stepwise cleavage of the

pro part of promelittin by dipeptidylpeptidase. IV. Evidence for a new type of precursor-product conversion. *Eur. J. Biochem.* **111,** 49–58.
16. Julius D., Blair L., Brake A., Sprague G., and Thorner J. (1983) Yeast α factor is processed from a larger precursor polypeptide: The essential role of a membrane-bound dipeptidyl aminopeptidase. *Cell* **32,** 839–852.
17. Hoffmann W., Richter K., Hutticher A., and Kreil G. (1983) Biosynthesis of Peptides in Amphibian Skin, in *Biochemical and Clinical Aspects of Neuropeptides.* (Koch G. and Richter D., eds.), Academic, New York.
18. Tatemoto K. and Mutt V. (1980) Isolation of two novel candidate hormones using a chemical method for finding naturally occurring polypeptides. *Nature* **285,** 417–418.
19. Andreu D., Aschauer H., Kreil G., and Merrifield R. B. (1985) Solid-phase synthesis of PYL[a] and isolation of its natural counterpart, PGL[a] (PYL[a] – (4 – 24) ) from skin secretion of *Xenopus laevis. Eur. J. Biochem.* **149,** 531–535.
20. Richter K., Aschauer H., and Kreil G. (1985) Peptides secreted by the skin of *Xenopus laevis. Peptides* (in press).
21. Gibson B. W., Poulter L., and Williams D. H. (1985) A mass spectrometric assay for novel peptides: Application to *Xenopus laevis* skin secretions. *Peptides* (in press).
22. Araki K., Tachibana S., Uchiyama M., Nakajima T., and Yasuhara T. (1973) Isolation and structure of a new active peptide "Xenopsin" on the smooth muscle, especially on a strip of fundus from a rat stomach, from the skin of *Xenopus laevis. Chem. Pharm. Bull.* (Tokyo) **21,** 2801–2804.
23. Carraway R. and Leeman S. E. (1975) The amino acid sequence of a hypothalamic peptide, neurotensin. *J. Biol. Chem.* **250,** 1907–1911.
24. Goedert M., Sturmey N., Williams B. J., and Emson P. C. (1984) The comparative distribution of xenopsin- and neurotensin-like immunoreactivity in *Xenopus laevis* and rat tissues. *Brain Res.* **308(2),** 273–280.
25. Yasuhara T. and Nakajima T. (1975) Occurrence of thyrotropin releasing hormone in frog skin. *Chem. Pharm. Bull.* **23,** 3301–3303.
26. Jackson I. M. D. and Reichlin S. (1977) Thyrotropin-releasing hormone: Abundance in the skin of the frog, *Rana pipiens. Science* **198,** 414–415.
27. Bennett G. W., Marsden C. A., Clothier R. M., Waters A. D., and Balls M. (1982) Co-existence of thyrotropin-releasing hormone and 5-hydroxytryptamine in the skin of *Xenopus laevis. Comp. Biochem. Physiol.* **72C** (no. 2), 257–561.
28. Grimmelikhuijzen C. J., Sundler F., and Rehfeld J. F. (1980) Gastrin/

CCK-like immunoreactivity in the nervous system of coelenterates. *Histochemistry* ***69,*** 61–68.

29. Rehfeld J. F. (1981) Four basic characteristics of the gastrin-cholecystokinin system (editorial review). *Am. J. Physiol.* **240,** G255–G256.
30. Morley J. E. (1982) The ascent of cholecystokinin (CCK)—from gut to brain (minireview). *Life Sci.* ***30,*** 479–493.

# Chapter 17

# cDNA Encoding Precursors of the Bee-Venom Peptides Melittin and Secapin

R. VLASAK, I. MALEC, AND G. KREIL

## 1. INTRODUCTION

The venom gland of honeybees is a Y-shaped, tubular structure that contains highly polyploid cells. The secretion produced by the gland cells is collected in a central duct and stored in the venom sac. Whereas the large venom gland of queen bees operates at maximal capacity in newly emerged animals, venom production in worker bees starts slowly after emergence and then increases gradually over a period of about 2 wk (*1*). Worker bee venom is available in large quantities and its constituents have been studied in great detail (*2,3,*). In addition to small amounts of biogenic amines, the enzymes and peptides listed in Table 1 have been isolated from this venom. Even though worker and queen bees are genetically identical, their venoms differ markedly. Phospholipase $A_2$, which is barely detectable in queen bee venom (*4*), is one of the differences noted (*5*). However, in both cases, melittin is the main venom constituent. This peptide was discovered through it lytic action on cells and liposomes. The primary structure of melittin, which is composed of 26 amino acids, was established by Habermann and Jentsch (*6*). The interaction of this peptide with natural and artificial phospholipid bilayers was studied extensively by various biophysical and biochemical methods (*see* ref. *7* for a brief summary), and the crystal structure of the melittin tetramer was elucidated (*8*). The amino acid sequence of the peptides mentioned in Table 1 and of phospholipase $A_2$ were also determined (*2,9,10*).

The biosynthesis of melittin has been studied in some detail. Translation of total poly(A)-RNA isolated from venom glands of newly emerged queen bees in cell-free systems yields prepromelittin, the sequence of which was determined (*11,12*). The liberation of the final venom constituent is a complex process involving three different reactions. First, the signal sequence is cleaved, a reaction that has also been observed in vitro with signal peptidase prepared from rat liver (*13*) or *E. coli* (*14*). The second reaction is the formation of the carboxy-terminal amide with concomitant loss of the extra Gly residue found in the product in vitro (*12*). Finally, promelittin is activated by stepwise cleavage of dipeptides, starting from the amino end (*15*). A dipeptidylaminopeptidase (DAP) was detected in venom glands of queen bees, which catalyzed this late processing reaction. The end product melittin is stored in the venom sac, which has a chitin wall and is thus resistant to the action of the lytic peptide.

## 2. CLONING AND SEQUENCE ANALYSIS OF MELITTIN mRNA

From the venom composition it was possible to anticipate that melittin mRNA should be the main poly(A)-RNA in venom glands. Total mRNA was isolated from venom glands of young queen bees, transcribed into double-stranded cDNA, and inserted into the *Pst*I and Cla site of the plasmid pBR322. Because no specific probe was available, recombinant plasmids were prepared from different clones and sequenced. One of the first, clone pBM22, contained an insert coding for part of the melittin precursor. Subsequently, a clone with a larger insert was found, and its sequence was also determined. The nucleotide sequences of this clone, termed pUM13/4, and of pBM22 are shown in Fig. 1 (*16*). The insert in clone pUM13/4 is composed of 374 nucleotides, excluding the poly(A)-tail. Starting at the 5′ end, there is a very purine-rich sequence of 52 nucleotides, followed by the first ATG codon and an open reading frame coding, for a total of 70 amino acids. The 3′ end contains 112 nucleotides, including the modified polyadenylation sequence AATATAAA. Clone pBM22 has a shorter insert that starts at the fourth melittin codon; its sequence is identical to the corresponding part in clone pUM13/4.

```
                                         1
                                         AGCGAATTAACAGAATTAACAGGAAGGAAGGAAGGAAGCGATCGGAGAAATC

ATG AAA TTC TTA GTC AAC GTT GCC CTT GTT TTT ATG GTC GTG TAC ATT TCT TAC ATC TAT GCG
Met-Lys-Phe-Leu-Val-Asn-Val-Ala-Leu-Val-Phe-Met-Val-Val-Tyr-Ile-Ser-Tyr-Ile-Tyr-Ala-
                                                                                   ↑
GCC CCT GAA CCG GAA CCG GCA CCA GAG CCA GAG GCG GAG GCA GAC GCG GAG GCA GAT CCG GAA
-Ala-Pro-Glu-Pro-Glu-Pro-Ala-Pro-Glu-Pro-Glu-Ala-Glu-Ala-Asp-Ala-Glu-Ala-Asp-Pro-Glu-

                *→
GCG GGA ATT GGA GCA GTT CTG AAG GTA TTA ACC ACA GGA TTG CCC GCC CTC ATA AGT TGG ATT
-Ala-Gly-Ile-Gly-Ala-Val-Leu-Lys-Val-Leu-Thr-Thr-Gly-Leu-Pro-Ala-Leu-Ile-Ser-Trp-Ile-

AAA CGT AAG AGG CAA CAG GGT TAG TCGGATCCATCGATGCCGATTTATCGATCTATCGAATCGTCGAAAAATCTT
-Lys-Arg-Lys-Arg-Gln-Gln-Gly ///

                                                          374
ATTGCAACTTGAAGTAAACATGTATACATGCTGATAATATAAATTTTCTCATTCATTCpolyA
```

Fig. 1. Nucleotide sequence of the cDNA present in clone pUM13/4 and amino acid sequence of prepromelittin. Only the encoding strand with the poly(dA) tail is shown. The sequence of melittin is underlined, the peptide bond hydrolized by signal peptidase is marked by an arrow. Clone pBM22 starts at (* →) (from ref. *16*).

The amino acid sequence obtained by translating the open reading frame in clone pUM13/4 is identical to the one determined for prepromelittin by peptide analysis and sequential Edman degradation. The nucleotide sequence also demonstrates that the mRNA can only code for a polypeptide composed of 70 amino acids, and that prepromelittin is indeed the primary translation product.

Using Southern blot analysis, it was shown that the insert of clone pUM13/4 hybridized mainly to a single *Eco*RI fragment of honeybee DNA with a length of about 3000 base pairs (bp). This indicates that the bee genome of about $0.35 \times 10^9$ bp may contain only a single gene for prepromelittin.

## 3. SEQUENCE ANALYSIS OF THE CLONED mRNA AND PREPROSECAPIN

Translation of total mRNA from venom glands of young queen bees yields prepromelittin and an unknown polypeptide with a molecular weight of about 22,000 as the only major products (*17*). The latter is not present in translates of the corresponding mRNA from worker bees. Cell-free translation in the presence of radioactive His, which does not occur in prepromelittin, revealed the presence of an additional major product that comigrates with the melittin precursor. It seemed likely that the mRNA coding for these two polypeptides should also be present in our cDNA library.

During our search for the melittin clones, two were found that contained inserts coding for the precursor of secapin (*17*). The sequences of these two clones, pUM9/24 and pBMC1, are shown in Fig. 2. Their nucleotide sequences are identical except for the 3′ end, where one contains an additional segment of about 100 nucleotides. It appears that these clones correspond to two different mRNAs generated through the use of different polyadenylation signals during transcription or mRNA maturation. Similar variants were described in a number of other cases (*18,19*).

Both inserts contain an open reading frame of 77 codons, starting with an ATG about 70 bases from the 5′ end. The predicted polypeptide appears to contain an unusually long signal peptide that probably terminates at Ala-32 (*see* ref. 20). The last 25 amino

```
                1    *→
                TTGAATCAAGAAAAAATTACAAGAAATTGAAATTAAATATTAAGGAGTTAAATCAACATTAAAGAAGAATT

ATG AAG AAC TAT TCA AAA AAT GCA ACA CAC TTA ATT ACG GTT CTT CTA TTC AGC TTT GTT GTT ATA
Met-Lys-Asn-Tyr-Ser-Lys-Asn-Ala-Thr-His-Leu-Ile-Thr-Val-Leu-Leu-Phe-Ser-Phe-Val-Val-Ile-

CTT TTA ATT ATT CCA TCA AAA TGT GAA GCC GTT AGC AAT GAT AGG CAA CCA TTG GAA GCA CGA TCT
-Leu-Leu-Ile-Ile-Pro-Ser-Lys-Cys-Glu-Ala-Val-Ser-Asn-Asp-Arg-Gln-Pro-Leu-Glu-Ala-Arg-Ser-
                                        ↑

GCT GAT TTA GTC CCG GAA CCA AGA TAC ATT ATT GAT GTT CCT CCT AGA TGT CCT CCA GGT TCT AAA
-Ala-Asp-Leu-Val-Pro-Glu-Pro-Arg-Tyr-Ile-Ile-Asp-Val-Pro-Pro-Arg-Cys-Pro-Pro-Gly-Ser-Lys-

TTC ATT AAG AAC AGA TGT AGA GTC ATA GTG CCT TAA ATTCGTATGAACTTGCGAGAAGTATATTCAAAAAATAAA
-Phe-Ile-Lys-Asn-Arg-Cys-Arg-Val-Ile-Val-Pro ///

                                                                                ←*
ATATTTTAAATTGATATTTTAAATTAAAGAGTTGCATGAATTAAAAATAATTATTATTAAATAAATATATATATATAAATATGGGAT

TTATAAATAGGAATTTCTTATTAAATACAAAGAAATTTATAATAATTGTGTGTAAAAAATAATAAATAAATAAAAAAAAATTTAAAT

  523
GATGTpolyA
```

Fig. 2. Nucleotide sequence of the cDNA present in clone pUM9/24 and amino acid sequence of preprosecapin. The sequence of secapin is underlined. Except for the last four amino acids, this sequence is identical to the structure determined by others (*10,21*). Clone pBMC1 starts at (* →) and has its poly(dA) tail at (← *). The postulated end of the signal peptide is marked by an arrow (from ref. *17*).

acids in the sequence correspond almost exactly to the structure of secapin, as determined by two groups (*10,21*). These published sequences differed from ours at the carboxyl end. Recently, the sequence derived from the cDNA clones has independently been corroborated (*22*).

Secapin is a minor constituent of worker bee venom (*see* Table 1), discovered by Gauldie et al. (*3*), for which no physiological action has yet been found. The presence of secapin in queen bee venom has not been investigated, but it is assumed that the polypeptide labeled with His, which comigrates with prepromelittin, is, in fact, preprosecapin. Evidence has been presented that queen bee venom has a higher content of secapin than worker bee venom (*17*).

From the structure of the secapin precursor, it is evident that its activation must proceed via a mechanism different from that found for promelittin. A sequence with Pro or Ala at every other position, as in the pro-part of promelittin, which can be hydrolyzed by the DAP present in venom glands, is not found in prosecapin. Therefore, instead, the liberation of secapin must involve cleavage after a single Arg residue. Such cleavages have also been found in the precursors of several hormones, such as somatostatin (*23*), vasopressin-neurophysin II (*24*), and vasoactive intestinal peptide (*25*), even though these are much rarer than processing at pairs of basic amino acids. It appears likely that the liberation of secapin is catalyzed by an enzyme with

TABLE 1
Composition of Worker Bee Venom

| Constituent | Percent of dry weight |
|---|---|
| Enzymes | |
| Phospholipase $A_2$ | 10–15 |
| Hyaluronidase | 2 |
| Acid phosphatase | <1 |
| Peptides | |
| Melittin | 45–60 |
| Apamin | 2–3 |
| MCD—peptide | 2 |
| Secapin | 1 |
| Tertiapin | <1 |

a trypsin-like specificity that may be homologous with other prohormone-cleaving proteases. This processing may well take place inside the cells of the venom gland, as opposed to the activation of melittin, which has to occur outside the cell by a very specific, late-acting mechanism. It will be interesting to determine which one of the two activation routes, the melittin- or the secapin-type, is used for the processing of other venom constituents.

On Southern blots, the secapin clones also hybridized to an *Eco*RI fragment of about 3000 bp. This raises the interesting possibility that the genes for melittin and secapin are in close proximity in the honeybee genome. We are currently trying to isolate these genes from a bee DNA library to test this assumption.

## ACKNOWLEDGMENTS

This work was supported by grants 4174 and 4907 from the Austrian Fonds zur Förderung der wissenschaftlichen Forschung.

## REFERENCES

1. Bachmayer H., Kreil G., and Suchanek G. (1972) Synthesis of promelittin and melittin in the venom gland of queen and worker bees patterns observed during maturation. *J. Insect Physiol.* **18,** 1515–1521.
2. Habermann E. (1972) Bee and wasp venoms. *Science* **177,** 314–318.
3. Gauldie J., Hanson J. M., Rumjanek F. D., Shipolini R. A., and Vernon C. A. (1976) The peptide components of bee venom. *Eur. J. Biochem.* **61,** 369–376.
4. Marz R., Mollay C., Kreil G., and Zelger J. (1981) Queen bee *Apis mellifera* venom contains much less phospholipase than worker-bee venom. *Insect Biochem.* **11,** 685–690.
5. Owen M. D. (1979) Relation between age and hyaluronidase activity in the venom of queen and worker honey bees (*Apis mellifera L*). *Toxicon* **17,** 94–98.
6. Habermann E. and Jentsch J. (1967) Sequenzanalyse des Melittins aus den tryptischen und peptischen Spaltstücken. *Hoppe Seylers Z. Physiol. Chem.* **348,** 37–50.
7. Posch M., Rakusch U., Mollay C., and Laggner P. (1983) Cooperative effects in the interaction between melittin and phospha-

tidylcholine model membranes: Studies by temperature scanning and densitometry. *J. Biol. Chem.* **258,** 1761–1766.

8. Terwilliger T. C. and Eisenberg D. (1982) The structure of melittin. II. Interpretation of the structure. *J. Biol. Chem.* **257,** 6016–6022.
9. Shipolini R. A., Callewaert G. L., Cottrell R. C., and Vernon C. A. (1971) The primary sequence of phospholipase-A from bee venom. *FEBS Lett.* **17,** 39–40.
10. Gauldie J., Hanson J. M., Shipolini R. A., and Vernon C. A. (1978) The structures of some peptides from bee venom. *Eur. J. Biochem.* **83,** 405–410.
11. Suchanek G., Kreil G., and Hermodson M. A. (1978) Amino acid sequence of honeybee prepromelittin synthesized in vitro. *Proc. Natl. Acad. Sci. USA* **75,** 701–704.
12. Suchanek G. and Kreil G. (1977) Translation of melittin messenger RNA in vitro yields a product terminating with glutaminylglycine rather than with glutaminamide. *Proc. Natl. Acad. Sci. USA* **74,** 975–978.
13. Mollay C., Vilas U., and Kreil G. (1982) Cleavage of honeybee prepromelittin by an endoprotease from rat liver microsomes: Identification of intact signal peptide. *Proc. Natl. Acad. Sci. USA* **79,** 2260–2263.
14. Mollay C., Vilas U., and Wickner W. (1983) Unpublished experiments.
15. Kreil G., Haiml L., and Suchanek G. (1980) Stepwise cleavage of the pro part of promelittin by dipepidyl peptidase IV: Evidence for a new type of precursor–product conversion. *Eur. J. Biochem.* **111,** 49–58.
16. Vlasak R., Unger-Ullmann C., Kreil G., and Frischauf A-M. (1983) Nucleotide sequence of cloned cDNA coding for honeybee prepromelittin. *Eur. J. Biochem.* **135,** 123–126.
17. Vlasak R. and Kreil G. (1984) *Eur. J. Biochem.* **145,** 279–282.
18. Lagace L., Chandra T., Woo S. L. C., and Means A. R. (1983) Identification of multiple species of calmodulin messenger RNA using a full length complementary DNA. *J. Biol. Chem.* **258,** 1684–1688.
19. Parnes J. R., Robinson R. R., and Seidman J. G. (1983) Multiple mRNA species with distinct 3′ termini are transcribed from the $\beta_2$-microglobulin gene. *Nature,* **302,** 449–452.
20. von Heijne G. (1983) Patterns of amino acids near signal–sequence cleavage sites. *Eur. J. Biochem.* **133,** 17–21.
21. Kudelin A. B., Martynov V. I., Kudelina I. A., and Miroshnikov A. I. (1978) *Abstr., 15th Eur. Pept. Symp.* Gdansk.
22. Lui L. K. and Vernon C. A. (1984) *J. Chem. Res. S.* 10–11.
23. Goodman R. H., Jacobs J. W., Dee P. C., and Habener J. F. (1982)

Somatostatin-28 encoded in a cloned cDNA obtained from a rat medullary thyroid carcinoma. *J. Biol. Chem.* **257,** 1156–1159.

24. Land H., Schütz G., Schmale H., and Richter D. (1982) Nucleotide sequence of cloned cDNA encoding bovine arginine vasopressin-neurophysin II precursor. *Nature* **295,** 299–303.
25. Itoh N., Obata K., Yanaihara N., and Okamoto H. (1983) Human preprovasoactive intestinal polypeptide contains a novel PHI-27-like peptide, PHM-27. *Nature* **304,** 547–550.

# Chapter 18

# Yeast α-Factor Genes

JANET KURJAN

## 1. INTRODUCTION

Each of the two haploid mating types (**a** and α) in the yeast *Saccharomyces cerevisiae* secretes an oligopeptide pheromone that plays a role in the mating process (reviewed in refs. *1,2*). Cells of **a** mating type secrete an 11-amino acid oligopeptide called **a**-factor (*3*) and cells of α mating type secrete a 13-amino acid oligopeptide called α-factor (*4,5*). The amino acid sequence of active α-factor is Trp-His-Trp-Leu-Gln-Leu-Lys-Pro-Gly-Gln-Pro-Met-Tyr. A portion of the α-factor isolated from cells lacks the amino-terminal Trp residue, but shows the same level of activity as that of full-length α-factor (*6–8*).

Several lines of evidence indicate that the sex pheromones are involved in the mating process. Exposure to the pheromone of the opposite mating type results in arrest in the G1 phase of the cell cycle, just prior to the initiation of DNA synthesis (*5,9*). It is at this point of the cell cycle, termed Start, that cells of the two mating types are able to mate with one another (*10*). A second response is the induction of cell surface proteins that promote agglutination of cells of opposite mating types, an important prelude to conjugation (reviewed in ref. *11*). Additional evidence supporting the role of **a**- and α-factor in mating arises from genetic analysis. For example, many mating-defective (sterile) mutants show defects in pheromone production and/or response

(*12–14*). Also, all of the mutants that have been isolated by selection for α-factor resistance (i.e., mutants defective in the response to α-factor) are also sterile (*15,16*). However, genetic analysis of mutations affecting mating or α-factor production has failed to identify an α-factor structural gene. Recently, two α-factor structural genes, *MF*α1 and *MF*α2, have been identified by cloning (*17,18*). The existence of two α-factor genes may explain the difficulty in obtaining α-factor mutations.

DNA sequencing of *MF*α1 and *MF*α2 indicates that α-factor is encoded within large precursor proteins of 165 amino acids for *MF*α1 and 120 amino acids for *MF*α2 (*17,18*). The *MF*α1 and *MF*α2 precursors show a strong similarity in overall structure and, although not identical in amino acid sequence, a fairly high level of homology. These precursors show several features in common with precursors of secreted peptides from higher organisms. Both the *MF*α1 and *MF*α2 gene products contain a signal sequence at the amino-terminus. This is followed by a region of approximately 60 amino acids within which are three putative glycosylation sites, which I will refer to as the pro region. The carboxy-terminal portion of the precursors consists of tandem repeats of α-factor (or α-factor-like) sequences, each preceded by a spacer sequence of six to eight amino acids. *MF*α1 contains four spacer-α-factor repeats and *MF*α2 contains two spacer-α-factor repeats. One of the *MF*α2 repeats encodes a peptide that differs from α-factor by two amino acid substitutions.

The sequence of the spacer sequences preceding each of the α-factor repeats suggests that several steps are required for the processing of mature α-factor from the precursor proteins and that these steps are similar to steps involved in peptide processing in higher eukaryotes. Maturation of the carboxy-terminus of α-factor involves cleavage at a dibasic amino acid sequence similar to that involved in maturation of polypeptide hormones in higher organisms (*19,20*). Maturation of the amino-terminus of α-factor involves cleavage by a dipeptidyl aminopeptidase in a manner similar to the processing of promelittin, the major component of honeybee venom (*21,22*).

This article will focus on the cloning and structure of the two α-factor structural genes and the precursor processing of the α-factor precursor proteins.

## 2. CLONING OF *MFα1* USING A GENETIC TECHNIQUE

In recent years, it has become straightforward to clone virtually any yeast gene in which there is a mutation, by virtue of the ability of the cloned structural gene to complement the mutation in yeast. The lack of α-factor mutations made cloning of an α-factor structural gene by complementation impossible. Instead, a somewhat more circuitous genetic technique was used (*17*). This technique involved the isolation of a clone that when present in yeast in high copy number is able to partially alleviate an α-factor defect resulting from a regulatory mutation. In order to explain this technique in more detail, it is necessary to explain features of the mating type locus (*MAT*) that are involved in the regulation of genes unlinked to *MAT* that are involved in the mating process.

*MAT***a** and *MAT*α are the alleles of the mating type locus responsible for the **a** and α cell types, respectively. *MAT*α contains two complementation groups, *MAT*α1 and *MAT*α2, both of which are proposed to be involved in the regulation of unlinked genes involved in the mating process (*23*). The analysis of *mat*α1$^-$ and *mat*α2$^-$ mutants led to a model for *MAT* regulation, as diagrammed in Fig. 1. This model has been confirmed and extended by more recent molecular experiments (*24,25*). *MAT*α1 is a positive regulator of α-specific genes; that is, genes (not at *MAT*) that are required for mating in α but not in **a** cells. Therefore, *mat*α1$^-$ strains are sterile because of the lack of α-specific products essential for the mating process. *MAT*α2 is a negative regulator of **a**-specific genes. A *mat*α2$^-$ mutation results in a defect in this negative regulator, thus allowing the production of **a**-specific gene products. The *MAT*α1 product is still functional, allowing expression of α-specific gene products. Therefore, the *mat*α2$^-$ defect leads to simultaneous production of both **a**- and α-specific gene products. It has been proposed that the sterility and lack of α-factor production by *mat*α2$^-$ strains is caused by an antagonism between **a**- and α-specific products (*14,26*). In particular, it has been proposed that the lack of α-factor secretion in the *mat*α2$^-$ strain is a result of the simultaneous production of α-factor, an α-specific gene product, and barrier, an *a*-specific product that is involved in α-factor degradation (*27–30*). This proposal was

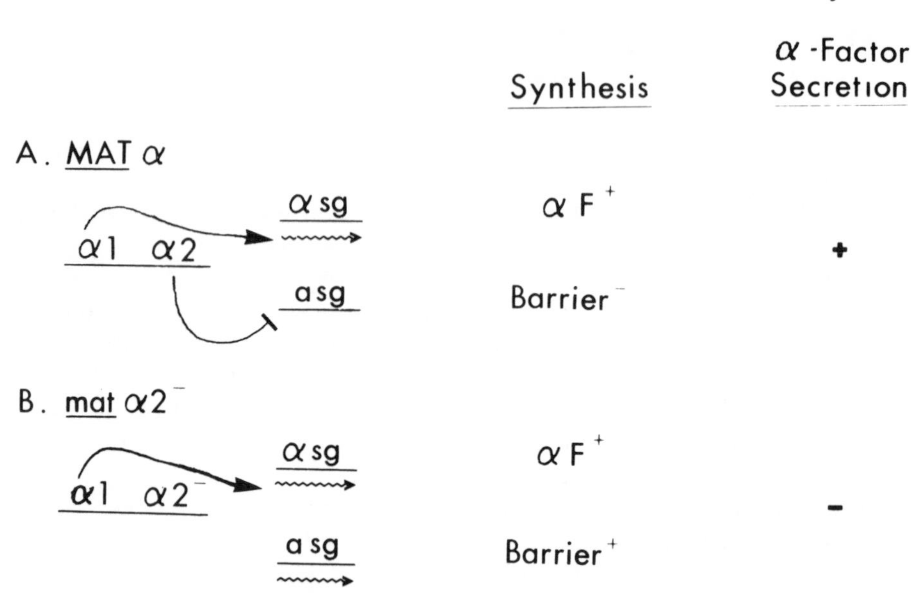


Fig. 1. Regulation of α-factor production by the mating-type locus. The mating-type locus is depicted on the left for a *MAT*α cell (A) and a *mat*α2⁻ cell (B). *MAT*α consists of two complementation groups. *MAT*α1 is a positive regulator of unlinked α-specific genes (αsg, including the α-factor structural genes) leading to expression of these genes, which is indicated by an arrow. *MAT*α2 is a negative regulator of **a**-specific genes (asg, including barrier, an activity involved in α-factor degradation). *MAT*α cells produce α-factor, but no barrier activity, leading to secretion of α-factor (A). In *mat*α2⁻ cells (B), α-factor is produced, as a result of the functional *MAT*α1 product. The defect in *mat*α2⁻ allows expression of barrier that degrades the α-factor. The *mat*α2⁻ cells are therefore defective in secretion of α-factor.

confirmed by isolation of a mutation in the *mat*α2⁻ strain (*bar*1⁻) that allows α-factor secretion but not mating (*26*). When present in a *MAT*α strain, the *bar*1⁻ mutation results in a defect in the barrier property (*26*), most likely because of a defect in an α-specific proteolytic activity.

In addition to confirming that the α-factor defect in *mat*α2⁻ cells is caused by the simultaneous production of **a**- and α-specific products, this result suggested a possible scheme for the isolation of the α-factor structural gene(s) (*17*). Presence in a *mat*α2⁻ strain of an α-factor structural gene on a high copy number plasmid might lead to a sufficient level of α-factor overproduction that some of the pheromone might escape degradation by the barrier protease. Therefore, a yeast clone bank (*31*) in YEp13 (a high copy number yeast–*E. coli* shuttle vector) (*32*) was screened in a *matα2*$^-$ strain for a clone that secretes α-factor. Secretion of α-factor was assayed utilizing a mutation (*sst2*$^-$) that results in greatly increased levels of sensitivity to the sex pheromones (*33*). Strains to be tested for α-factor secretion are replica plated to a thin lawn of *MAT***a** *sst2*$^-$ cells. Colonies that secrete α-factor inhibit the growth of the surrounding *MAT***a** *sst2*$^-$ cells and therefore produce a "halo" of no growth. This assay (modified from ref. *34*) is called the "halo" assay. A clone that produces a small halo on the *MAT***a** *sst2*$^-$ lawn was found. The plasmid responsible for this phenotype, p69A, was isolated and tested for other properties that would be expected of an α-factor structural gene clone (*17*). In particular, p69A was shown to result in the overproduction of α-factor when present in a *MATα* strain. Since the amino acid sequence of α-factor was known, the gene resulting in α-factor production was sequenced and shown to contain an α-factor coding sequence (*17*).

## 3. STRUCTURE OF *MFα1*

p69A contains a yeast DNA insert of approximately 4 kilobases (kb) (*17*). The restriction map is shown in Fig. 2. p69A was subcloned and the subclones were tested by the halo assay in or-

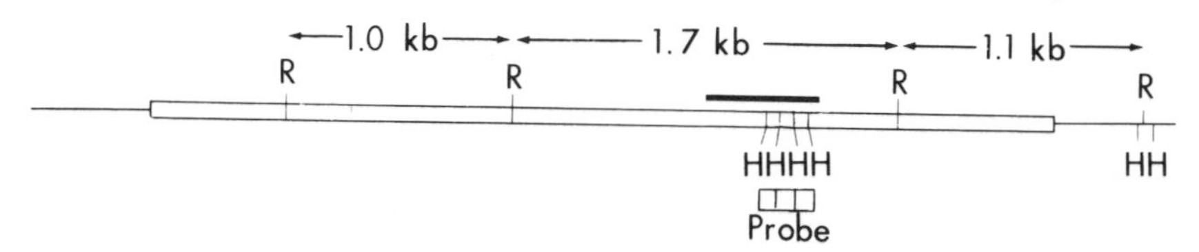


Fig. 2. Restriction map of *MFα1* and probe utilized for isolation of *MFα2*. The yeast DNA insert of p69A is shown as a rectangle. The flanking regions from YEp13 are shown as lines. The restriction sites for *Eco*RI (R) and *Hin*dIII (H) cleavage are shown. The "halo" activity of p69A was shown to be present on the 1.7-kb *Eco*RI fragment and is disrupted by the four *Hin*d III sites. The *MFα1* coding region is shown by a heavy bar. Cleavage of the 1.7-kb *Eco*RI fragment by *Hin*dIII generates three 63-bp *Hin*dIII fragments that contain only spacer and α-factor sequences. These fragments were labeled with $^{32}$P-deoxynucleotides and used as a probe to isolate *MFα2*.

der to determine the location of the gene responsible for the halo phenotype. The gene is present on the 1.7-kb *Eco*RI fragment and is interrupted by the four *Hin*dIII sites within this fragment. DNA sequencing of the region surrounding the *Hin*dIII sites showed that this fragment encodes a putative α-factor precursor of 165 amino acids (Figs. 3 and 4). The 20–22 amino acids at the amino-terminus of this precursor resemble the signal sequences found at the start of most secreted proteins (reviewed in ref. *35,36*). The signal sequence is followed by a pro region of approximately 60 amino acids, within which there are three putative glycosylation sites (Asn-*X*-Thr) (*37*). The carboxy-terminal half of the precursor consists of four tandem repeats of α-factor preceded by a region of six to eight amino acids called the "spacer" region. The amino acid sequence of the spacers is Lys-Arg-(Glu-Ala)-$^{Glu}/_{Asp}$-Ala-Glu-Ala. The first spacer is missing the two amino acids shown in parentheses, and the fifth residue is Glu in the first two spacers and Asp in the third and fourth spacers. The four α-factor repeats show a few nucleotide differences, but all code for the mature α-factor sequence.

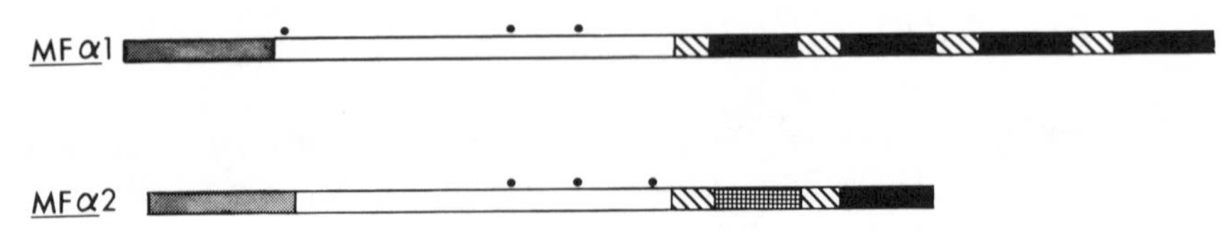


Fig. 3. Structure of *MF*α1 and *MF*α2 precursor proteins. The precursor proteins encoded in the *MF*α1 and *MF*α2 genes are diagrammed. *MF*α1 encodes a precursor of 165 amino acids, and *MF*α2 encodes a precursor of 120 amino acids. Both contain a typical signal sequence at the amino-terminus (shown by the dotted region). This is followed by a pro region of approximately 60 amino acids within which there are three putative N-glycosylation sites (indicated by large dots). The carboxy-terminal portion of the precursor consists of tandem repeats of spacer (shown by diagonal lines) and α-factor (solid or grid patterns) sequences. *MF*α1 contains four repeats, and *MF*α2 contains two repeats. The *MF*α1 spacers have the sequence Lys-Arg-(Glu-Ala)-$^{Glu}/_{Asp}$-Ala-Glu-Ala-and the *MF*α2 spacer sequences are Lys-Arg-Glu-Ala-$^{Val}/_{Asn}$-Ala-Asp-Ala-. All of the *MF*α1 and the second *MF*α2 α-factor repeat encode the mature α-factor sequence Trp-His-Trp-Leu-Gln-Leu-Lys-Pro-Gly-Gln-Pro-M et-Tyr. The first *MF*α2 repeat differs from this sequence by two amino acid substitutions: $Gln_5$ →Asn and $Lys_7$ →Arg.

## 4. CLONING OF *MF*α1 AND *MF*α2 BY HYBRIDIZATION

A second α-factor structural gene has been identified using two approaches (*18,38*). Singh et al. (*18*) synthesized oligonucleotides with sequences that would be likely to be homologous to a region of the α-factor coding sequence based on amino acid sequence. Two oligonucleotides (CC$^{A}_{G}$TTGGTTACATG and CC$^{A}_{G}$GTTGGATACATG) were synthesized. These oligonucleotide sequences were chosen as most likely to be homologous to the region of an α-factor structural gene coding for the last five amino acids of α-factor (Gly-Gln-Pro-Met-Tyr), based on preferred codon usage in yeast (*39*). The oligonucleotides were labeled and used to screen a yeast clone bank. Clones corresponding to *MF*α1 and a second gene, called *MF*α2, were isolated. The second approach took advantage of homology between *MF*α1 and *MF*α2. (*38*). As shown in Fig. 2, cleavage of p69A with *Hin*dIII generates three 63-bp *Hin*dIII fragments that consist only of spacer and α-factor sequences. These fragments were labeled and used as a probe against a Southern blot of yeast DNA. In addition to the *MF*α1 fragments, this probe hybridized well to an additional fragment, a large *Eco*RI, and a 1.8-kb *Hin*dIII fragment, and faintly to several other bands that have not been further characterized at this time. Clones containing the 1.8-kb *Hin*dIII fragment were isolated from a yeast clone bank and shown to contain the *MF*α2 gene by DNA sequencing.

The *MF*α1 genes isolated by the two groups were isolated from different yeast strains and contain several polymorphisms (*17,18*). Four out of five nucleotide differences are outside the coding region. Two nucleotides, 7 and 8 base pairs (bp) upstream from the ATG start codon, differ by CC-TT transitions. There are additional differences 106 and 143 bp 3′ to the stop codon; an A-C transversion and a C-T transition. Within the coding region there is a single polymorphism, a C-T transition that results in a Leu-Ser difference in the pro region. The function, if any, of this region of t he precursor is unknown and this amino acid difference is likely to have no effect on function. Within the region of *MF*α2 that has been sequenced by both groups, no polymorphisms are found.

The initial cloning method for *MF*α1 required that the gene be expressed when present in yeast in a high-copy-number plasmid

```
MFα1         CAACAGGTTT TGGATAAGTA CATATATAAG AGGGCCTTTT GTTCCCATCA AAAATGTTAC TGTTCTTACG
MFα2         CGCAGAGACG AGGGCTTATA TGTATAAAAG CTGTCCTTGA TTCTGGTGTA GTTTGAGGTG TCCTTCCTAT    5' Flanking
                                                                                    TT           Region
 ATTCATTTAC GATTCAAGAA TAGTTCAAAC AAGAAGATTA CAAACTATCA ATTTCATACA CAATATAAAC GACCAAAAGA
 ATCTGTTTTT ATATTCTATA TAATGGATAA TTACTACCAT CACCTGCATC AAATTCCAGT AAATTCACAT ATTGGAGAAA
________________________________________________________________________________________________

MFα2     A   C AT   T  CC    CTC    TTT A      ——  G   C GTT   T  TC —— A
MFα1 ATG AGA TTT CCT TCA ATT TTT —— ACT GCA GTT TTA TTC GCA GCA TCC TCC GCA TTA GCT GCT      Signal
     Met Arg Phe Pro Ser Ile Phe —— Thr Ala Val Leu Phe Ala Ala Ser Ser Ala Leu Ala Ala      Sequence
MFα2     Lys     Ile     Thr     Leu     Phe Ile     ——          Val     Val —— Thr
________________________________________________________________________________________________

     —— —— —— —— GT T C  T   A   T  TC   T  G G G  A   C   G   C A T  T   A
     CCA GTC AAC ACT ACA ACA GAA GAT GAA ACG GCA CAA ATT CCG GCT GAA GCT GTC ATC GGT TAC
     Pro Val ASN THR THR Thr Glu Asp Glu Thr Ala Gln Ile Pro Ala Glu Ala Val Ile Gly Tyr
     —— —— —— —— Ser Ser Asp Glu Asp Ile         Val                 Ile

      TG      C   G   T      CAT  C A A    T    A        C AGT     GCT  C GCC  GT
     TCA GAT TTA GAA GGG GAT TTC GAT GTT GCT GTT TTG CCA TTT TCC AAC AGC ACA AAT AAC GGG     Proregion
     Ser Asp Leu Glu Gly Asp Phe Asp Val Ala Val Leu Pro Phe Ser ASN SER THR Asn Asn Gly
     Leu     Phe Gly         His     Ile     Phe                     ALA     Ala Ser

     C            C   C   C           T GAG ——   G      GAA      G C  AAC ACC A        CG
     TTA TTG TTT ATA AAT ACT ACT ATT GCC AGC ATT GCT GCT —— AAA GAA GAA GGG GTA TCT TTG GAT
     Leu Leu Phe Ile ASN THR THR Ile Ala Ser Ile Ala Ala —— Lys Glu Glu Gly Val Ser Leu Asp
                                         Glu ——         Glu         Gln ASN THR THR     Ala
________________________________________________________________________________________________

                     GTT GCC   C             C       A A T T G  G    A
     AAA AGA GAG GCT —— —— GAA GCT TGG CAT TGG TTG CAA CTA AAA CCT GGC CAA CCA ATG TAC
     Lys Arg Glu Ala —— —— Glu Ala Trp His Trp Leu Gln Leu Lys Pro Gly Gln Pro Met Tyr
                     Val Ala Asp                     Asn     Arg
```

```
                  G      A C       T           C       T         C       A
AAG AGA GAA GCC GAA GCT GAA GCT TGG CAT TGG CTG CAA CTA AAG CCT GGC CAA CCA ATG TAC     Spacer-
Lys Arg Glu Ala Glu Ala Glu Ala Trp His Trp Leu Gln Leu Lys Pro Gly Gln Pro Met Tyr    α-Factor
                Asn     Asp                                                             Repeats
AAA AGA GAA GCC GAC GCT GAA GCT TGG CAT TGG CTG CAA CTA AAG CCT GGC CAA CCA ATG TAC
Lys Arg Glu Ala Asp Ala Glu Ala Trp His Trp Leu Gln Leu Lys Pro Gly Gln Pro Met Tyr

AAA AGA GAA GCC GAC GCT GAA GCT TGG CAT TGG TTG CAG TTA AAA CCC GGC CAA CCA ATG TAC
Lys Arg Glu Ala Asp Ala Glu Ala Trp His Trp Leu Gln Leu Lys Pro Gly Gln Pro Met Tyr
```

```
MFα1 TAA GCCCGACTGA TAACAACAGT GTAGATGTAA CAAAGTCGAC TTTGTTCCCA CTGTACTTTT AGCTCGTACA
MFα2 TGA AAAATGACCC TAAACTACTT CTAAACCCTC TCGATTTCTT TCACGTTCAT ACAACACCTA GTTTTATTTA   3' Flanking
                                       A                                         C          Region
AAATACAATA TACTTTTCAT TTCTCCGTAA ACAACCTGTT TTCCCATGTA ATATCCTTTT CTATTTTTCG TTTCGTTACC
TTTTCTTTTC AATCTGAGTA GTTGAGTTTT CGATCACTCA CATAGAACTA TTTTTTGCCA TTTAAATAAA GTATTCTCTC
```

Fig. 4. Nucleotide and amino acid sequence of *MFα1* and *MFα2*. The nucleotide sequences for 150 nucleotides of the coding strand of the 5′-flanking regions of *MFα1* and *MFα2* are shown. The TATA boxes are underlined. The nucleotide and amino acid sequences of the coding region of *MFα1* are shown divided into the signal sequence, pro region, and spacer–α-factor repeat regions. Positions at which *MFα2* differs from *MFα1* are indicated above the *MFα1* sequence for nucleotide differences and below the *MFα1* sequence for amino acid differences. The sequence shown is that determined by Kurjan and Herskowitz (*17*). The sequence determined by Singh et al. (*18*) contains a TTA (Leu) codon at the 22nd amino acid of the proregion. The nucleotide sequences for 150 nucleotides of the 3′-flanking sequences of *MFα1* and *MFα2* are shown. Sequences that show homology to sequences found in the 3′-region of other yeast genes are underlined. The positions in the 5′- and 3′-flanking regions that are polymorphic in the two *MFα1* genes that have been sequenced (*17,18*) are indicated.

(*17*). The *MF*α2 clones were tested in a similar manner (*18,38*). When present in a *mat*α2$^{-}$ strain on YEp13, *MF*α2 results in the production of a small halo, approximately equivalent to that produced with *MF*α1. The results of the two groups differ about the phenotype of a *MAT*α strain carrying *MF*α2. The same clone as that tested in *mat*α2$^{-}$ resulted in, at most, a slight increase in α-factor production in the *MAT*α strain. However, the *MF*α2 clone of Singh et al. (*18*) resulted in a significant increase in α-factor production in a *MAT*α strain. This difference could be a result of a difference in the vector used, in the *MAT*α strain, or in the *MF*α2 gene itself. The latter possibility seems unlikely, however, because the genes isolated by the two groups are identical within the regions that have been sequenced.

## 5. STRUCTURE OF *MF*α2 AND COMPARISON OF *MF*α1 AND *MF*α2

*MF*α2 encodes a putative α-factor precursor of 120 amino acids that shows a considerable degree of similarity to the *MF*α1 precursor (*18,38*). These precursors are diagrammed in Fig. 3 and the sequences are shown in Fig. 4. Both precursors contain a signal sequence, a pro region, and tandem spacer–α-factor repeats. The difference in length between the two precursors (165 amino acids for *MF*α1 and 120 amino acids for *MF*α2) is accounted for mainly by the presence of only two spacer–α-factor repeats in the *MF*α2 precursor vs four in *MF*α1. Another important difference is that, whereas all four α-factor repeats in *MF*α1 encode the mature α-factor sequence that has been isolated from yeast cells, only the second α-factor repeats in *MF*α2 encodes a α-factor. The first repeat encodes a peptide that differs from α-factor by two amino acid substitutions. The spacer sequences in *MF*α2 also show some differences from the *MF*α1 spacers. The *MF*α2 spacers have the sequence Lys-Arg-Glu-Ala-Val/Asn-Ala-Asp-Ala-, as opposed to the sequence of Lys-Arg-(Glu-Ala)-Glu/Asp-Ala-Glu-Ala-in *MF*α1. The possible effect of these differences on precursor processing are discussed below.

The amount of homology between the *MF*α1 and *MF*α2 coding regions is considerable, although the level varies between the different regions of the precursors. The homology is greatest in the regions of spacer–α-factor repeats. The two genes differ at only 15% of the positions in this region, both in terms of amino acid sequence and nucleotide sequence (a deletion is counted as a single difference). The greatest number of differences are found in the signal sequences, with 52% of the amino acids and 41% of the base pairs showing differences. Both signal sequences are typical and are likely to function as such. Both precursors contain a pro region of about 60 amino acids of unknown function that contain three putative glycosylation sites. Two of these sites are at identical positions and the third is at quite different positions in the two precursors (see Fig. 3). Overall, there is a fairly high degree of homology within this region; 41–43% of the amino acids and 38–40% of the nucleotides differ (the range of percentages is a result of the single polymorphism found within the two *MF*α1 genes that have been sequenced). The considerable similarity in structure and sequence between the two precursors suggests that the two genes originated by a gene-duplication event. However, the noncoding regions of the two genes show no homology.

The noncoding regions of both genes show features similar to other yeast genes. There are TATA boxes (*40*) at position −128 (upstream from the ATG initiation codon) in *MF*α1 and −128 and −64 in *MF*α2 (*18*). Many yeast genes do not contain the consensus polyadenylation sequence AATAAA found downstream from most protein-encoding genes in eukaryotes. However, a sequence similar to the consensus sequence TAAATAAAG is found downstream of most yeast genes (*41*). The sequences TAACAAAG and TAAACAAC are found 28 and 98 bp, respectively, downstream from the stop codon in *MF*α1, and the sequence TAAATAAA is found 133 bp downstream from the stop codon in *MF*α2. Another sequence that has been implicated in the process of transcription termination in yeast, (*42*) TAG . . . TAGT/TATCT . . . TTT is not present in the 3′-flanking region of *MF*α1 (*42*). A sequence similar to this consensus sequence, TAC-------TAGTTTT-TTT-TTTT-TTTT is found about 55 bp downstream from the stop codon in *MF*α2. Both genes show a tendency to use codons that are preferred for yeast genes, although the level of preferred codons is somewhat less than that found in many genes in yeast (*39*). For those amino acids for which a preferred codon (or codons) has been determined (excluding those amino acids for which there is a single codon), 63% of the codons in *MF*α1 and

71% of the codons in *MF*α2 are preferred. The stop codon present in *MF*α1 is TAA and in *MF*α2, TGA. In both cases, a second stop codon is present eight codons downstream from the first stop codon.

## 6. PRECURSOR PROCESSING

The sequences of the precursor proteins encoded in *MF*α1 and *MF*α2 show certain similarities to precursors of secreted peptides from higher organisms, suggesting that processing of the prepropoly-α-factor precursor utilizes mechanisms similar to the mechanisms involved in precursor processing in higher organisms. The processing pathway for α-factor maturation is shown in Fig. 5. The majority of this pathway was initially speculative, being based on features in common with other processed peptides. Recently, most of the steps were confirmed for processing of the *MF*α1 precursor by experimental results (*19–21,43,44*), as described below.

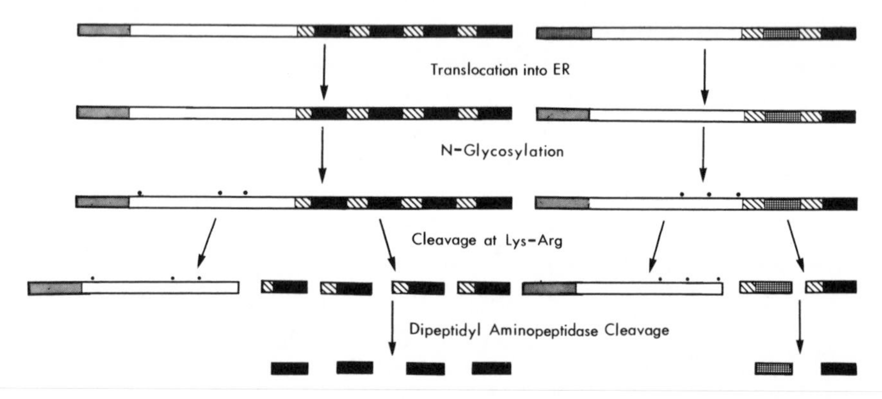


Fig. 5. α-Factor-precursor processing pathway. The signal sequences are indicated by a dotted rectangle, the pro region by an open rectangle, and the putative glycosylation sites by large dots. The spacer is indicated by diagonal lines, and α-factor sequences are indicated by solid rectangles or a grid for the first *MF*α2 α-factor sequence. Details of this processing pathway are described in the text.

At the amino-terminus, both precursors contain a sequence that resembles the signal sequences of most secreted proteins (*35,36*). It seems probable that this sequence directs translation to the endoplasmic reticulum (ER), leading to translocation of the precursor into the ER. Surprisingly, experiments using yeast mutants defective in secretion indicate that the *MFα1* signal sequence is not removed (*44*). The three glycosylation sequences within the *MFα1* precursor are glycosylated during the secretion process (*37,43,44*).

Processing of the propoly-α-factor to the mature pheromone must require two different mechanisms. Whereas polypeptide hormones for which the precursor sequence is known are flanked on both sides by basic amino acid pairs (for example, *see* refs. *35,45–48*), the α-factor sequences contain such a sequence only at the carboxy-terminus. Recent results have identified a membrane-bound proteinase that cleaves after a basic amino acid pair similar or identical to that found in the *MFα1* and *MFα2* spacer peptides (*19,20*). The *KEX2* gene is required for this activity and is likely to be the structural gene encoding the endopeptidase (*20*). Cleavage after the Lys-Arg sequence should produce an α-factor processing intermediate with several extra amino acids at the amino-terminus and the two basic amino acids at the carboxy-terminus. A membrane-bound carboxypeptidase activity was identified that may be involved in removal of the two basic amino acids from the carboxy-terminus of α-factor (*19*).

The processing steps described above should yield α-factor peptides with four to six additional amino acids attached to the amino-terminus, (Glu-Ala)-Glu/Asp-Ala-Glu-Ala- for *MFα1*, and Glu-Ala-Val/Asn-Ala-Asp-Ala- for *MFα2*. Clearly, a different mechanism is required to process the amino-terminus of α-factor. Such a mechanism was suggested by the work of Kreil et al., on the protein melittin, the major component of honeybee venom (*22*). Melittin is produced as a preproprotein. The pro part of the precursor consists of a sequence quite similar to that of the spacer peptides in the α-factor precursors, particularly those in *MFα1*. The pro region of promelittin consists of 11 repeats of the dipeptide *X*-*Y*, where *X* = Glu, Asp, or Ala; and *Y* = Ala or Pro. These spacer sequences in the *MFα1* precursor contain two to three repeats of the dipeptide *X*-*Y*, where *X* = Glu or Asp; and *Y* = Ala. The *MFα2* spacers are similar, except that the *X* residue at position 5 of the spacer is Val or Asn. Kreil and coworkers showed that promelittin is processed by a DAP that removes dipeptides of the *X-Y* sequence found in melittin from the amino-terminus (*22*). Promelittin can be processed to the mature melittin by a DAP from pig kidney or from queenbee venom glands. Suarez-Rendueles et al. (*49*) showed that yeast contains a DAP activity that efficiently releases *P*-nitroaniline (*p*NA) from the substrate *X*-Pro-*p*NA when $X$ = Ala, and considerably less efficiently when $X$ = Gly ($X$ = Glu or Asp was not tested). The similarity of the α-factor-spacer sequences to the promelittin sequences processed by the dipeptidyl aminopeptidases suggested that the amino terminus of α-factor is likely to be processed in a similar manner (*17*). Recent results confirmed that this mechanism is involved in the maturation of α-factor (*21*).

The involvement of a DAP in the maturation of the amino-terminus of α-factor was elucidated by experiments utilizing the α-specific sterile mutant, *ste*13$^-$ (*21*). The *ste*13$^-$ mutation results in both a mating defect and a defect in α-factor secretion in *MAT*α cells, but has no effect on mating or **a**-factor secretion in *MAT***a** cells. Julius et al. (*21*) also showed that *ste*13$^-$ strains are defective in a membrane-bound, heat-stable DAP (a second, heat-labile DAP is not affected), and that strains containing the wild-type *STE*13 gene on a high copy number plasmid show significantly elevated levels of the heat-stable DAP. These results suggest that the DAP associated with the *STE*13 gene is responsible for maturation of the amino-terminus of mature α-factor.

Processing the amino-terminus of α-factor from the *MF*α2 precursor requires removal of *X*-Ala dipeptides in which *X* is different from the amino acids found in promelittin or in *MF*α1 pro-α-factor. The sequences Val-Ala and Asn-Ala are present in the first and second spacer-α-factor repeats of *MF*α2, respectively. It was shown that DAPs from different sources have different specificities with respect to the dipeptide cleaved (*50*). The honeybee DAP described above shows a relatively low rate of cleavage of the sequence *X*-Gly (a dipeptide that is not present in promelittin), whereas a DAP isolated from frog-skin secretion cleaves *X*-Gly with high efficiency. This enzyme is likely to be involved in the processing of the frog-skin peptide cerulein, which contains the dipeptide Asp-Gly in the pro part of the cerulein precursor. It is possible that the yeast DAP involved in α-factor processing may remove the dipeptides Val-Ala and Asn-Ala present in *MF*α2 with a different efficiency than the more common Glu/Asp-Ala dipeptide.

## 7. FUNCTION OF *MF*α1 VS *MF*α2

The existence of at least two α-factor-encoding genes brings up the question of the role of the two genes in α-factor production and mating. To determine the relative importance of these two genes with respect to these processes, mutant analysis was undertaken (*38*). This utilized the one-step gene-disruption technique, developed by Rothstein (*51*), which allows the construction of a totally nonfunctional gene. This technique involves inserting a DNA fragment that contains a selectable yeast marker into the gene to be mutated. The fragment may be inserted into the coding region of the gene or may actually replace the entire gene, as diagrammed in Fig. 6. Both types of construction should result in a nonfunctional gene product. The mutant gene constructed in vitro is then used to replace the wild-type gene in yeast, and the phenotype of the mutant strain is tested.

Several gene disruptions were made in *MF*α1 and *MF*α2 individually, and the double mutants were also constructed (*38*). The *mf*α1$^-$ mutations result in a significant reduction in α-factor secretion. The *mf*α2$^-$ mutations have no detectable effect on α-factor secretion, as assayed by the halo assay. The *mf*α1$^-$ *mf*α2$^-$ double mutant secretes no detectable α-factor. These results suggest that these two genes are the only (functional) α-factor structural genes and that *MF*α1 plays a significantly more important role than *MF*α2 does in α-factor production. Because of the different numbers of α-factor repeats in *MF*α1 and *MF*α2, it might be expected that *MF*α1 would produce more α-factor than *MF*α2. However, the level of α-factor produced in the *mf*α1$^-$ mutants is considerably lower than the level expected if the two genes were expressed at an equivalent level. This difference could be caused by differential transcription, messenger stability, translation, or precursor processing of the *MF*α2 gene product with respect to the *MF*α1 product.

The mating phenotypes of the mutants were also assayed (*38*). The proportion of cells that are able to mate in wild-type and mu-

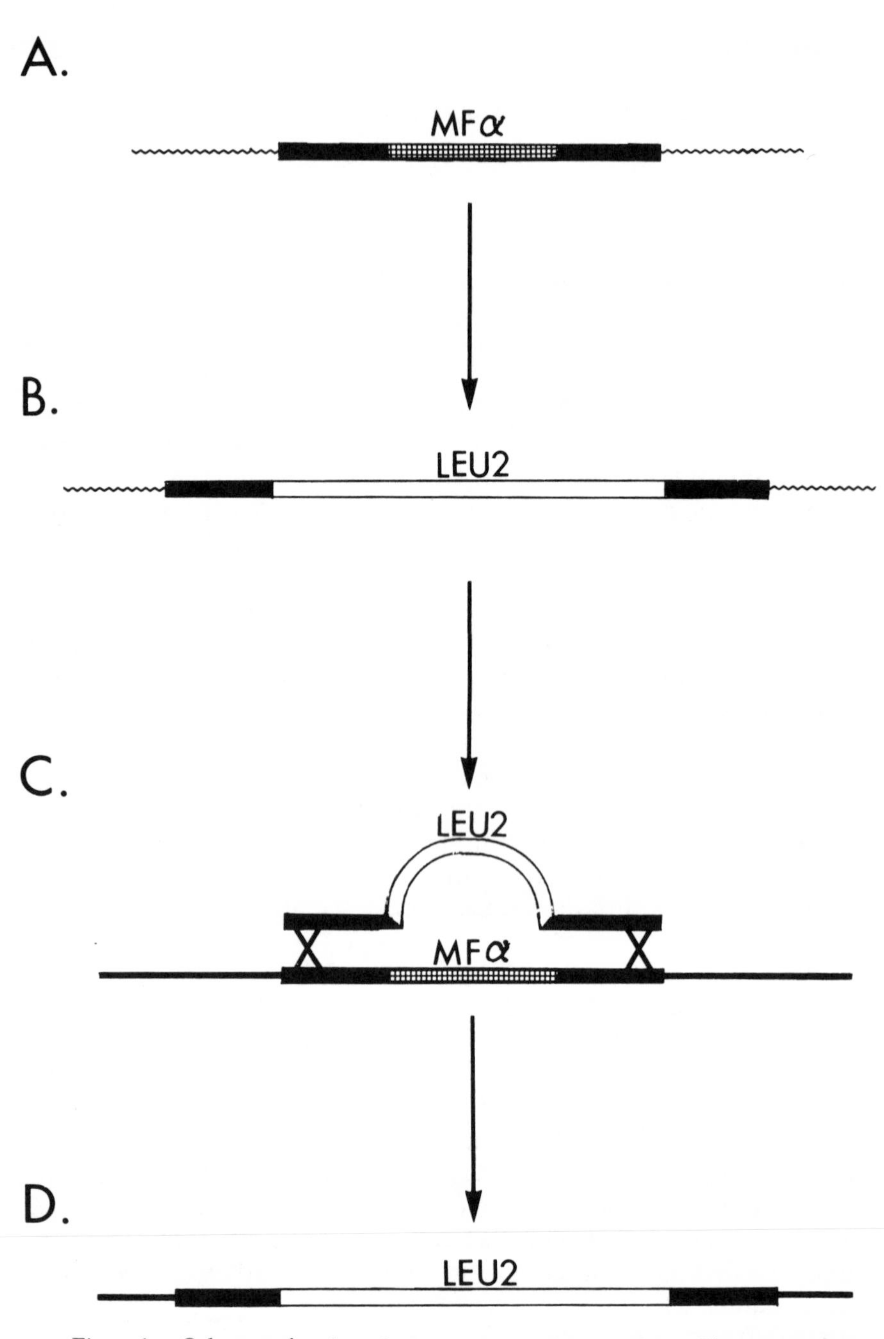


Fig. 6. Scheme for in vitro mutant construction. The one-step gene-disruption technique (*51*) used to mutate *MF*α1 and *MF*α2 is illustrated. The α-factor coding sequence is indicated by a cross-hatched rectangle; the *MF*α flanking sequences by a solid rectangle; the *LEU2* fragment by an open rectangle; flanking plasmid sequences by a wavy line; and flanking genomic sequences by a solid line. (A) A plasmid

tant strains were compared. The *mfα2⁻* mutations had little or no effect on mating efficiency. The *mfα1⁻* mutations did have an effect in *MATα* cells, however, although the extent of the effect varies depending on the particular mutant construction. All except one of the *mfα1⁻* mutants, including a mutant in which the entire *MFα1* coding region is deleted, show a 2.5- to 4-fold reduction in mating efficiency in comparison to the parental strain. Because the total deletion mutant shows this phenotype, it represents the null phenotype for *MFα1*. One mutant shows a 30-fold reduction in comparison to the wild-type, which corresponds to a 10-fold inhibition of α-mating in comparison to the null mutants. This mutant contains an insertion of a DNA fragment containing the yeast *URA3* at the end of the first spacer sequence, and therefore should produce a hybrid protein consisting of the *MFα1* signal sequence, pro region, and first spacer sequence, followed by several amino acids encoded by the end of the *URA3* fragment (until a stop codon is reached). The mating phenotype of this mutant suggests that the hybrid protein is inhibitory to α-mating. The double mutants (*mfα1⁻ mfα2⁻*) resulted in a total mating defect in *MATα* cells, indicating that the *MFα1* and *MFα2* gene products are essential for α mating.

## 8. SUMMARY

Yeast contains two α-factor structural genes, called *MFα1* and *MFα2* (*38*). Both genes encode large precursor proteins, of 165 and 120 amino acids, respectively. There is a significant degree of

---

Fig. 6. (*cont.*) containing an *MFα* restriction fragment is shown. (B) A plasmid in which the *MFα* coding sequences have been replaced by a restriction fragment containing a selectable yeast marker (*LEU2* in the case diagrammed) is shown. The *LEU2* gene is flanked by *MFα* flanking sequences. (C) An *MFα⁺ leu2⁻* strain is transformed with linear DNA that has been released by restriction endonuclease cleavage from the plasmid containing the mutant gene, with the result that the linear fragment contains the *LEU2* gene and the *MFα* flanking sequences. (D) Replacement of the wild-type genomic sequence with the mutant *mfα* gene containing the *LEU2* insert can be obtained by selection for Leu⁺ transformants. This results in the total deletion of the *MFα* gene. Alternatively, plasmid (B) can be constructed in such a way that the selectable marker is inserted into the *MFα* coding region without deletion of the entire gene. This should also result in a nonfunctional *mfα* gene because of the large insertion within the coding sequence.

homology between the two precursors, suggesting that the two genes originated by a gene-duplication event. Both genes contain a signal sequence at the amino-terminus that is followed by a block of about 60 amino acids of unknown function, within which there are three putative glycosylation sites. The α-factor sequences are present in tandem repeats at the carboxy-terminal portion of the precursor. Each of the α-factor sequences is preceded by a 6–8 amino acid segment called the spacer sequence. *MFα1* contains four tandem spacer–α-factor repeats, all of which encode mature α-factor. *MFα2* contains two spacer-α-factor repeats. The first repeat contains a sequence that differs from mature α-factor by two amino acid substitutions. The second repeat contains the mature α-factor sequence. Processing of the α-factor precursors involves mechanisms similar to those used for processing of secreted peptides in higher eukaryotes. The carboxy-terminus of the α-factor peptide is processed at the dibasic Lys-Arg sequence in a manner similar to that found for mammalian peptide hormones (*19,20*). The *KEX2* gene is required for the Lys-Arg-cleaving endopeptidase activity (*20*). The amino-terminus of α-factor is processed by a DAP encoded (or regulated) by the α-specific gene *STE13*, in a manner similar to the processing of promelittin in honeybees (*21*). Differences in the spacer regions of the two α-factor precursors may result in differences in the efficiency of processing the two precursor proteins. Replacement of the wild-type α-factor genes with genes mutated in vitro indicates that *MFα1* and *MFα2* are the only functional α-factor structural genes, that *MFα1* is responsible for the majority of α-factor production, and that at least one of the α-factor genes is required for α mating (*38*).

## ACKNOWLEDGMENTS

I would like to thank Rodney Rothstein, Geoffrey Zubay, Mark Dworkin, Eva Dworkin, Chris Dietzel, and John Hill for comments on the manuscript and Cindy Neary for typing the manuscript.

## REFERENCES


1. Manney T. R., Duntze W., and Betz R. (1981) The Isolation, Characterization and Physiological Effects of the *Saccharomyces cerevisiae* sex pheromones, in *Sexual Interactions in Eukaryotic Microbes,* (O'Day D. O. and Horgen P. A., eds.) Academic, New York.
2. Thorner J. (1981) Pheromonal Regulation of Development in *Saccharomyces cerevisiae,* in *The Molecular Biology of the Yeast Saccharomyces: Life Cycle and Inheritance,* (Strathern J. N., Jones E. W., and Broach J. R., eds.) Cold Spring Harbor Laboratory, Cold Spring Harbor, New York.
3. Betz R., Duntze W., and Manney T. R. (1981) Hormonal control of gametogenesis in the yeast *Saccharomyces cerevisiae. Gamete Res.* **4,** 571–584.
4. Levi J. D. (1956) Mating reaction in yeast. *Nature* **177,** 753–754.
5. Duntze W., MacKay V. L., and Manney T. R. (1970) *Saccharomyces cerevisiae:* A diffusible sex factor. *Science* **168,** 1472–1473.
6. Sakurai A., Sakata K., Tamura S., Aizawa K., Yanagishima N., and Shimoda C. (1976) Structure of the peptidyl factor inducing sexual agglutination in *S. cerevisiae. Agric. Biol. Chem.* **40,** 1057–1058.
7. Stötzler D. and Dunne W. (1976) Isolation and characterization of four related peptides exhibiting α-factor activity from *Saccharomyces cerevisiae. Eur. J. Biochem.* **65,** 257–262.
8. Stötzler D., Kiltz H., and Duntze W. (1976) Primary structure of α-factor peptides from *Saccharomyces cerevisiae. Eur. J. Biochem.* **69,** 397–400.
9. Bücking-Throm E., Duntze W., Hartwell L. H., and Manney T. R. (1973) Reversible arrest of haploid yeast cells at the initiation of DNA synthesis by a diffusible sex factor. *Exp. Cell Res.* **76,** 99–110.
10. Hartwell L. H. (1973) Synchronization of haploid yeast cell cycles, a prelude to conjugation. *Exp. Cell Res.* **76,** 111–117.
11. Yanagishima N. and Yoshida K. (1981) Sexual Interactions in *Saccharomyces cerevisiae* With Special Reference to the Regulation of Sexual Agglutinability, in *Sexual Interactions in Eukaryotic Microbes.* (O'Day D. O. and Horgen P. A., eds.), Academic, New York.
12. Mackay V. L. and Manney T. R. (1974) Mutations affecting sexual conjugation and related processes in *Saccharomyces cerevisiae.* I. Isolation and phenotypic characterization of nonmating mutants. *Genetics* **76,** 255–271.
13. Liebowitz M. J. and Wickner R. B. (1976) A chromosomal gene required for killer plasmid expression, mating, and spore maturation in *Saccharomyces cerevisiae. Proc. Natl. Acad. Sci. USA* **73,** 2061–2065.

14. Sprague G. F., Jr., Rine J., and Herskowitz I. (1981) Control of yeast cell type by the mating type locus. II. Genetic interactions between *MAT*α and unlinked α-specific *STE* genes. *J. Mol. Biol.* **153,** 323–335.
15. Manney T. R. and Woods V. (1976) Mutants of *Saccharomyces cerevisiae* resistant to the α mating-type factor. *Genetics* **82,** 639–644.
16. Hartwell L. H. (1980) Mutants of *Saccharomyces cerevisiae* unresponsive to cell division control by polypeptide mating hormone. *J. Cell Biol.* **85,** 811–822.
17. Kurjan J. and Hershowitz I. (1982) Structure of a yeast pheromone gene (*MF*α): A putative α-factor precursor contains four tandem copies of mature α-factor. *Cell* **30,** 933–943.
18. Singh A., Chen E. Y., Lugovoy J. M., Chang C. N., Hitzeman R. A., and Seeburg P. H. (1983) *Saccharomyces cerevisiae* contains two discrete genes coding for the α-factor pheromone. *Nucleic Acids Res.* **11,** 4049–4063.
19. Achstetter T. and Wolf D. H. (1985) Hormone processing and membrane-bound proteinases in yeast. *EMBO J.* **4,** 173–177.
20. Julius D., Brake A., Blair L., Kunisawa R., and Thorner J. (1984) Isolation of the putative structural gene for the lysine-arginine-cleaving endopeptidase required for processing of yeast prepro-α-factor. *Cell* **37,** 1075–1089.
21. Julius D., Blair L., Brake A., Sprague G., and Thorner J. (1983) Yeast α factor is processed from a larger precursor polypeptide: The essential role of a membrane-bound dipeptidyl aminopeptidase. *Cell* **32,** 839–852.
22. Kreil G., Haiml L., and Suchanek G. (1980) Stepwise cleavage of the pro part of promelittin by dipeptidylpeptidase IV. *Eur. J. Biochem.* **111,** 49–58.
23. Strathern J., Hicks J., and Herskowitz I. (1981) Control of cell type in yeast by the mating type locus: The α1-α2 hypothesis. *J. Mol. Biol.* **147,** 357–372.
24. Jensen R., Sprague G. F., Jr., and Herskowitz I. (1983) Regulation of yeast mating-type interconversion: Feedback control of *HO* gene expression by the mating-type locus. *Proc. Natl. Acad. Sci. USA* **80,** 3035–3039.
25. Sprague G. F., Jr., Jensen R., and Herskowitz I. (1983) Control of yeast cell type by the mating type locus: Positive regulation of the α-specific *STE*3 gene by the *MAT*α1 product. *Cell* **32,** 409–415.
26. Sprague G. F., Jr. and Herskowitz I. (1981) Control of yeast cell type by the mating locus. I. Identification and control of expression of the *a*-specific gene, *BAR*1. *J. Mol. Biol.* **153,** 305–321.
27. Hicks J. B. and Herskowitz I. (1976) Evidence for a new diffusible element of mating pheromones in yeast. *Nature* **260,** 246–248.
28. Tanaka T. and Kita H. (1977) Degradation of mating factor by

α-mating type cells of *Saccharomyces cerevisiae*. *J. Biochem.* **82,** 1689–1693.

29. Maness P. F. and Edelman G. M. (1978) Inactivation and chemical alteration of mating factor α by cells and spheroplasts of yeast. *Proc. Natl. Acad. Sci. USA* **75,** 1304–1308.
30. Ciejek E. and Thorner J. (1979) Recovery of *S. cerevisiae* cells from G1 arrest by α factor pheromone requires endopeptidase action. *Cell* **18,** 623–635.
31. Nasmyth K. A. and Tatchell K. (1980) The structure of transposable yeast mating type loci. *Cell* **19,** 753–764.
32. Broach J. R., Strathern J. N., and Hicks J. B. (1979) Transformation in yeast: Development of a hybrid cloning vector and isolation of the *CAN*1 gene. *Gene* **8,** 121–133.
33. Chan R. K. and Otte C. A. (1982) Isolation and genetic analysis of *Saccharomyces cerevisiae* mutants supersensitive to G1 arrest by α-factor and α-factor pheromones. *Mol. Cell. Biol.* **2,** 11–20.
34. Fink G. R. and Styles C. A. (1972) Curing of a killer factor in *Saccharomyces cerevisiae. Proc. Natl. Acad. Sci. USA* **69,** 2846–2849.
35. Steiner D. F., Quinn P. S., Chan S. J., Marsh J., and Trager H. S. (1980) Processing mechanisms in the biosynthesis of proteins. *Ann. NY Acad. Sci.* **343,** 1–16.
36. Kreil G. (1981) Transfer of proteins across membranes. *Ann. Rev. Biochem.* **50,** 317–348.
37. Struck D. K., Lennarz W. J., and Brew K. (1978) Primary structural requirements for enzymatic formation of the *N*-glycosidic bond in glycoproteins. *J. Biol. Chem.* **253,** 5786–5794.
38. Kurjan J. (1985) α-Factor structural gene mutations in *Saccharomyces cerevisiae:* Effects on α-factor production and mating. *Mol. Cell. Biol.* **5,** 787–796.
39. Bennetzen J. L. and Hall B. D. (1982) Codon selection in yeast. *J. Biol. Chem.* **257,** 3026–3031.
40. Breathnach R. and Chambon P. (1981) Organization and expression of eucaryotic split genes coding for proteins. *Ann. Rev. Biochem.* **50,** 349–383.
41. Bennetzen J. L. and Hall B. D. (1982) The primary structure of the *Saccharomyces cerevisiae* gene for alcohol dehydrogenase I. *J. Biol. Chem.* **257,** 3018–3025.
42. Zaret K. S. and Sherman F. (1982) DNA sequence required for efficient transcription termination in yeast. *Cell* **28,** 563–573.
43. Achstetter T. and Wolf D. H. (1985) Hormone processing and membrane-bound proteinases in yeast. *EMBO J.* **4,** 173–177.
44. Julius D., Schekman R., and Thorner J. (1984) Glycosylation and processing of prepro-α-factor through the yeast secretory pathway. *Cell* **36,** 309–318.

45. Nakanishi S., Inoue A., Kita T., Nakamura M., Chang A. C. Y., Cohen S. N., and Numa S. (1979) Nucleotide sequence of cloned cDNA for bovine corticotropin-β lipotropin precursor. *Nature* **278,** 423–427.
46. Stern A. S., Jones B. N., Shively J. E., Stern S., and Udenfriend S. (1981) Two adrenal opioid polypeptides: Proposed intermediates in the processing of proenkephalin. *Proc. Natl. Acad. Sci. USA* **78,** 1962–1966.
47. Comb M., Seeburg P. H., Adelman J., Eiden L., and Herbert E. (1982) Primary structure of the human met- and leu-enkephalin precursor and its mRNA. *Nature* **295,** 663–666.
48. Lund P. K., Goodman R. H., Dee P. C., and Habener J. F. (1982) Pancreatic preproglucagon DNA contains two glucagon-related coding sequences arrayed in tandem. *Proc. Natl. Acad. Sci. USA* **79,** 345–349.
49. Suarez-Rendueles M. P., Schwencke J., Garcia-Alvarez M., and Gascon S. (1981) A new X-prolyl-dipeptidyl aminopeptidase from yeast associated with a particulate fraction. *FEBS Lett.* **131,** 296–300.
50. Hoffmann W., Bach T. C., Seliger H., and Kreil G. (1983) Biosynthesis of caerulein in the skin of *Xenopus laevis:* Partial sequences of precursors as deduced from cDNA clones. *EMBO J.* **2,** 111–114.
51. Rothstein R. J. (1983) One-step gene disruption in yeast. *Meth. Enzym.* **101,** 202–211.

# Index